HIGHER MATHEMATICS.

A TEXT-BOOK FOR

CLASSICAL AND ENGINEERING COLLEGES.

EDITED BY

MANSFIELD MERRIMAN,
PROFESSOR OF CIVIL ENGINEERING IN LEHIGH UNIVERSITY,

AND

ROBERT S. WOODWARD,
PROFESSOR OF MECHANICS IN COLUMBIA UNIVERSITY.

FIRST EDITION.

FIRST THOUSAND.

NEW YORK:

JOHN WILEY & SONS.

LONDON: CHAPMAN & HALL, LIMITED.

1896.

PREFACE.

IN the early part of this century it was possible for an industrious student to acquire a comprehensive if not minute knowledge of the entire realm of mathematical science. The more eminent minds of that time, like Lagrange, Laplace, and Gauss, were about equally familiar with all branches of pure and applied mathematics. Since that epoch the tendency has been constantly towards specialization; and additions to pure theory along with extensions of applications have been made with increasing rapidity, until now the mere quantity of information available presents a formidable obstacle to the simultaneous attainment of the breadth and depth of knowledge which characterized the mathematician of a generation ago. It would appear, however, that this obstacle is due to the bewildering mass of details rather than to any considerable increase in the number of fundamental principles. Hence the student who seeks to gain a comprehensive view of the mathematics of the present day needs most of all that sort of guidance which fixes his attention on essentials and prevents him from wasting valuable time and energy in the pursuit of non-essentials.

During the past twenty years a marked change of opinion has occurred as to the aims and methods of mathematical instruction. The old ideas that mathematical studies should be pursued to discipline the mind, and that such studies were ended when an elementary course in the calculus had been covered, have for the most part disappeared. In our best classical and engineering colleges the elementary course in calculus is now given in the sophomore year, while lectures and seminary work in pure mathematics are continued during the junior and senior years. It is with the hope of meeting the existing demand for a suitable text to be used in such upper-class work that the editors enlisted the cooperation of the authors in the task of bringing together the chapters of this book. It was the intention of the editors to include a chapter on elliptic integrals and functions; much to their regret, however, it was found impracticable to obtain the manuscript in time for publication. Notwithstanding this omission the volume contains about one-fifth more matter than was originally contemplated.

Each chapter, so far as it goes, is complete in itself, and is intended primarily to give a clear idea of the leading principles of the subject treated. While the authors have been guided by general instructions issued by the editors, each has been free to follow his own plan of treatment. It will be found that certain chapters adopt the formal method usual in text-books, while others employ what may be called the historical and intuitive method. A glance at the table of contents will show that the chapters of the work present a considerable variety of subjects, thus affording teachers and students an opportunity to select such topics as may be suited to their time and tastes. Numerous problems are given for solution, numerical examples of the application of theory to physical science are freely introduced, and the foot-notes set forth much suggestive matter of a historical and critical nature.

THE EDITORS.

JUNE 30, 1896.

CONTENTS.

Chapter III. PROJECTIVE GEOMETRY.

By George Bruce Halsted,

Professor of Mathematics in the University of Texas.

Chapter IV. HYPERBOLIC FUNCTIONS.

By James McMahon,

Associate Professor of Mathematics in Cornell University.

CHAPTER V. HARMONIC FUNCTIONS.

By William E. Byerly,
Professor of Mathematics in Harvard University.

CHAPTER VI.   FUNCTIONS OF A COMPLEX VARIABLE.

By Thomas S. Fiske,
Adjunct Professor of Mathematics in Columbia University.

Chapter VII. DIFFERENTIAL EQUATIONS.

By W. Woolsey Johnson,
Professor of Mathematics in United States Naval Academy.

Chapter VIII. GRASSMANN'S SPACE ANALYSIS.

By Edward W. Hyde,
Professor of Mathematics in the University of Cincinnati.

CHAPTER IX.   VECTOR ANALYSIS AND QUATERNIONS.

By ALEXANDER MACFARLANE,
Lecturer in Electrical Engineering in Lehigh University.

CHAPTER X.   PROBABILITY AND THEORY OF ERRORS.

By ROBERT S. WOODWARD,
Professor of Mechanics in Columbia University.

CHAPTER XI.   HISTORY OF MODERN MATHEMATICS.

By DAVID EUGENE SMITH,
Professor of Mathematics in Michigan State Normal School.

HIGHER MATHEMATICS

Chapter I.

THE SOLUTION OF EQUATIONS.

By Mansfield Merriman,

Professor of Civil Engineering in Lehigh University.

Art. 1. Introduction.

In this Chapter will be presented a brief outline of methods, not commonly found in text-books, for the solution of an equation containing one unknown quantity. Graphic, numeric, and algebraic solutions will be given by which the real roots of both algebraic and transcendental equations may be obtained, together with historical information and theoretic discussions.

An algebraic equation is one that involves only the operations of arithmetic. It is to be first freed from radicals so as to make the exponents of the unknown quantity all integers; the degree of the equation is then indicated by the highest exponent of the unknown quantity. The algebraic solution of an algebraic equation is the expression of its roots in terms of the literal coefficients; this is possible, in general, only for linear, quadratic, cubic, and quartic equations, that is, for equations of the first, second, third, and fourth degrees. A numerical equation is an algebraic equation having all its coefficients real numbers, either positive or negative. For the four degrees above mentioned the roots of numerical equations may be computed from the formulas for the algebraic solutions, unless they fall under the so-called irreducible case wherein real quantities are expressed in imaginary forms.

An algebraic equation of the nth degree may be written with all its terms transposed to the first member, thus:

$$x^n + a_1 x^{n-1} + a_2 x^{n-2} + \ldots + a_{n-1} x + a_n = 0,$$

and, for brevity, the first member will be called $f(x)$ and the equation be referred to as $f(x) = 0$. The roots of this equation are the values of x which satisfy it, that is, those values of x that reduce $f(x)$ to 0. When all the coefficients $a_1, a_2, \ldots a_n$ are real, as will always be supposed to be the case, Sturm's theorem gives the number of real roots, provided they are unequal, as also the number of real roots lying between two assumed values of x, while Horner's method furnishes a convenient process for obtaining the values of the roots to any required degree of precision.

A transcendental equation is one involving the operations of trigonometry or of logarithms, as, for example, $x + \cos x = 0$, or $a^{2x} + x b^x = 0$. No general method for the literal solution of these equations exists ; but when all known quantities are expressed as real numbers, the real roots may be located and computed by tentative methods. Here also the equation may be designated as $f(x) = 0$, and the discussions in Arts. 2–5 will apply equally well to both algebraic and transcendental forms. The methods to be given are thus, in a sense, more valuable than Sturm's theorem and Horner's process, although for algebraic equations they may be somewhat longer. It should be remembered, however, that algebraic equations higher than the fourth degree do not often occur in physical problems, and that the value of a method of solution is to be measured not merely by the rapidity of computation, but also by the ease with which it can be kept in mind and applied.

Prob. 1.   Reduce the equation $(a + x)^{\frac{1}{2}} + (a - x)^{\frac{2}{3}} = 2b$ to an equation having the exponents of the unknown quantity all integers.

Art. 2. Graphic Solutions.

Approximate values of the real roots of two simultaneous algebraic equations may be found by the methods of plane analytic geometry when the coefficients are numerically expressed. For example, let the given equations be

$$x^2 + y^2 = a^2, \qquad x^2 - bx = y^2 - cy,$$

the first representing a circle and the second a hyperbola. Drawing two rectangular axes OX and OY, the circle is described from O with the radius a. The coordinates of the center of the hyperbola are found to be $OA = \tfrac{1}{2}b$ and $AC = \tfrac{1}{2}c$, while its diameter $BD = \sqrt{b^2 - c^2}$, from which the two branches may be described.

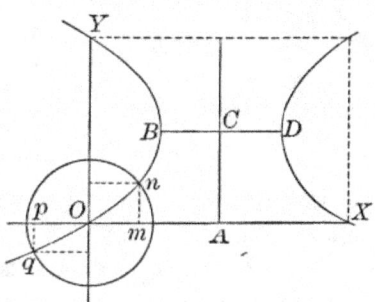

The intersections of the circle with the hyperbola give the real values of x and y. If $a = 1$, $b = 4$, and $c = 3$, there are but two real values for x and two real values for y, since the circle intersects but one branch of the hyperbola ; here Om is the positive and Op the negative value of x, while mn is the positive and pq the negative value of y. When the radius a is so large that the circle intersects both branches of the hyperbola there are four real values of both x and y.

By a similar method approximate values of the real roots of an algebraic equation containing but one unknown quantity may be graphically found. For instance, let the cubic equation $x^3 + ax - b = 0$ be required to be solved.* This may be written as the two simultaneous equations

$$y = x^3, \qquad y = -ax + b,$$

* See Proceedings of the Engineers' Club of Philadelphia, 1884, Vol. IV, pp. 47–49

and the graph of each being plotted, the abscissas of their points of intersection give the real roots of the cubic. The

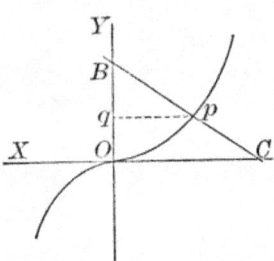

curve $y = x^3$ should be plotted upon cross-section paper by the help of a table of cubes; then OB is laid off equal to b, and OC equal to a/b, taking care to observe the signs of a and b. The line joining B and C cuts the curve at p, and hence qp is the real root of $x^3 + ax - b = 0$. If the cubic equation have three real roots the straight line BC will intersect the curve in three points.

Some algebraic equations of higher degrees may be graphically solved in a similar manner. For the quartic equation $z^4 + Az^2 + Bz - C = 0$, it is best to put $z = A^{\frac{1}{2}}x$, and thus reduce it to the form $x^4 + x^2 + bx - c = 0$; then the two equations to be plotted are

$$y = x^4 + x^2, \qquad y = -bx + c,$$

the first of which may be drawn once for all upon cross-section paper, while the straight line represented by the second may be drawn for each particular case, as described above.*

This method is also applicable to many transcendental equations; thus for the equation $Ax - B\sin x = 0$ it is best to write $ax - \sin x = 0$; then $y = \sin x$ is readily plotted by help of a table of sines, while $y = ax$ is a straight line passing through the origin. In the same way $a^x - x^2 = 0$ gives the curve represented by $y = a^x$ and the parabola represented by $y = x^2$, the intersections of which determine the real roots of the given equation.

Prob. 2. Devise a graphic solution for finding approximate values of the real roots of the equation $x^4 + ax^3 + bx^2 + cx + d = 0$.

Prob. 3. Determine graphically the number and the approximate values of the real roots of the equation arc $x - 8\sin x = 0$. (Ans.—Six real roots. $x = \pm 159°$, $\pm 430°$, and $\pm 456°$.)

---

* For an extension of this method to the determination of imaginary roots, see Phillips and Beebe's Graphic Algebra, New York, 1882.

ART. 3. THE REGULA FALSI.

One of the oldest methods for computing the real root of an equation is the rule known as "regula falsi," often called the method of double position.* It depends upon the principle that if two numbers x_1 and x_2 be substituted in the expression $f(x)$, and if one of these renders $f(x)$ positive and the other renders it negative, then at least one real root of the equation $f(x) = 0$ lies between x_1 and x_2. Let the figure represent a part of the real graph of the equation $y = f(x)$. The point X, where the curve crosses the axis of abscissas, gives a real root OX of the equation $f(x) = 0$. Let OA and OB be inferior and superior limits of the root OX which are determined either by trial or by the method of Art. 5. Let Aa and Bb be the values of $f(x)$ corresponding to these limits. Join ab, then the intersection C of the straight line ab with the axis OB gives an approximate value OC for the root. Now compute

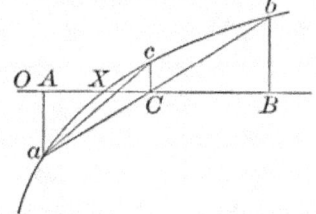

Cc and join ac, then the intersection D gives a value OD which is closer still to the root OX.

Let x_1 and x_2 be the assumed values OA and OB, and let $f(x_1)$ and $f(x_2)$ be the corresponding values of $f(x)$ represented by Aa and Bb, these values being with contrary signs. Then from the similar triangle AaC and BbC the abscissa OC is

$$x_3 = \frac{x_2 f(x_1) - x_1 f(x_2)}{f(x_1) - f(x_2)} = x_1 + \frac{(x_2 - x_1)f(x_1)}{f(x_1) - f(x_2)} = x_2 + \frac{(x_2 - x_1)f(x_2)}{f(x_1) - f(x_2)}.$$

By a second application of the rule to x_1 and x_3, another value x_4 is computed, and by continuing the process the value of x can be obtained to any required degree of precision.

As an example let $f(x) = x^5 + 5x^2 + 7 = 0$. Here it may be found by trial that a real root lies between -2 and -1.8.

* This originated in India, and its first publication in Europe was by Abraham ben Esra, in 1130. See Matthiesen, Grundzüge der antiken und modernen Algebra der litteralen Gleichungen, Leipzig. 1878.

For $x_1 = -2$, $f(x_1) = -5$, and for $x_2 = -1.8$, $f(x_2) = +4.304$; then by the regula falsi there is found $x_3 = -1.90$ nearly. Again, for $x_3 = -1.90$, $f(x_3) = +0.290$, and these combined with x_1 and $f(x_1)$ give $x_4 = -1.906$, which is correct to the third decimal.

As a second example let $f(x) = \arcsin x - \sin x - 0.5 = 0$. Here a graphic solution shows that there is but one real root, and that the value of it lies between $85°$ and $86°$. For $x_1 = 85°$, $f(x_1) = -0.01266$, and for $x_2 = 86°$, $f(x_2) = +0.00342$; then by the rule $x_3 = 85°\,44'$, which gives $f(x_3) = -0.00090$. Again, combining the values for x_2 and x_3, there is found $x_4 = 85°\,47'$. which gives $f(x_4) = -0.00009$. Lastly, combining the values for x_2 and x_4, there is found $x_5 = 85°\,47'.\dot{4}$, which is as close an approximation as can be made with five-place tables.

In the application of this method it is to be observed that the signs of the values of x and $f(x)$ are to be carefully regarded, and also that the values of $f(x)$ to be combined in one operation should have opposite signs. For the quickest approximation the values of $f(x)$ to be selected should be those having the smallest numerical values.

Prob. 4. Compute by the regula falsi the real roots of $x^5 - 0.25 = 0$. Also those of $x^2 + \sin 2x = 0$.

ART. 4. NEWTON'S APPROXIMATION RULE.

Another useful method for approximating to the value of the real root of an equation is that devised by Newton in 1666.*

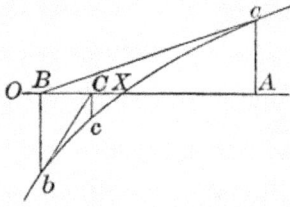

If $y = f(x)$ be the equation of a curve, OX in the figure represents a real root of the equation $f(x) = 0$. Let OA be an approximate value of OX, and Aa the corresponding value of $f(x)$. At a let aB be drawn tangent to the curve; then OB is another approximate value of OX.

---

* See Analysis per equationes numero terminorum infinitas, p. 269, Vol. I of Horsely's edition of Newton's works (London, 1779), where the method is given in a somewhat different form.

Let Bb be the value of $f(x)$ corresponding to OB, and at b let the tangent bC be drawn; then OC is a closer approximation to OX, and thus the process may be continued.

Let $f'(x)$ be the first derivative of $f(x)$; or, $f'(x) = d\,f(x)/dx$. For $x = x_1 = OA$ in the figure, the value of $f(x_1)$ is the ordinate Aa, and the value of $f'(x_1)$ is the tangent of the angle aBA; this tangent is also Aa/AB.   Hence $AB = f(x_1)/f'(x_1)$, and accordingly OB and OC are found by

$$x_2 = x_1 - \frac{f(x_1)}{f'(x_1)}, \qquad x_3 = x_2 - \frac{f(x_2)}{f'(x_2)},$$

which is Newton's approximation rule.   By a third application to x_3 the closer value x_4 is found, and the process may be continued to any degree of precision required.

For example, let $f(x) = x^5 + 5x^2 + 7 = 0$.   The first derivative is $f'(x) = 5x^4 + 10x$.   Here it may be found by trial that -2 is an approximate value of the real root.   For $x_1 = -2$ $f(x_1) = -5$, and $f'(x_1) = 60$, whence by the rule $x_2 = -1.92$. Now for $x_2 = -1.92$ are found $f(x_2) = -0.6599$ and $f'(x_2) = 29.052$, whence by the rule $x_3 = -1.906$, which is correct to the third decimal.

As a second example let $f(x) = x^2 + 4 \sin x = 0$.   Here the first derivative is $f'(x) = 2x + 4\cos x$.   An approximate value of x found either by trial or by a graphic solution is $x = -1.94$, corresponding to about $-111° 09'$.   For $x_1 = -1.94$, $f(x_1) = 0.03304$ and $f'(x_1) = -5.323$, whence by the rule $x_2 = -1.934$.   By a second application $x_3 = -1.9328$, which corresponds to an angle of $-110° 54\frac{1}{2}'$.

In the application of Newton's rule it is best that the assumed value of x_1 should be such as to render $f(x_1)$ as small as possible, and also $f'(x_1)$ as large as possible.   The method will fail if the curve has a maximum or minimum between a and b.   It is seen that Newton's rule, like the regula falsi, applies equally well to both transcendental and algebraic equations, and moreover that the rule itself is readily kept in mind by help of the diagram.

Prob. 5. Compute by Newton's rule the real roots of the algebraic equation $x^4 - 7x + 6 = 0$. Also the real roots of the transcendental equation $\sin x + \text{arc } x - 2 = 0$.

Art. 5. Separation of the Roots.

The roots of an equation are of two kinds, real roots and imaginary roots. Equal real roots may be regarded as a special class, which lie at the limit between the real and the imaginary. If an equation has p equal roots of one value and q equal roots of another value, then its first derivative equation has $p - 1$ roots of the first value and $q - 1$ roots of the second value, and thus all the equal roots are contained in a factor common to both primitive and derivative. Equal roots may hence always be readily detected and removed from the given equation. For instance, let $x^4 - 9x^2 + 4x + 12 = 0$, of which the derivative equation is $4x^3 - 18x + 4 = 0$; as $x - 2$ is a factor of these two equations, two of the roots of the primitive equation are $+ 2$.

The problem of determining the number of the real and imaginary roots of an algebraic equation is completely solved by Sturm's theorem. If, then, two values be assigned to x the number of real roots between those limits is found by the same theorem, and thus by a sufficient number of assumptions limits may be found for each real root. As Sturm's theorem is known to all who read these pages, no applications of it will be here given, but instead an older method due to Hudde will be presented which has the merit of giving a comprehensive view of the subject, and which moreover applies to transcendental as well as to algebraic equations.*

If any equation $y = f(x)$ be plotted with values of x as abscissas and values of y as ordinates, a real graph is obtained whose intersections with the axis OX give the real roots of the

* Devised by Hudde in 1659 and published by Rolle in 1690. See Œuvres de Lagrange, Vol. VIII, p. 190.

equal ion $f(x) = $ o.   Thus in the figure the three points marked X give three values OX for three real roots. The curve which represents $y = f(x)$ has points of maxima and minima marked A, and inflection points marked B. Now let the first deriva-

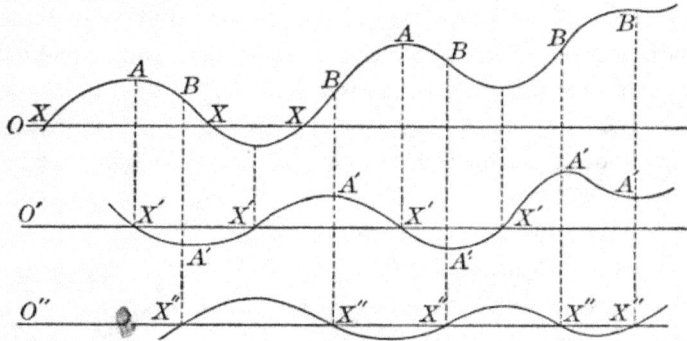

tive equation $dy/dx = f'(x)$ be formed and be plotted in the same manner on the axis $O'X'$. The condition $f'(x) = $ o gives the abscissas of the points A, and thus the real roots $O'X'$ give limits separating the real roots of $f(x) = $ o. To ascertain if a real root OX lies between two values of $O'X'$ these two values are to be substituted in $f(x)$: if the signs of $f(x)$ are unlike in the two cases, a real root of $f(x) = $ o lies between the two limits; if the signs are the same, a real root does not lie between those limits.

In like manner if the second derivative equation, that is, $d^2y/dx^2 = f''(x)$, be plotted on $O''X''$, the intersections give limits which separate the real roots of $f'(x) = $ o. It is also seen that the roots of the second derivative equation are the abscissas of the points of inflection of the curve $y = f(x)$.

To illustrate this method let the given equation be the quintic $f(x) = x^5 - 5x^3 + 6x + 2 = $ o. The first derivative equation is $f'(x) = 5x^4 - 15x^2 + 6 = $ o, the roots of which are approximately $- 1.59, - 0.69, + 0.69, + 1.59$. Now let each of these values be substituted for x in the given quintic, as also the values $- \infty$, o, and $+ \infty$, and let the corresponding values of $f(x)$ be determined as follows:

$$x = -\infty, \quad -1.59, \quad -0.69, \quad 0, \quad +0.69, \quad +1.59, \quad +\infty;$$
$$f(x) = -\infty, \quad +2.4, \quad -0.6, +2, \quad +4.7, \quad +1.6, \quad +\infty.$$

Since $f(x)$ changes sign between $x_0 = -\infty$ and $x_1 = -1.59$, one real root lies between these limits; since $f(x)$ changes sign between $x_1 = -1.59$ and $x_2 = -0.69$, one real root lies between these limits; since $f(x)$ changes sign between $x_2 = -0.69$ and $x_3 = 0$, one real root lies between these limits; since $f(x)$ does not change sign between $x_3 = 0$ and $x_4 = \infty$, a pair of imaginary roots is indicated, the sum of which lies between $+0.69$ and ∞.

As a second example let $f(x) = e^x - e^{2x} - 4 = 0$. The first derivative equation is $f'(x) = e^x - 2e^{2x} = 0$, which has two roots $e^x = \frac{1}{2}$ and $e^x = 0$, the latter corresponding to $x = -\infty$. For $x = -\infty$, $f(x)$ is negative; for $e^x = \frac{1}{2}$, $f(x)$ is negative; for $x = +\infty$, $f(x)$ is negative. The equation $e^x - e^{2x} - 4 = 0$ has, therefore, no real roots.

When the first derivative equation is not easily solved, the second, third, and following derivatives may be taken until an equation is found whose roots may be obtained. Then, by working backward, limits may be found in succession for the roots of the derivative equations until finally those of the primative are ascertained. In many cases, it is true, this process may prove lengthy and difficult, and in some it may fail entirely; nevertheless the method is one of great theoretical and practical value.

Prob. 6. Show that $e^x + e^{-3x} - 4 = 0$ has two real roots, one positive and one negative.

Prob. 7. Show that $x^6 + x + 1 = 0$ has no real roots; also that $x^6 - x - 1 = 0$ has two real roots, one positive and one negative.

ART. 6.   NUMERICAL ALGEBRAIC EQUATIONS.

An algebraic equation of the nth degree may be written with all its terms transposed to the first member, thus:

$$x^n + a_1 x^{n-1} + a_2 x^{n-2} + \ldots + a_{n-1} x + a_n = 0;$$

and if all the coefficients and the absolute term are real numbers, this is commonly called a numerical equation. The first member may for brevity be denoted by $f(x)$ and the equation itself by $f(x) = 0$.

The following principles of the theory of algebraic equations with real coefficients, deduced in text-books on algebra, are here recapitulated for convenience of reference:

(1) If x_1 is a root of the equation, $f(x)$ is divisible by $x - x_1$; and conversely, if $f(x)$ is divisible by $x - x_1$, then x_1 is a root of the equation.

(2) An equation of the nth degree has n roots and no more.

(3) If $x_1, x_2, \ldots x_n$ are the roots of the equation, then the product $(x - x_1)(x - x_2) \ldots (x - x_n)$ is equal to $f(x)$.

(4) The sum of the roots is equal to $-a_1$; the sum of the products of the roots, taken two in a set, is equal to $+a_2$; the sum of the products of the roots, taken three in a set, is equal to $-a_3$; and so on. The product of all the roots is equal to $-a_n$ when n is odd, and to $+a_n$ when n is even.

(5) The equation $f(x) = 0$ may be reduced to an equation lacking its second term by substituting $y - a_1/n$ for x.*

(6) If an equation has imaginary roots, they occur in pairs of the form $p \pm qi$ where i represents $\sqrt{-1}$.

(7) An equation of odd degree has at least one real root whose sign is opposite to that of a_n.

(8) An equation of even degree, having a_n negative, has at least two real roots, one being positive and the other negative.

(9) A complete equation cannot have more positive roots than variations in the signs of its terms, nor more negative roots than permanences in signs. If all roots be real, there are as many positive roots as variations, and as many negative roots as permanences.†

(10) In an incomplete equation, if an even number of terms, say $2m$, are lacking between two other terms, then it has at least $2m$

---

* By substituting $y^2 + py + q$ for x, the quantities p and q may be determined so as to remove the second and third terms by means of a quadratic equation, the second and fourth terms by means of a cubic equation, or the second and fifth terms by means of a quartic equation.

† The law deduced by Harriot in 1631 and by Descartes in 1639.

imaginary roots; if an odd number of terms, say $2m + 1$, are lacking between two other terms, then it has at least either $2m + 2$ or $2m$ imaginary roots, according as the two terms have like or unlike signs.*

(11) Sturm's theorem gives the number of real roots, provided that they are unequal, as also the number of real roots lying between two assumed values of x.

(12) If a_r is the greatest negative coefficient, and if a_s is the greatest negative coefficient after x is changed into $- x$, then all real roots lie between the limits $a_r + 1$ and $- (a_s + 1)$.

(13) If a_h is the first negative and a_r the greatest negative coefficient, then $a_r^{\frac{1}{n-h}} + 1$ is a superior limit of the positive roots. If a_k be the first negative and a_s the greatest negative coefficient after x is changed into $- x$, then $a_s^{\frac{1}{n-k}} + 1$ is a numerically superior limit of the negative roots.

(14) Inferior limits of the positive and negative roots may be found by placing $x = z^{-1}$ and thus obtaining an equation $f(z) = 0$ whose roots are the reciprocals of $f(x) = 0$.

(15) Horner's method, using the substitution $x = z - r$ where r is an approximate value of x_1, enables the real root x_1 to be computed to any required degree of precision.

The application of these principles and methods will be familiar to all who read these pages. Horner's method may be also modified so as to apply to the computation of imaginary roots after their approximate values have been found.†
The older method of Hudde and Rolle, set forth in Art. 5, is however one of frequent convenient application, for such algebraic equations as actually arise in practice. By its use, together with principles (13) and (14) above, and the regula falsi of Art. 3, the real roots may be computed without any assumptions whatever regarding their values.

For example, let a sphere of diameter D and specific gravity

---

* Established by Du Gua; see Memoirs Paris Academy, 1741, pp. 435–494.

† Sheffler, Die Auflösung der algebraischen und transzendenten Gleichungen, Braunschweig, 1859; and Jelink, Die Auflösung der höheren numerischen Gleichungen, Leipzig, 1865.

g float in water, and let it be required to find the depth of immersion. The solution of the problem gives for the depth x the cubic equation

$$x^3 - \tfrac{3}{2}Dx^2 + \tfrac{1}{2}D^3g = 0.$$

As a particular case let $D = 2$ feet and $g = 0.65$; then the equation

$$x^3 - 3x^2 + 2.6 = 0$$

is to be solved. The first derivative equation is $3x^2 - 6x = 0$ whose roots are o and 2. Substituting these, there is found one negative root, one positive root less than 2, and one positive root greater than 2. The physical aspect of the question excludes the first and last root, and the second is to be computed. By (13) and (14) an inferior limit of this root is about 0.5, so that it lies between 0.5 and 2. For $x_1 = 0.5$, $f(x_1) = + 1.975$, and for $x_2 = 2$, $f(x_2) = -1.4$; then by the regula falsi $x_3 = 1.35$. For $x_3 = 1.35$, $f(x_3) = - 0.408$, and combining this with x, the regula falsi gives $x_4 = 1.204$ feet, which, except in the last decimal, is the correct depth of immersion of the sphere.

Prob. 8. The diameter of a water-pipe whose length is 200 feet and which is to discharge 100 cubic feet per second under a head of 10 feet is given by the real root of the quintic equation $x^5 - 38x - 101 = 0$. Find the value of x.

ART. 7. TRANSCENDENTAL EQUATIONS.

Rules (1) to (15) of the last article have no application to trigonometrical or exponential equations, but the general principles and methods of Arts. 2–5 may be always used in attempting their solution. Transcendental equations' may have one, many, or no real roots, but those arising from problems in physical science must have at least one real root. Two examples of such equations will be presented.

A cylinder of specific gravity g floats in water, and it is required to find the immersed arc of the circumference. If this be expressed in circular measure it is given by the transcedental equation

$$f(x) = x - \sin x - 2\pi g = 0.$$

The first derivative equation is $1 - \cos x = 0$, whose root is any even multiple of 2π. Substituting such multiples in $f(x)$ it is found that the equation has but one real root, and that this lies between 0 and 2π; substituting $\frac{1}{2}\pi$, $\frac{3}{4}\pi$, and π for x, it is further found that this root lies between $\frac{3}{4}\pi$ and π.

As a particular case let $g = 0.424$, and for convenience in using the tables let x be expressed in degrees; then

$$f(x) = x - 57°.2958 \sin x - 152°.64.$$

Now proceeding by the regula falsi (Art. 3) let $x_1 = 180°$ and $x_2 = 135°$, giving $f(x_1) = + 27°.36$ and $f(x_2) = -58°.16$, whence $x_3 = 166°$. For $x_3 = 166°$, $f(x_3) = -0°.469$, and hence $166°$ is an approximate value of the root. Continuing the process, x is found to be $166°.237$, or in circular measure $x = 2.9014$ radians.

As a second example let it be required to find the horizontal tension of a catenary cable whose length is 22 feet, span 20 feet, and weight 10 pounds per linear foot, the ends being suspended from two points on the same level. If l be the span, s the length of the cable, and z a length of the cable whose weight equals the horizontal tension, the solution of the problem leads to the transcendental equation $s = \left(e^{\frac{l}{2z}} - e^{-\frac{l}{2z}} \right) z$, or inserting the numerical values,

$$f(z) = 22 - \left(e^{\frac{10}{z}} - e^{-\frac{10}{z}} \right) z = 0$$

is the equation to be solved. The first derivative equation is

$$f'(z) = - \left(e^{\frac{10}{z}} - e^{-\frac{10}{z}} \right) + \frac{10}{z} \left(e^{\frac{10}{z}} + e^{-\frac{10}{z}} \right) = 0,$$

and this substituted in $f(z)$ shows that one real root is less than about 20. Assume $z_1 = 15$, then $f(z_1) = 0.486$ and $f'(z_1) = 0.206$, whence by Newton's rule (Art. 4) $z_2 = 13$ nearly. Next for $z_2 = 13$, $f(z_2) = -0.0298$ and $f'(z_2) = 0.322$, whence $z_3 = 13.1$. Lastly for $z_3 = 13.1$ $f(z_3) = 0.0012$ and $f'(z_3) = 0.3142$, whence $z_4 = 13.096$, which is a sufficiently close approximation. The horizontal tension in the given catenary is hence 130.96 pounds.*

---

* Since $e^{\theta} - e^{-\theta} = 2 \sinh \theta$, this equation may be written $110 - 10 \sinh \theta$, where $\theta = 10z^{-1}$, and the solution may be expedited by the help of tables of hyperbolic functions. See Chapter IV.

Prob. 9. Show that the equation $3 \sin x - 2x - 5 = 0$ has but one real root, and compute its value.

Prob. 10. Find the number of real roots of the equation $2x + \log x - 10\,000 = 0$, and show that the value of one of them is $x = 4995.74$.

ART. 8. ALGEBRAIC SOLUTIONS.

Algebraic solutions of complete algebraic equations are only possible when the degree n is less than 5. It frequently happens, moreover, that the algebraic solution cannot be used to determine numerical values of the roots as the formulas expressing them are in irreducible imaginary form. Nevertheless the algebraic solutions of quadratic, cubic, and quartic equations are of great practical value, and the theory of the subject is of the highest importance, having given rise in fact to a large part of modern algebra.

The solution of the quadratic has been known from very early times, and solutions of the cubic and quartic equations were effected in the sixteenth century. A complete investigation of the fundamental principles of these solutions was, however, first given by Lagrange in 1770.* This discussion showed, if the general equation of the nth degree, $f(x) = 0$, be deprived of its second term, thus giving the equation $f(y) = 0$, that the expression for the root y is given by

$$y = \omega s_1 + \omega^2 s_2 + \ldots + \omega^{n-1} s_{n-1},$$

in which n is the degree of the given equation, ω is, in succession, each of the nth roots of unity, $1, \epsilon, \epsilon^2, \ldots \epsilon^{n-1}$, and $s_1, s_2, \ldots s_{n-1}$ are the so-called elements which in soluble cases are determined by an equation of the $n - 1^{\text{th}}$ degree. For instance, if $n = 3$ the equation is of the third degree or a cubic, the three values of ω are

$$\omega_1 = 1, \quad \omega = -\tfrac{1}{2} + \tfrac{1}{2}\sqrt{-3} = \epsilon, \quad \omega = -\tfrac{1}{2} - \tfrac{1}{2}\sqrt{-3} = \epsilon^2,$$

* Memoirs of Berlin Academy. 1769 and 1770: reprinted in Œuvres de Lagrange (Paris, 1868), Vol. II, pp. 539–562. See also Traité de la résolution des équations numeriques, Paris, 1798 and 1808.

and the three roots are expressed by

$$y_1 = s_1 + s_2, \qquad y_2 = \epsilon s_1 + \epsilon^2 s_2, \qquad y_3 = \epsilon^2 s_1 + \epsilon s_2,$$

in which s_1^3 and s_2^3 are found to be the roots of a quadratic equation (Art. 9).

The n values of ω are the n roots of the binomial equation $\omega^n - 1 = 0$. If n be odd, one of these is real and the others are imaginary; if n be even, two are real and $n - 2$ are imaginary.* Thus the roots of $\omega^2 - 1 = 0$ are $+ 1$ and $- 1$; those of $\omega^3 - 1 = 0$ are given above; those of $\omega^4 - 1 = 0$ are $+ 1, + i, - 1$, and $- i$ where i is $\sqrt{-1}$. For the equation $\omega^5 - 1 = 0$ the real root is $+ 1$, and the imaginary roots are denoted by $\epsilon, \epsilon^2, \epsilon^3, \epsilon^4$; to find these let $\omega^5 - 1 = 0$ be divided by $\omega - 1$, giving

$$\omega^4 + \omega^3 + \omega^2 + \omega + 1 = 0,$$

which being a reciprocal equation can be reduced to a quadratic, and the solution of this furnishes the four values,

$$\epsilon = -\tfrac{1}{4}(1 - \sqrt{5} + \sqrt{-10 - 2\sqrt{5}}), \quad \epsilon^2 = -\tfrac{1}{4}(1 + \sqrt{5} + \sqrt{-10 + 2\sqrt{5}}),$$

$$\epsilon^4 = -\tfrac{1}{4}(1 - \sqrt{5} - \sqrt{-10 - 2\sqrt{5}}), \quad \epsilon^3 = -\tfrac{1}{4}(1 + \sqrt{5} - \sqrt{-10 + 2\sqrt{5}}),$$

where it will be seen that $\epsilon . \epsilon^4 = 1$ and $\epsilon^2 . \epsilon^3 = 1$, as should be the case, since $\epsilon^5 = 1$.

In order to solve a quadratic equation by this general method let it be of the form

$$x^2 + 2ax + b = 0,$$

and let x be replaced by $y - a$, thus reducing it to

$$y^2 - (a^2 - b) = 0.$$

Now the two roots of this are $y_1 = + s_1$ and $y_2 = - s_1$, whence the product of $(y - s_1)$ and $(y + s_1)$ is

$$y^2 - s^2 = 0.$$

Thus the value of s^2 is given by an equation of the first degree,

---

* The values of ω are, in short, those of the n "vectors" drawn from the center which divide a circle of radius unity into n equal parts, the first vector $\omega_1 = 1$ being measured on the axis of real quantities. See Chapter X.

$s^2 = a^2 - b$; and since $x = -a + y$, the roots of the given equation are

$$x_1 = -a + \sqrt{a^2 - b}, \qquad x_2 = -a - \sqrt{a^2 - b},$$

which is the algebraic solution of the quadratic.

The equation of the $n - 1^{\text{th}}$ degree upon which the solution of the equation of the nth degree depends is called a resolvent. If such a resolvent exists, the given equation is algebraically solvable; but, as before remarked, this is only the case for quadratic, cubic, and quartic equations.

Prob. 11. Show that the six 6th roots of unity are $+1$, $+\frac{1}{2}(1 + \sqrt{-3})$, $-\frac{1}{2}(1 - \sqrt{-3})$, -1, $-\frac{1}{2}(1 + \sqrt{-3})$, $-\frac{1}{2}(1 - \sqrt{-3})$.

ART. 9. THE CUBIC EQUATION.

All methods for the solution of the cubic equation lead to the result commonly known as Cardan's formula.* Let the cubic be

$$x^3 + 3ax^2 + 3bx + 2c = 0, \tag{1}$$

and let the second term be removed by substituting $y - a$ for x, giving the form,

$$y^3 + 3By + 2C = 0, \tag{1'}$$

in which the values of B and C are

$$B = -a^2 + b, \qquad C = a^3 - \tfrac{3}{2}ab + c. \tag{2}$$

Now by the Lagrangian method of Art. 8 the values of y are

$$y_1 = s_1 + s_2, \qquad y_2 = \epsilon s_1 + \epsilon^2 s_2, \qquad y_3 = \epsilon^2 s_1 + \epsilon s_2,$$

in which ϵ and ϵ^2 are the imaginary cube roots of unity. Forming the products of the roots, and remembering that $\epsilon^3 = 1$ and $\epsilon^2 + \epsilon^2 + 1 = 0$, there are found

$$y_1 y_2 + y_1 y_3 + y_2 y_3 = -3 s_1 s_2 = +3B,$$
$$y_1 y_2 y_3 = s_1^3 + s_2^3 = -2C.$$

For the determination of s_1 and s_2 there are hence two equations from which results the quadratic resolvent $s^6 + 2Cs^3 - B^3 = 0$, and thus

$$s_1 = (-C + \sqrt{B^3 + C^2})^{\frac{1}{3}}, \quad s_2 = (-C - \sqrt{B^3 + C^2})^{\frac{1}{3}}. \tag{3}$$

* Deduced by Ferreo in 1515, and first published by Cardan in 1545.

One of the roots of the cubic in y therefore is

$$y_1 = (-C + \sqrt{B^3 + C^2})^{\frac{1}{3}} + (-C - \sqrt{B^3 + C^2})^{\frac{1}{3}},$$

and this is the well-known formula of Cardan.

The algebraic solution of the cubic equation (1) hence consists in finding B and C by (2) in terms of the given coefficients, and then by (3) the elements s_1 and s_2 are determined. Finally,

$$x_1 = -a + (s_1 + s_2),$$
$$x_2 = -a - \tfrac{1}{2}(s_1 + s_2) + \tfrac{1}{2}\sqrt{-3}(s_1 - s_2), \qquad (4)$$
$$x_3 = -a - \tfrac{1}{2}(s_1 + s_2) - \tfrac{1}{2}\sqrt{-3}(s_1 - s_2),$$

which are the algebraic expressions of the three roots.

When $B^3 + C^2$ is negative the numerical solution of the cubic is not possible by these formulas, as then both s_1 and s_2 are in irreducible imaginary form. This, as is well known, is the case of three real roots, $s_1 + s_2$ being a real, while $s_1 - s_2$ is a pure imaginary.* When $B^3 + C^2$ is 0 the elements s_1 and s_2 are equal, and there are two equal roots, $x_2 = x_3 = -a + C^{\frac{1}{3}}$, while the other root is $x_1 = -a - 2C^{\frac{1}{3}}$.

When $B^3 + C^2$ is positive the equation has one real and two imaginary roots, and formulas (2), (3), and (4) furnish the numerical values of the roots of (1). For example, take the cubic

$$x^3 - 4.5x^2 + 12x - 5 = 0,$$

whence by comparison with (1) are found $a = -1.5$, $b = +4$, $c = -2.5$. Then from (2) are computed $B = 1.75$, $C = +3.125$. These values inserted in (3) give $s_1 = +0.9142$, $s_2 = -1.9142$; thus $s_1 + s_2 = -1.0$ and $s_1 - s_2 = +2.8284$. Finally, from (4)

$$x_1 = 1.5 - 1.0 = +0.5,$$
$$x_2 = 1.5 + 0.5 + 1.4142\sqrt{-3} = 2 + 2.4495i,$$
$$x_3 = 1.5 + 0.5 - 1.4142\sqrt{-3} = 2 - 2.4495i,$$

which are the three roots of the given cubic.

* The numerical solution of this case is possible whenever the angle whose cosine is $-C/\sqrt{-B^3}$ can be geometrically trisected.

Prob. 12. Compute the roots of $x^3 - 2x - 5 = 0$. Also the roots of $x^3 + 0.6x^2 - 5.76x + 4.32 = 0$.

Prob. 13. A cone has its altitude 6 inches and the diameter of its base 5 inches. It is placed with vertex downwards and one fifth of its volume is filled with water. If a sphere 4 inches in diameter be then put into the cone, what part of its radius is immersed in the water? (Ans. 0.5459 inches).

ART. 10. THE QUARTIC EQUATION.

The quartic equation was first solved in 1545 by Ferrari, who separated it into the difference of two squares. Lagrange in 1637 resolved it into the product of two quadratic factors. Tschirnhausen in 1683 removed the second and fourth terms. Euler in 1732 and Lagrange in 1767 effected solutions by assuming the form of the roots. All these methods lead to cubic resolvents, the roots of which are first to be found in order to determine those of the quartic.

The methods of Euler and Lagrange, which are closely similar, first reduce the quartic to one lacking the second term,

$$y^4 + 6By^2 + 4Cy + D = 0;$$

and the general form of the roots being taken as

$$y_1 = + \sqrt{s_1} + \sqrt{s_2} + \sqrt{s_3}, \qquad y_3 = - \sqrt{s_1} + \sqrt{s_2} - \sqrt{s_3},$$
$$y_2 = + \sqrt{s_1} - \sqrt{s_2} - \sqrt{s_3}, \qquad y_4 = - \sqrt{s_1} - \sqrt{s_2} + \sqrt{s_3},$$

the values s_1, s_2, s_3, are shown to be the roots of the resolvent,

$$s^3 + 3Bs^2 + \tfrac{1}{4}(9B^2 - D)s - \tfrac{1}{4}C^2 = 0.$$

Thus the roots of the quartic are algebraically expressed in terms of the coefficients of the quartic, since the resolvent is solvable by the process of Art. 9.

Whatever method of solution be followed, the following final formulas, deduced by the author in 1892, will result.* Let the complete quartic equation be written in the form

$$x^4 + 4ax^3 + 6bx^2 + 4cx + d = 0. \tag{1}$$

* See American Journal Mathematics, 1892, Vol. XIV, pp. 237–245.

First, let g, h, and k be determined from

$$g = a^2 - b, \quad h = b^2 + c^2 - 2abc + dg, \quad k = \tfrac{4}{3}ac - b^2 - \tfrac{1}{3}d. \quad (2)$$

Secondly, let l be obtained by

$$l = \tfrac{1}{2}(h + \sqrt{h^2 + k^3})^{\frac{1}{3}} + \tfrac{1}{2}(h - \sqrt{h^2 + k^3})^{\frac{1}{3}} \qquad (3)$$

Thirdly, let u, v, and w be found from

$$u = g + l, \quad v = 2g - l, \quad w = 4u^2 + 3k - 12gl. \qquad (4)$$

Then the four roots of the quartic equation are

$$\left. \begin{aligned}
x_1 &= -a + \sqrt{u} + \sqrt{v} + \sqrt{w}, \\
x_2 &= -a + \sqrt{u} - \sqrt{v} + \sqrt{w}, \\
x_3 &= -a - \sqrt{u} + \sqrt{v} - \sqrt{w}, \\
x_4 &= -a - \sqrt{u} - \sqrt{v} - \sqrt{w},
\end{aligned} \right\} \qquad (5)$$

in which the signs are to be used as written provided that $2a^3 - 3ab + c$ is a negative number; but if this is positive all radicals except $\sqrt{w}$ are to be reversed in sign.

These formulas not only serve for the complete theoretic discussion of the quartic (1), but they enable numerical solutions to be made whenever (3) can be computed, that is, whenever $h^2 + k^3$ is positive. For this case the quartic has two real and two imaginary roots. If there be either four real roots or four imaginary roots $h^2 + k^3$ is negative, and the irreducible case arises where convenient numerical values cannot be obtained, although they are correctly represented by the formulas.

As an example let a given rectangle have the sides p and q, and let it be required to find the length of an inscribed rectangle whose width is m. If x be this length, this is a root of the quartic equation

$$x^4 - (p^2 + q^2 + 2m^2)x^2 + 4pqmx - (p^2 + q^2 - m^2)m^2 = 0,$$

and thus the problem is numerically solvable by the above formulas if two roots are real and two imaginary. As a special case let $p = 4$ feet, $q = 3$ feet, and $m = 1$ foot; then

$$x^4 - 27x^2 + 48x - 24 = 0.$$

By comparison with (1) are found $a = 0$, $b = -4\frac{1}{2}$, $c = +12$, and $d = -24$. Then from (2), $g = +4\frac{1}{2}$, $h = -\frac{441}{8}$, and $k = +\frac{49}{4}$. Thus $h^2 + k^3$ is positive, and from (3) the value of l is -3.6067. From (4) are now found, $u = +0.8933$, $v = 12.6067$, and $w = +161.20$. Then, since c is positive, the values of the four roots are, by (5),

$$x_1 = -0.945 - \sqrt{12.607 + 12.697} = -5.975 \text{ feet,}$$
$$x_2 = -0.945 + \sqrt{12.607 + 12.697} = +4.085 \text{ feet,}$$
$$x_3 = -0.945 + \sqrt{12.607 - 12.697} = +0.945 + 0.30i,$$
$$x_4 = -0.945 - \sqrt{12.607 - 12.697} = +0.945 - 0.30i,$$

the second of which is evidently the required length. Each of these roots closely satisfies the given equation, the slight discrepancy in each case being due to the rounding off at the third decimal.*

Prob. 14. Compute the roots of the equation $x^4 + 7x + 6 = 0$. (Ans. -1.388, -1.000, $1.194 \pm 1.701i$.)

Art. 11. Quintic Equations.

The complete equation of the fifth degree is not algebraically solvable, nor is it reducible to a solvable form. Let the equation be

$$x^5 + 5ax^4 + 5bx^3 + 5cx^2 + 5dx + 2e = 0,$$

and by substituting $y - a$ for x let it be reduced to

$$y^5 + 5By^3 + 5Cy^2 + 5Dy + 2E = 0.$$

The five roots of this are, according to Art. 8,

$$y_1 = s_1 + s_2 + s_3 + s_4,$$
$$y_2 = \epsilon s_1 + \epsilon^2 s_2 + \epsilon^3 s_3 + \epsilon^4 s_4,$$
$$y_3 = \epsilon^2 s_1 + \epsilon^4 s_2 + \epsilon s_3 + \epsilon^3 s_4,$$
$$y_4 = \epsilon^3 s_1 + \epsilon s_2 + \epsilon^4 s_3 + \epsilon^2 s_4,$$
$$y_5 = \epsilon^4 s_1 + \epsilon^3 s_2 + \epsilon^2 s_3 + \epsilon s_4,$$

in which ϵ, ϵ^2, ϵ^3. ϵ^4 are the imaginary fifth roots of unity. Now if the several products of these roots be taken there will be

* This example is known by civil engineers as the problem of finding the length of a strut in a panel of the Howe truss.

found, by (4) of Art. 6, four equations connecting the four elements s_1, s_2, s_3, and s_4, namely,

$$- B = s_1 s_4 + s_2 s_3,$$
$$- C = s_1^2 s_3 + s_2^2 s_1 + s_3^2 s_4 + s_4^2 s_2,$$
$$- D = s_1^3 s_2 + s_2^3 s_4 + s_3^3 s_1 + s_4^3 s_3 - s_1^2 s_4^2 - s_2^2 s_3 + s_1 s_2 s_3 s_4,$$
$$- 2E = s_1^5 + s_2^5 + s_3^5 + s_4^5 + 5(s_1^2 s_2^2 s_4 + s_1^2 s_3^2 s_2 + s_2^2 s_4^2 s_3 + s_3^2 s_4^2 s_1)$$
$$- 5(s_1^2 s_3^2 s_4 + s_2^2 s_1 s_3 + s_3^2 s_2 s_4 + s_4^2 s_1 s_2);$$

but the solution of these leads to an equation of the 120th degree for s, or of the 24th degree for s^5. However, by taking $s_1 s_4 - s_2 s_3$, or $s_1^5 + s_2^5 + s_3^5 + s_4^5$ as the unknown quantity, a resolvent of the 6th degree is obtained, and all efforts to find a resolvent of the fourth degree have proved unavailing.

Another line of attack upon the quintic is in attempting to remove all the terms intermediate between the first and the last. By substituting $y^2 + py + q$ for x, the values of p and q may be determined so as to remove the second and third terms by a quadratic equation, or the second and third by a cubic equation, or the second and fourth by a quartic equation, as was first shown by Tschirnhausen in 1683. By substituting $y^3 + py^2 + qy + r$ for x, three terms may be removed, as was shown by Bring in 1786. By substituting $y^4 + py^3 + qy^2 + ry + t$ for x it was thought by Jerrard in 1833 that four terms might be removed, but Hamilton showed later that this leads to equations of a degree higher than the fourth.

In 1826 Abel gave a demonstration that the algebraic solution of the general quintic is impossible, and later Galois published a more extended investigation leading to the same conclusion.* Although these discussions are complex, and not devoid of points of doubt,† they have been generally accepted as conclusive. Moreover, the fact that the quintic is still unsolved in spite of the enormous amount of work done upon it during the past two centuries, is strong evidence that the problem is an impossible one.

* See Jordan's Traité des substitutions et des équations algébriques, 1870.

† See Kronecker, Verhandlungen der Berliner Akademie, 1853, p. 38; also Cockle, Philosophical Magazine, 1854, Vol. VII, p. 134.

There are, however, numerous special forms of the quintic whose algebraic solution is possible. The oldest of these is the quintic of De Moivre,

$$y^5 + 5By^3 + 5B^2y + 2E = 0,$$

which is solved at once by making $s_2 = s_3 = 0$ in the element equations; then $-B = s_1 s_4$ and $-2E = s_1{}^5 + s_4{}^5$, from which s_1 and s_4 are found, and $y_1 = s_1 + s_4$, or

$$y_1 = (-E + \sqrt{B^5 + E^2})^{\frac{1}{5}} + (-E - \sqrt{B^5 + E^2})^{\frac{1}{5}},$$

while the other roots are $y_2 = \epsilon s_1 + \epsilon^4 s_4$, $y_3 = \epsilon^2 s_1 + \epsilon^3 s_4$, $y_4 = \epsilon^3 s_1 + \epsilon^2 s_4$, and $y_5 = \epsilon^4 s_1 + \epsilon s_4$. If $B^5 + E^2$ be negative, this quintic has five real roots; if positive, there are one real and four imaginary roots.

When any relation, other than those expressed by the four element equations, exists between s_1, s_2, s_3, s_4, the quintic is solvable algebraically. As an infinite number of such relations may be stated, it follows that there are an infinite number of solvable quintics. In each case of this kind, however, the coefficients of the quintic are also related to each other by a certain equation of condition.

The complete solution of the quintic in terms of one of the roots of its resolvent sextic was made by McClintock in 1884.[*] By this method $s_1{}^5$, $s_2{}^5$, $s_3{}^5$, and $s_4{}^5$ are expressed as the roots of a quartic in terms of a quantity t which is the root of a sextic whose coefficients are rational functions of those of the given quintic. Although this has great theoretic interest, it is, of course, of little practical value for the determination of numerical values of the roots.

By means of elliptic functions the complete quintic can, however, be solved, as was first shown by Hermite in 1858. For this purpose the quintic is reduced by Jerrard's transformation to the form $x^5 + 5dx + 2e = 0$, and to this form can also be reduced the elliptic modular equation of the sixth degree. Other solutions by elliptic functions were made by

[*] American Journal of Mathematics, 1886, Vol. VIII, pp. 49-83.

Kronecker in 1861 and by Klein in 1884.* These methods, though feasible by the help of tables, have not yet been systematized so as to be of practical advantage in the numerical computation of roots.

Prob. 15. If the relation $s_1 s_4 = s_2 s_3$ exists, between the elements show that $s_1^5 + s_2^5 + s_3^5 + s_4^5 = -2E$.

Prob. 16. Compute the roots of $y^3 + 10y^2 + 20y + 6 = 0$, and also those of $y^3 - 10y^2 + 20y + 6 = 0$.

ART. 12. TRIGONOMETRIC SOLUTIONS.

When a cubic equation has three real roots the most convenient practical method of solution is by the use of a table of sines and cosines. If the cubic be stated in the form (1) of Art. 9, let the second term be removed, giving

$$y^3 + 3By + 2C = 0.$$

Now suppose $y = 2r \sin \theta$, then this equation becomes

$$8 \sin^3 \theta + 6\frac{B}{r^2} \sin \theta + 2\frac{C}{r^3} = 0,$$

and by comparison with the known trigonometric formula

$$8 \sin^3 \theta - 6 \sin \theta + 2 \sin 3\theta = 0,$$

there are found for r and $\sin 3\theta$ the values

$$r = \sqrt{-B}, \qquad \sin 3\theta = C/\sqrt{-B^3},$$

in which B is always negative for the case of three real roots (Art. 9). Now $\sin 3\theta$ being computed, 3θ is found from a table of sines, and then θ is known. Thus,

$$y_1 = 2r \sin \theta, \quad y_2 = 2r \sin (120° + \theta), \quad y_3 = 2r \sin (240° + \theta),$$

are the real roots of the cubic in y.†

---

* For an outline of these transcendental methods, see Hagen's Synopsis der höheren Mathematik, Vol. I, pp. 339–344.

† When B^3 is negative and numerically less than C^2, as also when B^3 is positive, this solution fails, as then one root is real and two are imaginary. In this case, however, a similar method of solution by means of hyperbolic sines is possible. See Grunert's Archiv für Mathematik und Physik, Vol. xxxviii, pp. 48–76.

For example, the depth of flotation of a sphere whose diameter is 2 feet and specific gravity 0.65, is given by the cubic equation $x^3 - 3x^2 + 2.6 = 0$ (Art. 6). Placing $x = y + 1$ this reduces to $y^3 - 3y + 0.6 = 0$, for which $B = -1$ and $C = +0.3$. Thus $r = 1$ and $\sin 3\theta = +0.3$. Next from a table of sines, $3\theta = 17° 27'$, and accordingly $\theta = 5° 49'$. Then

$$y_1 = 2 \sin \quad 5° 49' = +0.2027,$$
$$y_2 = 2 \sin 125° 49' = +1.6218,$$
$$y_3 = 2 \sin 245° 49' = -1.8245.$$

Adding 1 to each of these, the values of x are

$$x_1 = +1.203 \text{ feet}, \quad x_2 = +2.622 \text{ feet}, \quad x_3 = -0.825 \text{ feet};$$

and evidently, from the physical aspect of the question, the first of these is the required depth. It may be noted that the number 0.3 is also the sine of $162° 11'$, but by using this the three roots have the same values in a different order.

When the quartic equation has four real roots its cubic resolvent has also three real roots. In this case the formulas of Art. 10 will furnish the solution if the three values of l be obtained from (3) by the help of a table of sines. The quartic being given, g, h, and k are found as before, and the value of k will always be negative for four real roots. Then

$$r = \sqrt{-k}, \qquad \sin 3\theta = -h/r^3,$$

and 3θ is taken from a table; thus θ is known, and the three values of l are

$$l_1 = r \sin \theta, \qquad l_2 = r \sin(120° + \theta), \qquad l_3 = r \sin(240° + \theta).$$

Next the three values of u, of v, and of w are computed, and those selected which give u, w, and $v - \sqrt{w}$ all positive quantities. Then (5) gives the required roots of the quartic.

As an example, take the case of the inscribed rectangle in Art. 10, and let $p = 4$ feet, $q = 3$ feet, $m = \sqrt{13}$ feet; then the quartic equation is

$$x^4 - 51x^2 + 48\sqrt{13}\, x - 156 = 0.$$

Here $a = 0$. $b = -8\frac{1}{2}$, $c = +12\sqrt{13}$, and $d = -156$. Next $g = +8\frac{1}{2}$, $h = -\frac{54\frac{1}{2}}{8}$, and $k = -\frac{8\frac{1}{4}}{4}$. The trigonometric work now begins; the value of r is found to be $+4\frac{1}{2}$, and that of $\sin 3\theta$ to be $+0.7476$; hence from the table $3\theta = 48°\ 23'$, and $\theta = 16°\ 07'\ 40''$. The three values of l are then computed by logarithmic tables, and found to be,

$$l_1 = +1.250, \qquad l_2 = +3.1187, \qquad l_3 = -4.3687.$$

Next the values of u, v, and w are obtained, and it is seen that only those corresponding to l_1 will render all quantities under the radicals positive; these quantities are $u = 9.75$, $v = 15.75$, and $w = 192.0$. Then the four roots of the quartic are

$$x_1 = -8.564, \quad x_2 = +2.319, \quad x_3 = +1.746, \quad x_4 = +4.499 \text{ feet,}$$

of which only the second and third belong to inscribed rectangles, while the first and fourth belong to rectangles whose corners are on the sides of the given rectangle produced.

Trigonometric solutions of the quintic equation are not possible except for the binomial $x^5 \pm a$, and the quintic of De Moivre. The general trigonometric expression for the root of a quintic lacking its second term is $y = 2r_1 \cos \theta_1 + 2r_2 \cos \theta_2$, and to render a solution possible, r_1 and r_2, as well as $\cos \theta_1$ and $\cos \theta_2$, must be found; but these in general are roots of equations of the sixth or twelfth degree: in fact r_1^2 is the same as the function $s_1 s_4$ of Art. 11, and r_2^2 is the same as $s_2 s_3$. Here $\cos \theta_1$ and $\cos \theta_2$ may be either circular or hyperbolic cosines, depending upon the signs and values of the coefficients of the quintic.

Trigonometric solutions are possible for any binomial equation, and also for any equation which expresses the division of an angle into equal parts. Thus the roots of $x^6 + 1 = 0$ are $\cos m\ 30° \pm i \sin m\ 30°$, in which m has the values 1, 2, and 3. The roots of $x^5 - 5x^3 + 5x - 2 \cos 5\ \theta = 0$ are $2 \cos (m\ 72° + \theta)$ where m has the values 0, 1, 2, 3, and 4.

Prob. 17. Compute by a trigonometric solution the four roots of the quartic $x^4 + 4x^3 - 24x^2 - 76x - 29 = 0$. (Ans. -6.734, -1.550, $+0.262$, $+4.022$).

Prob. 18. Give a trigonometric solution of the quintic equation $x^5 - 5bx^3 + 5b^2x - 2e = 0$ for the case of five real roots. Compute the roots when $b = 1$ and $e = 0.752798$. (Ans. -1.7940, -1.3952, 0.2864, 0.9317, 1.9710.)

Art. 13.   Real Roots by Series.

The value of x in any algebraic equation may be expressed as an infinite series. Let the equation be of any degree, and by dividing by the coefficient of the term containing the first power of x let it be placed in the form

$$a = x + bx^2 + cx^3 + dx^4 + ex^5 + fx^6 + \ldots$$

Now let it be assumed that x can be expressed by the series

$$x = a + ma^2 + na^3 + pa^4 + qa^5 + \ldots$$

By inserting this value of x in the equation and equating the coefficients of like powers of a, the values of m, n, etc., are found, and then

$$x = a - ba^2 + (2b^2 - c)a^3 - (5b^3 - 5bc + d)a^4 + (14b^4 - 21b^2c + 6bd + 3c^2 - e)a^5$$
$$- (42b^5 - 84b^3c + 28b^2d + 28bc^2 - 7be - 7cd + f)a^6 + \ldots,$$

is an expression of one of the roots of the equation. In order that this series may converge rapidly it is necessary that a should be a small fraction.*

To apply this to a cubic equation the coefficients d, e, f, etc., are made equal to o, For example, let $x^3 - 3x + 0.6 = 0$; this reduced to the given form is $0.2 = x - \frac{1}{3}x^3$, hence $a = 0.2$, $b = 0$, $c = -\frac{1}{3}$, and then

$$x = 0.2 + \tfrac{1}{3} \cdot 0.2^3 + \tfrac{1}{3} \cdot 0.2^5 + \text{etc.} = +0.20277,$$

which is the value of one of the roots correct to the fourth decimal place. This equation has three real roots, but the series gives only one of them; the others can, however, be found if their approximate values are known. Thus, one root is about $+1.6$, and by placing $x = y + 1.6$ there results an equation in y whose root by the series is found to be $+0.0218$, and hence $+1.6218$ is another root of $x^3 - 3x + 0.6 = 0$.

* This method is given by J. B. Mott in The Analyst, 1882. Vol. IX, p. 104.

Cardan's expression for the root of a cubic equation can be expressed as a series by developing each of the cube roots by the binomial formula and adding the results. Let the equation be $y^3 + 3By + 2C = 0$, whose root is, by Art. 9,

$$y = (-C + \sqrt{B^3 + C^2})^{\frac{1}{3}} + (-C - \sqrt{B^3 + C^2})^{\frac{1}{3}},$$

then this development gives the series,

$$y = 2(-C)^{\frac{1}{3}}\left(1 - \frac{2}{2}r - \frac{2.5.8}{2.3.4}r^2 - \frac{2.5.8.11.14}{2.3.4.5.6}r^3 - \ldots\right),$$

in which r represents the quantity $(B^3 + C^2)/3C^2$. If $r = 0$ the equation has two equal roots and the third root is $2(-C)^{\frac{1}{3}}$. If r is numerically greater than unity the series is divergent, and the solution fails. If r is numerically less than unity and sufficiently small to make a quick convergence, the series will serve for the computation of one real root. For example, take the equation $x^3 - 6x + 6 = 0$, where $B = -2$ and $C = 3$; hence $r = 1/81$, and one root is

$$y = -2.8845(1 - 0.01235 - 0.00051 - 0.00032 -) = -2.846,$$

which is correct to the third decimal. In comparatively few cases, however, is this series of value for the solution of cubics.

Many other series for the expression of the roots of equations, particularly for trinomial equations, have been devised. One of the oldest is that given by Lambert in 1758, whereby the root of $x^n + ax - b = 0$ is developed in terms of the ascending powers of b/a. Other solutions were published by Euler and Lagrange. These series usually give but one root, and this only when the values of the coefficients are such as to render convergence rapid.

Prob. 19. Consult Euler's Anleitung zur Algebra (St. Petersburg, 1771), pp. 143–150, and apply his method of series to the solution of a quartic equation.

ART. 14. COMPUTATION OF ALL ROOTS.

A comprehensive and valuable method for the solution of equations by series was developed by McClintock, in 1894, by means of his Calculus of Enlargement.* By this method all the roots, whether real or imaginary, may be computed from a single series. The following is a statement of the method as applied to trinomial equations:

Let $x^n = nAx^{n-k} + B^n$ be the given trinomial equation. Substitute $x = By$ and thus reduce the equation to the form $y^n = nay^{n-k} + 1$ where $a = A/B^k$. Then if B^n is positive, the roots are given by the series

$$y = \omega + \omega^{1-k}a + \omega^{1-2k}(1 - 2k + n)a^2/2!$$
$$+ \omega^{1-3k}(1 - 3k + n)(1 - 3k + 2n)a^3/3!$$
$$+ \omega^{1-4k}(1 - 4k + n)(1 - 4k + 2n)(1 - 4k + 3n)a^4/4! + \cdots,$$

in which ω represents in succession each of the roots of unity. If, however, B^n is negative, the given equation reduces to $y^n = nay^{n-k} - 1$, and the same series gives the roots if ω be taken in succession as each of the roots of -1.

In order that this series may be convergent the value of a^n must be numerically less than $k^{-k}(n - k)^{k-n}$; thus for the quartic $y^4 = 4ax + 1$, where $n = 4$ and $k = 3$, the value of a must be less than $27^{-\frac{1}{4}}$.

To apply this method to the cubic equation $x^3 = 3Ax \pm B^2$, place $n = 3$ and $k = 2$, and put $y = Bx$. It then becomes $y^3 = 3ay \pm 1$ where $a = A/B^2$, and the series is

$$y = \omega + \omega^2 a - \tfrac{1}{3}\omega a^2 + \tfrac{1}{3}\omega^2 a^3 + \cdots,$$

in which the values to be taken for ω are the cube roots of 1 or -1, as the case may be. For example, let $x^3 - 2x - 5 = 0$. Placing $y = 5^{\frac{1}{3}}x$, this reduces to $y^3 = 0.684\,y + 1$. Here $a = 0.228$, and as this is less than $4^{-\frac{1}{3}}$ the series is convergent. Making $\omega = 1$, the first root is

$$y = 1 + 0.2280 - 0.0039 + 0.0009 = 1.2250.$$

*See Bulletin of American Mathematical Society, 1894, Vol. I, p. 3; also American Journal of Mathematics, 1895, Vol. XVII, pp. 89-110.

Next making $\omega = -\frac{1}{2} + \frac{1}{2}\sqrt{-3}$, ω^2 is $-\frac{1}{2} - \frac{1}{2}\sqrt{-3}$, and the corresponding root is found to be

$$y = -0.6125 + 0.3836\sqrt{-3}.$$

Again, making $\omega = -\frac{1}{2} - \frac{1}{2}\sqrt{-3}$ the third root is found to be the conjugate imaginary of the second. Lastly, multiplying each value of y by $5^{\frac{1}{3}}$,

$$x = 2.095, \qquad x = -1.047 \pm 1.136\sqrt{-1},$$

which are very nearly the roots of $x^3 - 2x - 5 = 0$.

In a similar manner the cubic $x^3 + 2x + 5 = 0$ reduces to $y^3 = -0.684y - 1$, for which the series is convergent. Here the three values of ω are, in succession, -1, $\frac{1}{2} + \frac{1}{2}\sqrt{-3}$, $-\frac{1}{2} + \frac{1}{2}\sqrt{-3}$, and the three roots are $y = -0.777$ and $y = 0.388 \pm 1.137i$.

When all the roots are real, the method as above stated fails because the series is divergent. The given equation can, however, be transformed so as to obtain $n - k$ roots by one application of the general series and k roots by another. As an example, let $x^3 - 243x + 330 = 0$. For the first application this is to be written in the form

$$x = \frac{x^3}{243} + \frac{330}{243},$$

for which $n = 1$ and $k = -2$. To make the last term unity place $x = \frac{330}{243}y$, and the equation becomes

$$y = \frac{330^3}{243}y^3 + 1,$$

whence $a = 330^3/3.243^3$. These values of n, k, and a are now inserted in the above general value of y, and ω made unity: thus $y = 0.9983$, whence $x_1 = 1.368$ is one of the roots. For the second application the equation is to be written

$$x^3 = -\frac{330}{243}x^{-1} + 243,$$

for which $n = 2$ and $k = 3$. Placing $x = 243^{\frac{1}{3}}y$, this becomes

$$y^2 = -\frac{340}{243^{\frac{2}{3}}}y^{-1} + 1,$$

whence $a = -110/243^{\frac{2}{3}}$, and the series is convergent. These values of n, k, and a are now inserted in the formula for y, and ω is made $+1$ and -1 in succession, thus giving two values for y, from which $x_2 = 14.86$ and $x_3 = -16.22$ are the other roots of the given cubic.

McClintock has also given a similar and more general method applicable to other algebraic equations than trinomials. The equation is reduced to the form $y^n = na \cdot \phi y \pm 1$, where $na \cdot \phi y$ denotes all the terms except the first and the last. Then the values of y are expressed by the series

$$y = \omega + \omega^{1-n}\phi\omega \cdot a + \omega^{1-n}\frac{d}{d\omega}\omega^{1-n}(\phi\omega)^2 \cdot \frac{a^2}{2!} +$$

$$+ \left(\omega^{1-n}\frac{d}{d\omega}\right)^2 \omega^{1-n}(\phi\omega)^3 \cdot \frac{a^3}{3!} + \dots,$$

in which the values of ω are to be taken as before. The method is one of great importance in the theory of equations, as it enables not only the number of real and imaginary roots to be determined, but also gives their values when the convergence of the series is secured.

Prob. 20. Compute by the above method all the roots of the quartic $x^4 + x + 10 = 0$.

Art. 15. Conclusion.

While this Chapter forms a supplement to the theory of equations as commonly given in college text-books, yet the brief space allotted to it has prevented the discussion and development of many interesting branches. Chief among these is the topic of complex or imaginary roots, particularly of their graphical representation and their numerical computation. Although such roots rarely, if ever, are required in the solution of problems in physical science, their determination is a matter of much theoretic interest. It may be mentioned, however, that both the regula falsi and Newton's approximation rule may, by a slight modification, be adapted to the computation of these imaginary roots, approximate values of them being first obtained by trial.

A method of solution of numerical algebraic equations, which may be called a logarithmic process, was published by Gräffe in 1837, and exemplified by Encke in 1841.* It consists in deriving from the given equation another equation whose roots are high powers of those of the given one, the coefficients of the latter then easily furnishing the real roots and the moduluses of the imaginary roots. . The method, although little known, is without doubt one of high practical values, as logarithmic tables are used throughout; moreover, Encke states that the time required to completely solve an equation of the seventh degree with six imaginary roots, as accurately as can be done with seven-place tables, is less than three hours.

The algebraic solutions of the quadratic, cubic, and quartic equations are valid not only for real coefficients, but also for imaginary ones. In the latter case the imaginary roots do not necessarily occur in pairs. The method of McClintock has the great merit that it is applicable also to equations with imaginary coefficients; it constitutes indeed the only general method by which the roots in such cases can be computed.

Prob. 21. Compute by McClintock's series the roots of the equation $x^3 - ix - 1 = 0$.

Prob. 22. Solve the equation $\cos x \cosh x + 1 = 0$, and also the equation $x - e^x = 0$. (For answers see Crelle's Journal für Mathematik, 1841, Vol. XXII, pp. 1–62.)

* See Crelle's Journal für Mathematik, 1841, pp. 193–248.

Chapter II.

DETERMINANTS.

By Laenas Gifford Weld,

Professor of Mathematics in the State University of Iowa.

Art. 1. Introduction.

As early as 1693 Leibnitz arrived at some vague notions regarding the functions which we now know as determinants. His researches in this subject, the first account of which is contained in his correspondence with De L'Hospital, resulted simply in the statement of some rather clumsy rules for eliminating the unknowns from systems of linear equations, and exerted no influence whatever upon subsequent investigations in the same direction. It was over half a century later, in 1750, that Gabriel Cramer first formulated an intelligible and general definition of the functions, based upon the recognition of the two classes of permutations, as presently to be set forth.

Though Cramer failed to recognize, even to the same extent as Leibnitz, the importance of the functions thus defined, the development of the subject from this time on has been almost continuous and often rapid. The name "determinant" is due to Gauss, who, with Vandermonde, Lagrange, Cauchy, Jacobi, and others, ranks among the great pioneers in this development.

Within recent years the theory of determinants has come into very general use, and has, in the hands of such mathematicians as Cayley and Sylvester, led to results of the greatest interest and importance, both through the study of special forms of the functions themselves and through their applications.*

* A list of writings on Determinants is given by Muir in Quarterly Journal of Mathematics, 1881, Vol. XVIII, pp. 110-149.

Art. 2. Permutations.

The various orders in which the elements of a group may be arranged in a row are called their permutations.

Any two elements, as a and b, may be arranged in two orders: ab and ba. A third, as c, may be introduced into each of these two permutations in three ways: before either element, or after both; thus giving $3 \times 2 = 6$ permutations of the three elements. In like manner an additional element may be introduced into each of the permutations of i elements in $(i + 1)$ ways: before any one of them, or after all. Hence, in general, if P_i denote the number of permutations of i elements, $P_{i+1} = (i + 1)P_i$. Now, $P_3 = 3 \times 2 \times 1 = 3!$; hence $P_4 = 4 \times 3! = 4!$; and, n being any integer,

$$P_n = n(n - 1)(n - 2) \ldots 1 = n!.$$

That is, the number of permutations of n elements is $n!$.

For all integral values of n greater than unity, $n!$ is an even number.

If the elements of any group be represented by the different letters, $a, b, c, \ldots$, the alphabetical order will be considered as the *natural order* of the elements. If represented by the same letter with different indices, thus:

$$a_1, a_2, a_3, \ldots; \quad \text{or thus:} \quad a', a'', a''', \ldots,$$

the natural order of the elements is that in which the indices form a continually increasing series.

Any two elements, whether adjacent or not, standing in their natural order in a permutation constitute a permanence; standing in an order which is the reverse of the natural, an inversion. Thus, in the permutation *daecb*, the permanences are *de, ae, ab, ac*; the inversions, *da, dc, db, ec, cb, cb*.

The permutations of the elements of a group are divided into two classes, viz.: even or positive permutations, in which the number of inversions is even; and odd or negative permutations, in which the number of inversions is odd.

When the elements are arranged in the natural order the number of inversions is zero—an even number.

Thus, the even or positive permutations of the elements a_1, a_2, a_3 are

$$a_1 \, a_2 \, a_3, \quad a_2 \, a_3 \, a_1, \quad a_3 \, a_1 \, a_2;$$

while the odd or negative permutations are

$$a_3 \, a_2 \, a_1, \quad a_1 \, a_3 \, a_2, \quad a_2 \, a_1 \, a_3.$$

ART. 3. INTERCHANGE OF TWO ELEMENTS.

It will now be shown that if, in any permutation of the elements of a group, two of the elements be interchanged the class of the permutation will be changed.

Let q and s be the elements in question. Then, representing collectively all the elements which precede these two by P, those which fall between them by R, and those which follow by T, any permutation of the group may be written

$$PqRsT.$$

Of the elements R, supposed to be r in number, let represent

h the number of an order higher than q,
i " " " " " lower " q,
j " " " " " lower " s,
k " " " " " higher " s.

It is evident that no change in the order of the elements qRs can affect their relations to the elements of either P or T. Then, passing from the order $PqRsT$ to the order

$$PRqsT$$

changes the number of inversions by $(h - i)$; and passing from this to the order

$$PsRqT$$

again changes the number of inversions by $(j - k) \pm 1$, the $\begin{Bmatrix} \text{plus} \\ \text{minus} \end{Bmatrix}$ sign being used as q is of $\begin{Bmatrix} \text{lower} \\ \text{higher} \end{Bmatrix}$ order than s. The total change in the number of inversions due to the interchange of the two elements in question is, therefore,

$$h - i + j - k \pm 1.$$

But since $i = r - h$ and $k = r - j$, this may be written

$$2(h + j - r) \pm 1,$$

which is an odd number for all admissible values of h, j, and r. Hence, the interchange of any two elements in a permutation changes the number of inversions by an odd number, thus changing the class of the permutation.

Art. 4. Positive and Negative Permutations.

Of all the permutations of the elements of a group, one half are even and one half odd.

To prove this, write out all the permutations. Now choose any two of the elements and interchange them in each permutation. The result will be the same set of permutations as before, only differently arranged. But each $\begin{Bmatrix} \text{even} \\ \text{odd} \end{Bmatrix}$ permutation of the old set has been converted into an $\begin{Bmatrix} \text{odd} \\ \text{even} \end{Bmatrix}$ one in the new. Hence, in either set, there are as many even permutations as odd; that is, one half are even and one half odd.

Prob. 1. Classify the following permutations:

(1) $b c d e a$;　　　(2) III V I II IV;　　　(3) $k n i m l j$;
(4) $a'' a^v a' a^{iv} a'''$; (5) $\beta \epsilon \gamma \zeta \alpha \delta$;　　(6) $5 2 4 1 3$;
(7) $x_1 x_3 x_0 x_4 x_2 x_5$; (8) F. Tu. M. Th. W.; (9) $\mu \kappa \nu \iota \lambda$.

Prob. 2. Derive the formula for the number of permutations of n elements taken m at a time. (Ans. $n!/(n - m)!$.)

Prob. 3. How many combinations of m elements arranged in the natural order may be selected from a group of n elements? (Ans. $n!/m!(n - m)!$.)

Prob. 4. Show that $0! = 1$.

Art. 5. The Determinant Array.

Assume n^2 elements arranged in n vertical ranks or columns, and n horizontal ranks or rows, thus:

$$a_1' \; a_1'' \ldots a_1^{(n)}$$
$$a_2' \; a_2'' \ldots a_2^{(n)}$$
$$\cdots \cdots \cdots \cdots \cdots$$
$$a_n' \; a_n'' \ldots a_n^{(n)}.$$

In this array all the elements in the same column have the same superscript, and those in the same row the same subscript. The columns being arranged in order from left to right, and the rows likewise in order from the top row downward, the position of any element of the array is shown at once by its indices. Thus, a_5''' is in the third column and the fifth row of the above array.

The diagonal passing through the elements $a_1', a_2'', \ldots a_n^{(n)}$ is called the principal diagonal of the array; that passing through $a_n', a_{n-1}'', \ldots a_1^{(n)}$, the secondary diagonal. The position occupied by the element a_1' is designated as the leading position.

ART. 6.   Determinant as Function of n^2 Elements.

The array just considered, inclosed between two vertical bars, thus :

$$\begin{vmatrix} a_1' & a_1'' & \ldots & a_1^{(n)} \\ a_2' & a_2'' & \ldots & a_2^{(n)} \\ \ldots & \ldots & \ldots & \ldots \\ a_n' & a_n'' & \ldots & a_n^{(n)} \end{vmatrix}$$

is used in analysis to represent a certain function of its n^2 elements called their determinant.*   This function may be defined as follows:

Write down the product of the elements on the principal diagonal, taking them in the natural order; thus :

$$a_1' a_2'' a_3''' \ldots a_n^{(n)}.$$

This product is called the principal term of the determinant. Now permute the subscripts in this principal term in every possible way, leaving the superscripts undisturbed.   To such of the $n!$ resulting terms as involve the even permutations of the subscripts give the positive sign; to those involving the odd

* This notation was first employed by Cauchy in 1815.   See Dostor's Théorie des déterminants, Paris, 1877.

permutations, the negative sign. The algebraic sum of all the terms thus formed is the determinant represented by the given array.

ART. 7. EXAMPLES OF DETERMINANTS.

Applying the process above explained to the array of four elements gives

$$\begin{vmatrix} a_1' & a_1'' \\ a_2' & a_2'' \end{vmatrix} \equiv a_1'a_2'' - a_2'a_1''. \tag{1}$$

As an example of a determinant of nine elements, with its expansion, may be written

$$\begin{vmatrix} a_1' & a_1'' & a_1''' \\ a_2' & a_2'' & a_2''' \\ a_3' & a_3'' & a_3''' \end{vmatrix} \equiv + a_1'a_2''a_3''' + a_2'a_3''a_1''' + a_3'a_1''a_2''' \\ - a_3'a_2''a_1''' - a_1'a_3''a_2''' - a_2'a_1''a_3'''. \tag{2}$$

It is evident, from the mode of its formation, that each term of the expansion of a determinant contains one, and only one, element from each column and each row of the array.

It follows that every complete determinant is a homogeneous function of its elements. The degree of this function, with respect to its elements, is called the order of the determinant. Thus, (1) and (2) are of the second and third order respectively.

The definition of a determinant given in the preceding article is once more illustrated by the following example of a determinant of the fourth order with its complete development:

$$\begin{vmatrix} a_1 & b_1 & c_1 & d_1 \\ a_2 & b_2 & c_2 & d_2 \\ a_3 & b_3 & c_3 & d_3 \\ a_4 & b_4 & c_4 & d_4 \end{vmatrix} \equiv \begin{aligned} &+ a_1b_2c_3d_4 - a_1b_2c_4d_3 - a_1b_3c_2d_4 + a_1b_3c_4d_2 \\ &+ a_1b_4c_2d_3 - a_1b_4c_3d_2 - a_2b_1c_3d_4 + a_2b_1c_4d_3 \\ &+ a_2b_3c_1d_4 - a_2b_3c_4d_1 - a_2b_4c_1d_3 + a_2b_4c_3d_1 \\ &+ a_3b_1c_2d_4 - a_3b_1c_4d_2 - a_3b_2c_1d_4 + a_3b_2c_4d_1 \\ &+ a_3b_4c_1d_2 - a_3b_4c_2d_1 - a_4b_1c_2d_3 + a_4b_1c_3d_2 \\ &+ a_4b_2c_1d_3 - a_4b_2c_3d_1 - a_4b_3c_1d_2 + a_4b_3c_2d_1 \end{aligned} \tag{3}$$

It will be noticed that, in this case, the columns are ranked alphabetically instead of by the numerical values of a series of indices.

ART. 8.   NOTATIONS.

Besides the notations already employed, the following is very extensively used :

$$\begin{vmatrix} a_{11} & a_{12} & \cdots & a_{1n} \\ a_{21} & a_{22} & \cdots & a_{2n} \\ \cdots & \cdots & \cdots & \cdots \\ a_{n1} & a_{n2} & \cdots & a_{nn} \end{vmatrix}.$$

This is called the double-subscript notation ; the first subscript indicating the rank of the row, the second that of the column. Thus the element a_{23} is in the second row and the third column. The letters are sometimes omitted, the elements being thus represented by the double subscripts alone.*

Instead of writing out the array in full, it is customary, when the elements are merely symbolic, to write only the principal term and enclose it between vertical bars.   This is called the umbral notation.   Thus, the determinant of the nth order is written

$$| \, a_1' \, a_2'' \, \ldots \, a_n^{(n)} \, | \; ;$$

or, using double subscripts,

$$| \, a_{11} \, a_{22} \, \ldots, \, a_{nn} \, | \, .$$

These last two forms are sometimes still further abridged to

$$| \, a_1^{(n)} \, | \quad \text{and} \quad | \, a_{1,n} \, | \, ,$$

respectively.

Prob. 5. Write out the developments of the following determinants:

(1) $\begin{vmatrix} a_1 & b_1 \\ a_2 & b_2 \end{vmatrix}$;   (2) $\begin{vmatrix} p' & p'' \\ q' & q'' \end{vmatrix}$;   (3) $\begin{vmatrix} p' & q' \\ p'' & q'' \end{vmatrix}$;   (4) $\begin{vmatrix} a & b \\ \alpha & \beta \end{vmatrix}$;

* Leibnitz indicated the elements of a determinant in this same manner, though he made no use of the array.

(5) $\left|\,a_1 b_2 c_3\,\right|$; (6) $\begin{vmatrix} p' & p'' & p''' \\ q' & q'' & q''' \\ r' & r'' & r''' \end{vmatrix}$; (7) $\begin{vmatrix} p' & q' & r' \\ p'' & q'' & r'' \\ p''' & q''' & r''' \end{vmatrix}$; (8) $\begin{vmatrix} a & b & c \\ \alpha & \beta & \gamma \\ x & y & z \end{vmatrix}$;

(9) $\left|\,11,\ 22\,\right|$; (10) $\left|\,a_{1,3}\,\right|$; (11) $\left|\,l_0 m_1 n_2\,\right|$; (12) $\left|\,a_{11} a_{22} a_{33} a_{44}\,\right|$.

Prob. 6. How many terms are there in the development of the determinant $\left|\,a_1{}^{vi}\,\right|$?

In the above determinant tell the signs of the terms :

$$(1)\ a_6' a_2'' a_1''' a_4^{iv} a_6^v a_3^{vi}; \qquad (2)\ a_1' a_3'' a_2''' a_6^{iv} a_6^v a_4^{vi};$$

$$(3)\ a_6' a_4'' a_6''' a_1^{iv} a_2^v a_2^{vi}.$$

Prob. 7. Show that in the expansion of any determinant, all of whose elements are positive, one half the terms are positive and one half negative.

Prob. 8. In determinants of what orders is the term containing the elements on the secondary diagonal (called the secondary term) positive ?

Prob. 9. What is the order of the determinant whose secondary term contains 10 inversions ? 36 inversions ?

Prob. 10. In the expansion of a determinant of the nth order, how many terms contain the leading element ?

ART. 9. SECOND AND THIRD ORDERS.

Simple rules will now be given for writing out the expansions of determinants of the second and third orders directly from the arrays by which they are represented.

To expand a determinant of the second order, write the product of the elements on the principal diagonal minus the product of those on the secondary diagonal, thus :

$$\begin{vmatrix} a & b \\ c & d \end{vmatrix} \equiv ad - bc.$$

Likewise, $\begin{vmatrix} -9 & 5 \\ -2 & \frac{1}{3} \end{vmatrix} = -3 + 10 = 7.$

The following method is applicable to determinants of the third order :*

---

* This method was first given by Sarrus, and is often called the rule of Sarrus; see Finck's Éléments d'Algèbre, 1846, p. 95.

Beneath the square array let the first two rows be repeated in order, as shown in the figure. Now write down six terms, each the product of the three elements lying along one of the six oblique lines parallel to the diagonals of the original square. Give to those terms whose elements lie on lines parallel to the principal diagonal the positive sign; to the others, the negative sign. The result is the required expansion. Ap-

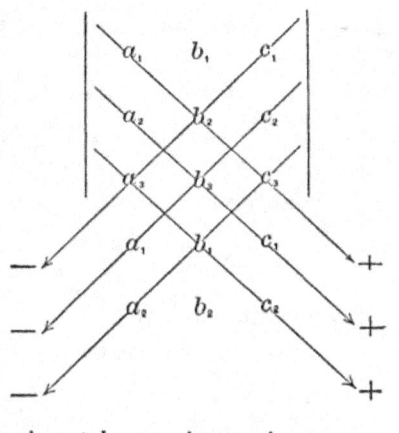

plying the method to the determinant just written gives

$$|a_1b_2c_3| = a_1b_2c_3 + a_2b_3c_1 + a_3b_1c_2 - a_3b_2c_1 - a_1b_3c_2 - a_2b_1c_3.$$

After a little practice the repetition of the first two rows will be dispensed with.

The above methods are especially useful in expanding determinants whose elements are not marked with indices, or in evaluating those having numerical elements. No such simple methods can be given for developing determinants of higher orders, but it will be shown later that these can always be resolved into determinants of the third or second order.

Prob. 11. Develop the following determinants:

(1) $\begin{vmatrix} a & h & g \\ h & b & f \\ g & f & c \end{vmatrix}$;

(2) $\begin{vmatrix} 0 & -n & -m \\ n & 0 & -l \\ m & l & 0 \end{vmatrix}$;

(3) $\begin{vmatrix} A & c & b \\ c & B & a \\ b & a & C \end{vmatrix}$;

(4) $\begin{vmatrix} x_0 & y_0 & 1 \\ x_1 & y_1 & 1 \\ x_2 & y_2 & 1 \end{vmatrix}$;

(5) $\begin{vmatrix} 1 & P & Q \\ 0 & \cos\alpha & \sin\beta \\ 0 & \sin\alpha & \cos\beta \end{vmatrix}$;

(6) $\begin{vmatrix} \cos\alpha & \sin\beta \\ \sin\alpha & \cos\beta \end{vmatrix}$;

(7) $\begin{vmatrix} 1 & \cos\alpha \\ \cos\alpha & 1 \end{vmatrix}$;

(8) $\begin{vmatrix} 1 & \sqrt{-1} \\ 4 & \sqrt{-2} \end{vmatrix}$;

(9) $\begin{vmatrix} a & b & c \\ c & a & b \\ b & c & a \end{vmatrix}$.

Prob. 12. Evaluate the following:

(1) $\begin{vmatrix} 1 & 2 & 3 \\ 3 & 1 & 2 \\ 2 & 3 & 1 \end{vmatrix}$;

(2) $\begin{vmatrix} -2 & -2\frac{1}{2} \\ 0 & -2 & 0 \\ 12 & 2 & 1 \end{vmatrix}$;

(3) $\begin{vmatrix} -1 & -\sqrt{-1} & -\sqrt{-1} \\ \sqrt{-1} & -1 & -\sqrt{-1} \\ \sqrt{-1} & \sqrt{-1} & -1 \end{vmatrix}$.

(Ans. 18; 16; 2.)

ART. 10. INTERCHANGE OF ROWS AND COLUMNS.

Any term in the development of the determinant $|a_1^{(n)}|$ may be written

$$\pm a_h' \, a_i'' \, a_j''' \ldots a_l^{(n)},$$

in which $hij \ldots l$ is some permutation of the subscripts $1, 2, 3, \ldots n$. Designate by u the number of inversions in $hij \ldots l$. Also, let v be the number of interchanges of two elements necessary to bring the given term into the form

$$\pm a_1^{(p)} \, a_2^{(q)} \, a_3^{(r)} \ldots a_n^{(t)},$$

in which the subscripts are arranged in the natural order, while $pqr \ldots t$ is a certain permutation of the superscripts $', '', ''', \ldots^{(n)}$.

This permutation is even or odd according as v is even or odd. But u and v are obviously of the same class; that is, both are even or both odd. Hence the permutations $hij \ldots l$ and $pqr \ldots t$ are of the same class; and the term will have the same sign, whether the sign be determined by the class of the permutation of the subscripts when the superscripts stand in the natural order, or by the class of the permutation of the superscripts when the order of the subscripts is natural.

It follows that the same development of the determinant array will be obtained if, instead of proceeding as indicated in Art. 6, the superscripts of the principal term be permuted, the subscripts being left in the natural order, and the sign of each of the resulting terms written in accordance with the class of the permutations of its superscripts.

Passing from one of these methods of development to the other amounts to the same thing as changing each column of the array into a row of the same rank, and *vice versa*. Hence, a determinant is not altered by changing the columns into corresponding rows and the rows into corresponding columns. Thus:

$$\begin{vmatrix} a_1' & a_1'' & \ldots & a_1^{(n)} \\ a_2' & a_2'' & \ldots & a_2^{(n)} \\ \hdotsfor{4} \\ a_n' & a_n'' & \ldots & a_n^{(n)} \end{vmatrix} \equiv \begin{vmatrix} a_1' & a_2' & \ldots & a_n' \\ a_1'' & a_2'' & \ldots & a_n'' \\ \hdotsfor{4} \\ a_1^{(n)} & a_2^{(n)} & \ldots & a_n^{(n)} \end{vmatrix}.$$

Whatever theorem, therefore, is demonstrated with reference to the rows of a determinant is also true with reference to the columns.

The rows and columns of a determinant array are alike called lines.

Art. 11. Interchange of Two Parallel Lines.

If any two parallel lines of a determinant be interchanged, the determinant will be changed only in sign.

For, interchanging any two parallel lines of a determinant array amounts to the same thing as interchanging, in every term of the expansion, the indices which correspond to these lines. Since this changes the class of each permutation of the indices in question from odd to even or from even to odd, it changes the sign of each term of the expansion, and therefore that of the whole determinant.

It follows from the above that if any line of a determinant be passed over m parallel lines to a new position in the array the new determinant will be equal to the original one multiplied by $(-1)^m$.

The element $a_k^{(s)}$ may be brought to the leading position by passing the kth row over the $(k-1)$ preceding rows, and the sth column over the $(s-1)$ preceding columns. This being done the determinant is multiplied by

$$(-1)^{k-1} \cdot (-1)^{s-1} = (-1)^{k+s},$$

which changes its sign or not according as $(k+s)$ is odd or even.

The position occupied by $a_k^{(s)}$ is called a positive position when $(k+s)$ is even; a negative position when $(k+s)$ is odd.

Art. 12. Two Identical Parallel Lines.

A determinant in which any two parallel lines are identical is equal to zero.

For the interchange of these two parallel lines, while it

changes the sign of the determinant, will in no way alter its value. The value then, if finite, can only be zero.

ART. 13. MULTIPLYING BY A FACTOR.

Multiplying each element of a line of a determinant by a given factor multiplies the determinant by that factor.

Since each term of the development contains one and only one element from the line in question (Art. 7), then multiplying each element of this line by the given factor multiplies each term of the development, and therefore the whole determinant, by the same factor.

It follows that, if the elements of any line of a determinant contain a common factor, this factor may be canceled and written outside the array as a factor of the whole determinant; thus:

$$\begin{vmatrix} a_{11} & . . & m\,a_{1i} & . . . & a_{1n} \\ a_{21} & . . & m\,a_{2i} & . . . & a_{2n} \\ . & . & . . . . & \\ a_{n1} & . . & m\,a_{ni} & . . . & a_{nn} \end{vmatrix} = m \begin{vmatrix} a_{11} & a_{22} & . . . & a_{nn} \end{vmatrix} .$$

A determinant in which the elements of any line have a common ratio to the corresponding elements of any parallel line is equal to zero. For this common ratio may be written outside the array, which will then have two identical lines. Its value is therefore zero (Art. 12).

A determinant having a line of zeros is equal to zero.

ART. 14. A LINE OF POLYNOMIAL ELEMENTS.

A determinant having a line of elements each of which is the sum of two or more quantities can be expressed as the sum of two or more determinants.

$$\text{Let} \quad \begin{vmatrix} a_1 & (b_1 - b_1' + b_1'' \pm \cdots) & c_1 \cdots \\ a_2 & (b_2 - b_2' + b_2'' \pm \cdots) & c_2 \cdots \\ a_3 & (b_3 - b_3' + b_3'' \pm \cdots) & c_3 \cdots \\ . . . . \end{vmatrix} \equiv \Delta \qquad (1)$$

be such a determinant. Then, if

$$B_i \equiv b_i - b_i' + b_i'' \pm \cdots ,$$

any term of the expansion of the determinant Δ is

$$\pm a_h\, B_i\, c_j \ldots = \pm a_h\, b_i\, c_j \ldots \mp a_h\, b_i'\, c_j \ldots$$
$$\pm a_h\, b_i''\, c_j \ldots \pm \ldots \qquad (2)$$

The terms in the expansion of Δ are obtained by permuting the subscripts $h,\, i,\, j,\, \ldots$ of $a_h\, B_i\, c_j \ldots$. But permuting at the same time the subscripts of the terms in the second member of (2), and giving to each term thus obtained its proper sign, there results

$$\Delta \equiv |\, a_1 B_2 c_3 \ldots\, | = |\, a_1 b_2 c_3 \ldots\, | - |\, a_1 b_2' c_3 \ldots\, | + |\, a_1 b_2'' c_3 \ldots\, | \pm \ldots,$$

which proves the theorem.

ART. 15. COMPOSITION OF PARALLEL LINES.

If each element of a line of a determinant be multiplied by a given factor and the product added to the corresponding element of any parallel line, the value of the determinant will not be changed ; thus:

$$\begin{vmatrix} a_{11}\, a_{12}\, a_{13} \ldots a_{1n} \\ \cdots\cdots\cdots \\ a_{n1}\, a_{n2}\, a_{n3} \ldots a_{nn} \end{vmatrix} = \begin{vmatrix} a_{11}\, a_{12}\, (a_{13} + ma_{11}) \ldots a_{1n} \\ \cdots\cdots\cdots\cdots \\ a_{n1}\, a_{n2}\, (a_{n3} + ma_{n1}) \ldots a_{nn} \end{vmatrix}.$$

This will appear upon resolving the second member into two determinants (Art. 14), one of which will be the given determinant, while the other, upon removal of the given factor, will vanish because of having two identical lines.

In like manner any number of parallel lines may be combined without changing the value of the determinant, care being taken not to modify in any way the elements to which are added multiples of corresponding elements from other parallel lines. For example, $|\, a_{1,n}\, |$ is equivalent to

$$\begin{vmatrix} a_{11} & (la_{11} + a_{12} - ma_{13} + \ldots) & a_{12} \ldots a_{1n} \\ & (l(a_{21}+\lambda a_{11})+(a_{22}+\lambda a_{12}) & \\ (a_{21}+\lambda a_{11}) & & (a_{22}+\lambda a_{13}) \ldots (a_{2n}+\lambda a_{1n}) \\ & -m(a_{23} + \lambda a_{13}) + \ldots) & \\ \cdots\cdots\cdots\cdots\cdots\cdots\cdots & & \cdots\cdots\cdots \\ a_{n1} & (la_{n1} + a_{n2} - ma_{n3} + \ldots) & a_{n3} \ldots a_{nn} \end{vmatrix}.$$

ART. 16. BINOMIAL FACTORS.

A determinant which is a rational integral function of a and of b, such that if b is substituted for a the determinant vanishes, contains $(a - b)$ as a factor. For example,

$$\Delta \equiv \begin{vmatrix} a^2 - p^2 & a - q & a + r \\ b^2 - p^2 & b - q & b + r \\ p & q & r \end{vmatrix}$$

is divisible by $(a - b)$.

To prove this, let the expansion of any such determinant be written in the form

$$\Delta = m_0 + m_1 a + m_2 a^2 + \dots,$$

the coefficients $m_0, m_1, m_2, \dots$ being independent of a. Now when b is substituted for a the determinant vanishes. Hence,

$$0 = m_0 + m_1 b + m_2 b^2 + \dots$$

Subtracting this from the preceding gives

$$\Delta = m_1(a - b) + m_2(a^2 - b^2) + \dots$$

This being divisible by $(a - b)$, the theorem is proven.

Prob. 13. Prove the following without expansion :

(1) $\begin{vmatrix} 0 & -x & x \\ my & 0 & -y \\ -mnz & nz & 0 \end{vmatrix} = 0;$
(2) $\begin{vmatrix} 0 & c - b \\ -c & 0 & a \\ b & -a & 0 \end{vmatrix} = 0;$

(3) $\begin{vmatrix} b + c & a & a \\ b & c + a & b \\ c & c & a + b \end{vmatrix} = 2 \begin{vmatrix} 0 & c & b \\ c & 0 & a \\ b & a & 0 \end{vmatrix};$

(4) $\begin{vmatrix} b^2 + c^2 & a & a \\ a & & \\ b & \dfrac{c^2 + a^2}{b} & b \\ c & c & \dfrac{a^2 + b^2}{c} \end{vmatrix} = 2 \begin{vmatrix} 0 & c & b \\ c & 0 & a \\ b & a & 0 \end{vmatrix};$

(5) $\begin{vmatrix} a & \sin A & b - c \\ b & \sin B & c - a \\ c & \sin C & a - b \end{vmatrix} = 0$, the elements referring to the triangle ABC.

Prob. 14. Prove that

$$\begin{vmatrix} 1 & x-a & y-b \\ 1 & x_1-a & y_1-b \\ 1 & x_2-a & y_2-b \end{vmatrix} = \begin{vmatrix} 1 & x & y \\ 1 & x_1 & y_1 \\ 1 & x_2 & y_2 \end{vmatrix} = \begin{vmatrix} 1 & x & y \\ 0 & x_1-x & y_1-y \\ 0 & x_2-x & y_2-y \end{vmatrix}.$$

Prob. 15. Find the value of θ in the equation

$$\begin{vmatrix} \sin\theta & \sin\theta & 0 \\ 1 & 0 & 1 \\ 0 & \cos\theta & \cos\theta \end{vmatrix} = 0. \quad (\text{Ans. } \theta = \pi/4.)$$

Prob. 16. Show that the proportion $a:b::l:m$ may be written in the form $\begin{vmatrix} a & b \\ l & m \end{vmatrix} = 0$; and from the properties of this determinant prove the common theorems in proportion.

Prob. 17. Show that the determinant $\begin{vmatrix} ab & c^2 & c^2 \\ a^2 & bc & a^2 \\ b^2 & b^2 & ca \end{vmatrix}$ contains the factor $(bc + ca + ab)$.

Prob. 18. Resolve the following determinants into factors:*

(1) $\begin{vmatrix} 1 & a & a^2 \\ 1 & b & b^2 \\ 1 & c & c^2 \end{vmatrix}$;　　(2) $\begin{vmatrix} 1 & a & a^2 & a^3 \\ 1 & b & b^2 & b^3 \\ 1 & c & c^2 & c^3 \\ 1 & d & d^2 & d^3 \end{vmatrix}$;　　(3) $\begin{vmatrix} 1 & a_1 & a_1^2 \dots a_1^{n-1} \\ 1 & a_2 & a_2^2 \dots a_2^{n-1} \\ \dots\dots\dots\dots \\ 1 & a_n & a_n^2 \dots a_n^{n-1} \end{vmatrix}$;

(4) $\begin{vmatrix} 1 & 1 & 1 & 1 \\ a & b & c & d \\ a^2 & b^2 & c^2 & d^2 \\ a^4 & b^4 & c^4 & d^4 \end{vmatrix}$;　　(5) $\begin{vmatrix} 1 & 1 & 1 \\ a & b & c \\ a^3 & b^3 & c^3 \end{vmatrix}$;　　(6) $\begin{vmatrix} 1 & 1 & 1 \\ a & b & c \\ a^4 & b^4 & c^4 \end{vmatrix}$.

ART. 17.　CO-FACTORS; MINORS.

The terms of $\Delta \equiv |a_1^{(n)}|$ which contain the element a_1' may be obtained by expanding the determinant

$$\begin{vmatrix} a_1' & 0 & 0 & \dots 0 \\ a_2' & a_2'' & a_2''' & \dots a_2^{(n)} \\ \dots\dots\dots\dots\dots \\ a_n' & a_n'' & a_n''' & \dots a_n^{(n)} \end{vmatrix}. \tag{1}$$

For, in writing out this expansion each term is formed by taking one, and only one, element from each column and each

* These determinants belong to an important class known as alternates. See Hanus' Elements of Determinants, Boston, 1888, pp. 187–201.

row of the array (Art. 7). If, therefore, in selecting the elements for any term, any other element than a_1' be taken from the first column, the one taken from the first row must be zero. Hence, the only terms which do not vanish are those which contain the element a_1'.

Moreover, in the terms of the expansion of (1) which do not vanish, a_1' is multiplied by $(n-1)$ elements chosen one from each column and each row of

$$\begin{vmatrix} a_2'' & a_2''' & \dots & a_2^{n)} \\ \dots\dots\dots\dots \\ a_n'' & a_n''' & \dots & a_n^{(n)} \end{vmatrix}. \tag{2}$$

There are $(n-1)!$ such terms, any one of which may be written $\pm\, a_1' a_i'' a_j''' \dots a_l^{(n)}$; the sign being determined by the class of the permutation of the n subscripts $1, i, j, \dots l$. But since this is of the same class as the permutation of the $(n-1)$ subscripts $i, j, \dots l$, the sign of any term, $\pm\, a_1' a_i'' a_j''' \dots a_l^{(n)}$, of the expansion of (1) is the same as the sign of the corresponding term, $a_i'' a_j''' \dots a_l^{(n)}$, of the expansion of (2). Hence,

$$\begin{vmatrix} a_1' & 0 & 0 & \dots 0 \\ a_2' & a_2'' & a_2''' \dots a_2^{(n)} \\ \dots\dots\dots\dots\dots \\ a_n' & a_n'' & a_n''' \dots a_n^{(n)} \end{vmatrix} = a_1' \begin{vmatrix} a_2'' & a_2''' \dots a_2^{(n)} \\ \dots\dots\dots\dots \\ a_n'' & a_n''' \dots a_n^{(n)} \end{vmatrix}. \tag{3}$$

The determinant (2) is called the co-factor or complement of the element a_1' in the determinant $|a_1^{(n)}|$. It is obtained from this determinant by deleting the first column and the first row.

The co-factor of any element $a_k^{(s)}$ may be found in the same manner upon transposing this element to the leading position. But by this transposition the sign of the determinant will be changed or not according as $a_k^{(s)}$ occupies a negative or a positive position (Art. 11). Hence, to find the co-factor of any element $a_i^{(k)}$ of the determinant $|a_1^{(n)}|$, delete the row and the column to which the element belongs, giving the resulting determinant the $\left\{\begin{array}{l} \text{positive} \\ \text{negative} \end{array}\right\}$ sign when $(k+s)$ is $\left\{\begin{array}{l} \text{even} \\ \text{odd} \end{array}\right\}$.

The co-factor thus obtained is represented by the symbol

$$A_k^{(s)};$$

the sign-factor of which, $(-1)^{k+s}$, is intrinsic, i.e., included in the symbol itself, which is accordingly written as positive. The co-factors of the various elements of $|a_{11}a_{22}a_{33}|$ are as follows:

$A_{11} \equiv \begin{vmatrix} a_{22} & a_{23} \\ a_{32} & a_{33} \end{vmatrix};$ $A_{11} \equiv - \begin{vmatrix} a_{21} & a_{23} \\ a_{31} & a_{33} \end{vmatrix};$ $A_{13} \equiv \begin{vmatrix} a_{21} & a_{22} \\ a_{31} & a_{32} \end{vmatrix};$

$A_{21} \equiv - \begin{vmatrix} a_{12} & a_{13} \\ a_{32} & a_{33} \end{vmatrix};$ $A_{22} \equiv \begin{vmatrix} a_{11} & a_{13} \\ a_{31} & a_{33} \end{vmatrix};$ $A_{23} \equiv - \begin{vmatrix} a_{11} & a_{12} \\ a_{31} & a_{32} \end{vmatrix};$

$A_{31} \equiv \begin{vmatrix} a_{12} & a_{13} \\ a_{22} & a_{23} \end{vmatrix};$ $A_{32} \equiv - \begin{vmatrix} a_{11} & a_{13} \\ a_{21} & a_{23} \end{vmatrix};$ $A_{33} \equiv \begin{vmatrix} a_{11} & a_{12} \\ a_{21} & a_{22} \end{vmatrix}.$

The result obtained by deleting the kth row and the sth column of $\varDelta \equiv |a_1^{(n)}|$ is called the minor of the determinant with respect to the element $a_k^{(s)}$, and is written $\varDelta_{(k}^{(s}$. This minor is the same as the co-factor of the same element without its sign-factor; thus:

$$A_k^{(s)} = (-1)^{k+s}\varDelta_{(k}^{(s}.$$

Similarly $\varDelta_{(h,k}^{(p,s}$ is the result obtained by deleting the hth and kth rows and the pth and sth columns of $\varDelta$, and is called a second minor of the given determinant. Minors of still lower orders are obtained in a similar manner, and expressed by a similar notation. The kth minors are determinants of the order $(n-k)$.

ART. 18.   DEVELOPMENT IN TERMS OF CO-FACTORS.

The $(n-1)!$ terms of $|a_1^{(n)}|$ which contain $a_k^{(s)}$ are represented in the aggregate by $a_k^{(s)}A_k^{(s)}$ (Eq. 3, Art. 17). In like manner the groups of terms containing the successive elements $a_k{}', a_k{}'', \ldots a_k^{(n)}$ are respectively

$$a_k{}'A_k{}', \quad a_k{}''A_k{}'', \ldots a_k^{(n)}A_k^{(n)}.$$

Each one of these n groups includes $(n-1)!$ terms of the determinant $|a_1^{(n)}|$, no one of which is found in any other group.   In all of them, then, there are $n \times (n-1)!$ or $n!$ different terms of the determinant, which is the whole number. Hence,

$$| a_1^{(n)} | = a_k' A_k' + a_k'' A_k'' + \ldots + a_k^{(n)} A_k^{(n)}. \qquad (1)$$

Similarly (Art. 10),

$$| a_1^{(n)} | = a_1^{(s)} A_1^{(s)} + a_2^{(s)} A_2^{(s)} + \ldots + a_n^{(s)} A_n^{(s)}. \qquad (2)$$

Any determinant may, by means of either (1) or (2), be resolved into determinants of an order one lower.   Since, in these formulas $A_k', \ldots A_k^{(n)}$ or $A_1^{(s)}, \ldots A_n^{(s)}$ are themselves determinants, they may be resolved into determinants of an order still one lower in the same manner.   By continuing the process any determinant may ultimately be expressed in terms of determinants of the third or second order, which may be easily expanded by methods already given (Art. 9).

For example, let it be required to develop the determinant $\varDelta \equiv | a_1, b_2, c_3, d_4 |$.   Applying formula (1), letting $k = 1$, gives

$$\varDelta = a_1 \begin{vmatrix} b_2, c_2, d_2 \\ b_3, c_3, d_3 \\ b_4, c_4, d_4 \end{vmatrix} - b_1 \begin{vmatrix} a_2, c_2, d_2 \\ a_3, c_3, d_3 \\ a_4, c_4, d_4 \end{vmatrix} + c_1 \begin{vmatrix} a_2, b_2, d_2 \\ a_3, b_3, d_3 \\ a_4, b_4, d_4 \end{vmatrix} - d_1 \begin{vmatrix} a_2, b_2, c_2 \\ a_3, b_3, c_3 \\ a_4, b_4, c_4 \end{vmatrix}.$$

Upon a second application of the same formula this becomes

$$\varDelta = a_1 b_2 \begin{vmatrix} c_3, d_3 \\ c_4, d_4 \end{vmatrix} - a_1 c_2 \begin{vmatrix} b_3, d_3 \\ b_4, d_4 \end{vmatrix} + a_1 d_2 \begin{vmatrix} b_3, c_3 \\ b_4, c_4 \end{vmatrix}$$

$$- a_2 b_1 \begin{vmatrix} c_3, d_3 \\ c_4, d_4 \end{vmatrix} + b_2 c_1 \begin{vmatrix} a_3, d_3 \\ a_4, d_4 \end{vmatrix} - b_1 d_2 \begin{vmatrix} a_3, c_3 \\ a_4, c_4 \end{vmatrix}$$

$$+ a_2 c_1 \begin{vmatrix} b_3, d_3 \\ b_4, d_4 \end{vmatrix} - b_2 c_1 \begin{vmatrix} a_3, d_3 \\ a_4, d_4 \end{vmatrix} + c_1 d_2 \begin{vmatrix} a_3, b_3 \\ a_4, b_4 \end{vmatrix}$$

$$- a_2 d_1 \begin{vmatrix} b_3, c_3 \\ b_4, c_4 \end{vmatrix} + b_2 d_1 \begin{vmatrix} a_3, c_3 \\ a_4, c_4 \end{vmatrix} - c_2 d_1 \begin{vmatrix} a_3, b_3 \\ a_4, b_4 \end{vmatrix} .$$

The complete development may be written out directly from the above.   It is given in Eq. 3, Art. 7.

Prob. 19. Develop the following determinants:

$$(1) \begin{vmatrix} 1 & x & 1 & y \\ x & 1 & y & 1 \\ 1 & y & 1 & x \\ y & 1 & x & 1 \end{vmatrix}; \qquad (2) \begin{vmatrix} a & x & y & a \\ x & 0 & 0 & y \\ y & 0 & 0 & x \\ a & y & x & a \end{vmatrix}; \qquad (3) \begin{vmatrix} 0 & q & r & s \\ p & 0 & r & s \\ p & q & 0 & s \\ p & q & r & 0 \end{vmatrix}.$$

(Ans. $(x - y)^2\big((x + y)^2 - 4\big)$; $(x^2 - y^2)^2$; $3pqrs$.)

Prob. 20. Find the values of the following determinants:

$$(1) \begin{vmatrix} 1 & 2 & 3 & 4 \\ 2 & 3 & 4 & 1 \\ 3 & 4 & 1 & 2 \\ 4 & 1 & 2 & 3 \end{vmatrix}; \qquad (2) \begin{vmatrix} 0 & 1 & 0 & 2 \\ 1 & 0 & 2 & 0 \\ 0 & 2 & 0 & 1 \\ 2 & 0 & 1 & 0 \end{vmatrix}; \qquad (3) \begin{vmatrix} 3 & 5 & 3 & 1 \\ 6 & 6 & -1 & 1 \\ 9 & -3 & 5 & 1 \\ 8 & 3 & 0 & 1 \end{vmatrix};$$

$$(4) \begin{vmatrix} 0 & 1 & 1 & 1 \\ 1 & 0 & 1 & 1 \\ 1 & 1 & 0 & 1 \\ 1 & 1 & 1 & 0 \end{vmatrix}; \qquad (5) \begin{vmatrix} 3 & 3 & 3 & 3 \\ 3 & 2 & 2 & 2 \\ 2 & 2 & 1 & 1 \\ 1 & 1 & 1 & 0 \end{vmatrix}; \qquad (6) \begin{vmatrix} 0 & 0 & 0 & 3 \\ 1 & 0 & 0 & 2 \\ 0 & 1 & 0 & 1 \\ 0 & 0 & 1 & 0 \end{vmatrix}.$$

(Ans. 160; 9; 0; 3; 3; 3.)

Prob. 21. Obtain the determinants in Exs. 5 and 6 of the preceding problem from that in Ex. 4.

Prob. 22. Evaluate $\begin{vmatrix} 0 & 1 & 1 & \ldots \\ 1 & 0 & 1 & \ldots \\ 1 & 1 & 0 & \ldots \\ \ldots & \ldots & \ldots \end{vmatrix}$, of the nth order.

(Ans. $(n - 1)(- 1)^{n-1}$.)

Prob. 23. Show that $\begin{vmatrix} a & b & c & d \\ -b & a & -d & c \\ -c & d & a & -b \\ -d & -c & b & a \end{vmatrix} = (a^2 + b^2 + c^2 + d^2)^2.$

ART. 19. THE ZERO FORMULAS.

If in the determinant $| a_1^{(n)} |$ the hth and kth rows be supposed identical, the elements a_k', a_k'', $\ldots a_k^{(n)}$ in the formula (1) of the last article may be replaced by a_h', a_h'', $\ldots a_h^{(n)}$ respectively. But in this case the value of the determinant is zero (Art. 12). Hence, in reference to the determinant $| a_1^{(n)} |$, h and k being different subscripts,

$$a_h'A_k' + a_h''A_k'' + \ldots + a_h^{(n)}A_k^{(n)} = 0.$$

Similarly, p and s being different superscripts,

$$a_1^{(p)}A_1^{(s)} + a_2^{(p)}A_2^{(s)} + \ldots + a_n^{(p)}A_n^{(s)} = 0.$$

Art. 20. Cauchy's Method of Development.

It is frequently desirable to expand a determinant with reference to the elements of a given row and column.

Let the determinant be $\Delta \equiv \mid a_1^{(n)} \mid$, and the given row and column the hth and pth respectively. Then is $A_h^{(p)}$ the co-factor of $a_h^{(p)}$, the element at the intersection of the two given lines. The co-factor of any element $a_k^{(s)}$ of $A_h^{(p)}$ will be designated by $B_k^{(s)}$, this being a determinant of the order $(n-2)$. The required expansion may now be obtained by means of the following formula, due to Cauchy:

$$\mid a_1^{(n)} \mid = a_h^{(p)}A_h^{(p)} - \Sigma a_h^{(s)}a_k^{(p)}B_k^{(s)}, \tag{1}$$

in which $k = 1, 2, \ldots h - 1, h + 1, \ldots n$, and $s = 1, 2, \ldots p - 1, p + 1, \ldots n$, successively.

To prove this, consider that $B_k^{(s)}$ is the aggregate of all terms of the expansion of Δ which contain the product $a_h^{(p)}a_k^{(s)}$. These terms are included in $a_h^{(p)}A_h^{(p)}$. Now, every term in the expansion which does not contain $a_h^{(p)}$ must contain some other element $a_k^{(s)}$ from the hth row and also some other element $a_k^{(p)}$ from the pth column, and thus contains the product $a_h^{(s)}a_k^{(p)}$. But this product differs from $a_h^{(p)}a_k^{(s)}$ only in the order of the superscripts; and is, therefore, in the expansion of Δ, multiplied by an aggregate of terms differing in sign only from that multiplying $a_h^{(p)}a_k^{(s)}$. Hence, $-B_k^{(s)}$ is the coefficient of $a_h^{(s)}a_k^{(p)}$ in the required expansion.

In the formula $a_h^{(p)}A_h^{(p)}$ gives $(n-1)!$ terms of Δ. There are also $(n-1)^2$ such aggregates as $-a_h^{(s)}a_k^{(p)}B_k^{(s)}$, each containing $(n-2)!$ terms. The formula therefore gives $(n-1)! + (n-1)^2(n-2)! = n!$ terms, which is the complete expansion.

When the expansion is required with reference to the elements of the first column and the first row the formula, written explicitly, becomes

$$|a_1^{(n)}| = a_1'A_1' - a_2'a_1''B_2'' - a_2'a_1'''B_2''' - \ldots - a_2'a_1^{(n)}B_2^{(n)}$$
$$- a_3'a_1''B_3'' - a_3'a_1'''B_3''' - \ldots - a_3'a_1^{(n)}B_3^{(n)}$$
$$\ldots\ldots\ldots\ldots\ldots\ldots\ldots\ldots\ldots$$
$$- a_n'a_1''B_n'' - a_n'a_1'''B_n''' - \ldots - a_n'a_1^{(n)}B_n^{(n)}, \quad (2)$$

in which $B_k^{(s)}$ has intrinsically the sign $(-1)^{k+s}$.

Cauchy's formula is particularly useful in expanding determinants which have been bordered; such as

$$- Q = \begin{vmatrix} 0 & u_1 & u_2 & u_3 \\ u_1 & a_{11} & a_{12} & a_{13} \\ u_2 & a_{21} & a_{22} & a_{23} \\ u_3 & a_{31} & a_{32} & a_{33} \end{vmatrix}. \quad (3)$$

Applying formula (2) to this determinant gives

$$- Q = - u_1^2 \begin{vmatrix} a_{22} & a_{23} \\ a_{32} & a_{33} \end{vmatrix} + u_1 u_2 \begin{vmatrix} a_{21} & a_{23} \\ a_{31} & a_{33} \end{vmatrix} - u_3 u_1 \begin{vmatrix} a_{21} & a_{22} \\ a_{31} & a_{32} \end{vmatrix}$$
$$+ u_1 u_2 \begin{vmatrix} a_{12} & a_{13} \\ a_{32} & a_{33} \end{vmatrix} - u_2^2 \begin{vmatrix} a_{11} & a_{13} \\ a_{31} & a_{33} \end{vmatrix} + u_3 u_2 \begin{vmatrix} a_{11} & a_{12} \\ a_{31} & a_{32} \end{vmatrix}$$
$$- u_3 u_1 \begin{vmatrix} a_{12} & a_{13} \\ a_{22} & a_{23} \end{vmatrix} + u_2 u_3 \begin{vmatrix} a_{11} & a_{13} \\ a_{21} & a_{23} \end{vmatrix} - u_3^2 \begin{vmatrix} a_{11} & a_{12} \\ a_{21} & a_{22} \end{vmatrix}.$$

Letting $a_{ks} \equiv a_{sk}$, and writing A_{11}, A_{12}, $\ldots$ for the co-factors of the elements of $|a_{11}a_{22}a_{33}|$, the above becomes

$$Q = A_{11}u_1^2 + A_{22}u_2^2 + A_{33}u_3^2 + 2A_{23}u_2u_3 + 2A_{31}u_3u_1 + 2A_{12}u_1u_2.$$

Prob. 24. Develop the following determinants by Cauchy's formula:

(1) $-\begin{vmatrix} a & h & g & u \\ h & b & f & v \\ g & f & c & w \\ u & v & w & 0 \end{vmatrix}$; (2) $\begin{vmatrix} 0 & yz & zx & xy \\ yz & 0 & 1 & 1 \\ zx & 1 & 0 & 1 \\ xy & 1 & 1 & 0 \end{vmatrix}$; (3) $\begin{vmatrix} 0 & 1 & 1 & 1 \\ 1 & 0 & xy & zx \\ 1 & xy & 0 & yz \\ 1 & zx & yz & 0 \end{vmatrix}$;

$$(4) \begin{vmatrix} -1-x & 1 & 1 \\ 1-y & -1 & 1 \\ x & 0 & y & z \\ 1-z & 1 & -1 \end{vmatrix}; \quad (5) \begin{vmatrix} 1 & 1 & 1 & x \\ x & y & z & 0 \\ 1 & 1 & 1 & y \\ 1 & 1 & 1 & z \end{vmatrix}; \quad (6) \begin{vmatrix} 0 & a & b \\ -a & \sin A & \sin B \\ -b & -\cos A & \cos B \end{vmatrix}.$$

ART. 21. DIFFERENTIATION OF DETERMINANTS.

By the formula (1) of Art. 18

$$\Delta \equiv |y_{1,n}| = Y_{k1} y_{k1} + Y_{k2} y_{k2} + \ldots Y_{kn} y_{kn}. \tag{1}$$

Considering the elements of the determinant as independent variables and differentiating with respect to y_{ks} gives

$$\delta_{ks}\Delta = Y_{ks}\, dy_{ks}, \quad \text{or} \quad Y_{ks} = \frac{\delta\Delta}{dy_{ks}}. \tag{2}$$

Substituting in (1),

$$\Delta \equiv |y_{1,n}| = y_{k1}\frac{\delta\Delta}{dy_{k1}} + y_{k2}\frac{\delta\Delta}{dy_{k2}} + \ldots + y_{kn}\frac{\delta\Delta}{dy_{kn}}. \tag{3}$$

Similarly

$$\Delta \equiv |y_{1,n}| = y_{1s}\frac{\delta\Delta}{dy_{1s}} + y_{2s}\frac{\delta\Delta}{dy_{2s}} + \ldots + y_{ns}\frac{\delta\Delta}{dy_{ns}}. \tag{4}$$

Again differentiating (1), this time with respect to all the elements of the kth row, there results

$$\delta_k\Delta = Y_{k1}dy_{k1} + Y_{k2}dy_{k2} + \ldots + Y_{kn}dy_{kn}. \tag{5}$$

In the total differential of Δ there are obviously n such expressions as (5), each of which may be obtained from Δ by replacing the elements of some one of the rows by their differentials; thus:

$$d\Delta = \begin{vmatrix} dy_{11} \ldots dy_{1n} \\ y_{21} \cdots y_{2n} \\ \cdots \cdots \cdots \\ y_{n1} \cdots y_{nn} \end{vmatrix} + \begin{vmatrix} y_{11} \cdots y_{1n} \\ dy_{21} \ldots dy_{2n} \\ \cdots \cdots \cdots \\ y_{n1} \cdots y_{nn} \end{vmatrix} + \ldots + \begin{vmatrix} y_{11} \cdots y_{1n} \\ y_{21} \cdots y_{2n} \\ \cdots \cdots \cdots \\ dy_{n1} \ldots dy_{nn} \end{vmatrix}. \tag{6}$$

If all the elements are functions of one independent variable x, then, representing $\dfrac{dy_{ks}}{dx}$ by y_{ks}',

$$\frac{d\Delta}{dx} = \begin{vmatrix} y_{11}' \cdots y_{1n}' \\ y_{21} \cdots y_{2n} \\ \cdots \cdots \cdots \\ y_{n1} \cdots y_{nn} \end{vmatrix} + \begin{vmatrix} y_{11} \cdots y_{1n} \\ y_{21}' \cdots y_{2n}' \\ \cdots \cdots \cdots \\ y_{n1} \cdots y_{nn} \end{vmatrix} + \ldots + \begin{vmatrix} y_{11} \cdots y_{1n} \\ y_{21} \cdots y_{2n} \\ \cdots \cdots \cdots \\ y_{n1}' \cdots y_{nn}' \end{vmatrix}. \tag{7}$$

Prob. 25. Show that Cauchy's formula may be written

$$\Delta \equiv |a_1^{(n)}| = a_h^{(p)} \frac{\delta \Delta}{da_h^{(p)}} - \Sigma a_k^{(p)} a_h^{(s)} \frac{\delta^2 \Delta}{da_h^{(p)} da_k^{(s)}}.$$

ART. 22.　RAISING THE ORDER.

Since, in the expansion of the determinant (1) of Art. 17 the elements $a_2', \ldots a_n'$ do not appear, these may be replaced by any quantities whatever, as $Q, \ldots T$, without changing the value of the determinant; thus:

$$\begin{vmatrix} a_1' & 0 & 0 & \ldots & 0 \\ a_2' & a_2'' & a_2''' & \ldots & a_2^{(n)} \\ \ldots & \ldots & \ldots & & \\ a_n' & a_n'' & a_n''' & \ldots & a_n^{(n)} \end{vmatrix} = \begin{vmatrix} a_1' & 0 & 0 & \ldots & 0 \\ Q & a_2'' & a_2''' & \ldots & a_2^{(n)} \\ \ldots & \ldots & \ldots & & \\ T & a_n'' & a_n''' & \ldots & a_n^{(n)} \end{vmatrix}.$$

Similarly,

$$\begin{vmatrix} a_1' & 0 & 0 & \ldots & 0 \\ a_2' & a_2'' & 0 & \ldots & 0 \\ a_3' & a_3'' & a_3''' & \ldots & a_3^{(n)} \\ \ldots & \ldots & \ldots & & \\ a_n' & a_n'' & a_n''' & \ldots & a_n^{(n)} \end{vmatrix} = a_1' a_2'' \begin{vmatrix} a_3''' & \ldots & a_3^{(n)} \\ \ldots & \ldots & \\ a_n''' & \ldots & a_n^{(n)} \end{vmatrix} = \begin{vmatrix} a_1' & 0 & 0 & \ldots & 0 \\ Q & a_2'' & 0 & \ldots & 0 \\ R & L & a_3''' & \ldots & a_3^{(n)} \\ \ldots & \ldots & \ldots & & \\ T & N & a_n''' & \ldots & a_n^{(n)} \end{vmatrix},$$

in which $Q, R, \ldots T$ and $L, \ldots N$ are any quantities whatever. Finally,

$$\begin{vmatrix} a_1' & 0 & \ldots 0 & 0 \\ a_2' & a_2'' & \ldots 0 & 0 \\ \ldots & \ldots & \ldots & \\ a_{n-1}' & a_{n-1}'' & \ldots a_{n-1}^{(n-1)} & 0 \\ a_n & a_n'' & \ldots a_n^{(n-1)} & a_n^{(n)} \end{vmatrix} = a_1' a_2'' \ldots a_{n-1}^{(n-1)} a_n^{(n)} = \begin{vmatrix} a_1' & 0 & \ldots 0 & 0 \\ Q & a_2'' & \ldots 0 & 0 \\ \ldots & \ldots & \ldots & \\ S & M & \ldots a_{n-1}^{(n-1)} & 0 \\ T & N & \ldots C & a_n^{(n)} \end{vmatrix};$$

that is, if all the elements on one side of the principal diagonal are zeros the determinant is equal to its principal term, and the elements on the other side of this diagonal may be replaced by any quantities whatever.

By what precedes,

$$\begin{vmatrix} a_1' & \ldots & a_1^{(n)} \\ \ldots & \ldots & \\ a_n' & \ldots & a_n^{(n)} \end{vmatrix} = \begin{vmatrix} 1 & 0 & \ldots & 0 \\ Q & a_1' & \ldots & a_1^{(n)} \\ \ldots & \ldots & \ldots & \\ T & a_n' & \ldots & a_n^{(n)} \end{vmatrix}$$

Hence, a determinant of the nth order may be expressed as a determinant of the order $(n + 1)$ by bordering it above by a row (to the left by a column) of zeros, to the left by a column (above by a row) of elements chosen arbitrarily, and writing 1 at the intersection of the lines thus added. By continuing this process any determinant may be expressed as a determinant of any higher order.

Prob. 26. If all the elements on one side of the secondary diagonal are zeros, what is the value of the determinant?

Prob. 27. Develop the determinant $\begin{vmatrix} a & h & g & u & 0 \\ h & b & f & v & 0 \\ g & f & c & w & 0 \\ u & v & w & 0 & t \\ 0 & 0 & 0 & t & s \end{vmatrix}$.

Prob. 28. A determinant in which $a_k^{(s)} = -a_s^{(k)}$ and $a_k^{(k)} = 0$ is said to be skew-symmetric. Prove that every skew-symmetric determinant of odd order is equal to zero.

ART. 23.　SOLUTION OF LINEAR EQUATIONS.

Of the many analytical processes giving rise to determinants the simplest and most common is the solution of systems of simultaneous linear equations. Thus, solving the equations

$$a_1'x' + a_1''x'' = \kappa_1, \atop a_2'x' + a_2''x'' = \kappa_2,$$

by the methods of ordinary algebra gives:

$$x' = \frac{\kappa_1 a_2'' - \kappa_2 a_1''}{a_1'a_2'' - a_2'a_1''}, \quad x'' = \frac{a_1'\kappa_2 - a_2'\kappa_1}{a_1'a_2'' - a_2'a_1''}.$$

In the notation of determinants these are written:

$$x' = \begin{vmatrix} \kappa_1 & a_1'' \\ \kappa_2 & a_2'' \end{vmatrix} / \begin{vmatrix} a_1' & a_1'' \\ a_2' & a_2'' \end{vmatrix}, \quad x'' = \begin{vmatrix} a_1' & \kappa_1 \\ a_2' & \kappa_2 \end{vmatrix} / \begin{vmatrix} a_1' & a_1'' \\ a_2' & a_2'' \end{vmatrix}.$$

It will be noted that the two fractions expressing the values of x' and x'' have a common denominator, this being the determinant whose elements are the coefficients of the unknowns arranged in the same order as in the given equations. The numerator of the fraction giving the value of x' is formed from this denominator by replacing each coefficient of x' by the corresponding absolute term.   Similarly for x''.

The difficulty of solving systems of linear equations by the ordinary processes of elimination increases rapidly as the number of equations is increased.   The law of formation of the roots explained above is however, capable of generalization, being equally applicable to all complete linear systems, as will now be shown.

Let such a system be written

$$\left.\begin{array}{l} a_1'x' + a_1''x'' + \ldots + a_1^{(n)}x^{(n)} = \kappa_1, \\ a_2'x' + a_2''x'' + \ldots + a_2^{(n)}x^{(n)} = \kappa_2, \\ \cdots\cdots\cdots\cdots\cdots\cdots\cdots\cdots\cdots\cdots \\ a_n'x' + a_n''x'' + \ldots + a_n^{(n)}x^{(n)} = \kappa_n. \end{array}\right\} . \tag{1}$$

Now form the determinant of the coefficients of these equations ; thus :

$$D \equiv \begin{vmatrix} a_1'a_1'' \ldots a_1^{(n)} \\ a_2'a_2'' \ldots a_2^{(n)} \\ \cdots\cdots\cdots\cdots \\ a_n'a_n'' \ldots a_n^{(n)} \end{vmatrix},$$

and let $A_k^{(s)}$ be the co-factor of $a_k^{(s)}$ in this determinant.   The function

$$a_1^{(p)}A_1^{(s)} + a_2^{(p)}A_2^{(s)} + \ldots + a_n^{(p)}A_n^{(s)}$$

is equal to D when $p = s$ (Art. 18) ; to zero when p and s are different superscripts (Art. 19).   Then, multiplying the given equations by $A_1^{(s)}$, $A_2^{(s)}$, $\ldots A_n^{(s)}$ respectively, the sum of the resulting equations is a linear equation in which the coefficient of $x^{(s)}$ is equal to D, while those of all the other unknowns vanish.   The sum is, therefore,

$$Dx^{(s)} = \kappa_1 A_1^{(s)} + \kappa_2 A_2^{(s)} + \ldots + \kappa_n A_n^{(s)}. \tag{2}$$

But the second member of this equation is what D becomes upon replacing the coefficients $a_1^{(s)}$, $a_2^{(s)}$, $\ldots a_n^{(s)}$ of the unknown $x^{(s)}$ by the absolute terms κ_1, κ_2, $\ldots \kappa_n$ in order.   Hence,

$$x^{(s)} = \begin{vmatrix} a_1' & \dots & a_1^{(s-1)} & \kappa_1 a_1^{(s+1)} & \dots & a_1^{(n)} \\ a_2' & \dots & a_2^{(s-1)} & \kappa_2 a_2^{(s+1)} & \dots & a_2^{(n)} \\ \dots & \dots & \dots & \dots & \dots & \dots \\ a_n' & \dots & a_n^{(s-1)} & \kappa_n a_n^{(s+1)} & \dots & a_n^{(n)} \end{vmatrix} \Big/ \begin{vmatrix} a_1' & a_1'' & \dots & a_1^{(n)} \\ a_2' & a_2'' & \dots & a_2^{(n)} \\ \dots & \dots & \dots & \dots \\ a_n' & a_n'' & \dots & a_n^{(n)} \end{vmatrix} . \tag{3}$$

This result may be stated as follows :

(*a*) The common denominator of the fractions expressing the values of the unknowns in a system of n linear equations involving n unknown quantities is the determinant of the coefficients, these being written in the same order as in the given equations. (*b*) The numerator of the fraction giving the value of any one of the unknowns is a determinant, which may be formed from the determinant of the coefficients by substituting for the column made up of the coefficients of the unknown in question a column whose elements are the absolute terms of the equations taken in the same order as the coefficients which they displace.

Prob. 29. Solve the following systems of equations :

(1) $3x + 5y = 21$, $6x + 2y = 15$;

(2) $\dfrac{x}{3} + \dfrac{3y}{2} = 5$, $\dfrac{2x}{3} + y = 6$;

(3) $3x + y + 2z = 50$, $x + 2y - 3z = 15$, $2x + 2y - 3z = 25$;

(4) $\dfrac{1}{y} + \dfrac{1}{z} = p$, $\dfrac{1}{z} + \dfrac{1}{x} = q$, $\dfrac{1}{x} + \dfrac{1}{y} = r$;

(5) $\dfrac{w}{3} + \dfrac{x}{5} + \dfrac{y}{7} + \dfrac{z}{9} = 2800$, $\dfrac{w}{5} + \dfrac{x}{7} + \dfrac{y}{9} + \dfrac{z}{11} = 2144$,

$\dfrac{w}{7} + \dfrac{x}{9} + \dfrac{y}{11} + \dfrac{z}{13} = 1744$, $\dfrac{w}{9} + \dfrac{x}{11} + \dfrac{y}{13} + \dfrac{z}{15} = 1472$.

Prob. 30. Show that the three right lines

$$y = x + 1, \quad y = -2x + 16, \quad y = 3x - 9,$$

intersect in a common point.

ART. 24.   CONSISTENCE OF LINEAR SYSTEMS.

When the number of given equations is greater than the number of unknowns their consistency with one another must obviously depend upon some relation among the coefficients. This relation will now be investigated for the case of $(n+1)$ linear equations involving n unknowns.   Let the equations be

$$\left.\begin{array}{l} a_1'x' \;+\ldots+\; a_1^{(n)}x^{(n)} \;= \kappa_1, \\ \ldots\ldots\ldots\ldots\ldots\ldots\ldots \\ a_n'x' \;+\ldots+\; a_n^{(n)}x^{(n)} \;= \kappa_n, \\ a_{n+1}'x' +\ldots+\; a_{n+1}^{(n)}x^{(n)} = \kappa_{n+1}. \end{array}\right\}$$

If the above equations are consistent the values of the unknowns obtained by solving any n of them must satisfy the remaining equation.   Solving the first n equations by the method of the preceding article, substituting the values of x', x'', $\ldots x^{(n)}$ thus obtained in the last equation, and clearing of fractions, the result reduces to (Art. 18)

$$E = \begin{vmatrix} a_1' & \ldots a_1^{(n)} & \kappa_1 \\ \ldots & \ldots\ldots\ldots & \\ a_n' & \ldots a_n^{(n)} & \kappa_n \\ a_{n+1}' & \ldots a_{n+1}^{(n)} & \kappa_{n+1} \end{vmatrix} = 0,$$

which is the condition to be fulfilled by the coefficients in order that the given equations may be consistent.

Hence the condition of consistency for a set of linear equations involving a number of unknowns one less than the number of equations is that the determinant of the coefficients and absolute terms, written in the same order as in the given equations, shall be zero.   This determinant is called the resultant* or eliminant of the equations.   Thus the equations

$$x+y-z=0, \; x-y+z=2, \; -x+y+z=4, \; x+y+z=6$$

are consistent, for the reason that

$$\begin{vmatrix} 1 & 1 & -1 & 0 \\ 1 & -1 & 1 & 2 \\ -1 & 1 & 1 & 4 \\ 1 & 1 & 1 & 6 \end{vmatrix} = 0.$$

---

* This term was introduced by Laplace in 1772.   The term eliminant is due to Sylvester.

ART. 25. THE MATRIX.

Assume r linear equations involving n unknowns, r being greater than n as follows:

$$\left.\begin{array}{c} a_1'x' + \ldots + a_1^{(n)}x^{(n)} = \kappa_1, \\ \ldots\ldots\ldots\ldots\ldots\ldots\ldots\ldots \\ a_n'x' + \ldots + a_n^{(n)}x^{(n)} = \kappa_n, \\ \ldots\ldots\ldots\ldots\ldots\ldots\ldots \ldots \\ a_r'x' + \ldots + a_r^{(n)}x^{(n)} = \kappa_r. \end{array}\right\}$$

The consistency of these equations requires that every determinant of the order $(n + 1)$, formed by selecting $(n + 1)$ rows from the array whose elements are the coefficients and absolute terms written in order, shall be zero.

If the elements of the array fulfill this condition the fact is expressed thus:

$$\left\| \begin{array}{ccc} a_1' & \ldots a_n' & \ldots a_r' \\ \ldots\ldots\ldots\ldots\ldots\ldots \\ a_1^{(n)} & \ldots a_n^{(n)} & \ldots a_r^{(n)} \\ \kappa_1 & \ldots \kappa_n & \ldots \kappa_r \end{array} \right\| = 0;$$

the change of rows into columns being purely arbitrary. The above expression is called a rectangular array, or a matrix.

ART. 26. HOMOGENEOUS LINEAR SYSTEMS.

Let the equations of the given system be both linear and homogeneous; thus:

$$\left.\begin{array}{c} a_1'\, x' + \ldots + a_1^{(n)}x^{(n)} = 0, \\ \ldots\ldots\ldots\ldots\ldots\ldots\ldots \\ a_n'\, x' + \ldots + a_n^{(n)}x^{(n)} = 0. \end{array}\right\} \tag{1}$$

Representing the determinant of the coefficients by E, the general solution, as given by the formula (3) of Art. 23, is

$$x^{(s)} = 0/E.$$

That is, all the unknowns are equal to zero, and the equations have no other solution than this unless

$$E = 0.$$

But in this case the value of each unknown is obtainable only in the indeterminate form 0/0. The ratios of the unknowns may be readily obtained, however. For, dividing each equation through by any one of these, as $x^{(s)}$, the system (1) becomes

$$a_1' \frac{x'}{x^{(s)}} + \ldots + a_1^{(s-1)} \frac{x^{(s-1)}}{x^{(s)}} + a_1^{(s+1)} \frac{x^{(s+1)}}{x^{(s)}} + \ldots + a_1^{(n)} \frac{x^{(n)}}{x^{(s)}} = -a_1^{(s)},$$
$$\cdots\cdots\cdots\cdots\cdots\cdots\cdots\cdots\cdots\cdots\cdots\cdots\cdots$$
$$a_n' \frac{x'}{x^{(s)}} + \ldots + a_n^{(s-1)} \frac{x^{(s-1)}}{x^{(s)}} + a_n^{(s+1)} \frac{x^{(s+1)}}{x^{(s)}} + \ldots + a_n^{(n)} \frac{x^{(n)}}{x^{(s)}} = -a_n^{(s)}. \qquad (2)$$

Now the condition $E = 0$ establishes the consistency of the n equations (2) involving the $(n-1)$ unknown ratios (Art. 24),

$$\frac{x'}{x^{(s)}}, \quad \cdots \quad \frac{x^{(s-1)}}{x^{(s)}}, \quad \frac{x^{(s+1)}}{x^{(s)}}, \quad \cdots \quad \frac{x^{(n)}}{x^{(s)}}.$$

Hence, if $E = 0$ the given equations (1) are consistent; that is, the values of the above $(n-1)$ ratios obtained by solving any $(n-1)$ of them will satisfy the remaining equation. Any n quantities having among themselves the ratios thus determined will satisfy the given equations. Thus, if $x_0', x_0'', \ldots x_0^{(n)}$ are n such quantities, so also are $\lambda x_0', \lambda x_0'', \ldots \lambda x_0^{(n)}$, λ being any factor whatever.

The determinant E of the coefficients of the given homogeneous linear equations is called the resultant or eliminant of the system.

When the number of equations is greater than the number of unknowns the conditions of consistency are expressible in the form of a rectangular array, as in Art. 25.

As an example, consider the five equations

$$2x - 3y + z = 0, \quad 4x - y - z = 0, \quad -7x + 3y + z = 0,$$
$$x + y - z = 0, \quad 5x - 5y + z = 0.$$

Dividing each of the first two equations by z and solving for the two unknowns $\frac{x}{z}$ and $\frac{y}{z}$ gives

$$\frac{x}{z} = \begin{vmatrix} -1 & -3 \\ 1 & -1 \end{vmatrix} \Big/ \begin{vmatrix} 2 & -3 \\ 4 & -1 \end{vmatrix} = \frac{2}{5}, \quad \frac{y}{z} = \begin{vmatrix} 2 & -1 \\ 4 & 1 \end{vmatrix} \Big/ \begin{vmatrix} 2 & -3 \\ 4 & -1 \end{vmatrix} = \frac{3}{5},$$

or $\qquad\qquad x : y : z :: 2 : 3 : 5;$

and any three quantities having these ratios will satisfy the two equations, as 10, 15, and 25. That the third equation is consistent with the first two is shown by the vanishing of the determinant

$$\begin{vmatrix} 2 - 3 & 1 \\ 4 - 1 - 1 \\ - 7 & 3 & 1 \end{vmatrix} = \text{o}.$$

If all the equations are consistent the determinant of the coefficients of any three of them must vanish; that is,

$$\begin{Vmatrix} 2 & 4 - 7 & 1 & 5 \\ - 3 - 1 & 3 & 1 - 5 \\ 1 - 1 & 1 - 1 & 1 \end{Vmatrix} = \text{o}.$$

ART. 27.   CO-FACTORS IN A ZERO DETERMINANT.

If, in the preceding article, $E = \text{o}$, it follows from Arts. 18 and 19 that

$$a_1{}' A_k{}' + a_1{}'' A_k{}'' + \ldots + a_1{}^{(n)} A_k{}^{(n)} = \text{o},$$
$$\cdots\cdots\cdots\cdots\cdots\cdots\cdots$$
$$a_k{}' A_k{}' + a_k{}'' A_k{}'' + \ldots + a_k{}^{(n)} A_k{}^{(n)} = \text{o} = E,$$
$$\cdots\cdots\cdots\cdots\cdots\cdots\cdots$$
$$a_n{}' A_k{}' + a_n{}'' A_k{}'' + \ldots + a_n{}^{(n)} A_k{}^{(n)} = \text{o}.$$

These equations obviously give for the ratios

$$\frac{A_k{}'}{A_k{}^{(s)}}, \; \cdots \; \frac{A_k{}^{(s-1)}}{A_k{}^{(s)}}, \; \frac{A_k{}^{(s+1)}}{A_k{}^{(s)}}, \; \cdots \; \frac{A_k{}^{(n)}}{A_k{}^{(s)}}$$

values which are identical with those obtained for the ratios

$$\frac{x'}{x^{(s)}}, \; \cdots \; \frac{x^{(s-1)}}{x^{(s)}}, \; \frac{x^{(s+1)}}{x^{(s)}}, \; \cdots \; \frac{x^{(n)}}{x^{(s)}}$$

from the equations (1) of Art. 26. It follows that $x', x'', \ldots x^{(n)}$ are proportional to $A_k{}', A_k{}'', \ldots A_k{}^{(n)}$, whatever the value of k. Thus, giving to k the successive values 1, 2, ... n, there result

$$x' : x'' : \ldots : x^{(n)} :: A_1{}' : A_1{}'' : \ldots : A_1{}^{(n)}$$
$$:: A_2{}' : A_2{}'' : \ldots : A_2{}^{(n)}$$
$$\cdots\cdots\cdots\cdots\cdots$$
$$:: A_n{}' : A_n{}'' : \ldots : A_n{}^{(n)}.$$

Hence, when a determinant is equal to zero, the co-factors of the elements of any line are proportional to the co-factors of the corresponding elements of any parallel line.

ART. 28.   SYLVESTER'S METHOD OF ELIMINATION.[*]

Let it be required to eliminate the unknown from the two equations

$$a_3 x^3 + a_2 x^2 + a_1 x + a_0 = 0,$$

$$b_2 x^2 + b_1 x + b_0 = 0.$$

This will be done by what is called the dialytic method, the invention of which is due to Sylvester. Multiplying the first of the given equations by x, and the second by x and x^2 successively, the result is a system of five equations, viz.:

$$\left.\begin{array}{l} a_3 x^3 + a_2 x^2 + a_1 x + a_0 = 0, \\ a_3 x^4 + a_2 x^3 + a_1 x^2 + a_0 x = 0, \\ b_2 x^2 + b_1 x + b_0 = 0, \\ b_2 x^3 + b_1 x^2 + b_0 x = 0, \\ b_2 x^4 + b_1 x^3 + b_0 x^2 = 0. \end{array}\right\}$$

The eliminant of these five equations, involving the four unknowns x, x^2, x^3, and x^4 is (Art. 24)

$$E \equiv \begin{vmatrix} 0 & a_3 & a_2 & a_1 & a_0 \\ a_3 & a_2 & a_1 & a_0 & 0 \\ 0 & 0 & b_2 & b_1 & b_0 \\ 0 & b_2 & b_1 & b_0 & 0 \\ b_2 & b_1 & b_0 & 0 & 0 \end{vmatrix} = 0.$$

If the given equations be not consistent this determinant will not vanish.

The above method is a general one. Thus, let the two given equations be

$$a_m x^m + \ldots + a_1 x + a_0 = 0,$$

$$b_n x^n + \ldots \ldots + b_1 x + b_0 = 0.$$

Multiplying the first equation $(n-1)$ times in succession by x, and the second $(m-1)$ times, $(m+n)$ equations are

---

[*] Philosophical Magazine, 1840, and Crelle's Journal, Vol. XXI.

obtained which involve as unknowns the first $(m + n - 1)$ powers of x. The eliminant of these equations is a determinant of the order $(m + n)$, which is of the nth degree in terms of the coefficients of the equation of the mth degree, and *vice versa*. The law of formation of the eliminant is obvious.

The same method may be used in eliminating one or both the variables from a pair of homogeneous equations.

As an example, let it be required to eliminate the variables from the equations

$$2x^3 - 5x^2y - 9y^3 = 0 \quad \text{and} \quad 3x^2 - 7xy - 6y^2 = 0.$$

Dividing the first by y^3, and multiplying by $\dfrac{x}{y}$; the second by y^2, and multiplying by $\dfrac{x}{y}$ twice in succession, there result, in all, five equations involving $\dfrac{x}{y}, \dfrac{x^2}{y^2}, \dfrac{x^3}{y^3}$, and $\dfrac{x^4}{y^4}$. Eliminating these four ratios gives

$$E = \begin{vmatrix} 0 & 2 & -5 & 0 & -9 \\ 2 & -5 & 0 & -9 & 0 \\ 0 & 0 & 3 & -7 & -6 \\ 0 & 3 & -7 & -6 & 0 \\ 3 & -7 & -6 & 0 & 0 \end{vmatrix},$$

the vanishing of which shows that the two given equations are consistent.

Prob. 31. Test the consistency of each of the following systems of equations:

(1) $x + y + 2z = 9$, $x + y - z = 0$, $2x - y + z = 3$, $x - 3y + 2z = 1$;

(2) $x - y - 2z = 0$, $x - 2y + z = 0$, $2x - 3y - z = 0$;

(3) $2x^2y - xy^2 = 0$, $8x^3y + 8xy^3 - 5y^4 = 0$.

Prob. 32. Find the ratios of the unknowns in the equations

$$2x + y - 2z = 0, \quad 4w - y - 4z = 0, \quad 2w + x - 5v + z = 0.$$

Prob. 33. In the equations

$$a_k'x' + \ldots + a_k^{(n)}x^{(n)} + a_k^{(n+1)}x^{(n+1)} = 0, \quad [k = 1, 2, \ldots n]$$

prove that $x' : \ldots : x^{(n)} : x^{(n+1)} :: M' : \ldots : M^{(n)} : M^{(n+1)}$, where

$(-1)^{i-1}M^{(i)}$ is the determinant obtained by deleting the ith column from the rectangular array

$$M \equiv \begin{vmatrix} a_1' \dots a_1^{(n)} a_1^{(n+1)} \\ \dots\dots\dots\dots \\ a_n' \dots a_n^{(n)} a_n^{(n+1)} \end{vmatrix}.$$

Prob. 34. From $\dfrac{lx+vy+\mu z}{p} = \dfrac{vx+my+\lambda z}{q} = \dfrac{\mu x+\lambda y+nz}{r}$,

deduce $\dfrac{x}{\begin{vmatrix} v & \mu & p \\ m & \lambda & q \\ \lambda & n & r \end{vmatrix}} = -\dfrac{y}{\begin{vmatrix} l & \mu & p \\ v & \lambda & q \\ \mu & n & r \end{vmatrix}} = \dfrac{z}{\begin{vmatrix} l & v & p \\ v & m & q \\ \mu & \lambda & r \end{vmatrix}}.$

Prob. 35. Show that the three straight lines $a'x + b'y + c' = 0$, $a''x + b''y + c'' = 0$, and $a'''x + b'''y + c''' = 0$, are concurrent when $|\,a'b''c'''\,| = 0$.

Prob. 36. Prove that the medians of a triangle are concurrent.

Prob. 37. Show that the points (x_0, y_0), (x_1, y_1), and (x_2, y_2) are collinear when $\begin{vmatrix} x_0 & y_0 & 1 \\ x_1 & y_1 & 1 \\ x_2 & y_2 & 1 \end{vmatrix} = 0.$

Prob. 38. Write the conditions that all the points (x_1, y_1), (x_2, y_2), ... (x_n, y_n) shall be collinear in the form of a matrix.

Prob. 39. Obtain the equation of a right line through (x_1, y_1) and (x_2, y_2) in the form of a determinant.

Prob. 40. Show that the equation $\begin{vmatrix} x & y & z & 1 \\ x_1 & y_1 & z_1 & 1 \\ x_2 & y_2 & z_2 & 1 \\ x_3 & y_3 & z_3 & 1 \end{vmatrix} = 0$

represents a plane through (x_1, y_1, z_1), (x_2, y_2, z_2), and (x_3, y_3, z_3).

ART. 29.　THE MULTIPLICATION THEOREM.

Let the two homogeneous linear equations

$$\left. \begin{array}{l} a_{11}x_1 + a_{12}x_2 = 0, \\ a_{21}x_1 + a_{22}x_2 = 0, \end{array} \right\} \tag{1}$$

be subjected to linear transformation by substituting

$$\left. \begin{array}{l} x_1 = b_{11}u_1 + b_{21}u_2, \\ x_2 = b_{12}u_1 + b_{22}u_2. \end{array} \right\} \tag{2}$$

The result of such transformation is

$$(a_{11}b_{11} + a_{12}b_{12})u_1 + (a_{11}b_{21} + a_{12}b_{22})u_2 = 0,$$
$$(a_{21}b_{11} + a_{22}b_{12})u_1 + (a_{21}b_{21} + a_{22}b_{22})u_2 = 0. \qquad (3)$$

The vanishing of the determinant

$$\begin{vmatrix} a_{11}b_{11} + a_{12}b_{12} & a_{11}b_{21} + a_{12}b_{22} \\ a_{21}b_{11} + a_{22}b_{12} & a_{21}b_{21} + a_{22}b_{22} \end{vmatrix} \cdot \qquad (4)$$

is the condition that the equations (3) may be consistent; that is, the condition that they may have solutions other than $u_1 = 0 = u_2$ (Art. 26). Now the equations (3) may be consistent because of the consistency of the equations (1), in which case the determinant

$$\begin{vmatrix} a_{11} & a_{12} \\ a_{21} & a_{22} \end{vmatrix} \qquad (5)$$

vanishes. Or, this condition failing, and the equations (1) thus having no solution other than $x_1 = 0 = x_2$, the equations (3) will still be consistent if the equations (2) are so; that is, if the determinant

$$\begin{vmatrix} b_{11} & b_{12} \\ b_{21} & b_{22} \end{vmatrix} \qquad (6)$$

vanishes. The vanishing of either of the determinants (5) or (6), therefore, causes the determinant (4) to vanish. It follows that (5) and (6) are factors of (4); and since their product and the determinant (4) are of the same degree with respect to the coefficients $a_{11}, \ldots, b_{11}, \ldots$, they are the only factors. Hence,

$$\begin{vmatrix} a_{11} & a_{12} \\ a_{21} & a_{22} \end{vmatrix} \cdot \begin{vmatrix} b_{11} & b_{12} \\ b_{21} & b_{22} \end{vmatrix} = \begin{vmatrix} a_{11}b_{11} + a_{12}b_{12} & a_{11}b_{21} + a_{12}b_{22} \\ a_{21}b_{11} + a_{22}b_{12} & a_{21}b_{21} + a_{22}b_{22} \end{vmatrix} \cdot \qquad (7)$$

The above method is equally applicable to the formation of the product of any two determinants of the same order. Hence results the following general formula:

$$| a_{11} a_{22} \ldots a_{nn} | \cdot | b_{11} b_{22} \ldots b_{nn} | =$$

$$\begin{vmatrix} a_{11}b_{11} + \ldots + a_{1n}b_{1n} & a_{11}b_{21} + \ldots + a_{1n}b_{2n} & \ldots & a_{11}b_{n1} + \ldots + a_{1n}b_{nn} \\ a_{21}b_{11} + \ldots + a_{2n}b_{1n} & a_{21}b_{21} + \ldots + a_{2n}b_{2n} & \ldots & a_{21}b_{n1} + \ldots + a_{2n}b_{nn} \\ \cdots & \cdots & \cdots & \cdots \\ a_{n1}b_{11} + \ldots + a_{nn}b_{1n} & a_{n1}b_{21} + \ldots + a_{nn}b_{2n} & \ldots & a_{n1}b_{n1} + \ldots + a_{nn}b_{nn} \end{vmatrix} \cdot \qquad (8)$$

The process indicated by this formula may be described as follows : *

To form the determinant $|p_{1,n}|$, which is the product of two determinants $|a_{1,n}|$ and $|b_{1,n}|$, first connect by plus signs the elements in the rows of both $|a_{1,n}|$ and $|b_{1,n}|$. Then place the first row of $|a_{1,n}|$ upon each row of $|b_{1,n}|$ in turn and let each two elements as they touch become products. This is the first row of $|p_{1,n}|$. Perform the same operation upon $|b_{1,n}|$ with the second row of $|a_{1,n}|$ to obtain the second row of $|p_{1,n}|$; and again with the third row of $|a_{1,n}|$ to obtain the third row of $|p_{1,n}|$; etc.

Any element of this product is

$$p_{ks} = a_{k_1}b_{s_1} + a_{k_2}b_{s_2} + \ldots + a_{kn}b_{sn}. \tag{9}$$

When the two determinants to be multiplied together are of different orders the one of lower order should be expressed as a determinant of the same order as the other (Art. 22), after which the above rule is applicable.

The product of two determinants may be formed by columns, instead of by rows as above. In this case the result is obtained in a different form. Thus the product of the determinants (5) and (6) by columns is

$$\begin{vmatrix} a_{11}b_{11} + a_{21}b_{21} & a_{12}b_{11} + a_{22}b_{12} \\ a_{11}b_{12} + a_{21}b_{22} & a_{12}b_{12} + a_{22}b_{22} \end{vmatrix}.$$

Prob. 41. Form the following products :

(1) $\begin{vmatrix} a & h & g \\ h & b & f \\ g & f & c \end{vmatrix} \cdot \begin{vmatrix} a & g \\ g & c \end{vmatrix}$;   (2) $\begin{vmatrix} b & f \\ f & c \end{vmatrix} \cdot \begin{vmatrix} a & g \\ g & c \end{vmatrix} \cdot \begin{vmatrix} a & h \\ h & b \end{vmatrix}$;

(3) $\begin{vmatrix} a_{11} & a_{12} & a_{13} \\ a_{21} & a_{22} & a_{23} \\ a_{31} & a_{32} & a_{33} \end{vmatrix} \cdot \begin{vmatrix} A_{11} & A_{12} & A_{13} \\ A_{21} & A_{22} & A_{23} \\ A_{31} & A_{32} & A_{33} \end{vmatrix}$;   (4) $\begin{vmatrix} a_1 & b_1 & c_1 \\ a_2 & b_2 & c_2 \\ a_3 & b_3 & c_3 \end{vmatrix} \cdot \begin{vmatrix} 0 & 1 & 1 \\ 1 & 0 & 1 \\ 1 & 1 & 0 \end{vmatrix}.$

Prob. 42. Generalize the last example (see Prob. 22, Art. 18).

Prob. 43. By forming the product

$$\begin{vmatrix} a + b\sqrt{-1} & -c + d\sqrt{-1} \\ c + d\sqrt{-1} & a - b\sqrt{-1} \end{vmatrix} \cdot \begin{vmatrix} j + k\sqrt{-1} & -l + m\sqrt{-1} \\ l + m\sqrt{-1} & j - k\sqrt{-1} \end{vmatrix}.$$

* Carr's Synopsis of Pure Mathematics, London, 1886, Article 570.

show that the product of two numbers, each the sum of four squares, is itself the sum of four squares.

ART. 30. PRODUCT OF TWO ARRAYS.

The process explained in the preceding article may be applied to form what is conventionally termed the product of two rectangular arrays. It will appear, however, that multiplying two such arrays together by columns leads to a result radically different from that obtained when the product is formed by rows.

Let the two rectangular arrays be

$$\begin{vmatrix} a_{11}a_{12}a_{13} \\ a_{21}a_{22}a_{23} \end{vmatrix} \text{ and } \begin{vmatrix} b_{11}b_{12}b_{13} \\ b_{21}b_{22}b_{23} \end{vmatrix}.$$

The product of these by columns is

$$\Delta \equiv \begin{vmatrix} a_{11}b_{11} + a_{21}b_{11} & a_{12}b_{11} + a_{22}b_{21} & a_{13}b_{11} + a_{23}b_{21} \\ a_{11}b_{12} + a_{21}b_{12} & a_{12}b_{12} + a_{22}b_{22} & a_{13}b_{12} + a_{23}b_{22} \\ a_{11}b_{13} + a_{21}b_{13} & a_{12}b_{13} + a_{22}b_{23} & a_{13}b_{13} + a_{23}b_{23} \end{vmatrix}.$$

The determinant Δ is plainly equal to zero, being the product of two determinants formed by adding a row of zeros to one of the given rectangular arrays and a row of elements chosen arbitrarily to the other.

In general, the product by columns of two rectangular arrays having m rows and n columns, m being less than n, is a determinant of the nth order whose value is zero.

Multiplying together the above rectangular arrays by rows, the result is

$$\Delta' \equiv \begin{vmatrix} a_{11}b_{11} + a_{12}b_{12} + a_{13}b_{13} & a_{11}b_{21} + a_{12}b_{22} + a_{13}b_{23} \\ a_{21}b_{11} + a_{22}b_{12} + a_{23}b_{13} & a_{21}b_{21} + a_{22}b_{22} + a_{23}b_{23} \end{vmatrix}$$

$$= \begin{vmatrix} a_{12}a_{13} \\ a_{22}a_{23} \end{vmatrix} \cdot \begin{vmatrix} b_{12}b_{13} \\ b_{22}b_{23} \end{vmatrix} + \begin{vmatrix} a_{11}a_{13} \\ a_{21}a_{23} \end{vmatrix} \cdot \begin{vmatrix} b_{11}b_{13} \\ b_{21}b_{23} \end{vmatrix} + \begin{vmatrix} a_{11}a_{12} \\ a_{21}a_{22} \end{vmatrix} \cdot \begin{vmatrix} b_{11}b_{12} \\ b_{21}b_{22} \end{vmatrix}.$$

In the same manner it may be shown that the product by rows of two rectangular arrays having m rows and n columns, m being less than n, is a determinant of the mth order, which may be expressed as the sum of the $n!/m!(n-m)!$ determinants formed from one of the arrays by deleting $(n - m)$ columns, each multiplied by the determinant formed by deleting the same columns from the other array.

ART. 31.   RECIPROCAL DETERMINANTS.

The determinant formed by replacing each element of a given determinant by its co-factor is called the reciprocal of the given determinant.*   Thus, the reciprocal of

$$\delta \equiv \begin{vmatrix} a_{11}a_{12} \dots a_{1n} \\ a_{21}a_{22} \dots a_{2n} \\ \dots \dots \dots \\ a_{n1}a_{n2} \dots a_{nn} \end{vmatrix} \text{ is } \varDelta \equiv \begin{vmatrix} A_{11}A_{12} \dots A_{1n} \\ A_{21}A_{22} \dots A_{2n} \\ \dots \dots \dots \\ A_{n1}A_{n2} \dots A_{nn} \end{vmatrix}.$$

The product of these two determinants is

$$\delta . \varDelta = \begin{vmatrix} a_{11}A_{11}+ \dots +a_{1n}A_{1n} & a_{11}A_{21}+ \dots +a_{1n}A_{2n} \dots \dots a_{11}A_{n1}+ \dots +a_{1n}A_{nn} \\ a_{21}A_{11}+ \dots +a_{2n}A_{1n} & a_{21}A_{21}+ \dots +a_{2n}A_{2n} \dots \dots a_{21}A_{n1}+ \dots +a_{2n}A_{nn} \\ \dots \dots \dots \dots \dots \dots \dots \dots \dots \dots \dots \dots \\ a_{n1}A_{11}+ \dots +a_{nn}A_{1n} & a_{n1}A_{21}+ \dots +a_{nn}A_{2n} \dots \dots a_{n1}A_{n1}+ \dots +a_{nn}A_{nn} \end{vmatrix}$$

Each element on the principal diagonal of this product is equal to δ (Art. 18), while all the other elements vanish (Art. 19).   Hence,

$$\delta . \varDelta = \begin{vmatrix} \delta & 0 \dots 0^{(n)} \\ 0 & \delta \dots 0 \\ \dots \dots \dots \\ 0_n & 0 \dots \delta \end{vmatrix} = \delta^n, \text{ or } \varDelta = \delta^{n-1}.$$

That is, the reciprocal of a determinant of the nth order is equal to its $(n - 1)^{\text{th}}$ power.

* The term reciprocal as here used has reference to the algebraic transformation concerned in the passage from point coördinates to line coördinates, called reciprocation.   The reciprocal of a determinant is also called the determinant adjugate.

CHAPTER III.

PROJECTIVE GEOMETRY.

By GEORGE BRUCE HALSTED,

Professor of Mathematics in the University of Texas.

ART. 1. THE ELEMENTS AND PRIMAL FORMS.*

1. A line determined by two points on it is called a 'straight.'

2. On any two points can be put one, but only one, straight, their 'join.'

3. A surface determined by three non-costraight points on it is called a 'plane.'

4. Any three points, not costraight, lie all on one and only one plane, their 'junction.'

5. If two points lie on a plane, so does their join.

6. The plane, the straight, and the point are the elements in projective geometry.

7. A straight is not to be considered as an aggregate of points. It is a monad, an atom, a simple positional concept as primal as the point. It is the 'bearer' of any points on it. It is the bearer of any planes on it.

8. Just so the plane is an element coeval with the point. It is the bearer of any points on it, or any straights on it.

9. A point is the bearer of any straight on it or any plane on it.

10. A point which is on each of two straights is called their 'cross.'

---

* This Chapter treats Projective Geometry entirely by the synthetic method. Metric relations are not considered, and nothing is borrowed from Analytic Geometry.

11. Planes all on the same point, or straights all with the same cross, are called 'copunctal.'

12. Any two planes lie both on one and only one straight, their 'meet.'

13. Like points with the same join, planes with the same meet are called costraight.

14. A plane and a straight not on it have one and only one point in common, their 'pass.'

15. Any three planes not costraight are copunctal on one and only one point, their 'apex.'

16. While these elements, namely, the plane, the straight, and the point, retain their atomic character, they can be united into compound figures, of which the primal class consists of three forms, the 'range,' the 'flat-pencil,' the 'axial-pencil.'

17. The aggregate of all points on a straight is called a 'point-row,' or 'range.'   If a point be common to two ranges, it is called their 'intersection.'

18. A piece of a range bounded by two points is called a 'sect.'

19. The aggregate of all coplanar, copunctal straights is called a 'flat-pencil.'   The common cross is called the 'pencil-point.'   The common plane is called the 'pencil-plane.'

20. A piece of a flat-pencil bounded by two of the straights, as 'sides,' is called an 'angle.'

21. The aggregate of all planes on a straight is called an 'axial-pencil,' or 'axial.'   Their common meet, the 'axis,' is their bearer.

22. A piece of the axial bounded by two of its planes, as sides, is called an 'axial angle.'

23. Angles are always pieces of the figure, not rotations.

24. No use is made of motion.   If a moving point is spoken of, it is to be interpreted as the mind shifting its attention.

25. When there can be no ambiguity of meaning, a figure in a pencil, though consisting only of some single elements of the complete pencil, may yet itself be called a pencil.   Just so, certain separate costraight points may be called a range.

ART. 2. PROJECTING AND CUTTING.

26. To 'project' from a fixed point M (the 'projection-vertex') a figure, the 'original,' composed of points B, C, D, etc., and straights b, c, d, etc., is to construct the 'projecting straights' $\overline{MB}$, $\overline{MC}$, $\overline{MD}$, and the 'projecting planes' $\overline{Mb}$, $\overline{Mc}$, $\overline{Md}$. Thus is obtained a new figure composed of straights and planes, all on M, and called an 'eject' of the original.

27. To 'cut' by a fixed plane μ (the 'picture-plane') a figure, the 'subject,' made up of planes β, γ, δ, etc., and straights b, c, d, etc., is to construct the meets $\overline{\mu\beta}$, $\overline{\mu\gamma}$, $\overline{\mu\delta}$, and the passes $\dot{\mu b}$, $\dot{\mu c}$, $\dot{\mu d}$. Thus is obtained a new figure composed of straights and points, all on μ, and called a 'cut' of the subject. If the subject is an eject of an original, the cut of the subject is an 'image' of the original.

28. Axial projection. To project from a fixed straight m (the 'projection-axis'), an original composed of points B, C, D, etc., is to construct the projecting planes $\overline{mB}$, $\overline{mC}$, $\overline{mD}$. Thus is obtained a new figure composed of planes all on the axis m, and called an 'axial-eject' of the original.

29. To cut by a fixed straight m (to 'transfix') a subject composed of planes β, γ, δ, etc., is to construct the passes $\dot{m\beta}$, $\dot{m\gamma}$, $\dot{m\delta}$. The cut obtained by transfixion is a range on the 'transversal' m.

30. Any two fixed primal figures are called 'projective' ($\overline{\wedge}$) when one can be derived from the other by any number of projectings and cuttings.*

ART. 3. ELEMENTS AT INFINITY.

31. It is assumed that for every element in either of the three primal figures there is always an element in each of the others.

---

*Pascal (1625–62) and Desargues (1593–1662) seem to have been the first to derive properties of conics from the properties of the circle by considering the fact that these curves lie in perspective on the surface of the cone.

32. On each straight is one and only one point ' at infinity,' or 'figurative' point. The others are 'proper' points. Any point going either way (moving in either 'sense') ever forward on a straight is at the same time approaching and receding from its point at infinity. The straight is thus a closed line compendent through its point at infinity.

33. 'Parallels' are straights on a common point at infinity.

34. Two proper points in it divide a range into a finite sect and a sect through the infinite. Its figurative point and a proper point in it divide a range into two sects to the infinite ('rays').

35. All the straights parallel to each other on a plane are on the same point at infinity, and so form a flat-pencil whose pencil-point is figurative. Such a pencil is called a 'parallel-flat-pencil.'

36. All points at infinity on a plane lie on one straight at infinity or figurative straight.* Its cross with any proper straight on the plane is the point at infinity on the proper straight.

37. Parallel-flat-pencils on the same plane have all a straight in common, namely, the straight at infinity on which are the figurative pencil-points of all these pencils.

38. Two planes whose meet is a straight at infinity are called parallel.

39. All the planes parallel to each other are on the same figurative straight, and so form an axial pencil whose axis is at infinity. Such an axial is called a parallel-axial.

40. All points at infinity and all straights at infinity lie on a plane at infinity or figurative plane. This plane at infinity is common to all parallel-axials, since it is on the axis of each.

Prob. 1. From each of the three primal figures generate the other two by projecting and cutting.

* This statement should not be interpreted as descriptive of the nature of infinity. In the Function Theory it is expedient to consider all points in a plane at infinity as coincident.

ART. 4.   CORRELATION AND DUALITY.

41. Two figures are called 'correlated' when every element of each is paired with one and only one element of the other. Correlation is a one-to-one correspondence of elements. The paired elements are called 'mates.'

42. Two figures correlated to a third are correlated to each other. For each element of the third has just one mate in each of the others, and these two are thus so paired as to be themselves mates.

43. On a plane, any theorem of configuration and determination, with its proof, gives also a like theorem with its proof, by simply interchanging point with straight, join with cross, sect with angle.*

This correlation of points with straights on a plane is termed a 'principle of duality.' Each of two figures or theorems so related is called the 'dual' of the other.†

Prob. 2. When two coplanar ranges m_1 and m' are correlated as cuts of a flat-pencil M, show that the figurative point P_1, or Q', of the one is mated, in general, to a proper point P', or Q_1, of the other.

Prob. 3. Give the duals of the following:

$1'$. Two coplanar straights determine a flat-pencil on their cross.

$2'$. Two coplanar flat-pencils determine a straight, their 'concur.'

3_1. Two points bound two 'explemental' sects.

Prob. 4. To draw a straight crossing three given straights, join the passes of two with a plane on the third. Give and solve dual.

ART. 5.   POLYSTIMS AND POLYGRAMS.

44_1. A 'polystim' is a system of n coplanar points ('dots'), with all the ranges they determine ('connectors'). Assume that no three dots are costraight.	$44'$. A 'polygram' is a system of n coplanar straights ('sides'), with all the flat-pencils they determine ('fans'). Assume that no three sides are copunctal.

---

*Culmann's Graphic Statics (Zürich, 1864) made extensive use of the theorems of duality. Reye's Geometrie der Lage (Hannover, 1856) was issued as a consequence of the Graphic Statics of Culmann.

†In Algebra the principle of duality consists in the interpretation of the same equation in different kinds of coordinates—point and line or point and plane coordinates.

In each dot intersect $(n-1)$ connectors, going through the remaining $(n-1)$ dots. So there are $n(n-1)/2$ connectors.

In each side concur $(n-1)$ fans, going through the remaining $(n-1)$ sides. So there are $n(n-1)/2$ fans.

45_1. For n greater than 3, the connectors will intersect in points other than the dots. Such intersections are called 'codots.'

$45'$. For n greater than 3, the fans will concur in straights other than the sides. Such concurs are called 'diagonals.'

46_1. There are $n(n-1)(n-2)(n-3)/8$ codots.

$46'$. There are $n(n-1)(n-2)(n-3)/8$ diagonals.

Proof of 46_1. In a polystim of n dots there are $n(n-1)/2$ connectors. These connectors intersect in

$$[n(n-1)/2][n(n-1)/2-1]/2 = n(n-1)(n^2-n-2)/8$$

points; i.e., the number of different combinations of $n(n-1)/2$ things, two at a time.

But some of these intersections are dots, and the remaining ones are codots. Now $(n-1)$ of these connectors meet at each dot. Therefore each dot is repeated $(n-1)(n-2)/2$ times; or the number of times the connectors intersect in points not codots, i.e. in dots, is $n(n-1)(n-2)/2$.

Therefore the number of codots is

$$n(n-1)(n^2-n-2)/8 - n(n-1)(n-2)/2$$
$$= [n(n-1)/8][n^2-n-2-4(n-2)]$$
$$= n(n-1)(n-2)(n-3)/8.$$

47_1. A set of n connectors may be selected in several ways so that two and only two contain each one of the n dots. Such a set of connectors is called a 'complete set' of connectors.

$47'$. A set of n fans may be selected in several ways so that two and only two contain each one of the n sides. Such a set of fans is called a 'complete set' of fans.

48_1. There are $(n-1)!/2$ complete sets of connectors.

$48'$. There are $(n-1)!/2$ complete sets of fans.

Proof of 48_1. In a polystim of n dots there are through any single dot $(n-1)$ connectors, and hence $(n-1)(n-2)/2$ pairs of connectors. Consider one such pair, as BC and BE.

The number of different sets (each of $n - 2$ connectors) from C to E through A, D, F, G, etc. [there being $(n - 3)$ such dots], is $(n - 3)!$, i.e. the number of permutations of $(n - 3)$ things. Hence the number of complete sets of connectors having the pair BC and BE is $(n - 3)!$ Therefore the whole number of complete sets of connectors is

$$(n - 1)(n - 2)[(n - 3)!]/2 = (n - 1)!/2.$$

49_1. In any complete set of connectors, when n is even, the first and the $(n/2+1)$th are called 'opposite'.

$49'$. In any complete set of fans, when n is even, the first and the $(n/2+1)$th are called 'opposite.'

50_1. A 'tetrastim' is a system of four dots with their six connectors. Each pair of opposite connectors intersect in a codot. These three codots determine the 'codot-tristim' of the tetrastim.

$50'$. A 'tetragram' is a system of four straights with their six fans. Each pair of opposite fans concur in a diagonal. These three diagonals determine the 'diagonal-trigram' of the tetragram.

51. Two correlated polystims whose paired dots and codots have their joins copunctal are called 'copolar.'

52. Two correlated polystims whose paired connectors intersect and have their intersections costraight are called 'coaxal.'

53. If two non-coplanar tristims be copolar, they are coaxal. For since AA' crosses BB', therefore AB and $A'B'$ intersect on the meet of the planes of the tristims.

54. If two non-coplanar tristims be coaxal, they are copolar. For since AB intersects $A'B'$, these four points are coplanar. The three planes $ABA'B'$, $ACA'C'$, $BCB'C'$ are copunctal. Hence so are their meets AA', BB', CC'.

55. By taking the angle between the planes evanescent, is seen that coplanar coaxal tristims are copolar; and then by reductio ad absurdum that coplanar copolar tristims are coaxal.

56. If two coplanar polystims are copolar and coaxal they are said to be 'complete plane perspectives.' Their pole and

axis are called the 'center of perspective' and the 'axis of perspective.'

57. If two coplanar tristims are copolar or coaxal, they are complete plane perspectives.

58. If two coplanar polystims are images of the same polystim from different projection vertices V_1, V_2, they are complete plane perspectives. For the joins of pairs of correlated points are all copunctal (on the pass of the straight $V_1 V_2$ with the picture plane), and the intersections of paired connectors are all costraight (on the meet of the picture plane and the plane of the original).

Prob. 5. In a hexastim there are 15 connectors and 45 codots. In a hexagram there are 15 fans and 45 diagonals.

Prob. 6. If the vertices of three coplanar angles are costraight, their sides make three tetragrams whose other diagonals are copunctal by threes four times. [Prove and give dual.]

Prob. 7. The corresponding sides of any two funiculars of a given system of forces cross on a straight parallel to the join of the poles of the two funiculars.

ART. 6. HARMONIC ELEMENTS.

59. Fundamental Theorem.—If two correlated tetrastims lie on different planes whose meet is on no one of the eight dots, and if five connectors of the one intersect their mates, then the tetrastims are coaxal. For the two pairs of tristims fixed by the five pairs of intersecting connectors being coaxal are copolar. Hence the sixth pair of connectors are coplanar.

60. If the tetrastims be coplanar, and if five intersections of pairs of correlated connectors are costraight, this the coplanar case can be made to depend upon the other by substituting for one of the tetrastims its image on a second plane meeting the first on the bearer of the five intersections.

61. If the axis m is a figurative straight, the theorem reads : If of two correlated tetrastims five pairs of mated connectors are parallel, so are the remaining pair.

62. Four costraight points are called 'harmonic points,' or

a 'harmonic range,' if the first and third are codots of a tetrastim while the other two are on the connectors through the
third codot.

63. By three costraight points and their order the fourth
harmonic point is uniquely determined. For if the three points

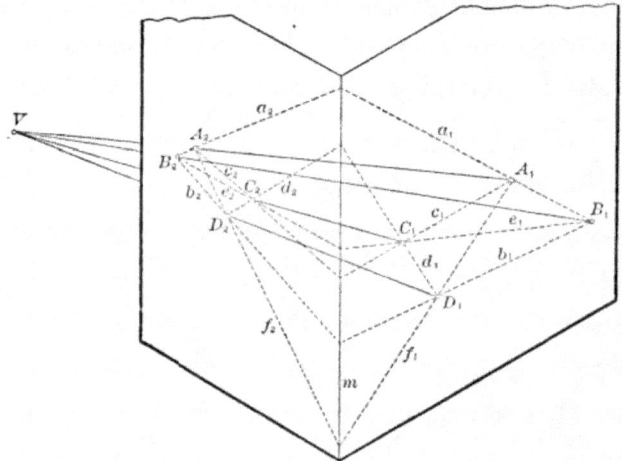

in order are A, B, C, draw any two straights through A, and a
third through B to cross these at K and M respectively. Join
CK, crossing AM at N. Join CM, crossing AK at L. Then the
join LN crosses the straight ABC, always at the same point D,
the fourth harmonic to ABC separated from B.

64. In projecting from a point not coplanar with it a
tetrastim defining a harmonic range, the four harmonic points
are projected by four coplanar straights, called 'harmonic
straights' or a 'harmonic flat-pencil.'

65. The four planes projecting harmonic points from an
axis not coplanar with them are called 'harmonic planes,' or a
'harmonic axial-pencil.'

66. Projecting or cutting a harmonic primal figure gives
always again a harmonic primal figure.

67. By three elements of a primal figure, given which is the
second, the fourth harmonic is completely determined.

68. Defining harmonic points by the tetrastim distinguishes

two points made codots from the other two. Yet it may be shown that the two pairs of points play identically the same rôle.

First, from the definition of four harmonic points each separated two may be interchanged without the points ceasing to

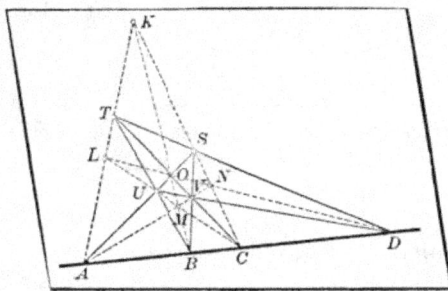

be harmonic [or, if $ABCD$ is a harmonic range, so is also $ADCB$, $CBAD$, and $CDAB$]. For the first and third remain codots.

Second, to prove that in a harmonic range the two pairs of separated points may be interchanged without the four points ceasing to be harmonic [or, if $ABCD$ is a harmonic range (and therefore $ADCB$, $CBAD$, and $CDAB$), then also is $BADC$, $DABC$, $BCDA$, and $DCBA$]: Through the third codot O draw the joins AO and CO. These determine on the connectors NK, KL, LM, and MN four new points, S, T, U, V, respectively. The tetrastim $KTOS$ has for two codots A and C, and has a connector though B; hence its remaining connector TS must pass though D. In like manner, the connector UV of the tetrastim $MVOU$ must pass through D, and a connector of each of the tetrastims $LUOT$ and $VNSO$ through B. Therefore B and D are codots of a tetrastim $STUV$ with the remaining connectors, one through A, one through C.

69. The separated points A and C are called 'conjugate points,' as also are B and D. Either two are said to be 'harmonic conjugates' with respect to the other two.

Prob. 8. To determine the join of a given point M with the inaccessible cross X of two given straights n and n'.

Through M draw any two straights crossing n at B and B', and n' at D and D'. Join DB and $D'B'$, crossing on A. Through A draw

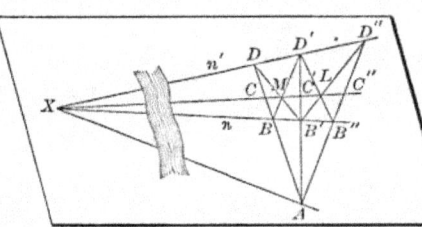

any third straight crossing n at B'' and n' at D''. Join $B'D''$ and $D'B''$, crossing at L. Then LM is the join required.

Proof. The tetrastim $XBMD$ makes $AB'C'D'$ a harmonic range, as $XB'LD'$ does $AB''C''D''$. But projecting $AB''C''D''$ from X, and cutting the eject by $AB'D'$ gives a harmonic range. Therefore C'', C', and X are costraight.*

Prob. 9. Through a given point to draw with the straight-edge a straight parallel to two given parallels.

Prob. 10. To determine the cross of a given straight m with the inconstructible join x of two given points N and N'. Join any two

points on m with N and N', giving b and b' on N, d and d' on N'. Join the crosses db and $d'b'$ by a. On a take any third point joining with N in b'' and with N' in d''.

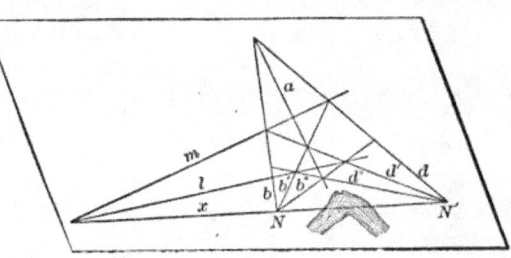

Join the crosses $b'd''$ and $d'b''$ by l. Then lm is the cross required. [From Prob. 8, by duality.]

Prob. 11. Cut four coplanar non-copunctal straights in a harmonic range.

Prob. 12. On a given straight determine a point from which the ejects of three given points form with the given straight a harmonic pencil.

ART. 7. PROJECTIVITY.

70. Two primal figures of three elements are always projective.—If one be a pencil, take its cut by a transversal. If the bearers of ABC and $A'B'C'$ be not coplanar, join AA', BB', CC', and cut these joins by a transversal, m. Then ABC and $A'B'C'$ are two cuts of the axial mAA', mBB', mCC'.

*Numerous problems in Surveying may be solved by the application of the preceding principles, but such application has not been found advantageous in practice. See Gillespie's Treatise on Land Surveying, New York, 1872.

If the bearers are coplanar, take on the join AA' any two projection vertices M and M'. Join MB and $M'B'$, crossing at B''; join MC and $M'C'$, crossing at C''. Join $B''C''$ crossing AA' at A''. Then ABC and $A'B'C'$ are images of $A''B''C''$.

71. If any four harmonic elements are taken in one of two projective figures, the four elements correlated to these are also harmonic. For both ejects and cuts of harmonic figures are themselves harmonic.

72. Two primal figures are projective if they are so correlated that to every four harmonic elements of the one are correlated always four harmonic elements of the other. For the same projectings and cuttings which derive $A'B'C'$ from ABC will give D_1 from D. Therefore $A'B'C'D_1$ is harmonic. But by hypothesis $A'B'C'D'$ is harmonic. Therefore D_1 is D'.

73. If two primal figures are projective, then to every consecutive order of elements of the one on a bearer corresponds a consecutive order of the correlated elements of the other on a bearer.

74. Two projective primal figures having three elements self-correlated are identical. For two self-correlated elements cannot bound an interval containing no such element, since they must harmonically separate one without it from one within.

75. Two ranges are called ' perspective ' if cuts of the same flat pencil.

Two flat pencils are perspective if cuts of the same axial pencil, or ejects of the same range. Two axials are perspective if ejects of the same flat pencil.

A range and a flat pencil, a range and an axial pencil, or a flat pencil and an axial are perspective if the first is a cut of the second.

76_1. If two projective ranges not costraight have a self-correlated point A, they are perspective.

$76'$. If two coplanar projective flat pencils not copunctal have a self-correlated straight a, they are perspective.

Let the join of any pair of correlated points BB' cross the join of any other pair CC' at V.

Projecting the two given ranges from V, their ejects are identical, since they are projective and have the three straights VA, VBB', VCC' self-correlated.

Let the cross of any pair of correlated straights bb' join the cross of any other pair cc' by m.

Cutting the two given flat pencils by m, their cuts are identical, since they are projective and have the three points ma, mbb', mcc' self-correlated.

ART. 8. CURVES OF THE SECOND DEGREE.

77₁. If two coplanar non-copunctal flat pencils are projective but not perspective, the crosses of correlated straights form a 'range of the second degree,' or 'conic range.'

77′. If two coplanar non-costraight ranges are projective but not perspective, the joins of correlated points form a 'pencil of the second class,' or 'conic pencil.'

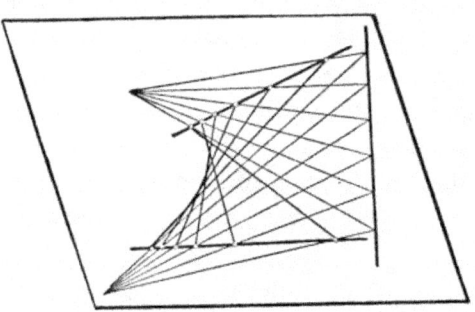

78₁. If two copunctual non-costraight axial pencils are projective but not perspective, the meets of correlated planes form a 'conic surface of the second order,' or 'cone.'

78′. If two copunctal non-coplanar flat pencils are projective but not perspective, the planes of correlated straights form a 'pencil of planes of the second class,' or 'cone of planes.'

79. All results obtained for the conic range or the conic pencil are interpretable for the cone or cone of planes, since the eject of a conic is a cone and the cut of a cone is a conic.

80₁. On the cross A of any pair of correlated straights a and a_1

80′. On the join a of any pair of correlated points A and A_1 of

of the projective flat pencils V

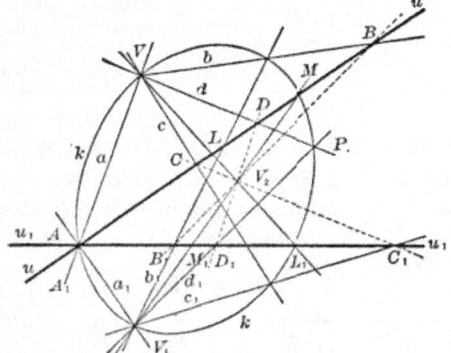

and V_1 draw two straights u and u_1.

The cuts ABC and $A_1B_1C_1$ being projective and having a pair of correlated points A, A_1 coincident, are perspective, both being cuts of the pencil on V_2, the cross of the joins BB_1 and CC_1.

Any straight d of V, crossing u at D, is then correlated to the join of V_1 with the cross D_1 of u_1 and the join DV_2. Any d crosses its d_1 so determined, at P, a point of the conic range k.

the projective ranges u and u_1 take two points V and V_1.

The ejects abc and $a_1b_1c_1$ being projective and having a pair of correlated straights a, a_1 coincident, are perspective, both being ejects of the range on u_2, the join of the crosses bb_1 and cc_1.

Any point D of u, joined with V by d, is then correlated to the cross of u_1 with the join d_1 of V_1 and the cross du_2.

Any D joined to its D_1 so determined, gives p a straight of the conic pencil K.

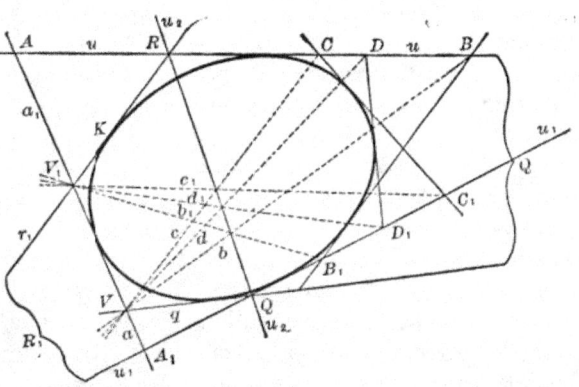

81₁. The pencil-points V, V_1 of the generating pencils pertain to the conic, since their join VV_1 is crossed by the element correlated to it in either pencil at its pencil-point.

81′. The bearers u, u_1 of the generating ranges pertain to the conic, since their cross uu_1 is joined to the element correlated to it in either range by its bearer.

82_1. The straight on V correlated to V_1V is called the 'tangent' at V. Every other straight on V is its join with a second point of the conic.

83_1. On any straight, as u, on any point A of the conic, its second element is its cross M with the join V_1V_2.

84_1. From the five given points VV_1AML, of k construct a sixth, P. The cross D of u with the join VP, and the cross D_1 of u_1 with the join V_1P are costraight with V_2. Therefore* the three opposite pairs in every complete set of connectors of a hexastim whose dots are in a conic intersect in three costraight codots whose bearer is called a 'Pascal straight.'

This hexastim has sixty Pascal straights, since it has sixty complete sets of connectors.

85_1. The ejects of the points of a conic from any two are projective.

86_1. By five of its points a conic is completely determined.

87_1. Instead of five points may be given the two pencil-points and three pairs of correlated straights. If one given straight is the join of the pencil-points, then four points and a tangent at one of them are given.

Thus by four of its points and the tangent at one of them a

$82'$. The point on u correlated to u_1u is called the 'contact' on u. Every other point on u is its cross with a second straight of the conic.

$83'$. On any point, as V, on any straight a of the conic, its second element is its join q with the cross u_1u_2.

$84'$. From the five given straights u, u_1, a, q, r_1, of K construct a sixth DD_1 or p. The join d of V with the cross up, and the join d_1 of V_1 with the cross u_1p are copunctal with u_2. Therefore † the three opposite pairs in every complete set of fans of a hexagram whose sides are in a conic concur in three copunctal diagonals whose bearer is called a 'Brianchon point.'

This hexagram has sixty Brianchon points, since it has sixty complete sets of fans.

$85'$. The cuts of the straights of a conic by any two are projective.

$86'$. By five of its straights a conic is completely determined.

$87'$. Instead of five straights may be given the two bearers and three pairs of correlated points.

If one given point is the cross of the bearers, then four straights and a contact point on one of them are given.

Thus by four of its straights and a contact-point on one of

---

* Pascal, 1640.   † Brianchon, 1806.

conic is completely determined.

88₁. By three of its points and the tangents at two of them the conic is completely determined.

89₁. Interpreting a pentastim as a hexastim with two dots coinciding gives: In every complete set of connectors of a pentastim whose dots are in a conic, two pairs of non-consecutive connectors determine by their two intersections a straight on which is the cross of the fifth connector with the tangent at the opposite dot.

them a conic is completely determined.

88'. By three of its straights and the contact-points on two of them the conic is completely determined.

89'. Interpreting a pentagram as a hexagram with two sides coinciding gives: In every complete set of fans of a pentagram whose sides are in a conic, two pairs of non-consecutive fans determine by their two concurs a point on which is the join of the fifth fan-point with the contact-point on the opposite side.

Thence follows the solution of the problems:

90₁. Given five points of a conic, to construct tangents at the points, using the ruler only.

91₁.* The hexastim with a pair of opposite connectors replaced by tangents gives: The intersections of the two opposite pairs in every complete set of connectors of a tetrastim with dots in a conic are both costraight with the crosses of the two pairs of tangents at opposite dots.

Or: A tetrastim with dots in a conic has each pair of codots costraight with a pair of fan-points of the tetragram of tangents at the dots.

90'. Given five straights of a conic, to find contact-points on the straights, using the ruler only.

91'. The hexagram with a pair of opposite fans replaced by contact-points gives: The concurs of the two opposite pairs in every complete set of fans of a tetragram with sides in a conic are both copunctal with the joins of the two pairs of contact-points on opposite sides.

Or: A tetragram with sides in a conic has each pair of diagonals copunctal with a pair of connectors of the tetrastim of contacts on the sides.

The figure for 91₁ and that for 91' are identical, and called Maclaurin's Configuration.   (See page 86.)

92₁. The tangents of a conic range are a conic pencil.

92'. The contact-points of a conic pencil are a conic range.

---

* Due to Maclaurin, 1748.

93. The points of a conic range may now be conceived as all on a curve, a 'conic curve,' their bearer. The straights of

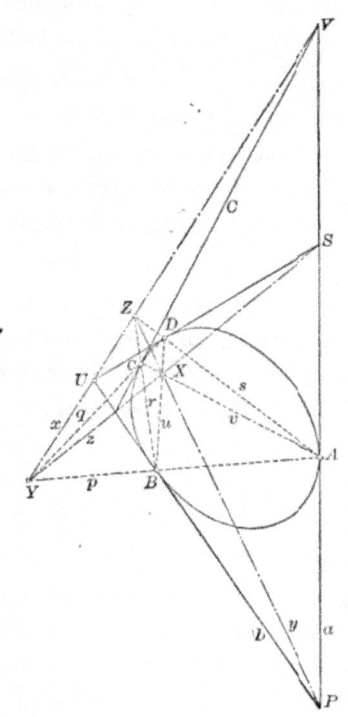

the corresponding conic pencil, tangents of this conic range, may now also be conceived as all on this same conic curve on which are their contact-points. Consequently the conic curve is dual to itself, and so the principle of duality on a plane receives an important extension.

94. It follows immediately from their generation that all conics are closed curves, though they may be compendent through one or two points at infinity. With two points at infinity the curve is called 'hyperbola ;' with one, 'parabola ;' with none, 'ellipse.' *

95. If a point has on it tangents to the curve, it is called 'without' the curve; if none, 'within' the curve. The contact-point on a tangent is 'on' the curve ; all other points on a tan-

---

* The generation shows that a straight cuts the curves in two points and that from any point two tangents to the curves may be drawn. Hence the curves are of the second order and of the second class, that is they are identical with the conics of analytic geometry. Analytically the equations $P + \lambda Q = 0$, $P' + \lambda Q' = 0$, where P, Q, P', Q' are linear functions of point coordinates, represent two projective pencils, the correlated rays corresponding to the same value of λ. Hence the locus of the intersection of correlated rays is represented by $PQ' - P'Q = 0$, a second-degree point equation. Projective ranges are represented by $R + \lambda S = 0$, $R' + \lambda S' = 0$, where R, S, R', S' are linear functions of line coordinates. The envelope of the joins of correlated points is represented by $RS' - R'S = 0$, a second-degree line equation.

The projective generation of conics is developed synthetically in Steiner's Theorie der Kegelschnitte, 1866, and in Chasles' Géométrie supérieure, 1852. For the analytic treatment see Clebsch, Geometrie, vol. I, 1876.

gent are without the curve. Every straight in its plane contains innumerable points without the curve, since the straight crosses every tangent.

Prob. 13. Given four points on a conic and the tangent at one of them, draw the tangent at another.

Prob. 14. If the n sides of a polygram rotate respectively about n fixed points not costraight, while $(n-1)$ of a complete set of fan-points glide respectively on $(n-1)$ fixed straights, then every remaining fan-point describes a conic.*

Prob. 15. In any tristim with dots on a conic the three crosses of the connectors with the tangents at the opposite dots are costraight.†

Prob. 16. If two given angles rotate about their fixed vertices so that one cross of their sides is on a straight, either of the other three crosses describes a conic.‡

Prob. 17. Construct a hyperbola from three given points, and straights on its figurative points.

ART. 9. POLE AND POLAR.

96. Taking every tangent to a conic as the dual to its own contact-point fixes as dual to any given point in the plane one particular straight, its 'polar,' of which the point is the 'pole.'

97. With reference to any given conic, to construct the polar of any given point in its plane. Put on the given point Z two secants crossing the curve, one at A and D, the other at B and C. The join of the other codots X and Y of $ABCD$ is the polar of Z. Varying either secant, as ZBC, does not change this polar, since on it must always be the cross S of the tangents at A and D, and also the point which D and A harmonically separate from Z (given by each of the variable tetrastims $BXCY$).

98. The join of any two codots of a tetrastim with dots on a conic is the polar of the third codot with respect to that

---

* Due to Braikenridge, 1735.

† From Pascal; dual from Brianchon.

‡ Given by Newton in Principia, Book I, lemma xxi, under the name of "the organic description" of a conic.

conic, and either codot is the pole of the join of the other two. Any point is harmonically separated from its polar by the conic.

99. To draw with ruler only the tangents to a conic from a point without, join it to the crosses of its polar with the conic.

100₁. Two points are called 'conjugate' with reference to a conic if one (and so each) is on the polar of the other.

100'. Two straights are called 'conjugate' with reference to a conic if one (and so each) is on the pole of the other.

101₁. All points on a tangent are conjugate to its contact-point.

101'. All straights on a contact-point are conjugate to its tangent.

102₁. The points of a range are projective to their conjugates on its bearer.

102'. The straights of a flat pencil are projective to their conjugates on its bearer.

103₁. With reference to a given conic, the 'kerncurve,' the polars of all points on a second conic make a conic pencil, whose bearer is the 'polarcurve' of the second conic.

103'. With reference to a given conic, the 'kerncurve,' the poles of all tangents on a second conic make a conic range, whose bearer is the 'polarcurve' of the second conic.

Prob. 18. Either diagonal of a circumscribed tetragram is the polar of the cross of the others.

Prob. 19. A pair of tangents from any point on a polar harmonically separate it from its pole.

Prob. 20. A pair of tangents are harmonic conjugates with respect to any pair of straights on their cross which are conjugate with respect to the conic.

ART. 10.   INVOLUTION.

104. If in a primal figure of four elements (a 'throw') first any two be interchanged, then the other two, the result is projective to the original.

[That is, $ABCD \,\overline{\wedge}\, BADC \,\overline{\wedge}\, CDAB \,\overline{\wedge}\, DCBA$.]

Let $ABCD$ be a throw on m. Project it from V. Cut this eject by a straight (m') on A. The cut is $AB'C'D'$. Now project $ABCD$ from C'. The cut of this latter eject by VB is

$B'BVH$.   Project $B'BVH$ from D and cut the eject by m'.   The
cut is $B'AD'C'$, which is perspective to $BADC$.

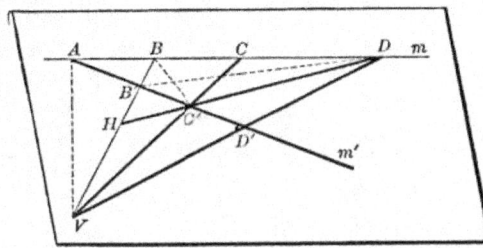

105. Two projective primal figures of the same kind of elements and both on the same bearer are called 'conjective.'
When in two conjective primal figures one particular element
has the same mate to whichever figure it be regarded as belonging, then every element has this property.

If $AA'BB'$ is projective to $A'AB'X$, then by § 104, $AA'BB'$
is projective to $AA'XB'$, and having three elements self-correlated, they are identical.

106. Two conjective figures such that the elements are
mutually paired ('coupled') form an 'Involution.' For example, the points of a range, and, on the same bearer, their conjugates with respect to a conic, form an involution.   Every
eject and every cut of an involution is an involution.

107. When two ranges are projective, the point at infinity
of either one is correlated to a point of the other called its
'vanishing point.'

108. When two conjective ranges form an involution the
two vanishing points coincide in a point called the 'center' of
the involution.

109. If two figures forming an involution have self-correlated elements, these are called the 'double' elements of the
involution.   An involution has at most two double elements;
for were three self-correlated, all would be self-correlated.

110. If a primal figure of four elements is projective with
a second made by interchanging two of these elements, they
harmonically separate the other two.

For project the range $ABCD$ from V and cut the eject by a straight on A. The cut $AB'C'D'$ is projective to $ABCD$,

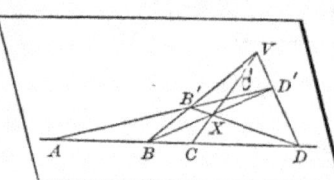

which by hypothesis is projective to $ADCB$. Therefore $ADCB$ is perspective to $AB'C'D'$. So $VC'C$ is on the cross X of the joins DB' and BD'. So B and D are codots of the tetrastim $VD'XB'$, while A and C are on the connectors through C', the third codot.

111. If an involution has two double elements these separate harmonically any two coupled elements. Let A and C be the double elements. Then $ABCB'$ is projective to $AB'CB$; therefore by § 110 $ABCB'$ is harmonic.

112. An involution is completely determined by two couples. For the projective correspondence $AA'B \ldots \bar{\wedge} A'AB' \ldots$ is completely determined by the three given pairs of correlated elements, and since among them is one couple, so are all correlated elements couples.

113. When there are double elements, then the elements of no couple are separated by those of another couple. Inversely, when the elements of one couple separate those of another, then the elements of every couple are separated by those of every other, and there are no double elements.

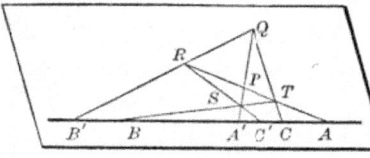

114$_1$. The three pairs of opposite connectors of a tetrastim are cut by any transversal in three couples of a point involution.*

114'. The three pairs of opposite fan-points of a tetragram are projected from any projection-vertex by three couples of an involution of straights.

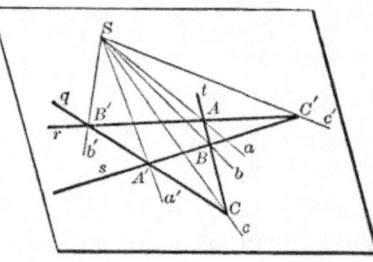

* Due to Desargues, 1639.

Let $QRST$ be a tetrastim of which the pairs of opposite connectors RT and QS, ST and QR, QT and RS are cut by any transversal respectively in A and A', B and B', C and C'. From the projection-vertex Q, the ranges $ATPR$ and $ACA'B'$ are perspective. But $ATPR$ and $ABA'C'$ are perspective from S. Therefore $ACA'B'$ is projective to $ABA'C'$, and therefore to $A'C'AB$ (§ 104). Since thus A and A' are coupled, so (§ 105) are B and B', and C and C'.

115. To construct the sixth point C' of an involution of which five points are given, draw through C any straight, on which take any two points Q and T. Join AT, $B'Q$ crossing at R. Join BT, $A'Q$ crossing at S. The join RS cuts the bearer of the involution in C'.

Prob. 21. Find the center O of a point involution of which two couples $AA'BB'$ are given.

Prob. 22. If two points M and N on m are harmonically separated by *two* pairs of opposite connectors of a tetrastim, then so are they by the third pair.

Prob. 23. To construct a conic which shall be on three given points, and with regard to which the couples of points of an involution on a given straight shall be conjugate points.

ART. 11.   PROJECTIVE CONIC RANGES.

116. Four points on a conic are called harmonic if they are projected from any (and so every) fifth point on the conic by four harmonic straights.

117. A conic and a primal figure or two conics are called projective when so correlated that every four harmonic elements of the one correspond to four harmonic elements of the other.

118. If a conic range and a flat pencil are projective, and every element of the one is on the correlated element of the other, they are called perspective. A conic is projected from every point on it by a flat pencil perspective to it. Inversely the pencil-point of every flat pencil perspective to a conic is on the conic.

119. Two conics are projective if flat pencils respectively perspective to them are projective. Therefore any three elements in one can be correlated to any three elements in the other, but this completely pairs all the elements.

120. Two different conic ranges on the same bearer have at most two self-correlated elements.

121. Two different coplanar conic ranges with a point V in common are projective if every two points costraight with V are correlated. For both are then perspective to the flat pencil on V. Every common point other than V is self-correlated; but V only when they have there a common tangent. They can have at most three self-correlated points.

122. If a flat pencil V and conic range k are coplanar and projective but not perspective, then at most three straights of the pencil are on their correlated points of the conic; but at least one.

For any flat pencil M perspective to k is projective to V, and with it determines in general a second conic range which must have in common with k every point which lies on its correlated straight of V. So if more than three straights of V were on their correlated points of k, the conics would be identical and V perspective to k.

Again, since every conic is compendent, and so divides its plane into two severed pieces, therefore the two different conics if they cross at their common point M must cross again, say at P. In this case the straights VP and MP are correlated, and so VP is on the point P correlated to it on k.

In case they do not cross at their common point M, the straight VM corresponds to the common tangent at M, and so to the point M correlated to it on k.

123. Two projective conic ranges on the same curve form an involution if a pair of points are doubly correlated. Besides the couple AA_1, let B and B_1 be any other two correlated points, so that AA_1B corresponds to A_1AB_1. The cross of AA_1 and BB_1 call U, and its polar u. Project AA_1B from B_1.

Project A_1AB_1 from B. The ejects $B_1(AA_1B)$ and $B(A_1AB_1)$ are projective, and having the straight B_1B (or BB_1) self-correlated, so are perspective. The crosses of their correlated elements are therefore costraight. But the cross of B_1A with its correlated straight BA_1 is known to be on u, the polar of U, the

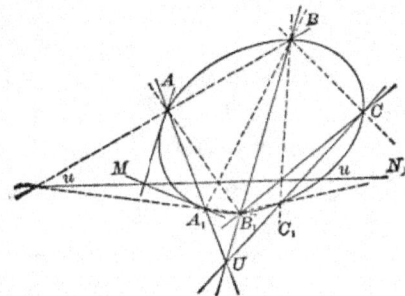

cross of AA_1 with BB_1. Likewise the cross of B_1A_1 with BA is on u. Therefore the point C_1 correlated to C is the cross of CU with the curve. So C and C_1 are coupled.

124. If two conic ranges form an involution, the joins of coupled points are all copunctal on the 'involutioncenter.'

125. Calling projective the conic pencils dual to projective conic ranges, if these ranges form an involution, so do the pencils, and the crosses of coupled tangents are all costraight on the 'involutionaxis.'

So two conic pencils forming an involution are cut by each of their straights in two ranges forming an involution. Two conic ranges forming an involution are projected from each of their points in two flat pencils, forming an involution.

126. If the involutioncenter lies without the conic bearer of an involution, it has two double elements where it is cut by the involutionaxis.

127. To construct the self-correlated points of two projective conic ranges on the same conic.—Let A, B, C be any three points of k, and A_1, B_1, C_1 their correlated points of k_1. The projective flat pencils $A(A_1B_1C_1)$ and $A_1(ABC)$ have AA_1 self-corresponding, hence they are perspective to a range on the join u of the cross of AB_1 and A_1B with the cross of AC_1

and A_1C. The crosses of the conic and this join u are the self-correlated points of k and k_1.

128. If the dots of a tetrastim are on a conic, the six points where a straight not on a dot cuts the conic and two pairs of opposite connectors form an involution.

For the two flat pencils in which the two crosses of m with the conic, P, P_1, and two opposite dots R, T, are projected from the other two dots Q, S, are projective, and consequently so are the cuts of these flat pencils by m; that is, $PBP_1A \barwedge PA_1P_1B_1$. But $PA_1P_1B_1 \barwedge P_1B_1PA_1$. Therefore $PBP_1A \barwedge P_1B_1PA_1$.

129₁. Conics on which are the dots of a tetrastim are cut by a transversal in points of an involution. At its double points the transversal is tangent to two of those conics.

129'. Copunctal tangents to conics on which are the sides of a tetragram form an involution. The double straights touch two of those conics at the pencil-point.

Prob. 24. The pairs of points in which a conic is cut by the straights of a pencil whose pencil-point is not on the conic form an involution.

ART. 12.   CENTER AND DIAMETER.

130. The harmonic conjugate of a point at infinity with respect to the end points of a finite sect is the 'center' of that sect.

131. The pole of a straight at infinity with respect to a certain conic is the 'center' of the conic.

132. The polar of any figurative point is on the centre of the conic, and is called a 'diameter.'

133. If a straight crosses a conic the sect between the crosses is called a 'chord.'

The center of a conic is the center of all chords on it.

134. The centers of chords on straights conjugate to a diameter are all on the diameter.

135. Two diameters are conjugate when each is the polar of the figurative point on the other.

136. The tangents at the crosses of a straight with a conic cross on the diameter which is a conjugate to that straight.

137. The joins of any point on the conic to the crosses of a diameter with the conic are parallel to two conjugate diameters.

138. Of two conjugate diameters, each is on the centers of the chords parallel to the other; and if one crosses the conic, the tangents at its crosses are parallel to the other diameter.

139. The center of an ellipse is within it, for its polar does not meet the curve, and so there are no tangents from it to the curve. The centre of a parabola is the contact point of the figurative straight. The centre of a hyperbola lies without the curve, since the figurative straight crosses the curve. The tangents from the center to the hyperbola are called 'asymptotes.' Their contact-points are the two points at infinity on the curve.

140. If a diameter which cuts the curve be given, the tangents at its crosses can be constructed with ruler only, and so however many chords on straights conjugate to the diameter.

141. Every flat pencil is an involution of conjugates with respect to a given conic. Hence the pairs of conjugate diameters of a conic form an involution.

If the conic is a hyperbola, the asymptotes are the double straights of the involution. Hence any two conjugate diameters of a hyperbola are harmonically separated by the asymptotes; and since the hyperbola lies wholly in one of the two explemental angles made by the asymptotes, one diameter cuts the curve, the other does not.

142. Any one pair of conjugate diameters of an ellipse is always separated by any other pair. Any one pair of conjugate diameters of a hyperbola is never separated by any other pair.

143. If a tangent to a hyperbola cuts the asymptotes at A and C, then the contact-point B is the center of the sect AC, since the tangent cuts the harmonic pencil made by the diameter through B, the conjugate diameter and the asymptotes, in the harmonic range $ABCD$ where D is at infinity. Just so the

center of any chord is the center of the costraight sect bounded by the asymptotes.

144. If a point is the center of two chords it is the center of the conic, for its polar is the figurative straight.

145. As many points as desired of a conic may be constructed by the ruler alone.

With the aid of one fixed conic all problems solvable by ruler and compasses can be solved by ruler alone, that is, by pure projective geometry. For example : Of two projective primal figures (say ranges) on the same bearer, given three pairs of correlated elements to find the self-corresponding elements, if there be any. Project the two ranges from any point V of the given conic. These ejects are cut by the conic in projective conic ranges. Of these determine the self-correlated points by § 127.

Project these from V. The ejects cut the bearer of the original ranges in the required self-correlated points.

Prob. 25. Find the crosses of a straight with a conic given only by five points.

Prob. 26. Given a conic and its center, find a point B such that for two given points A, C, the center of the sect AB shall be C.

Prob. 27. The join of the other extremities of two coinitial sects is parallel to the join of their centers.

Prob. 28. In an ellipse let A and B be crosses of conjugate diameters CA, CB with the curve. Through A' the cross of the diameter conjugate to CA with the curve draw a parallel to the join AB. Let it cut the curve again at B'. Then CB' is the diameter conjugate to CB.

Art. 13. Plane and Point Duality.

146_1. On a plane are ∞^2 points, a 'point-field.'

147_1. The ∞^1 planes of a single axial pencil have on them all the points of point-space; so there are just ∞^3 points.

Point-space is tridimensional.

$146'$. On a point are ∞^2 planes, a 'plane-sheaf.'

$147'$. The ∞^1 points of a single range have on them all the planes of plane-space; so there are just ∞^3 planes.

Plane-space is tridimensional.

148. With the straight as element, space is of four dimensions.

On a plane are ∞^2 straights, a 'straight-field.'

On a point are ∞^2 straights, a 'straight-sheaf.'

On a straight are ∞^1 planes, and so ∞^3 straights.

On a straight are ∞^1 points, and so ∞^3 straights.

On each of the ∞^2 points on a plane are the ∞^2 straights of a straight-sheaf; so there are just ∞^4 straights.

On each of the ∞^2 planes on a point are the ∞^2 straights of a straight-field; so there are just ∞^4 straights.

149_1. Two planes determine a straight, their meet.

$149'$. Two points determine a straight, their join.

150_1. Two planes determine an axial-pencil on their meet.

$150'$. Two points determine a range on their join.

151_1. Two bounding planes determine an axial angle.

$151'$. Two bounding points determine a sect.

152_1. A plane and a straight not on it determine a point, their pass.

$152'$. A point and a straight not on it determine a plane.

153_1. An axial pencil and a plane not on its bearer determine a flat pencil.

$153'$. A range and a point not on its bearer determine a flat pencil.

154_1. Three planes determine a point, their apex.

$154'$. Three points determine a plane, their junction.

155_1. Three planes determine a plane-sheaf.

$155'$. Three points determine a point-field.

156_1. Two coplanar straights are copunctal.

$156'$. Two copunctal straights are coplanar.

157. Any figure, or the proof of any theorem of configuration and determination, gives a dual figure or proves a dual theorem by simply interchanging point with plane. Thus all the pure projective geometry on a plane may be read as geometry on a point.

Prob. 29. If of straights copunctal in pairs not all are copunctal, then all are coplanar.

Prob. 30. On a given point put a straight to cut two given straights.

Prob. 31. If two triplets of planes $\alpha\beta\gamma$, $\alpha'\beta'\gamma'$ are such that the meets $\beta\gamma$ and $\beta'\gamma'$, $\gamma\alpha$ and $\gamma'\alpha'$, $\alpha\beta$ and $\alpha'\beta'$ lie on three planes α'', β'', γ'' which are costraight, then the meets $\alpha\alpha'$, $\beta\beta'$, $\gamma\gamma'$ are coplanar.

Prob. 32. Describe the figures in space dual to the polystim and the polygram.

ART. 14. RULED QUADRIC SURFACES.

158. The joins of the correlated points of two projective ranges whose bearers are not coplanar form a 'ruled system' of straights no two coplanar. For were two coplanar, then two points on the bearer m and two on the bearer m_1 would all four be on this plane, and so m and m_1 coplanar, contrary to hypothesis.

159. Let the straights n, n_1, n_2 be any three of the elements of a ruled system, and N_2 any point on n_2. Put a plane on N_2 and the straight n_1, and let its pass with n be called N. The straight NN_2 cuts n, n_1, n_2 all three. Projecting the generating ranges of the ruled system (on the bearers m and m_1) from the straight NN_2 (or m_2) as axis produces two projective axial pencils, which having three planes m_2n, m_2n_1, m_2n_2 self-corresponding, are identical. Therefore every pair of correlated points of the ranges on m and m_1 is coplanar with m_2; that is, m_2 cuts every element of the ruled system.

By varying the point N, ∞^1 straights are obtained, all cutting all the ∞^1 straights of the original ruled system and making on every two projective ranges. Of the straights so obtained no two cross, for that would make two of the first ruled system coplanar.

Either of these two systems may be considered as generating a 'ruled surface,' which is the bearer of both. Each of the two systems is completely determined by any three straights of the other, and therefore so is the ruled surface also. From the construction follows that the straights of either ruled system cut all the straights of the other in projective ranges. So any two straights of either system may be considered as bearers of projective ranges generating the other system, or indeed the ruled surface.

160. On each point of this ruled surface are two and only two straights lying wholly in the surface (one in each ruled

system). So a plane on one straight of the ruled surface is
also on another straight of this
surface.

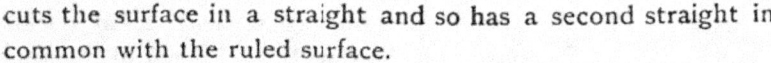

161. If in the two generating
projective ranges the point at
infinity of one is correlated to the
point at infinity of the other, the
ruled surface is called a 'hyperbolic-paraboloid.'

The join of these figurative
points is on the figurative plane.
Therefore the plane at infinity
cuts the surface in a straight and so has a second straight in
common with the ruled surface.

That a hyperbolic-paraboloid has two straights in common
with the plane at infinity may also be proved as follows:

Call the bearers of the generating ranges m and m_1, and let
n, n_1 be any two elements, and f the element at infinity. By
§ 159 the ruled surface may be considered as generated by the
straights on the three elements n, n_1, f. But all these straights
must be parallel to the same plane, namely, to any plane on f.
On f and each one of these straights put a plane ; these planes
make a parallel-axial-pencil, and cut any two of the original
elements in projective ranges with the figurative points correlated. Therefore the figurative straight joining the figurative
points of n and n_1 is wholly on the ruled surface.

162. From § 161 follows that all straights pertaining to the
same ruled system on a hyperbolic-paraboloid are parallel to
the same plane. Such planes are called 'asymptote-planes.'
A hyperbolic-paraboloid is completely determined by two non-
coplanar straights and an asymptote-plane cutting them. To
get an element cut the two given straights by any plane parallel to the asymptote-plane, and join the meets.

163. Three non-crossing straights, all parallel to the same
plane, completely determine a hyperbolic-paraboloid. Let m,
m_1, m_2 be the given straights. The passes of planes on m_2

with m and m_1 are projective ranges whose joins are a ruled system.

But from the hypothesis one of these planes is parallel to both m and m_1. Therefore their points at infinity are correlated and the ruled surface is a hyperbolic-paraboloid.

164. If two non-coplanar projective ranges be each axially projected from the bearer of the other, two projective axial pencils are formed, with those planes correlated on which are the correlated points of the ranges. If A, A_1 be correlated points, then the straight AA_1 is the meet of correlated planes. Thus two projective axial pencils with axes not coplanar generate a ruled system. If the whole figure be cut by a plane, this will cut these axial pencils in two projective flat pencils, and the conic generated by these will be the cut of the ruled surface. So every plane cuts it in a conic or a pair of straights. Hence no straight not wholly on the surface can cut it in more than two points. The surface is therefore of the second degree (quadric).

If the plane at infinity cuts the ruled surface in a pair of straights, it is a hyperbolic-paraboloid. If not, it is called a 'hyperboloid of one nappe,' a figure of which is here shown.

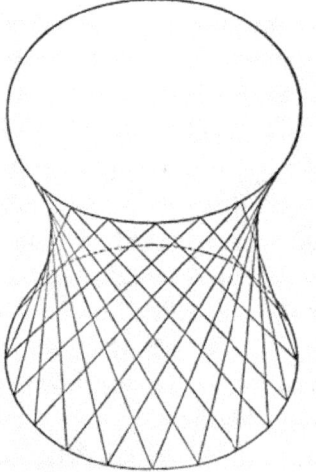

164½. Copunctal straights parallel to the generating elements of a hyperboloid of one nappe are on a cone. Copunctal straights parallel to the generating elements of a hyperbolic-paraboloid are on a system of two planes.

For the figurative plane cuts the hyperboloid of one nappe in a conic curve, but cuts the hyperbolic-paraboloid in two straights; and each of the copunctal straights goes to a point of the figurative cut.

165. Each straight in one ruled system of a hyperboloid of

one nappe is parallel to one, but only to one, straight in the other ruled system. Of the straights on a hyperbolic-paraboloid no two are parallel. Let n and n_1, any two elements of one ruled system, be the bearers of the generating ranges R and R_1. If V is the vanishing point of R, then the straight on V parallel to n_1 is an element of the other ruled system. But for the hyperbolic-paraboloid V is itself a figurative point.

166. Any straight of one ruled system on a ruled surface is called a ' guide-straight ' of the other ruled system.

167_1. A ruled system is cut by any two of its guide-straights in projective ranges.

$167'$. A ruled system is projected from any two of its guide-straights in projective axial pencils.

For if m, m_1, m_2 be any three guide-straights of the ruled system, the planes on m_1 cut m and m_1 in projective ranges the joins of whose correlated points are the elements of the ruled system. Again, if the points on m_2 be projected axially from m and m_1, the meets of the planes so correlated are the elements of the ruled system.

168. Four straights of a ruled system are called harmonic straights if they are cut in four harmonic points by one (and so by every) guide-straight. By three straights, no two coplanar, a fourth harmonic is determined lying in a ruled system with the given three and on a fourth harmonic point to any three costraight points of the given three.

169. A plane cutting the ruled surface in a straight m of one ruled system and consequently also in a straight n of the other ruled system has in common with the surface no point not on one of these straights. For any straight from such a point cutting both these straights would lie wholly on the ruled surface ; and so therefore would their whole plane, which is impossible. Any third straight coplanar with m and n on their cross has no second point in common with the surface and so is a tangent, and the plane of m and n is called tangent at their cross, the point mn.

The number of planes tangent to the ruled surface and on a given straight equals the number of points the straight has in common with the ruled surface, that is two; so the ruled surface is of the second class.

170. Project the two generating ranges of a ruled system from any projection-vertex V not on it. The eject consists of two copunctal projective flat pencils. The plane of any two correlated straights is on an element of the ruled system. All such planes form a cone of planes.

The points of contact of these planes with the ruled surface are a conic range. The planes tangent to a ruled surface at the points on its cut with a plane form a cone of planes.

171. The cut of a hyperbolic-paraboloid by a plane not on an element has on it the meets of the plane with the two figurative elements, and so is a hyperbola except when their cross is on the plane, in which case it is a parabola. The figurative plane is a tangent plane.

172. The planes tangent at the figurative points of a hyperboloid of one nappe are all proper planes, copunctal and forming a cone of planes tangent to the 'asymptote-cone' of the hyperboloid. Each element to the asymptote-cone is parallel to one element of each ruled system.

Any plane not on an element of the hyperboloid of one nappe cuts it in a hyperbola, parabola, or ellipse, according as it is parallel to two elements, one, or no element of the asymptote-cone, that is, according as it has in common with the figurative conic on the hyperboloid two points, one, or no point.

173. If an axial pencil and a ruled system are projective, they generate in general a 'twisted cubic curve,' which any plane cuts in one point at least and three at most. For a plane cuts the ruled system in a conic range perspective to it, of which in general three points at most lie on the corresponding planes of the pencil.

174. The ruled quadric surface is the only surface doubly

ruled.   The figure of two so united ruled systems is one of the most noteworthy discovered by the modern geometry.*

175. To find the straights crossing four given straights.— Let u_1, u_2, u_3, u_4 be the given straights.   Projecting the range R_1 on u_1 from the axes u_2 and u_3 gives two axial pencils, each perspective to R_1, and consequently projective.   The meets of their correlated planes are all the ∞^1 straights on u_1, u_2, u_3, and form a ruled system of which u_1, u_2, u_3 are guide-straights. The two projective axial-pencils cut the fourth straight u_4 in two 'conjective' ranges.   [Two projective primal figures of the same kind and on the same bearer are called conjective.]   If now a straight m of the ruled system crosses u_4, then the two correlated planes of which this straight m is the meet must cut u_4 in the same point, which consequently is a self-corresponding point of the two conjective ranges.   Since there are two such (the points common to u_4 and the ruled surface), so there are two straights (real or conjugate imaginary) crossing four given straights.   Their construction is shown to depend on that for the two self-correlated points of two conjective ranges.

This important problem in the four-dimensional space of straights, 'what is common to four straights?' is the analogue of the problem in the space of points, 'what is common to three points?' and its dual in the space of planes, 'what is common to three planes?'

It shows not only their fundamental diversity, but also, as compared to points-geometry and planes-geometry, the inherently quadratic character of straights-geometry.

Prob. 33. Find the straights cutting two given straights and parallel to a third.

Prob. 34. Three diagonals of a skew hexagram whose six sides are on a ruled surface are copunctal.

Prob. 35. If a flat pencil and a range not on parallel planes are projective, then straights on the points of the range parallel to the correlated straights of the pencil form one ruled system of a hyperbolic-paraboloid.

* See Monge, Journal de l'école polytechnique, Vol. I.

Prob. 36. What is the locus of a point harmonically separated from a given point by a ruled surface?

Art. 15.   Cross-Ratio.

176. Lindemann has shown how every one number, whether integer, fraction, or irrational, $+$ or $-$, may be correlated to one point of a straight, without making any use of measurement, without any comparison of sects by application of a unit sect.* He gets an analytic definition of the ' cross-ratio ' of four copunctal straights. Then this expression is applied to four costraight points. Then is deduced that the number previously attached to a point on a straight is the same as the cross-ratio of that point with three fixed points of the straight. Thus analytic geometry and metric geometry may be founded without using ratio in its old sense, involving measurement. Thus also the non-Euclidean geometries, that of Bolyai-Lobachévski in which the straight has two points at infinity, and that of Riemann in which the straight has no point at infinity, may be treated together with the limiting case of each between them, the Euclidean geometry, wherein the straight has one but only one point at infinity.

Relinquishing for brevity this pure projective standpoint and reverting to the old metric usages where an angle is an inclination, a sect is a piece of a straight, and any ratio is a number; distinguishing the sect AC from CA as of opposite ' sense,' so that $AC = - CA$, the ratio $[AC/BC]/[AD/BD]$ is called the cross-ratio of the range $ABCD$ and is written $[ABCD]$ where A and B, called conjugate points of the cross-ratio, may be looked upon as the extremities of a sect divided internally or externally by C and again by D.†

---

* Von Staudt in Beiträge zur Geometrie der Lage, 1856–60, determines the projective definition of number, and thus makes the metric geometry a consequence of projective geometry.

† The fundamental property of cross-ratio is stated in the Mathematical Collections of Pappus, about 400 B.C. The cross-ratio is the basis of Poncelet's Traité des propriétés projectives, 1822, which distinguishes sharply the projective and metric properties of curves.

177. If on $ABCD$ respectively be the straights $abcd$ co-punctal on V, then $AC/BC = \triangle AVC/\triangle BVC$

or $AC/BC = \frac{1}{2}AV.VC \sin(ac)/\frac{1}{2}BV.VC \sin(bc)$.

$AD/BD = \triangle AVD/\triangle BVD$

$= \frac{1}{2}AV.VD \sin(ad)/\frac{1}{2}BV.VD \sin(bd)$.

Therefore $[\dot{A}\dot{B}CD] = [\sin(ac)/\sin(bc)]/[\sin ad/\sin(bd)]$.

Thus as the cross-ratio of any flat pencil $V[abcd]$ or axial pencil $u(\alpha\beta\gamma\delta)$ may be taken the cross-ratio of the cut $ABCD$ on any transversal.

178. Two projective primal figures are 'equicross;' and inversely two equicross primal figures are projective.

179. As D approaches the point at infinity, AD/BD approaches 1. The cross-ratio $[\dot{A}\dot{B}CD]$ when D is figurative equals AC/BC.

180. Given three costraight points ABC, to find D so that $[\dot{A}\dot{B}CD]$ may equal a given number n (+ or −). On any straight on C take A' and B' such that $CA'/CB' = n$; A' and B' lying on the same side of C if n be positive, but on opposite sides if n be negative. Join AA', BB', crossing in V. The parallel to $A'B'$ on V will cut AB in the required D. For if D' be the point at infinity on $A'B'$, and $ABCD$ be projected from V, then $A'B'CD'$ is a cut of the eject; so

$$[\dot{A}\dot{B}CD] = [\dot{A}'\dot{B}'CD'] = A'C/B'C = n.$$

181. If $[\dot{A}\dot{B}CD] = [\dot{A}\dot{B}CD_1]$, then D_1 coincides with D.

182. If two figures be complete plane perspectives, four costraight points (or copunctal straights) in one are equicross with the correlated four in the other. Let O be the center of perspective. Let M and M' be any pair of correlated points of the two figures, N and N' another pair of correlated points lying on the straight OMM' whose cross with the axis of perspective is X. Then $[\dot{O}\dot{X}MN] = [\dot{O}\dot{X}M'N']$.

That is, $[OM/XM]/[ON/XN] = [OM'/XM']/[ON'/XN']$.

Therefore $[OM/XM]/[OM'/XM'] = [ON/XN]/[ON'/XN']$.

That is, $[\dot{O}\dot{X}MM'] = [\dot{O}\dot{X}NN']$; or the cross-ratio $[\dot{O}\dot{X}MM']$

is constant for all pairs of correlated points M and M' taken on a straight OX on the center of perspective.

Next let L and L' be another pair of correlated points and Y the cross of OLL' with the axis of perspective. Since LM and $L'M'$ cross on some point Z of the axis XY, therefore if $OXMM'$ be projected from Z, the cut of the eject by OY is $OYLL'$. So $[\dot{O}\dot{X}MM']=[\dot{O}\dot{Y}LL']$; or the cross-ratio $[\dot{O}\dot{X}MM']$ is constant for all pairs of correlated points.

It is called the 'parameter' of the correlation. When the parameter equals -1, the range $OXMM'$ is harmonic, and two correlated elements correspond doubly, are coupled, and the correlation is 'involutorial.'

183. When the correlation is involutorial and the center of perspective is the figurative point on a perpendicular to the axis of perspective, this is called the 'axis of symmetry,' and the complete plane perspectives are said to be 'symmetrical.'

184. When the correlation is involutorial and the axis of perspective is figurative, then the center of perspective is called the 'symcenter,' and the complete plane perspectives are said to be 'symcentral.'

Prob. 37. In a plane are given a parallelogram and any sect. With the ruler alone find the center of the sect and draw a parallel to it.

Prob. 38. The locus of a point such that its joins to four given points have a given cross-ratio is a conic on which are the points.

Prob. 39. If the sides of a trigram are tangent to a conic, the joins of two of its fan-points to any point on the polar of the third are conjugate with respect to the conic.

Prob. 40. If from any point of the sect between the contact-points of a pair of tangents to a parabola straights be drawn parallel to these tangents, the join of their proper crosses with the tangents will be a tangent.

Chapter IV.

HYPERBOLIC FUNCTIONS.

By James McMahon,

Assistant Professor of Mathematics in Cornell University.

Art. 1. Correspondence of Points on Conics.

To prepare the way for a general treatment of the hyperbolic functions a preliminary discussion is given on the relations between hyperbolic sectors. The method adopted is such as to apply at the same time to sectors of the ellipse, including the circle; and the analogy of the hyperbolic and circular functions will be obvious at every step, since the same set of equations can be read in connection with either the hyperbola or the ellipse.* It is convenient to begin with the theory of correspondence of points on two central conics of like species, i.e. either both ellipses or both hyperbolas.

To obtain a definition of corresponding points, let O_1A_1, O_1B_1 be conjugate radii of a central conic, and O_2A_2, O_2B_2 conjugate radii of any other central conic of the same species; let P_1, P_2 be two points on the curves; and let their coordinates referred to the respective pairs of conjugate directions be (x_1, y_1), (x_2, y_2); then, by analytic geometry,

$$\frac{x_1^2}{a_1^2} \pm \frac{y_1^2}{b_1^2} = 1, \qquad \frac{x_2^2}{a_2^2} \pm \frac{y_2^2}{b_2^2} = 1. \tag{1}$$

---

* The hyperbolic functions are not so named on account of any analogy with what are termed Elliptic Functions. "The elliptic integrals, and thence the elliptic functions, derive their name from the early attempts of mathematicians at the rectification of the ellipse. . . . To a certain extent this is a disadvantage; . . . because we employ the name hyperbolic function to denote cosh u, sinh u, etc., by analogy with which the elliptic functions would be merely the circular functions cos ϕ, sin ϕ, etc. . . ." (Greenhill, Elliptic Functions, p. 175.)

Now if the points P_1, P_2 be so situated that

$$\frac{x_1}{a_1} = \frac{x_2}{a_2}, \qquad \frac{y_1}{b_1} = \frac{y_2}{b_2}, \tag{2}$$

the equalities referring to sign as well as magnitude, then P_1, P_2 are called corresponding points in the two systems. If Q_1, Q_2 be another pair of correspondents, then the sector and tri-

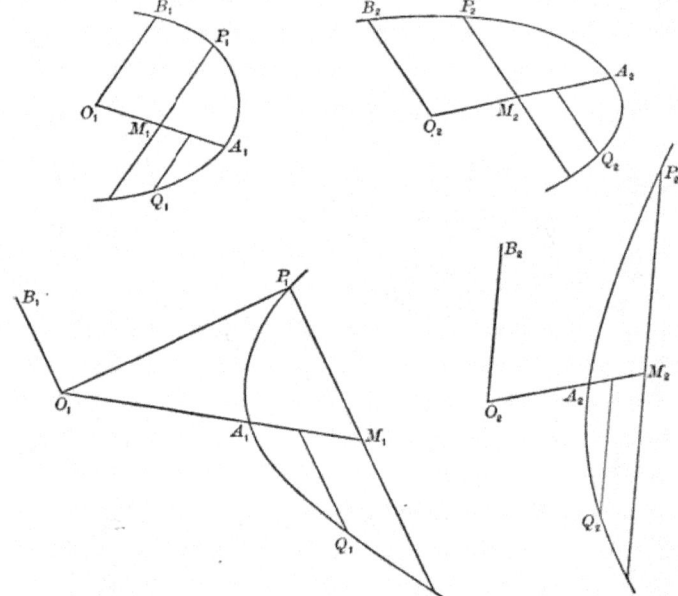

angle $P_1 O_1 Q_1$ are said to correspond respectively with the sector and triangle $P_2 O_2 Q_2$. These definitions will apply also when the conics coincide, the points P_1, P_2 being then referred to any two pairs of conjugate diameters of the same conic.

In discussing the relations between corresponding areas it is convenient to adopt the following use of the word "measure": The measure of any area connected with a given central conic is the ratio which it bears to the constant area of the triangle formed by two conjugate diameters of the same conic.

For example, the measure of the sector $A_1 O_1 P_1$ is the ratio

$$\frac{\text{sector } A_1 O_1 P_1}{\text{triangle } A_1 O_1 B_1}$$

and is to be regarded as positive or negative according as $A_1O_1P_1$ and $A_1O_1B_1$ are at the same or opposite sides of their common initial line.

ART. 2. AREAS OF CORRESPONDING TRIANGLES.

The areas of corresponding triangles have equal measures. For, let the coordinates of P_1, Q_1 be (x_1, y_1), (x_1', y_1'), and let those of their correspondents P_2, Q_2 be (x_2, y_2), (x_2', y_2'); let the triangles $P_1O_1Q_1$, $P_2O_2Q_2$ be T_1, T_2, and let the measuring triangles $A_1O_1B_1$, $A_2O_2B_2$ be K_1, K_2, and their angles ω_1, ω_2; then, by analytic geometry, taking account of both magnitude and direction of angles, areas, and lines,

$$\frac{T_1}{K_1} = \frac{\frac{1}{2}(x_1 y_1' - x_1' y_1) \sin \omega_1}{\frac{1}{2}a_1 b_1 \sin \omega_1} = \frac{x_1}{a_1}\frac{y_1'}{b_1} - \frac{x_1'}{a_1}\frac{y_1}{b_1};$$

$$\frac{T_2}{K_2} = \frac{\frac{1}{2}(x_2 y_2' - x_2' y_2) \sin \omega_2}{\frac{1}{2}a_2 b_2 \sin \omega_2} = \frac{x_2}{a_1}\frac{y_2'}{b_2} - \frac{x_2'}{a_2}\frac{y_2}{b_2}.$$

Therefore
$$\frac{T_1}{K_1} = \frac{T_2}{K_2}. \tag{3}$$

ART. 3. AREAS OF CORRESPONDING SECTORS.

The areas of corresponding sectors have equal measures. For conceive the sectors S_1, S_2 divided up into infinitesimal corresponding sectors; then the respective infinitesimal corresponding triangles have equal measures (Art. 2); but the given sectors are the limits of the sums of these infinitesimal triangles, hence

$$\frac{S_1}{K_1} = \frac{S_2}{K_2}. \tag{4}$$

In particular, the sectors $A_1O_1P_1$, $A_2O_2P_2$ have equal measures; for the initial points A_1, A_2 are corresponding points.

It may be proved conversely by an obvious reductio ad absurdum that if the initial points of two equal-measured sectors correspond, then their terminal points correspond.

Thus if any radii O_1A_1, O_2A_2 be the initial lines of two equal-measured sectors whose terminal radii are O_1P_1, O_2P_2, then P_1, P_2 are corresponding points referred respectively to the pairs of conjugate directions O_1A_1, O_1B_1, and O_2A_2, O_2B_2; that is,

$$\frac{x_1}{a_1} = \frac{x_2}{a_2}, \quad \frac{y_1}{b_1} = \frac{y_2}{b_2}.$$

Prob. 1. Prove that the sector $P_1O_1Q_1$ is bisected by the line joining O_1 to the mid-point of P_1Q_1. (Refer the points P_1, Q_1, respectively, to the median as common axis of x, and to the two opposite conjugate directions as axis of y, and show that P_1, Q_1 are then corresponding points.)

Prob. 2. Prove that the measure of a circular sector is equal to the radian measure of its angle.

Prob. 3. Find the measure of an elliptic quadrant, and of the sector included by conjugate radii.

Art. 4.   Characteristic Ratios of Sectorial Measures.

Let $A_1O_1P_1 = S_1$ be any sector of a central conic; draw P_1M_1 ordinate to O_1A_1, i.e. parallel to the tangent at A_1; let $O_1M_1 = x_1$, $M_1P_1 = y_1$, $O_1A_1 = a_1$, and the conjugate radius $O_1B_1 = b_1$; then the ratios x_1/a_1, y_1/b_1 are called the characteristic ratios of the given sectorial measure S_1/K_1. These ratios are constant both in magnitude and sign for all sectors of the same measure and species wherever these may be situated (Art. 3). Hence there exists a functional relation between the sectorial measure and each of its characteristic ratios.

Art. 5.   Ratios Expressed as Triangle-measures.

The triangle of a sector and its complementary triangle are measured by the two characteristic ratios. For, let the triangle $A_1O_1P_1$ and its complementary triangle $P_1O_1B_1$ be denoted by T_1, T_1'; then

$$\left. \begin{aligned} \frac{T_1}{K_1} &= \frac{\frac{1}{2}a_1y_1 \sin \omega_1}{\frac{1}{2}a_1b_1 \sin \omega_1} = \frac{y_1}{b_1}, \\[2mm] \frac{T_1'}{K_1} &= \frac{\frac{1}{2}b_1x_1 \sin \omega_1}{\frac{1}{2}a_1b_1 \sin \omega_1} = \frac{x_1}{a_1}. \end{aligned} \right\} \tag{5}$$

ART. 6. FUNCTIONAL RELATIONS FOR ELLIPSE.

The functional relations that exist between the sectorial measure and each of its characteristic ratios are the same for all elliptic, including circular, sectors (Art. 4). Let P_1, P_2 be corresponding points on an ellipse and a circle, referred to the conjugate directions 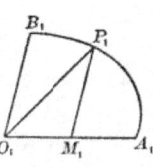 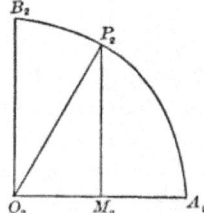 O_1A_1, O_1B_1, and O_2A_2, O_2B_2, the latter pair being at right angles; let the angle $A_2O_2P_2 = \theta$ in radian measure; then

$$\frac{S_2}{K_2} = \frac{\frac{1}{2}a_2^2\theta}{\frac{1}{2}a_2^2} = \theta. \tag{6}$$

$$\therefore \frac{x_2}{a_2} = \cos \frac{S_2}{K_2}, \quad \frac{y_2}{b_2} = \sin \frac{S_2}{K_2}; \qquad [a_2 = b_2$$

hence, in the ellipse, by Art. 3,

$$\frac{x_1}{a_1} = \cos \frac{S_1}{K_1}, \quad \frac{y_1}{b_1} = \sin \frac{S_1}{K_1}. \tag{7}$$

Prob. 4. Given $x_1 = \frac{1}{2}a_1$; find the measure of the elliptic sector $A_1O_1P_1$. Also find its area when $a_1 = 4$, $b_1 = 3$, $\omega = 60°$.

Prob. 5. Find the characteristic ratios of an elliptic sector whose measure is $\frac{1}{4}\pi$.

Prob. 6. Write down the relation between an elliptic sector and its triangle. (See Art. 5.)

ART. 7. FUNCTIONAL RELATIONS FOR HYPERBOLA.

The functional relations between a sectorial measure and its characteristic ratios in the case of the hyperbola may be written in the form

$$\frac{x_1}{a_1} = \cosh \frac{S_1}{K_1}, \quad \frac{y_1}{b_1} = \sinh \frac{S_1}{K_1};$$

and these express that the ratio of the two lines on the left is a certain definite function of the ratio of the two areas on the right. These functions are called by analogy the hyperbolic cosine ana the nyperbolic sine. Thus, writing u for S_1/K_1, the two equations

$$\frac{x_1}{a_1} = \cosh u, \quad \frac{y_1}{b_1} = \sinh u \tag{8}$$

serve to define the hyperbolic cosine and sine of a given sectorial measure u; and the hyperbolic tangent, cotangent, secant, and cosecant are then defined as follows:

$$\left.\begin{aligned}
\tanh u &= \frac{\sinh u}{\cosh u}, \quad \coth u = \frac{\cosh u}{\sinh u}, \\[2mm]
\operatorname{sech} u &= \frac{1}{\cosh u}, \quad \operatorname{csch} u = \frac{1}{\sinh u}.
\end{aligned}\right\} \tag{9}$$

The names of these functions may be read " h-cosine," " h-sine," " h-tangent," etc.

ART. 8. Relations between Hyperbolic Functions.

Among the six functions there are five independent relations, so that when the numerical value of one of the functions is given, the values of the other five can be found. Four of these relations consist of the four defining equations (9). The fifth is derived from the equation of the hyperbola

$$\frac{x_1^2}{a_1^2} - \frac{y_1^2}{b_1^2} = 1,$$

giving

$$\cosh^2 u - \sinh^2 u = 1. \tag{10}$$

By a combination of some of these equations other subsidiary relations may be obtained; thus, dividing (10) successively by $\cosh^2 u$, $\sinh^2 u$, and applying (9), give

$$\left.\begin{aligned}
1 - \tanh^2 u &= \operatorname{sech}^2 u, \\
\coth^2 u - 1 &= \operatorname{csch}^2 u.
\end{aligned}\right\} \tag{11}$$

Equations (9), (10), (11) will readily serve to express the value of any function in terms of any other. For example, when $\tanh u$ is given,

$$\coth u = \frac{1}{\tanh u}, \qquad \operatorname{sech} u = \sqrt{1 - \tanh^2 u},$$

$$\cosh u = \frac{1}{\sqrt{1 - \tanh^2 u}}, \qquad \sinh u = \frac{\tanh u}{\sqrt{1 - \tanh^2 u}},$$

$$\operatorname{csch} u = \frac{\sqrt{1 - \tanh^2 u}}{\tanh u}.$$

The ambiguity in the sign of the square root may usually be removed by the following considerations: The functions $\cosh u$, $\operatorname{sech} u$ are always positive, because the primary characteristic ratio x_1/a_1 is positive, since the initial line O_1A_1 and the abscissa O_1M_1 are similarly directed from O_1, on whichever branch of the hyperbola P_1 may be situated; but the functions $\sinh u$, $\tanh u$, $\coth u$, $\operatorname{csch} u$, involve the other characteristic ratio y_1/b_1, which is positive or negative according as y_1 and b_1 have the same or opposite signs, i.e., as the measure u is positive or negative; hence these four functions are either all positive or all negative. Thus when any one of the functions $\sinh u$, $\tanh u$, $\operatorname{csch} u$, $\coth u$, is given in magnitude and sign, there is no ambiguity in the value of any of the six hyperbolic functions; but when either $\cosh u$ or $\operatorname{sech} u$ is given, there is ambiguity as to whether the other four functions shall be all positive or all negative.

The hyperbolic tangent may be expressed as the ratio of two lines. For draw the tangent line $AC = t$; then

$$\tanh u = \frac{y}{b} : \frac{x}{a} = \frac{a}{b} \cdot \frac{y}{x}$$

$$= \frac{a}{b} \cdot \frac{t}{a} = \frac{t}{b}. \qquad (12)$$

The hyperbolic tangent is the measure of the triangle OAC. For

$$\frac{OAC}{OAB} = \frac{at}{ab} = \frac{t}{b} = \tanh u. \qquad (13)$$

Thus the sector AOP, and the triangles AOP, POB, AOC, are proportional to u, $\sinh u$, $\cosh u$, $\tanh u$ (eqs. 5, 13): hence

$$\sinh u > u > \tanh u. \qquad (14)$$

Prob. 7. Express all the hyperbolic functions in terms of sinh u. Given cosh $u = 2$, find the values of the other functions.

Prob. 8. Prove from eqs. 10, 11, that cosh $u >$ sinh u, cosh $u > 1$, tanh $u < 1$, sech $u < 1$.

Prob. 9. In the figure of Art. 1, let $OA=2$, $OB=1$, $AOB = 60°$, and area of sector $AOP = 3$; find the sectorial measure, and the two characteristic ratios, in the elliptic sector, and also in the hyperbolic sector; and find the area of the triangle AOP. (Use tables of cos, sin, cosh, sinh.)

Prob. 10. Show that coth u, sech u, csch u may each be expressed as the ratio of two lines, as follows: Let the tangent at P make on the conjugate axes OA, OB, intercepts $OS = m$, $OT = n$; let the tangent at B, to the conjugate hyperbola, meet OP in R, making $BR = l$; then

$$\coth u = l/a, \quad \text{sech } u = m/a, \quad \text{csch } u = n/b.$$

Prob. 11. The measure of segment AMP is sinh u cosh $u - u$. Modify this for the ellipse. Modify also eqs. 10–14, and probs. 8, 10.

ART. 9. VARIATIONS OF THE HYPERBOLIC FUNCTIONS.

Since the values of the hyperbolic functions depend only on the sectorial measure, it is convenient, in tracing their vari-

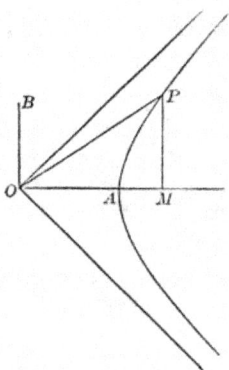

ations, to consider only sectors of one half of a rectangular hyperbola, whose conjugate radii are equal, and to take the principal axis OA as the common initial line of all the sectors. The sectorial measure u assumes every value from $- \infty$, through 0, to $+ \infty$, as the terminal point P comes in from infinity on the lower branch, and passes to infinity on the upper branch; that is, as the terminal line OP swings from the lower asymptotic position $y = - x$, to the upper one, $y = x$. It is here assumed, but is proved in Art. 17, that the sector AOP becomes infinite as P passes to infinity.

Since the functions cosh u, sinh u, tanh u, for any position of OP_1 are equal to the ratios of x, y, t, to the principal radius a, it is evident from the figure that

$$\cosh 0 = 1, \quad \sinh 0 = 0, \quad \tanh 0 = 0, \qquad (15)$$

and that as u increases towards positive infinity, $\cosh u$, $\sinh u$ are positive and become infinite, but $\tanh u$ approaches unity as a limit; thus

$$\cosh \infty = \infty, \quad \sinh \infty = \infty, \quad \tanh \infty = 1. \qquad (16)$$

Again, as u changes from zero towards the negative side, $\cosh u$ is positive and increases from unity to infinity, but $\sinh u$ is negative and increases numerically from zero to a negative infinite, and $\tanh u$ is also negative and increases numerically from zero to negative unity; hence

$$\cosh(-\infty) = \infty, \quad \sinh(-\infty) = -\infty, \quad \tanh(-\infty) = -1. \quad (17)$$

For intermediate values of u the numerical values of these functions can be found from the formulas of Arts. 16, 17, and are tabulated at the end of this chapter. A general idea of their manner of variation can be obtained from the curves in Art. 25, in which the sectorial measure u is represented by the abscissa, and the values of the functions $\cosh u$, $\sinh u$, etc., are represented by the ordinate.

The relations between the functions of $-u$ and of u are evident from the definitions, as indicated above, and in Art. 8. Thus

$$\left.\begin{array}{ll} \cosh(-u) = +\cosh u, & \sinh(-u) = -\sinh u, \\ \operatorname{sech}(-u) = +\operatorname{sech} u, & \operatorname{csch}(-u) = -\operatorname{csch} u, \\ \tanh(-u) = -\tanh u, & \coth(-u) = -\coth u. \end{array}\right\} \quad (18)$$

Prob. 12. Trace the changes in $\operatorname{sech} u$, $\coth u$, $\operatorname{csch} u$, as u passes from $-\infty$ to $+\infty$. Show that $\sinh u$, $\cosh u$ are infinites of the same order when u is infinite. (It will appear in Art. 17 that $\sinh u$, $\cosh u$ are infinites of an order infinitely higher than the order of u.)

Prob. 13. Applying eq. (12) to figure, page 114, prove $\tanh u$, $= \tan AOP$.

ART. 10.   ANTI-HYPERBOLIC FUNCTIONS.

The equations $\dfrac{x}{a} = \cosh u$, $\dfrac{y}{b} = \sinh u$, $\dfrac{t}{b} = \tanh u$, etc.,

may also be expressed by the inverse notation $u = \cosh^{-1}\dfrac{x}{a}$,

$u = \sinh^{-1}\dfrac{y}{b}$, $u = \tanh^{-1}\dfrac{t}{b}$, etc., which may be read: "u is
the sectorial measure whose hyperbolic cosine is the ratio x to
a," etc.; or "u is the anti-h.-cosine of x/a," etc.

Since there are two values of u, with opposite signs, that
correspond to a given value of $\cosh u$, it follows that if u be
determined from the equation $\cosh u = m$, where m is a given
number greater than unity, u is a two-valued function of m.
The symbol $\cosh^{-1} m$ will be used to denote the positive value
of u that satisfies the equation $\cosh u = m$. Similarly the
symbol $\operatorname{sech}^{-1} m$ will stand for the positive value of u that
satisfies the equation $\operatorname{sech} u = m$. The signs of the other
functions $\sinh^{-1} m$, $\tanh^{-1} m$, $\coth^{-1} m$, $\operatorname{csch}^{-1} m$, are the same
as the sign of m. Hence all of the anti-hyperbolic functions
of real numbers are one-valued.

Prob. 14. Prove the following relations:

$$\cosh^{-1} m = \sinh^{-1}\sqrt{m^2 - 1}, \quad \sinh^{-1} m = \pm \cosh^{-1}\sqrt{m^2 + 1},$$

the upper or lower sign being used according as m is positive or
negative. Modify these relations for $\sin^{-1}$, $\cos^{-1}$.

Prob. 15. In figure, Art. 1, let $OA = 2$, $OB = 1$, $AOB = 60°$; find
the area of the hyperbolic sector AOP, and of the segment AMP,
if the abscissa of P is 3. (Find $\cosh^{-1}$ from the tables for $\cosh$.)

ART. 11.   FUNCTIONS OF SUMS AND DIFFERENCES.

(a) To prove the difference-formulas

$$\left.\begin{array}{l} \sinh(u - v) = \sinh u \cosh v - \cosh u \sinh v, \\ \cosh(u - v) = \cosh u \cosh v - \sinh u \sinh v. \end{array}\right\} \quad (19)$$

Let OA be any radius of a hyperbola, and let the sectors AOP,
AOQ have the measures u, v; then $u - v$ is the measure of the
sector QOP. Let OB, OQ' be the radii conjugate to OA, OQ;
and let the coördinates of P, Q, Q' be (x_1, y_1), (x, y), (x', y')
with reference to the axes OA, OB; then

$$\sinh (u - v) = \sinh \frac{\text{sector } QOP}{K} = \frac{\text{triangle } QOP}{K} \quad \text{[Art. 5.}$$

$$= \frac{\frac{1}{2}(xy_1 - x_1y) \sin \omega}{\frac{1}{2}a_1b_1 \sin \omega} = \frac{y_1}{b_1} \frac{x}{a_1} - \frac{y}{b_1} \frac{x_1}{a_1}.$$

$$= \sinh u \cosh v - \cosh u \sinh v ;$$

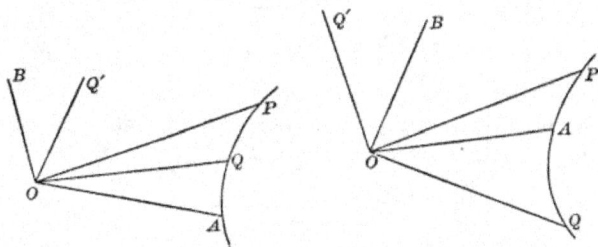

$$\cosh (u - v) = \cosh \frac{\text{sector } QOP}{K} = \frac{\text{triangle } POQ'}{K} \quad \text{[Art. 5.}$$

$$= \frac{\frac{1}{2}(xy' - y_1x') \sin \omega}{\frac{1}{2}a_1b_1 \sin \omega} = \frac{y'}{b_1} \frac{x_1}{a_1} - \frac{y_1}{b_1} \frac{x'}{a_1} ;$$

but $\qquad \dfrac{y'}{b_1} = \dfrac{x}{a_1}, \qquad \dfrac{x'}{a_1} = \dfrac{y}{b_1},$ $\hfill$ (20)

since Q, Q' are extremities of conjugate radii ; hence

$$\cosh (u - v) = \cosh u \cosh v - \sinh u \sinh v.$$

In the figures u is positive and v is positive or negative. Other figures may be drawn with u negative, and the language in the text will apply to all. In the case of elliptic sectors, similar figures may be drawn, and the same language will apply, except that the second equation of (20) will be $x'/a_1 = -y/b_1$; therefore

$$\sin (u - v) = \sin u \cos v - \cos u \sin v,$$
$$\cos (u - v) = \cos u \cos v + \sin u \sin v.$$

(b) To prove the sum-formulas

$$\left. \begin{array}{l} \sinh (u + v) = \sinh u \cosh v + \cosh u \sinh v, \\ \cosh (u + v) = \cosh u \cosh v + \sinh u \sinh v. \end{array} \right\} \quad (21)$$

These equations follow from (19) by changing v into $-v$,

and then for $\sinh(-v)$, $\cosh(-v)$, writing $-\sinh v$, $\cosh v$ (Art. 9, eqs. (18)).

(c) To prove that $\tanh(u \pm v) = \dfrac{\tanh u \pm \tanh v}{1 \pm \tanh u \tanh v}$. 　　(22)

Writing $\tanh(u \pm v) = \dfrac{\sinh(u \pm v)}{\cosh(u \pm v)}$, expanding and dividing numerator and denominator by $\cosh u \cosh v$, eq. (22) is obtained.

Prob. 16. Given $\cosh u = 2$, $\cosh v = 3$, find $\cosh(u + v)$.

Prob. 17. Prove the following identities:

1. $\sinh 2u = 2 \sinh u \cosh u$.
2. $\cosh 2u = \cosh^2 u + \sinh^2 u = 1 + 2 \sinh^2 u = 2 \cosh^2 u - 1$.
3. $1 + \cosh u = 2 \cosh^2 \tfrac{1}{2}u$, 　$\cosh u - 1 = 2 \sinh^2 \tfrac{1}{2}u$.
4. $\tanh \tfrac{1}{2}u = \dfrac{\sinh u}{1 + \cosh u} = \dfrac{\cosh u - 1}{\sinh u} = \left(\dfrac{\cosh u - 1}{\cosh u + 1}\right)^{\frac{1}{2}}$.
5. $\sinh 2u = \dfrac{2 \tanh u}{1 - \tanh^2 u}$, 　$\cosh 2u = \dfrac{1 + \tanh^2 u}{1 - \tanh^2 u}$.
6. $\sinh 3u = 3 \sinh u + 4 \sinh^3 u$, $\cosh 3u = 4 \cosh^3 u - 3 \cosh u$.
7. $\cosh u + \sinh u = \dfrac{1 + \tanh \tfrac{1}{2}u}{1 - \tanh \tfrac{1}{2}u}$.
8. $(\cosh u + \sinh u)(\cosh v + \sinh v) = \cosh(u + v) + \sinh(u + v)$.
9. Generalize (8); and show also what it becomes when $u = v = \ldots$
10. $\sinh^2 x \cos^2 y + \cosh^2 x \sin^2 y = \sinh^2 x + \sin^2 y$.
11. $\cosh^{-1} m \pm \cosh^{-1} n = \cosh^{-1}\left[mn \pm \sqrt{(m^2 - 1)(n^2 - 1)}\right]$.
12. $\sinh^{-1} m \pm \sinh^{-1} n = \sinh^{-1}\left[m \sqrt{1 + n^2} \pm n \sqrt{1 + m^2}\right]$.

Prob. 18. What modifications of signs are required in (21), (22), in order to pass to circular functions?

Prob. 19. Modify the identities of Prob. 17 for the same purpose.

ART. 12.　CONVERSION FORMULAS.

To prove that

$$\left.\begin{array}{l}
\cosh u_1 + \cosh u_2 = 2 \cosh \tfrac{1}{2}(u_1 + u_2) \cosh \tfrac{1}{2}(u_1 - u_2), \\[4pt]
\cosh u_1 - \cosh u_2 = 2 \sinh \tfrac{1}{2}(u_1 + u_2) \sinh \tfrac{1}{2}(u_1 - u_2), \\[4pt]
\sinh u_1 + \sinh u_2 = 2 \sinh \tfrac{1}{2}(u_1 + u_2) \cosh \tfrac{1}{2}(u_1 - u_2), \\[4pt]
\sinh u_1 - \sinh u_2 = 2 \cosh \tfrac{1}{2}(u_1 + u_2) \sinh \tfrac{1}{2}(u_1 - u_2).
\end{array}\right\} \quad (23)$$

From the addition formulas it follows that

$$\cosh (u + v) + \cosh (u - v) = 2 \cosh u \cosh v,$$

$$\cosh (u + v) - \cosh (u - v) = 2 \sinh u \sinh v,$$

$$\sinh (u + v) + \sinh (u - v) = 2 \sinh u \cosh v,$$

$$\sinh (u + v) - \sinh (u - v) = 2 \cosh u \sinh v,$$

and then by writing $u + v = u_1$, $u - v = u_2$, $u = \frac{1}{2}(u_1 + u_2)$, $v = \frac{1}{2}(u_1 - u_2)$, these equations take the form required.

Prob. 20. In passing to circular functions, show that the only modification to be made in the conversion formulas is in the algebraic sign of the right-hand member of the second formula.

Prob. 21. Simplify $\dfrac{\cosh 2u + \cosh 4v}{\sinh 2u + \sinh 4v}$, $\dfrac{\cosh 2u + \cosh 4v}{\cosh 2u - \cosh 4v}$.

Prob. 22. Prove $\sinh^2 x - \sinh^2 y = \sinh (x + y) \sinh (x - y)$.

Prob. 23. Simplify $\cosh^2 x \cosh^2 y \pm \sinh^2 x \sinh^2 y$.

Prob. 24. Simplify $\cosh^2 x \cos^2 y + \sinh^2 x \sin^2 y$.

ART. 13.   LIMITING RATIOS.

To find the limit, as u approaches zero, of

$$\frac{\sinh u}{u}, \quad \frac{\tanh u}{u},$$

which are then indeterminate in form.

By eq. (14), $\sinh u > u > \tanh u$; and if $\sinh u$ and $\tanh u$ be successively divided by each term of these inequalities, it follows that

$$1 < \frac{\sinh u}{u} < \cosh u,$$

$$\operatorname{sech} u < \frac{\tanh u}{u} < 1.$$

but when $u \doteq 0$, $\cosh u \doteq 1$, $\operatorname{sech} u = 1$, hence

$$\lim_{u \doteq 0} \frac{\sinh u}{u} = 1, \quad \lim_{u \doteq 0} \frac{\tanh u}{u} = 1. \qquad (24)$$

ART. 14. DERIVATIVES OF HYPERBOLIC FUNCTIONS.

To prove that

$$(a) \qquad \frac{d(\sinh u)}{du} = \cosh u,$$

$$(b) \qquad \frac{d(\cosh u)}{du} = \sinh u,$$

$$(c) \qquad \frac{d(\tanh u)}{du} = \operatorname{sech}^2 u,$$

$$(d) \qquad \frac{d(\operatorname{sech} u)}{du} = - \operatorname{sech} u \tanh u,$$

$$(e) \qquad \frac{d(\coth u)}{du} = - \operatorname{csch}^2 u,$$

$$(f) \qquad \frac{d(\operatorname{csch} u)}{du} = - \operatorname{csch} u \coth u.$$

$$(25)$$

(a) Let $y = \sinh u$,

$$\Delta y = \sinh (u + \Delta u) - \sinh u$$
$$= 2 \cosh \tfrac{1}{2}(2u + \Delta u) \sinh \tfrac{1}{2}\Delta u,$$
$$\frac{\Delta y}{\Delta u} = \cosh (u + \tfrac{1}{2}\Delta u) \frac{\sinh \tfrac{1}{2}\Delta u}{\tfrac{1}{2}\Delta u}.$$

Take the limit of both sides, as $\Delta u \doteq 0$, and put

$$\lim. \frac{\Delta y}{\Delta u} = \frac{dy}{du} = \frac{d(\sinh u)}{du},$$

$$\lim. \cosh (u + \tfrac{1}{2}\Delta u) = \cosh u,$$

$$\lim. \frac{\sinh \tfrac{1}{2}\Delta u}{\tfrac{1}{2}\Delta u} = 1; \qquad \text{(see Art. 13)}$$

$$\frac{d(\sinh u)}{du} = \cosh u.$$

(b) Similar to (a).

$$(c) \quad \frac{d(\tanh u)}{du} = \frac{d}{du} \cdot \frac{\sinh u}{\cosh u}$$
$$= \frac{\cosh^2 u - \sinh^2 u}{\cosh^2 u} = \frac{1}{\cosh^2 u} = \operatorname{sech}^2 u.$$

(d)   Similar to (c).

(e) $\dfrac{d(\operatorname{sech} u)}{du} = \dfrac{d}{du} \cdot \dfrac{1}{\cosh u} = -\dfrac{\sinh u}{\cosh^2 u} = -\operatorname{sech} u \tanh u.$

(f)   Similar to (e).

It thus appears that the functions $\sinh u$, $\cosh u$ reproduce themselves in two differentiations; and, similarly, that the circular functions $\sin u$, $\cos u$ produce their opposites in two differentiations.   In this connection it may be noted that the frequent appearance of the hyperbolic (and circular) functions in the solution of physical problems is chiefly due to the fact that they answer the question: What function has its second derivative equal to a positive (or negative) constant multiple of the function itself? (See Probs. 28–30.)   An answer such as $y = \cosh mx$ is not, however, to be understood as asserting that mx is an actual sectorial measure and y its characteristic ratio; but only that the relation between the numbers mx and y is the same as the known relation between the measure of a hyperbolic sector and its characteristic ratio; and that the numerical value of y could be found from a table of hyperbolic cosines.

Prob. 25. Show that for circular functions the only modifications required are in the algebraic signs of (b), (d).

Prob. 26. Show from their derivatives which of the hyperbolic and circular functions diminish as u increases.

Prob. 27. Find the derivative of $\tanh u$ independently of the derivatives of $\sinh u$, $\cosh u$.

Prob. 28. Eliminate the constants by differentiation from the equation $y = A \cosh mx + B \sinh mx$, and prove that $d^2y/dx^2 = m^2y$.

Prob. 29. Eliminate the constants from the equation

$$y = A \cos mx + B \sin mx,$$

and prove that $d^2y/dx^2 = -m^2y$.

Prob. 30. Write down the most general solutions of the differential equations

$$\frac{d^2y}{dx^2} = m^2y, \quad \frac{d^2y}{dx^2} = -m^2y, \quad \frac{d^4y}{dx^4} = m^4y.$$

ART. 15.   DERIVATIVES OF ANTI-HYPERBOLIC FUNCTIONS.

$$(a) \qquad \frac{d(\sinh^{-1} x)}{dx} = \frac{1}{\sqrt{x^2 + 1}},$$

$$(b) \qquad \frac{d(\cosh^{-1} x)}{dx} = \frac{1}{\sqrt{x^2 - 1}},$$

$$(c) \qquad \frac{d(\tanh^{-1} x)}{dx} = \frac{1}{1 - x^2}\bigg]_{x<1},$$

$$(d) \qquad \frac{d(\coth^{-1} x)}{dx} = -\frac{1}{x^2 - 1}\bigg]_{x>1}, \qquad\qquad (26).$$

$$(e) \qquad \frac{d(\operatorname{sech}^{-1} x)}{dx} = -\frac{1}{x\sqrt{1 - x^2}},$$

$$(f) \qquad \frac{d(\operatorname{csch}^{-1} x)}{dx} = -\frac{1}{x\sqrt{x^2 + 1}}.$$

(a) Let $\ u = \sinh^{-1} x,\ $ then $x = \sinh u,\ dx = \cosh u\, du$

$$= \sqrt{1 + \sinh^2 u}\, du = \sqrt{1 + x^2}\, du, \quad du = dx/\sqrt{1 + x^2}.$$

(b) Similar to (a).

(c) Let $\ u = \tanh^{-1} x,\ $ then $x = \tanh u,\ dx = \operatorname{sech}^2 u\, du$

$$= (1 - \tanh^2 u)du = (1 - x^2)du, \quad du = dx/1 - x^2.$$

(d) Similar to (c).

$$(e) \quad \frac{d(\operatorname{sech}^{-1} x)}{dx} = \frac{d}{dx}\left(\cosh^{-1}\frac{1}{x}\right) = \frac{-1}{x^2}\bigg/\left(\frac{1}{x^2} - 1\right)^{\frac12} = \frac{-1}{x\sqrt{1 - x^2}}.$$

(f) Similar to (e).

Prob. 31. Prove

$$\frac{d(\sin^{-1} x)}{dx} = \frac{1}{\sqrt{1 - x^2}}, \qquad \frac{d(\cos^{-1} x)}{dx} = -\frac{1}{\sqrt{1 - x^2}},$$

$$\frac{d(\tan^{-1} x)}{dx} = \frac{1}{1 + x^2}, \qquad \frac{d(\cot^{-1} x)}{dx} = -\frac{1}{1 + x^2}.$$

Prob. 32. Prove

$$d \sinh^{-1} \frac{x}{a} = \frac{dx}{\sqrt{x^2 + a^2}}, \qquad d \cosh^{-1} \frac{x}{a} = \frac{dx}{\sqrt{x^2 - a^2}},$$

$$d \tanh^{-1} \frac{x}{a} = \frac{a\,dx}{a^2 - x^2}\Big]_{x<a}, \qquad d \coth^{-1} \frac{x}{a} = -\frac{a\,dx}{x^2 - a^2}\Big]_{x>a}.$$

Prob. 33. Find $d(\operatorname{sech}^{-1} x)$ independently of $\cosh^{-1} x$.

Prob. 34. When $\tanh^{-1} x$ is real, prove that $\coth^{-1} x$ is imaginary, and conversely; except when $x = 1$.

Prob. 35. Evaluate $\dfrac{\sinh^{-1} x}{\log x}$, $\dfrac{\cosh^{-1} x}{\log x}$, when $x = \infty$.

ART. 16.   EXPANSION OF HYPERBOLIC FUNCTIONS.

For this purpose take Maclaurin's Theorem,

$$f(u) = f(0) + u f'(0) + \frac{1}{2!} u^2 f''(0) + \frac{1}{3!} u^3 f'''(0) + \cdots,$$

and put $f(u) = \sinh u$, $f'(u) = \cosh u$, $f''(u) = \sinh u$, $\ldots$,

then $f(0) = \sinh 0 = 0$, $f'(0) = \cosh 0 = 1$, $\ldots$;

hence $$\sinh u = u + \frac{1}{3!} u^3 + \frac{1}{5!} u^5 + \cdots; \qquad (27)$$

and similarly, or by differentiation,

$$\cosh u = 1 + \frac{1}{2!} u^2 + \frac{1}{4!} u^4 + \cdots. \qquad (28)$$

By means of these series the numerical values of $\sinh u$, $\cosh u$, can be computed and tabulated for successive values of the independent variable u. They are convergent for all values of u, because the ratio of the nth term to the preceding is in the first case $u^2/(2n - 1)(2n - 2)$, and in the second case $u^2/(2n - 2)(2n - 3)$, both of which ratios can be made less than unity by taking n large enough, no matter what value u may have.

From these series the following can be obtained by division :

$$\left.\begin{array}{l}
\tanh u = u - \tfrac{1}{3}u^3 + \tfrac{2}{15}u^5 + \tfrac{17}{315}u^7 + \cdots, \\[4pt]
\operatorname{sech} u = 1 - \tfrac{1}{2}u^2 + \tfrac{5}{24}u^4 - \tfrac{61}{720}u^6 + \cdots, \\[4pt]
u \coth u = 1 + \tfrac{1}{3}u^2 - \tfrac{1}{45}u^4 + \tfrac{2}{945}u^6 - \cdots, \\[4pt]
u \operatorname{csch} u = 1 - \tfrac{1}{6}u^2 + \tfrac{7}{360}u^4 - \tfrac{31}{15120}u^6 + \cdots.
\end{array}\right\} \qquad (29)$$

These four developments are seldom used, as there is no observable law in the coefficients, and as the functions tanh u, sech u, coth u, csch u, can be found directly from the previously computed values of cosh u, sinh u.

Prob. 36. Show that these six developments can be adapted to the circular functions by changing the alternate signs.

ART. 17.   EXPONENTIAL EXPRESSIONS.

Adding and subtracting (27), (28) give the identities

$$\cosh u + \sinh u = 1 + u + \frac{1}{2!}u^2 + \frac{1}{3!}u^3 + \frac{1}{4!}u^4 + \cdots = e^u,$$

$$\cosh u - \sinh u = 1 - u + \frac{1}{2!}u^2 - \frac{1}{3!}u^3 + \frac{1}{4!}u^4 - \cdots = e^{-u},$$

hence $\cosh u = \tfrac{1}{2}(e^u + e^{-u}),$ $\sinh u = \tfrac{1}{2}(e^u - e^{-u}),$

$$\left.\begin{array}{ll}
\tanh u = \dfrac{e^u - e^{-u}}{e^u + e^{-u}}, & \operatorname{sech} u = \dfrac{2}{e^u + e^{-u}}, \quad \text{etc.}
\end{array}\right\} \qquad (30)$$

The analogous exponential expressions for sin u, cos u are

$$\cos u = \frac{1}{2}(e^{ui} + e^{-ui}), \quad \sin u = \frac{1}{2i}(e^{ui} - e^{-ui}), \quad (i = \sqrt{-1})$$

where the symbol e^{ui} stands for the result of substituting ui for x in the exponential development

$$e^x = 1 + x + \frac{1}{2!}x^2 + \frac{1}{3!}x^3 + \cdots$$

This will be more fully explained in treating of complex numbers, Arts. 28, 29.

Prob. 37. Show that the properties of the hyperbolic functions could be placed on a purely algebraic basis by starting with equations (30) as their definitions ; for example, verify the identities :

$$\sinh(-u) = -\sinh u, \quad \cosh(-u) = \cosh u,$$

$$\cosh^2 u - \sinh^2 u = 1, \quad \sinh(u+v) = \sinh u \cosh v + \cosh u \sinh v,$$

$$\frac{d^2(\cosh mu)}{du^2} = m^2 \cosh mu, \quad \frac{d^2(\sinh mu)}{du^2} = m^2 \sinh mu.$$

Prob. 38. Prove $(\cosh u + \sinh u)^n = \cosh nu + \sinh nu$.

Prob. 39. Assuming from Art. 14 that $\cosh u$, $\sinh u$ satisfy the differential equation $d^2y/du^2 = y$, whose general solution may be written $y = Ae^u + Be^{-u}$, where A, B are arbitrary constants ; show how to determine A, B in order to derive the expressions for $\cosh u$, $\sinh u$, respectively. [Use eq. (15).]

Prob. 40. Show how to construct a table of exponential functions from a table of hyperbolic sines and cosines, and *vice versa*.

Prob. 41. Prove $u = \log_e(\cosh u + \sinh u)$.

Prob. 42. Show that the area of any hyperbolic sector is infinite when its terminal line is one of the asymptotes.

Prob. 43. From the relation $2\cosh u = e^u + e^{-u}$ prove

$$2^{n-1}(\cosh u)^n = \cosh nu + n\cosh(n-2)u + \tfrac{1}{2}n(n-1)\cosh(n-4)u + \dots,$$

and examine the last term when n is odd or even.

Find also the corresponding expression for $2^{n-1}(\sinh u)^n$.

Art. 18.   Expansion of Anti-Functions.

Since
$$\frac{d(\sinh^{-1} x)}{dx} = \frac{1}{\sqrt{1+x^2}} = (1+x^2)^{-\frac{1}{2}}$$

$$= 1 - \frac{1}{2}x^2 + \frac{1}{2}\frac{3}{4}x^4 - \frac{1}{2}\frac{3}{4}\frac{5}{6}x^6 + \dots,$$

hence, by integration,

$$\sinh^{-1} x = x - \frac{1}{2}\frac{x^3}{3} + \frac{1}{2}\frac{3}{4}\frac{x^5}{5} - \frac{1}{2}\frac{3}{4}\frac{5}{6}\frac{x^7}{7} + \dots, \quad (31)$$

the integration-constant being zero, since $\sinh^{-1} x$ vanishes with x. This series is convergent, and can be used in computation, only when $x < 1$. Another series, convergent when $x > 1$, is obtained by writing the above derivative in the form

$$\frac{d(\sinh^{-1} x)}{dx} = (x^2 + 1)^{-\frac{1}{2}} = \frac{1}{x}\left(1 + \frac{1}{x^2}\right)^{-\frac{1}{2}}$$

$$= \frac{1}{x}\left[1 - \frac{1}{2}\frac{1}{x^2} + \frac{1}{2}\frac{3}{4}\frac{1}{x^4} - \frac{1}{2}\frac{3}{4}\frac{5}{6}\frac{1}{x^6} + \cdots\right],$$

$$\therefore \sinh^{-1} x = C + \log x + \frac{1}{2}\frac{1}{2x^2} - \frac{1}{2}\frac{3}{4}\frac{1}{4x^4} + \frac{1}{2}\frac{3}{4}\frac{5}{6}\frac{1}{6x^6} - \cdots, \quad (32)$$

where C is the integration-constant, which will be shown in Art. 19 to be equal to $\log_e 2$.

A development of similar form is obtained for $\cosh^{-1} x$; for

$$\frac{d(\cosh^{-1} x)}{dx} = (x^2 - 1)^{-\frac{1}{2}} = \frac{1}{x}\left(1 - \frac{1}{x^2}\right)^{-\frac{1}{2}}$$

$$= \frac{1}{x}\left[1 + \frac{1}{2}\frac{1}{x^2} + \frac{1}{2}\frac{3}{4}\frac{1}{x^4} + \frac{1}{2}\frac{3}{4}\frac{5}{6}\frac{1}{x^6} + \cdots\right],$$

hence

$$\cosh^{-1} x = C + \log x - \frac{1}{2}\frac{1}{2x^2} - \frac{1}{2}\frac{3}{4}\frac{1}{4x^4} - \frac{1}{2}\frac{3}{4}\frac{5}{6}\frac{1}{6x^6} - \cdots, \quad (33)$$

in which C is again equal to $\log_e 2$ [Art. 19, Prob. 46]. In order that the function $\cosh^{-1} x$ may be real, x must not be less than unity; but when x exceeds unity, this series is convergent, hence it is always available for computation.

Again, $\qquad \dfrac{d(\tanh^{-1} x)}{dx} = \dfrac{1}{1 - x^2} = 1 + x^2 + x^4 + x^6 + \cdots,$

and hence $\qquad \tanh^{-1} x = x + \dfrac{1}{3}x^3 + \dfrac{1}{5}x^5 + \dfrac{1}{7}x^7 + \cdots, \quad (34)$

From (32), (33), (34) are derived:

$$\mathrm{sech}^{-1} x = \cosh^{-1}\frac{1}{x}$$

$$= C - \log x - \frac{x^2}{2.2} - \frac{1.3.x^4}{2.4.4} - \frac{1.3.5.x^6}{2.4.6.6} - \cdots; \quad (35)$$

$$\operatorname{csch}^{-1} x = \sinh^{-1} \frac{1}{x} = \frac{1}{x} - \frac{1}{2} \frac{1}{3x^3} + \frac{1}{2} \cdot \frac{3}{4} \frac{1}{5x^5} - \frac{1}{2} \frac{3}{4} \frac{5}{6} \frac{1}{7x^7} + \cdots,$$

$$= C - \log x + \frac{x^2}{2.2} - \frac{1.3 \cdot x^4}{2.4.4} + \frac{1.3.5 \cdot x^6}{2.4.6.6} - \cdots; \quad (36)$$

$$\coth^{-1} x = \tanh^{-1} \frac{1}{x} = \frac{1}{x} + \frac{1}{3x^3} + \frac{1}{5x^5} + \frac{1}{7x^7} + \cdots \quad (37)$$

Prob. 44. Show that the series for $\tanh^{-1} x$, $\coth^{-1} x$, $\operatorname{sech}^{-1} x$, are always available for computation.

Prob. 45. Show that one or other of the two developments of the inverse hyperbolic cosecant is available.

Art. 19. Logarithmic Expression of Anti-Functions.

Let $\qquad x = \cosh u$, then $\sqrt{x^2 - 1} = \sinh u$;

therefore $\qquad x + \sqrt{x^2 - 1} = \cosh u + \sinh u = e^u$,

and $\qquad u, = \cosh^{-1} x, = \log (x + \sqrt{x^2 - 1}). \qquad (38)$

Similarly, $\sinh^{-1} x = \log (x + \sqrt{x^2 + 1}). \qquad (39)$

Also $\qquad \operatorname{sech}^{-1} x = \cosh^{-1} \frac{1}{x} = \log \frac{1 + \sqrt{1 - x^2}}{x}, \qquad (40)$

$$\operatorname{csch}^{-1} x = \sinh^{-1} \frac{1}{x} = \log \frac{1 + \sqrt{1 + x^2}}{x}. \qquad (41)$$

Again, let $\qquad x = \tanh u = \frac{e^u - e^{-u}}{e^u + e^{-u}},$

therefore $\qquad \frac{1 + x}{1 - x} = \frac{e^u}{e^{-u}} = e^{2u},$

$$2u = \log \frac{1 + x}{1 - x}, \qquad \tanh^{-1} x = \tfrac{1}{2} \log \frac{1 + x}{1 - x}; \qquad (42)$$

and $\qquad \coth^{-1} x = \tanh^{-1} \frac{1}{x} = \tfrac{1}{2} \log \frac{x + 1}{x - 1}. \qquad (43)$

Prob. 46. Show from (39), (40), that, when $x \doteq \infty$,

$$\sinh^{-1} x - \log x \doteq \log 2, \qquad \cosh^{-1} x - \log x \doteq \log 2,$$

and hence show that the integration-constants in (32), (33) are each equal to $\log 2$.

Prob. 47. Derive from (42) the series for tanh^{-1}x given in (34).

Prob. 48. Prove the identities:

$$\log x = 2 \tanh^{-1}\frac{x-1}{x+1} = \tanh^{-1}\frac{x^2-1}{x^2+1} = \sinh^{-1}\tfrac{1}{2}(x-x^{-1}) = \cosh^{-1}\tfrac{1}{2}(x+x^{-1});$$

$$\log \sec x = 2 \tanh^{-1}\tan^2 \tfrac{1}{2}x; \quad \log \csc x = 2 \tanh^{-1}\tan^2(\tfrac{1}{4}\pi + \tfrac{1}{2}x);$$

$$\log \tan x = -\tanh^{-1}\cos 2x = -\sinh^{-1}\cot 2x = \cosh^{-1}\csc 2x.$$

ART. 20. THE GUDERMANIAN FUNCTION.

The correspondence of sectors of the same species was discussed in Arts. 1–4. It is now convenient to treat of the correspondence that may exist between sectors of different species.

Two points P_1, P_2, on any hyperbola and ellipse, are said to correspond with reference to two pairs of conjugates O_1A_1, O_1B_1, and O_2A_2, O_2B_2, respectively, when

$$x_1/a_1 = a_2/x_2, \tag{44}$$

and when y_1, y_2 have the same sign. The sectors $A_1O_1P_1$, $A_2O_2P_2$ are then also said to correspond. Thus corresponding sectors of central conics of different species are of the same sign and have their primary characteristic ratios reciprocal. Hence there is a fixed functional relation between their respective measures. The elliptic sectorial measure is called the gudermanian of the corresponding hyperbolic sectorial measure, and the latter the anti-gudermanian of the former. This relation is expressed by

$$S_2/K_2 = \text{gd } S_1/K_1$$

$$\text{or} \quad v = \text{gd } u, \quad \text{and} \quad u = \text{gd}^{-1}v. \tag{45}$$

ART. 21. CIRCULAR FUNCTIONS OF GUDERMANIAN.

The six hyperbolic functions of u are expressible in terms of the six circular functions of its gudermanian; for since

$$\frac{x_1}{a_1} = \cosh u, \quad \frac{x_2}{a_2} = \cos v, \qquad \text{(see Arts. 6, 7)}$$

in which u, v are the measures of corresponding hyperbolic and elliptic sectors,

hence
$$\left.\begin{array}{l} \cosh u = \sec v, \qquad [\text{eq. } (44)] \\ \sinh u = \sqrt{\sec^2 v - 1} = \tan v, \\ \tanh u = \tan v / \sec v = \sin v, \\ \coth u = \csc v, \\ \operatorname{sech} u = \cos v, \\ \operatorname{csch} u = \cot v. \end{array}\right\} \qquad (46)$$

The gudermanian is sometimes useful in computation; for instance, if $\sinh u$ be given, v can be found from a table of natural tangents, and the other circular functions of v will give the remaining hyperbolic functions of u. Other uses of this function are given in Arts. 22–26, 32–36.

Prob. 49. Prove that $\operatorname{gd} u = \sec^{-1}(\cosh u) = \tan^{-1}(\sinh u)$
$$= \cos^{-1}(\operatorname{sech} u) = \sin^{-1}(\tanh u),$$

Prob. 50. Prove $\quad \operatorname{gd}^{-1} v = \cosh^{-1}(\sec v) = \sinh^{-1}(\tan v)$
$$= \operatorname{sech}^{-1}(\cos v) = \tanh^{-1}(\sin v).$$

Prob. 51. Prove $\quad \operatorname{gd} 0 = 0, \ \operatorname{gd} \infty = \tfrac{1}{2}\pi, \quad \operatorname{gd}(-\infty) = -\tfrac{1}{2}\pi,$
$$\operatorname{gd}^{-1} 0 = 0, \ \operatorname{gd}^{-1}(\tfrac{1}{2}\pi) = \infty, \ \operatorname{gd}^{-1}(-\tfrac{1}{2}\pi) = -\infty.$$

Prob 52. Show that $\operatorname{gd} u$ and $\operatorname{gd}^{-1} v$ are odd functions of u, v.

Prob. 53. From the first identity in 4, Prob. 19, derive the relation $\tanh \tfrac{1}{2}u = \tan \tfrac{1}{2}v$.

Prob. 54. Prove
$$\tanh^{-1}(\tan u) = \tfrac{1}{2} \operatorname{gd} 2u, \text{ and } \tan^{-1}(\tanh x) = \tfrac{1}{2} \operatorname{gd}^{-1} 2x.$$

ART. 22. GUDERMANIAN ANGLE

If a circle be used instead of the ellipse of Art. 20, the gudermanian of the hyperbolic sectorial measure will be equal to the radian measure of the angle of the corresponding circular sector (see eq. (6), and Art. 2, Prob. 2). This angle will be called the gudermanian angle; but the gudermanian function v, as above defined, is merely a number, or ratio; and this number is equal to the radian measure of the gudermanian angle θ, which is itself usually tabulated in degree measure; thus

$$\theta = 180° v/\pi. \quad . \quad . \quad . \quad . \quad . \quad (47)$$

Prob. 55. Show that the gudermanian angle of u may be construct-ed as follows:

Take the principal radius OA of an equilateral hyperbola, as the

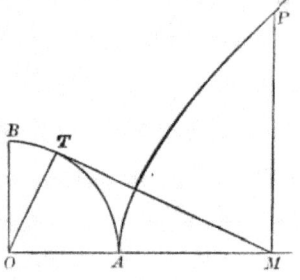

initial line, and OP as the terminal line, of the sector whose measure is u; from M, the foot of the ordinate of P, draw MT tangent to the circle whose diameter is the transverse axis; then AOT is the angle required.*

Prob. 56. Show that the angle θ never exceeds 90°.

Prob. 57. The bisector of angle AOT bisects the sector AOP (see Prob. 13, Art. 9, and Prob. 53, Art. 21), and the line AP.   (See Prob. 1, Art. 3.)

Prob. 58. This bisector is parallel to TP, and the points T, P are in line with the point diametrically opposite to A.

Prob. 59. The tangent at P passes through the foot of the ordinate of T, and intersects TM on the tangent at A.

Prob. 60. The angle APM is half the gudermanian angle.

ART. 23.   DERIVATIVES OF GUDERMANIAN AND INVERSE.

Let $\qquad\qquad v = \operatorname{gd} u,\quad u = \operatorname{gd}^{-1} v,$

then $\qquad\qquad \sec v = \cosh u,$

$$\sec v \tan v\, dv = \sinh u\, du,$$

$$\sec v\, dv = du,$$

therefore $\qquad d(\operatorname{gd}^{-1} v) = \sec v\, dv.$ $\qquad\qquad$ (48)

Again, $\qquad\qquad dv = \cos v\, du = \operatorname{sech} u\, du,$

therefore $\qquad d(\operatorname{gd} u) = \operatorname{sech} u\, du.$ $\qquad\qquad$ (49)

Prob. 61. Differentiate:

$$y = \sinh u - \operatorname{gd} u, \qquad\qquad y = \sin v + \operatorname{gd}^{-1} v,$$
$$y = \tanh u \operatorname{sech} u + \operatorname{gd} u, \qquad y = \tan v \sec v + \operatorname{gd}^{-1} v.$$

---

* This angle was called by Gudermann the longitude of u, and denoted by lu. His inverse symbol was $\mathfrak{L}$; thus $u = \mathfrak{L}(lu)$. (Crelle's Journal, vol. 6, 1830.) Lambert, who introduced the angle θ, named it the transcendent angle. (Hist. de l'acad. roy. de Berlin, 1761).   Hoüel (Nouvelles Annales, vol. 3, 1864) called it the hyperbolic amplitude of u, and wrote it amh u, in analogy with the amplitude of an elliptic function, as shown in Prob. 62. Cayley (Elliptic Functions, 1876) made the usage uniform by attaching to the angle the name of the mathematician who had used it extensively in tabulation and in the theory of elliptic functions of modulus unity.

Prob. 62. Writing the "elliptic integral of the first kind" in the form

$$u = \int_0^\phi \frac{d\phi}{\sqrt{1 - \kappa^2 \sin^2 \phi}},$$

κ being called the modulus, and ϕ the amplitude; that is,

$$\phi = \text{am } u, \ (\text{mod. } \kappa),$$

show that, in the special case when $\kappa = 1$,

$$u = \text{gd}^{-1} \phi, \qquad \text{am } u = \text{gd } u, \quad \sin \text{am } u = \tanh u,$$
$$\cos \text{am } u = \text{sech } u, \qquad \tan \text{am } u = \sinh u;$$

and that thus the elliptic functions $\sin \text{am } u$, etc., degenerate into the hyperbolic functions, when the modulus is unity.*

Art. 24. Series for Gudermanian and its Inverse.

Substitute for sech u, sec v in (49), (48) their expansions, Art. 16, and integrate, then

$$\text{gd } u = u - \tfrac{1}{6}u^3 + \tfrac{1}{24}u^5 - \tfrac{61}{5040}u^7 + \cdots \qquad (50)$$
$$\text{gd}^{-1}v = v + \tfrac{1}{6}v^3 + \tfrac{1}{24}v^5 + \tfrac{61}{5040}v^7 + \cdots \qquad (51)$$

No constants of integration appear, since gd u vanishes with u, and gd^{-1}v with v. These series are seldom used in computation, as gd u is best found and tabulated by means of tables of natural tangents and hyperbolic sines, from the equation

$$\text{gd } u = \tan^{-1}(\sinh u),$$

and a table of the direct function can be used to furnish the numerical values of the inverse function; or the latter can be obtained from the equation,

$$\text{gd}^{-1}v = \sinh^{-1}(\tan v) = \cosh^{-1}(\sec v).$$

To obtain a logarithmic expression for gd^{-1}v, let

$$\text{gd}^{-1}v = u, \quad v = \text{gd } u,$$

---

* The relation gd $u = \text{am } u$, (mod. 1), led Hoüel to name the function gd u, the hyperbolic amplitude of u, and to write it amh u (see note, Art. 22). In this connection Cayley expressed the functions tanh u, sech u, sinh u in the form sin gd u, cos gd u, tan gd u, and wrote them sg u, cg u, tg u, to correspond with the abbreviations sn u, cn u, dn u for sin am u, cos am u, tan am u. Thus tanh $u = \text{sg } u = \text{sn } u$, (mod. 1); etc.

It is well to note that neither the elliptic nor the hyperbolic functions received their names on account of the relation existing between them in a special case. (See foot-note, p. 107.)

therefore　　　$\sec v = \cosh u, \quad \tan v = \sinh u,$

$$\sec v + \tan v = \cosh u + \sinh u = e^u,$$

$$e^u = \frac{1 + \sin v}{\cos v} = \frac{1 - \cos (\tfrac{1}{2}\pi + v)}{\sin (\tfrac{1}{2}\pi + v)} = \tan (\tfrac{1}{4}\pi + \tfrac{1}{2}v),$$

$$u, = \mathrm{gd}^{-1}v, = \log_e \tan (\tfrac{1}{4}\pi + \tfrac{1}{2}v). \tag{52}$$

Prob. 63. Evaluate $\dfrac{\mathrm{gd}\,u - u}{u^3}\Big]_{u \doteq 0}, \quad \dfrac{\mathrm{gd}^{-1}v - v}{v^3}\Big]_{v \doteq 0}.$

Prob. 64. Prove that gd u − sin u is an infinitesimal of the fifth order, when $u \doteq 0$.

Prob. 65. Prove the relations

$$\tfrac{1}{4}\pi + \tfrac{1}{2}v = \tan^{-1}e^u, \quad \tfrac{1}{4}\pi - \tfrac{1}{2}v = \tan^{-1}e^{-u}.$$

ART. 25. GRAPHS OF HYPERBOLIC FUNCTIONS.

Drawing two rectangular axes, and laying down a series of points whose abscissas represent, on any convenient scale, successive values of the sectorial measure, and whose ordinates represent, preferably on the same scale, the corresponding values of the function to be plotted, the locus traced out by this series of points will be a graphical representation of the variation of the function as the sectorial meas-

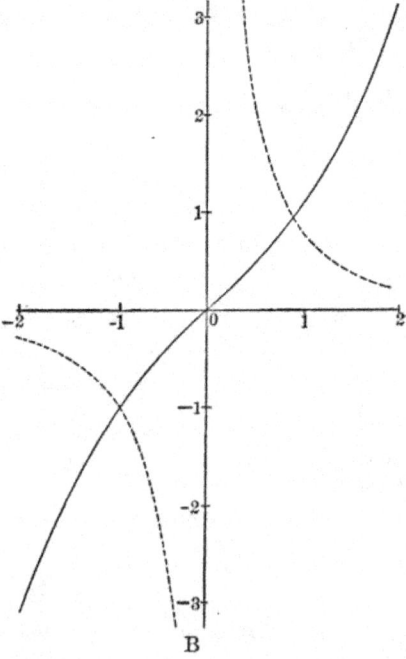

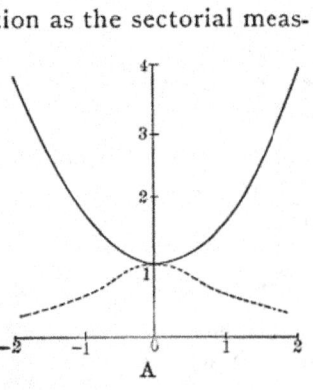

A　　　　　　　　　　　　　　　B

ure varies.   The equations of the curves in the ordinary cartesian notation are :

Fig.	Full Lines.	Dotted Lines.
A	$y = \cosh x,$	$y = \operatorname{sech} x$;
B	$y = \sinh x,$	$y = \operatorname{csch} x$;
C	$y = \tanh x,$	$y = \coth x$;
D	$y = \operatorname{gd} x.$	

Here x is written for the sectorial measure u, and y for the
numerical value of cosh u, etc.   It is thus to be noted that the
variables x, y are numbers, or ratios, and that the equation
$y = \cosh x$ merely expresses that the relation between the
numbers x and y is taken to be the same as the relation between a sectorial measure and its characteristic ratio.   The
numerical values of cosh u, sinh u, tanh u are given in the
tables at the end of this chapter for values of u between o and
4.   For greater values they may be computed from the developments of Art. 16.

The curves exhibit graphically the relations :

$$\operatorname{sech} u = \frac{\text{I}}{\cosh u}, \quad \operatorname{csch} u = \frac{\text{I}}{\sinh u}, \quad \coth u = \frac{\text{I}}{\tanh u};$$

$$\cosh u \not< \text{I}, \quad \operatorname{sech} u \not> \text{I}, \quad \tanh u \not> \text{I}, \quad \operatorname{gd} u \not< \tfrac{1}{2}\pi, \text{ etc.;}$$

$$\sinh (- u) = - \sinh u, \quad \cosh (- u) = \cosh u,$$

$$\tanh (- u) = - \tanh u, \quad \operatorname{gd} (- u) = - \operatorname{gd} u, \text{ etc.;}$$

$$\cosh o = \text{I}, \quad \sinh o = o, \quad \tanh o = o, \quad \operatorname{csch} (o) = \infty, \text{ etc.;}$$

$$\cosh (\pm \infty) = \infty, \quad \sinh (\pm \infty) = \pm \infty, \quad \tanh (\pm \infty) = \pm \text{I, etc.}$$

The slope of the curve $y = \sinh x$ is given by the equation
$dy/dx = \cosh x$, showing that it is always positive, and that
the curve becomes more nearly vertical as x becomes infinite.
Its direction of curvature is obtained from $d^2y/dx^2 = \sinh x$,
proving that the curve is concave downward when x is negative, and upward when x is positive.   The point of inflexion is
at the origin, and the inflexional tangent bisects the angle
between the axes.

The direction of curvature of the locus $y = \operatorname{sech} x$ is given by $d^2y/dx^2 = \operatorname{sech} x(2\tanh^2 x - 1)$, and thus the curve is con-

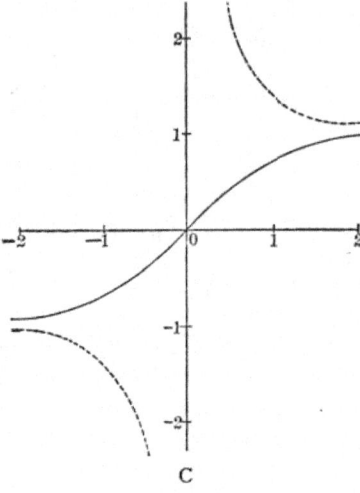

cave downwards or upwards according as $2\tanh^2 x - 1$ is negative or positive. The in-flexions occur at the points $x = \pm \tanh^{-1}.707, = \pm.881,$ $y = .707$; and the slopes of the inflexional tangents are $\pm 1/2$.

The curve $y = \operatorname{csch} x$ is asymptotic to both axes, but approaches the axis of x more rapidly than it approaches the axis of y, for when $x = 3, y$ is only .1, but it is not till $y = 10$

C

that x is so small as .1. The curves $y = \operatorname{csch} x, y = \sinh x$ cross at the points $x = \pm.881, y = \pm 1$.

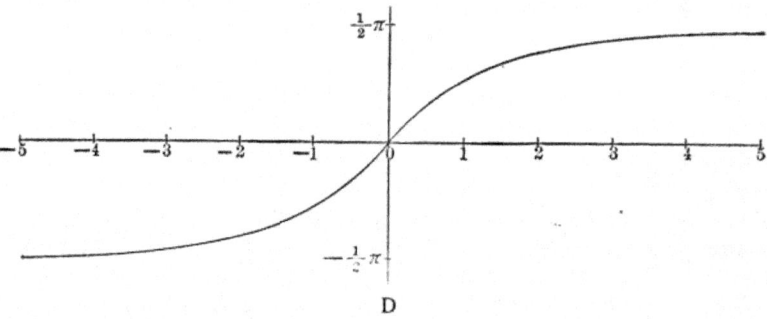

D

Prob. 66. Find the direction of curvature, the inflexional tangent, and the asymptotes of the curves $y = \operatorname{gd} x, y = \tanh x$.

Prob. 67. Show that there is no inflexion-point on the curves $y = \cosh x, y = \coth x$.

Prob. 68. Show that any line $y = mx + n$ meets the curve $y = \tanh x$ in either three real points or one. Hence prove that the equation $\tanh x = mx + n$ has either three real roots or one. From the figure give an approximate solution of the equation $\tanh x = x - 1$.

Prob. 69. Solve the equations: $\cosh x = x + 2$; $\sinh x = \frac{3}{2}x$; $\operatorname{gd} x = x - \frac{1}{2}\pi$.

Prob. 70. Show which of the graphs represent even functions, and which of them represent odd ones.

ART. 26. ELEMENTARY INTEGRALS.

The following useful indefinite integrals follow from Arts. 14, 15, 23:

Hyperbolic.	Circular.

1. $\displaystyle\int \sinh u\, du = \cosh u,$ $\qquad \displaystyle\int \sin u\, du = -\cos u,$

2. $\displaystyle\int \cosh u\, du = \sinh u,$ $\qquad \displaystyle\int \cos u\, du = \sin u,$

3. $\displaystyle\int \tanh u\, du = \log \cosh u,$ $\quad \displaystyle\int \tan u\, du = -\log \cos u,$

4. $\displaystyle\int \coth u\, du = \log \sinh u,$ $\quad \displaystyle\int \cot u\, du = \log \sin u,$

5. $\displaystyle\int \operatorname{csch} u\, du = \log \tanh \frac{u}{2},$ $\quad \displaystyle\int \csc u\, du = \log \tan \frac{u}{2},$

   $\qquad\qquad = -\sinh^{-1}(\operatorname{csch} u),$ $\qquad\qquad = -\cosh^{-1}(\csc u),$

6. $\displaystyle\int \operatorname{sech} u\, du = \operatorname{gd} u,$ $\qquad \displaystyle\int \sec u\, du = \operatorname{gd}^{-1} u,$

7. $\displaystyle\int \frac{dx}{\sqrt{x^2 + a^2}} = \sinh^{-1}\frac{x}{a},$* $\quad \displaystyle\int \frac{dx}{\sqrt{a^2 - x^2}} = \sin^{-1}\frac{x}{a},$

8. $\displaystyle\int \frac{dx}{\sqrt{x^2 - a^2}} = \cosh^{-1}\frac{x}{a},$ $\quad \displaystyle\int \frac{-dx}{\sqrt{a^2 - x^2}} = \cos^{-1}\frac{x}{a},$

9. $\displaystyle\int \frac{dx}{a^2 - x^2}\bigg]_{x<a} = \frac{1}{a}\tanh^{-1}\frac{x}{a},$ $\displaystyle\int \frac{dx}{a^2 + x^2} = \frac{1}{a}\tan^{-1}\frac{x}{a},$

* Forms 7–12 are preferable to the respective logarithmic expressions (Art. 19), on account of the close analogy with the circular forms, and also because they involve functions that are directly tabulated. This advantage appears more clearly in 13–20.

10. $\int \dfrac{-dx}{x^2-a^2}\bigg]_{x>a} = \dfrac{1}{a}\coth^{-1}\dfrac{x}{a}, \quad \int \dfrac{-dx}{a^2+x^2} = \dfrac{1}{a}\cot^{-1}\dfrac{x}{a},$

11. $\int \dfrac{-dx}{x\sqrt{a^2+x^2}} = \dfrac{1}{a}\operatorname{sech}^{-1}\dfrac{x}{a}, \quad \int \dfrac{dx}{x\sqrt{x^2-a^2}} = \dfrac{1}{a}\sec^{-1}\dfrac{x}{a},$

12. $\int \dfrac{-dx}{x\sqrt{a^2+x^2}} = \dfrac{1}{a}\operatorname{csch}^{-1}\dfrac{x}{a}, \quad \int \dfrac{-dx}{x\sqrt{x^2-a^2}} = \dfrac{1}{a}\csc^{-1}\dfrac{x}{a}.$

From these fundamental integrals the following may be derived:

13. $\int \dfrac{dx}{\sqrt{ax^2+2bx+c}} = \dfrac{1}{\sqrt{a}}\sinh^{-1}\dfrac{ax+b}{\sqrt{ac-b^2}}, \quad a \text{ positive}, ac>b^2;$

$$= \dfrac{1}{\sqrt{a}}\cosh^{-1}\dfrac{ax+b}{\sqrt{b^2-ac}}, \quad a \text{ positive}, ac<b^2;$$

$$= \dfrac{1}{\sqrt{-a}}\cos^{-1}\dfrac{ax+b}{\sqrt{b^2-ac}}, \quad a \text{ negative}.$$

14. $\int \dfrac{dx}{ax^2+2bx+c} = \dfrac{1}{\sqrt{ac-b^2}}\tan^{-1}\dfrac{ax+b}{\sqrt{ac-b^2}}, \quad ac>b^2;$

$$= \dfrac{-1}{\sqrt{b^2-ac}}\tanh^{-1}\dfrac{ax+b}{\sqrt{b^2-ac}}, \quad ac<b^2,\ ax+b<\sqrt{b^2-ac};$$

$$= \dfrac{-1}{\sqrt{b^2-ac}}\coth^{-1}\dfrac{ax+b}{\sqrt{b^2-ac}}, \quad ac<b^2,\ ax+b>\sqrt{b^2-ac};$$

Thus, $\int_4^5 \dfrac{dx}{x^2-4x+3} = -\coth^{-1}(x-2)\bigg]_4^5 = \coth^{-1}2-\coth^{-1}3$

$= \tanh^{-1}(.5)-\tanh^{-1}(.3333) = .5494-.3466 = .2028.*$

$\int_2^{2.5} \dfrac{dx}{x^2-4x+3} = -\tanh^{-1}(x-2)\bigg]_2^{2.5} = \tanh^{-1}0 - \tanh^{-1}(.5)$

$$= -.5494.$$

(By interpreting these two integrals as areas, show graphically that the first is positive, and the second negative.)

15. $\int \dfrac{dx}{(a-x)\sqrt{x-b}} = \dfrac{2}{\sqrt{a-b}}\tanh^{-1}\sqrt{\dfrac{x-b}{a-b}},$

---

*For $\tanh^{-1}(.5)$ interpolate between $\tanh(.54) = .4930$, $\tanh(.56) = .5080$ (see tables, pp. 162, 163); and similarly for $\tanh^{-1}(.3333)$.

or $\dfrac{-2}{\sqrt{b-a}}\tan^{-1}\sqrt{\dfrac{x-b}{b-a}}$, or $\dfrac{2}{\sqrt{a-b}}\coth^{-1}\sqrt{\dfrac{x--b}{a-b}}$;

the real form to be taken. (Put $x - b = z^2$, and apply 9, 10.)

16. $\displaystyle\int\dfrac{dx}{(a-x)\sqrt{b-x}}=\dfrac{2}{\sqrt{b-a}}\tanh^{-1}\sqrt{\dfrac{b-x}{b-a}}$,

or $\dfrac{2}{\sqrt{b-a}}\coth^{-1}\sqrt{\dfrac{b-x}{b-a}}$, or $\dfrac{-2}{\sqrt{a-b}}\tan^{-1}\sqrt{\dfrac{b-x}{a-b}}$;

the real form to be taken.

17. $\displaystyle\int(x^2 - a^2)^{\frac{1}{2}}dx = \dfrac{1}{2}x(x^2 - a^2)^{\frac{1}{2}} - \dfrac{1}{2}a^2\cosh^{-1}\dfrac{x}{a}$.

By means of a reduction-formula this integral is easily made to depend on 8. It may also be obtained by transforming the expression into hyperbolic functions by the assumption $x = a\cosh u$, when the integral takes the form

$$a^2\int\sinh^2 u\,du = \dfrac{a^2}{2}\int(\cosh 2u - 1)du = \dfrac{1}{4}a^2(\sinh 2u - 2u)$$
$$= \tfrac{1}{2}a^2(\sinh u\cosh u - u),$$

which gives 17 on replacing $a\cosh u$ by x, and $a\sinh u$ by $(x^2 - a^2)^{\frac{1}{2}}$. The geometrical interpretation of the result is evident, as it expresses that the area of a rectangular-hyperbolic segment AMP is the difference between a triangle OMP and a sector OAP.

18. $\displaystyle\int(a^2 - x^2)^{\frac{1}{2}}dx = \dfrac{1}{2}x(a^2 - x^2)^{\frac{1}{2}} + \dfrac{1}{2}a^2\sin^{-1}\dfrac{x}{a}$.

19. $\displaystyle\int(x^2 + a^2)^{\frac{1}{2}}dx = \dfrac{1}{2}x(x^2 + a^2)^{\frac{1}{2}} + \dfrac{1}{2}a^2\sinh^{-1}\dfrac{x}{a}$.

20. $\displaystyle\int\sec^3\phi\,d\phi = \int(1 + \tan^2\phi)^{\frac{1}{2}}d\tan\phi$

$$= \tfrac{1}{2}\tan\phi(1 + \tan^2\phi)^{\frac{1}{2}} + \tfrac{1}{2}\sinh^{-1}(\tan\phi)$$
$$= \tfrac{1}{2}\sec\phi\tan\phi + \tfrac{1}{2}\,gd^{-1}\phi.$$

21. $\displaystyle\int\operatorname{sech}^3 u\,du = \tfrac{1}{2}\operatorname{sech} u\tanh u + \tfrac{1}{2}\,gd\,u$.

Prob. 71. What is the geometrical interpretation of 18, 19?

Prob. 72. Show that $\displaystyle\int(ax^2 + 2bx + c)^{\frac{1}{2}}dx$ reduces to 17, 18, 19, respectively: when a is positive, with $ac < b^2$; when a is negative; and when a is positive, with $ac > b^2$.

Prob. 73. Prove $\int \sinh u \tanh u \, du = \sinh u - \operatorname{gd} u$,

$$\int \cosh u \coth u \, du = \cosh u + \log \tanh \frac{u}{2}.$$

Prob. 74. Integrate

$$(x^2 + 2x + 5)^{-\frac{1}{2}}dx, \quad (x^2 + 2x + 5)^{-1}dx, \quad (x^2 + 2x + 5)^{\frac{1}{2}}dx.$$

Prob. 75. In the parabola $y^2 = 4px$, if s be the length of arc measured from the vertex, and ϕ the angle which the tangent line makes with the vertical tangent, prove that the intrinsic equation of the curve is $ds/d\phi = 2p \sec^2 \phi$, $s = \frac{1}{2} \sec \phi \tan \phi + \frac{1}{2} \operatorname{gd}^{-1}\phi$.

Prob. 76. The polar equation of a parabola being $r = a \sec^2 \frac{1}{2}\theta$, referred to its focus as pole, express s in terms of θ.

Prob. 77. Find the intrinsic equation of the curve $y/a = \cosh x/a$, and of the curve $y/a = \log \sec x/a$.

Prob. 78. Investigate a formula of reduction for $\int \cosh^n x \, dx$; also integrate by parts $\cosh^{-1} x \, dx$, $\tanh^{-1} x \, dx$, $(\sinh^{-1} x)^2 dx$; and show that the ordinary methods of reduction for $\int \cos^m x \sin^n x \, dx$ can be applied to $\int \cosh^m x \sinh^n x \, dx$.

ART. 27.   FUNCTIONS OF COMPLEX NUMBERS.

As vector quantities are of frequent occurence in Mathematical Physics; and as the numerical measure of a vector in terms of a standard vector is a complex number of the form $x + iy$, in which x, y are real, and i stands for $\sqrt{-1}$; it becomes necessary in treating of any class of functional operations to consider the meaning of these operations when performed on such generalized numbers.* The geometrical definitions of $\cosh u$, $\sinh u$, given in Art. 7, being then no longer applicable, it is necessary to assign to each of the symbols

* The use of vectors in electrical theory is shown in Bedell and Crehore's Alternating Currents, Chaps. XIV-XX (first published in 1892). The advantage of introducing the complex measures of such vectors into the differential equations is shown by Steinmetz, Proc. Elec. Congress, 1893; while the additional convenience of expressing the solution in hyperbolic functions of these complex numbers is exemplified by Kennelly, Proc. American Institute Electrical Engineers, April 1895. (See below, Art. 37.)

$\cosh{(x + iy)}$, $\sinh{(x + iy)}$, a suitable algebraic meaning, which should be consistent with the known algebraic values of $\cosh x$, $\sinh x$, and include these values as a particular case when $y = 0$. The meanings assigned should also, if possible, be such as to permit the addition-formulas of Art. 11 to be made general, with all the consequences that flow from them.

Such definitions are furnished by the algebraic developments in Art. 16, which are convergent for all values of u, real or complex. Thus the definitions of $\cosh{(x + iy)}$, $\sinh{(x + iy)}$ are to be

$$
\left.
\begin{array}{l}
\cosh{(x + iy)} = 1 + \dfrac{1}{2!}(x + iy)^2 + \dfrac{1}{4!}(x + iy)^4 + \ldots, \\[2ex]
\sinh{(x + iy)} = (x + iy) + \dfrac{1}{3!}(x + iy)^3 + \ldots
\end{array}
\right\} \quad (52)
$$

From these series the numerical values of $\cosh{(x + iy)}$, $\sinh{(x + iy)}$ could be computed to any degree of approximation, when x and y are given. In general the results will come out in the complex form*

$$
\cosh{(x + iy)} = a + ib,
$$
$$
\sinh{(x + iy)} = c + id.
$$

The other functions are defined as in Art. 7, eq. (9).

Prob. 79. Prove from these definitions that, whatever u may be,

$$
\cosh{(-u)} = \cosh u, \qquad \sinh{(-u)} = -\sinh u,
$$

$$
\frac{d}{du}\cosh u = \sinh u, \qquad \frac{d}{du}\sinh u = \cosh u,
$$

$$
\frac{d^2}{du^2}\cosh mu = m^2 \cosh mu, \quad \frac{d^2}{du^2}\sinh mu = m^2 \sinh mu.\dagger
$$

---

* It is to be borne in mind that the symbols cosh, sinh, here stand for algebraic operators which convert one number into another; or which, in the language of vector-analysis change one vector into another, by stretching and turning.

† The generalized hyperbolic functions usually present themselves in Mathematical Physics as the solution of the differential equation $d^2\phi/du^2 = m^2\phi$, where ϕ, m, u are complex numbers, the measures of vector quantities. (See Art. 37.)

ART. 28.   ADDITION-THEOREMS FOR COMPLEXES.

The addition-theorems for $\cosh(u+v)$, etc., where u, v are complex numbers, may be derived as follows. First take u, v as real numbers, then, by Art. 11,

$$\cosh(u+v) = \cosh u \cosh v + \sinh u \sinh v;$$

hence $1 + \dfrac{1}{2!}(u+v)^2 + \ldots = \left(1 + \dfrac{1}{2!}u^2 + \ldots\right)\left(1 + \dfrac{1}{2!}v^2 + \ldots\right)$

$$+ \left(u + \dfrac{1}{3!}u^3 + \ldots\right)\left(v + \dfrac{1}{3!}v^3 + \ldots\right)$$

This equation is true when u, v are any real numbers. It must, then, be an algebraic identity. For, compare the terms of the rth degree in the letters u, v on each side. Those on the left are $\dfrac{1}{r!}(u+v)^r$; and those on the right, when collected, form an rth-degree function which is numerically equal to the former for more than r values of u when v is constant, and for more than r values of v when u is constant. Hence the terms of the rth degree on each side are algebraically identical functions of u and v.* Similarly for the terms of any other degree. Thus the equation above written is an algebraic identity, and is true for all values of u, v, whether real or complex. Then writing for each side its symbol, it follows that

$$\cosh(u+v) = \cosh u \cosh v + \sinh u \sinh v; \qquad (53)$$

and by changing v into $-v$,

$$\cosh(u-v) = \cosh u \cosh v - \sinh u \sinh v. \qquad (54)$$

In a similar manner is found

$$\sinh(u \pm v) = \sinh u \cosh v \pm \cosh u \sinh v. \qquad (55)$$

In particular, for a complex argument,

$$\left. \begin{aligned} \cosh(x \pm iy) &= \cosh x \cosh iy \pm \sinh x \sinh iy, \\ \sinh(x \pm iy) &= \sinh x \cosh iy \pm \cosh x \sinh iy. \end{aligned} \right\} \qquad (56)$$

---

* "If two rth-degree functions of a single variable be equal for more than r values of the variable, then they are equal for all values of the variable, and are algebraically identical."

Prob. 79. Show, by a similar process of generalization,* that if sin u, cos u, exp u† be defined by their developments in powers of u, then, whatever u may be,

$$\sin (u + v) = \sin u \cos v + \cos u \sin v,$$
$$\cos (u + v) = \cos u \cos v - \sin u \sin v,$$
$$\exp (u + v) = \exp u \exp v.$$

Prob. 80. Prove that the following are identities:

$$\cosh^2 u - \sinh^2 u = 1,$$
$$\cosh u + \sinh u = \exp u,$$
$$\cosh u - \sinh u = \exp (-u),$$
$$\cosh u = \tfrac{1}{2}[\exp u + \exp (-u)],$$
$$\sinh u = \tfrac{1}{2}[\exp u - \exp(-u)].$$

ART. 29.   FUNCTIONS OF PURE IMAGINARIES.

In the defining identities

$$\cosh u = 1 + \frac{1}{2!}u^2 + \frac{1}{4!}u^4 + \ldots,$$

$$\sinh u = u + \frac{1}{3!}u^3 + \frac{1}{5!}u^5 + \ldots,$$

put for u the pure imaginary iy, then

$$\cosh iy = 1 - \frac{1}{2!}y^2 + \frac{1}{4!}y^4 - \ldots = \cos y, \qquad (57)$$

$$\sinh iy = iy + \frac{1}{3!}(iy)^3 + \frac{1}{5!}(iy)^5 + \ldots$$

$$= i\left[y - \frac{1}{3!}y^3 + \frac{1}{5!}y^5 - \ldots\right] = i \sin y, \quad (58)$$

and, by division, $\tanh iy = i \tan y.$ \qquad (59)

* This method of generalization is sometimes called the principle of the " permanence of equivalence of forms." It is not, however, strictly speaking, a " principle," but a method; for, the validity of the generalization has to be demonstrated, for any particular form, by means of the principle of the algebraic identity of polynomials enunciated in the preceding foot-note. (See Annals of Mathematics, Vol. 6, p. 81.)

† The symbol exp u stands for "exponential function of u," which is identical with e^u when u is real.

These formulas serve to interchange hyperbolic and circular functions. The hyperbolic cosine of a pure imaginary is real, and the hyperbolic sine and tangent are pure imaginaries.

The following table exhibits the variation of sinh u, cosh u, tanh u, exp u, as u takes a succession of pure imaginary values.

u	sinh u	cosh u	tanh u	exp u
O	O	I	O	I
$\frac{1}{4}i\pi$	$.7i$	$.7$*	i	$.7(1+i)$
$\frac{1}{2}i\pi$	i	O	$\infty\ i$	i
$\frac{3}{4}i\pi$	$.7i$	$-.7$	$-i$	$.7(1-i)$
$i\pi$	O	-1	O	-1
$\frac{5}{4}i\pi$	$-.7i$	$-.7$	i	$-.7(1+i)$
$\frac{3}{2}i\pi$	$-i$	O	$\infty\ i$	$-i$
$\frac{7}{4}i\pi$	$-.7i$	$.7$	$-i$	$.7(1-i)$
$2i\pi$	O	I	O	I

* In this table .7 is written for $\frac{1}{2}\sqrt{2}, = .707 \ldots$.

Prob. 81. Prove the following identities :

$$\cos y = \cosh iy = \tfrac{1}{2}[\exp iy + \exp(-iy)],$$

$$\sin y = \frac{1}{i}\sinh iy = \frac{1}{2i}[\exp iy - \exp(-iy)],$$

$$\cos y + i \sin y = \cosh iy + \sinh iy = \exp iy,$$

$$\cos y - i \sin y = \cosh iy - \sinh iy = \exp(-iy),$$

$$\cos iy = \cosh y, \quad \sin iy = i \sinh y.$$

Prob. 82. Equating the respective real and imaginary parts on each side of the equation $\cos ny + i \sin ny = (\cos y + i \sin y)^n$, express $\cos ny$ in powers of $\cos y$, $\sin y$; and hence derive the corresponding expression for $\cosh ny$.

Prob. 83. Show that, in the identities (57) and (58), y may be replaced by a general complex, and hence that

$$\sinh(x \pm iy) = \pm i \sin(y \mp ix),$$

$$\cosh (x \pm iy) = \cos (y \mp ix),$$
$$\sin (x \pm iy) = \pm i \sinh (y \mp ix),$$
$$\cos (x \pm iy) = \cosh (y \mp ix).$$

Prob. 84. From the product-series for $\sin x$ derive that for $\sinh x$:

$$\sin x = x\left(1 - \frac{x^2}{\pi^2}\right)\left(1 - \frac{x^2}{2^2\pi^2}\right)\left(1 - \frac{x^2}{3^2\pi^2}\right) \cdots,$$

$$\sinh x = x\left(1 + \frac{x^2}{\pi^2}\right)\left(1 + \frac{x^2}{2^2\pi^2}\right)\left(1 + \frac{x^2}{3^2\pi^2}\right) \cdots$$

ART. 30. FUNCTIONS OF $x + iy$ IN THE FORM $X + iY$.

By the addition-formulas,

$$\cosh (x + iy) = \cosh x \cosh iy + \sinh x \sinh iy,$$
$$\sinh (x + iy) = \sinh x \cosh iy + \cosh x \sinh iy,$$

but $\qquad \cosh iy = \cos y, \quad \sinh iy = i \sin y,$

hence $\cosh (x + iy) = \cosh x \cos y + i \sinh x \sin y,$
$$\left. \sinh (x + iy) = \sinh x \cos y + i \cosh x \sin y. \right\} \qquad (60)$$

Thus if $\cosh (x + iy) = a + ib$, $\sinh (x + iy) = c + id$, then

$$a = \cosh x \cos y, \quad b = \sinh x \sin y,$$
$$\left. c = \sinh x \cos y, \quad d = \cosh x \sin y. \right\} \qquad (61)$$

From these expressions the complex tables at the end of this chapter have been computed.

Writing $\cosh z = Z$, where $z = x + iy$, $Z = X + iY$; let the complex numbers z, Z be represented on Argand diagrams, in the usual way, by the points whose coordinates are (x, y), (X, Y); and let the point z move parallel to the y-axis, on a given line $x = m$, then the point Z will describe an ellipse whose equation, obtained by eliminating y between the equations $X = \cosh m \cos y$, $Y = \sinh m \sin y$, is

$$\frac{X^2}{(\cosh m)^2} + \frac{Y^2}{(\sinh m)^2} = 1,$$

and which, as the parameter m varies, represents a series of confocal ellipses, the distance between whose foci is unity.

Similarly, if the point z move parallel to the x-axis, on a given line $y = n$, the point Z will describe an hyperbola whose equation, obtained by eliminating the variable x from the equations $X = \cosh x \cos n$, $Y = \sinh x \sin n$, is

$$\frac{X^2}{(\cos n)^2} - \frac{Y^2}{(\sin n)^2} = 1,$$

and which, as the parameter n varies, represents a series of hyperbolas confocal with the former series of ellipses.

These two systems of curves, when accurately drawn at close intervals on the Z plane, constitute a chart of the hyperbolic cosine; and the numerical value of $\cosh (m + in)$ can be read off at the intersection of the ellipse whose parameter is m with the hyperbola whose parameter is n.*

Prob. 85. Prove that, in the case of $\sinh (x + iy)$, the above two systems of curves are each turned through a right angle. Compare the chart of $\sin (x + iy)$, and also of $\cos (x + iy)$.

Prob. 86. Prove the identity $\tan (x + iy) = \dfrac{\sinh 2x + i \sin 2y}{\cosh 2x + \cos 2y}$.

Prob. 87. If $\cosh (x + iy)$, $= a + ib$, be written in the "modulus and amplitude" form as $r(\cos \theta + i \sin \theta)$, $= r \exp i\theta$, then

$$r^2 = a^2 + b^2 = \cosh^2 x - \sin^2 y = \cos^2 y - \sinh^2 x,$$
$$\tan \theta = b/a = \tanh x \tan y.$$

Prob. 88. Find the modulus and amplitude of $\sinh (x + iy)$, $\sin (x + iy)$, $\exp (x + iy)$.

Prob. 89. The functions $\sinh u$, $\cosh u$ have the pure imaginary period $2i\pi$; that is, $\sinh (u + 2i\pi) = \sinh u$, $\cosh (u + 2i\pi) = \cosh u$; also $\sinh (u + \frac{1}{2}i\pi) = i \cosh u$, $\cosh (u + \frac{1}{2}i\pi) = i \sinh u$, $\sinh (u + i\pi) = -\sinh u$, $\cosh (u + i\pi) = -\cosh u$.

Prob. 90. The functions $\cosh^{-1}m$, $\sinh^{-1}m$ have multiple values at intervals of $2i\pi$, but each has a unique value (called the principal value) in which the coefficient of i lies between 0 and π for the former, and between $-\frac{1}{2}\pi$ and $+\frac{1}{2}\pi$ for the latter.

* Such a chart is given by Kennelly, Proc. A. I. E. E., April 1895, and is used by him to obtain the numerical values of $\cosh (x + iy)$, $\sinh (x + iy)$, which present themselves as the measures of certain vector quantities in the theory of alternating currents. (See Art. 37.) The chart is constructed for values of x and of y between 0 and 1.2; but it is available for all values of y, on account of the periodicity of the functions.

ART. 31. THE CATENARY.

A flexible inextensible string is suspended from two fixed points, and takes up a position of equilibrium under the action of gravity. It is required to find the equation of the curve in which it hangs.

Let w be the weight of unit length, and s the length of arc AP measured from the lowest point A; then ws is the weight of the portion AP. This is balanced by the terminal tensions, T acting in the tangent line at P, and H in the horizontal tangent.   Resolving horizontally and vertically gives

$$T \cos \phi = H, \quad T \sin \phi = ws,$$

in which ϕ is the inclination of the tangent at P; hence

$$\tan \phi = \frac{ws}{H} = \frac{s}{c},$$

where c is written for H/w, the length whose weight is the constant horizontal tension ; therefore

$$\frac{dy}{dx} = \frac{s}{c}, \quad \frac{ds}{dx} = \sqrt{1 + \frac{s^2}{c^2}}, \quad \frac{dx}{c} = \frac{ds}{\sqrt{s^2 + c^2}},$$

$$\frac{x}{c} = \sinh^{-1} \frac{s}{c}, \quad \sinh \frac{x}{c} = \frac{s}{c} = \frac{dy}{dx}, \quad \frac{y}{c} = \cosh \frac{x}{c},$$

which is the required equation of the catenary, referred to an axis of x drawn at a distance c below A.

The following trigonometric method illustrates the use of the gudermanian: The "intrinsic equation," $s = c \tan \phi$, gives $ds = c \sec^2 \phi \, d\phi$; hence $dx, = ds \cos \phi, = c \sec \phi \, d\phi$; $dy, = ds \sin \phi, = c \sec \phi \tan \phi \, d\phi$; thus $x = c \operatorname{gd}^{-1} \phi$, $y = c \sec \phi$; whence $y/c = \sec \phi = \sec \operatorname{gd} x/c = \cosh x/c$; and

$$s/c = \tan \operatorname{gd} x/c = \sinh x/c.$$

Numerical Exercise.—A chain whose length is 30 feet is suspended from two points 20 feet apart in the same horizontal; find the parameter c, and the depth of the lowest point.

The equation $s/c = \sinh x/c$ gives $15/c = \sinh 10/c$, which, by putting $10/c = z$, may be written $1.5z = \sinh z$. By examining the intersection of the graphs of $y = \sinh z$, $y = 1.5z$, it appears that the root of this equation is $z = 1.6$, nearly. To find a closer approximation to the root, write the equation in the form $f(z) = \sinh z - 1.5z = 0$, then, by the tables,

$$f(1.60) = 2.3756 - 2.4000 = -.0244,$$

$$f(1.62) = 2.4276 - 2.4300 = -.0024,$$

$$f(1.64) = 2.4806 - 2.4600 = +.0206;$$

whence, by interpolation, it is found that $f(1.6221) = 0$, and $z = 1.6221$, $c = 10/z = 6.1649$. The ordinate of either of the fixed points is given by the equation

$$y/c = \cosh x/c = \cosh 10/c = \cosh 1.6221 = 2.6306,$$

from tables; hence $y = 16.2174$, and required depth of the vertex $= y - c = 10.0525$ feet.*

Prob. 91. In the above numerical problem, find the inclination of the terminal tangent to the horizon.

Prob. 92. If a perpendicular MN be drawn from the foot of the ordinate to the tangent at P, prove that MN is equal to the constant c, and that NP is equal to the arc AP. Hence show that the locus of N is the involute of the catenary, and has the property that the length of the tangent, from the point of contact to the axis of x, is constant. (This is the characteristic property of the tractory).

Prob. 93. The tension T at any point is equal to the weight of a portion of the string whose length is equal to the ordinate y of that point.

Prob. 94. An arch in the form of an inverted catenary† is 30 feet wide and 10 feet high; show that the length of the arch can be obtained from the equations $\cosh z - \dfrac{2}{3}z = 1$, $2s = \dfrac{30}{z}\sinh z$.

---

* See a similar problem in Chap. I, Art. 7.

† For the theory of this form of arch, see "Arch" in the Encyclopædia Britannica.

ART. 32. CATENARY OF UNIFORM STRENGTH.

If the area of the normal section at any point be made proportional to the tension at that point, there will then be a constant tension per unit of area, and the tendency to break will be the same at all points. To find the equation of the curve of equilibrium under gravity, consider the equilibrium of an element PP' whose length is ds, and whose weight is $g\rho\omega ds$, where ω is the section at P, and ρ the uniform density. This weight is balanced by the difference of the vertical components of the tensions at P and P', hence

$$d(T \sin \phi) = g\rho\omega ds,$$

$$d(T \cos \phi) = 0;$$

therefore $T \cos \phi = H$, the tension at the lowest point, and $T = H \sec \phi$. Again, if ω_0 be the section at the lowest point, then by hypothesis $\omega/\omega_0 = T/H = \sec \phi$, and the first equation becomes

$$Hd(\sec \phi \sin \phi) = g\rho\omega_0 \sec \phi ds,$$

or

$$c\, d \tan \phi = \sec \phi ds,$$

where c stands for the constant $H/g\rho\omega_0$, the length of string (of section ω_0) whose weight is equal to the tension at the lowest point; hence,

$$ds = c \sec \phi\, d\phi, \quad s/c = \text{gd}^{-1}\phi,$$

the intrinsic equation of the catenary of uniform strength.

Also $dx = ds \cos \phi = c\, d\phi$, $dy = ds \sin \phi = c \tan \phi\, d\phi$;

hence $x = c\phi$, $y = c \log \sec \phi$,

and thus the Cartesian equation is

$$y/c = \log \sec x/c,$$

in which the axis of x is the tangent at the lowest point.

Prob. 95. Using the same data as in Art. 31, find the parameter c and the depth of the lowest point. (The equation $x/c = \text{gd}\, s/c$ gives $10/c = \text{gd}\, 15/c$, which, by putting $15/c = z$, becomes gd $z = \frac{2}{3}z$. From the graph it is seen that z is nearly 1.7. If $f(z) = $ gd $z - \frac{2}{3}z$, then, from the tables of the gudermanian at the end of this chapter,

$$f(1.70) = 1.1780 - 1.1333 = + .0447,$$

$$f(1.75) = 1.1796 - 1.1667 = + .0129,$$

$$f(1.80) = 1.1804 - 1.2000 = - .0196,$$

whence, by interpolation, $z = 1.7698$ and $c = 8.4755$. Again, $y/c = \log_e \sec x/c$; but $x/c = 10/c = 1.1799$; and 1.1799 radians $= 67° 36' 29''$; hence $y = 8.4755 \times .41914 \times 2.3026 = 8.1798$, the required depth.)

Prob. 96. Find the inclination of the terminal tangent.

Prob. 97. Show that the curve has two vertical asymptotes.

Prob. 98. Prove that the law of the tension T, and of the section ω, at a distance s, measured from the lowest point along the curve, is

$$\frac{T}{H} = \frac{\omega}{\omega_0} = \cosh \frac{s}{c};$$

and show that in the above numerical example the terminal section is 2.85 times the minimum section.

Prob. 99. Prove that the radius of curvature is given by $\rho = c \cosh s/c$. Also that the weight of the arc s is given by $W = H \sinh s/c$, in which s is measured from the vertex.

ART. 33. THE ELASTIC CATENARY.

An elastic string of uniform section and density in its natural state is suspended from two points. Find its equation of equilibrium.

Let the element $d\sigma$ stretch into ds; then, by Hooke's law, $ds = d\sigma(1 + \lambda T)$, where λ is the elastic constant of the string; hence the weight of the stretched element ds, $= g\rho\omega d\sigma$, $= g\rho\omega ds/(1 + \lambda T)$. Accordingly, as before,

$$d(T \sin \phi) = g\rho\omega ds/(1 + \lambda T),$$

and $$T \cos \phi = H = g\rho\omega c,$$

hence $$cd(\tan \phi) = ds/(1 + \mu \sec \phi),$$

in which μ stands for λH, the extension at the lowest point;

therefore $ds = c(\sec^3 \phi + \mu \sec^3 \phi)d\phi,$

$$s/c = \tan \phi + \tfrac{1}{2}\mu(\sec \phi \tan \phi + \mathrm{gd}^{-1} \phi),$$

which is the intrinsic equation of the curve, and reduces to that of the common catenary when $\mu = 0$. The coordinates x, y may be expressed in terms of the single parameter ϕ by putting $dx = ds \cos \phi = c(\sec^2 \phi + \mu \sec^3 \phi)d\phi,$

$dy = ds \sin \phi = c(\sec^3 \phi + \mu \sec^3 \phi) \sin \phi \, d\phi.$ Whence

$$x/c = \mathrm{gd}^{-1} \phi + \mu \tan \phi, \quad y/c = \sec \phi + \tfrac{1}{2}\mu \tan^2 \phi.$$

These equations are more convenient than the result of eliminating ϕ, which is somewhat complicated.

ART. 34. THE TRACTORY.*

To find the equation of the curve which possesses the property that the length of the tangent from the point of contact to the axis of x is constant.

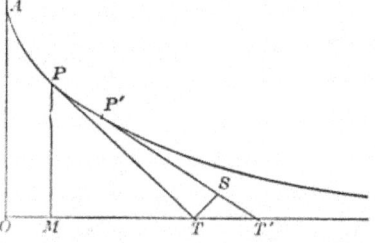

Let PT, $P'T'$ be two consecutive tangents such that $PT = P'T' = c$, and let $OT = t$; draw TS perpendicular to $P'T'$; then if $PP' = ds$, it is evident that ST' differs from ds by an infinitesimal of a higher order.   Let PT make an angle ϕ with OA, the axis of y; then (to the first order of infinitesimals) $PTd\phi = TS = TT' \cos \phi$; that is,

$$cd\phi = \cos \phi dt, \quad t = c \, \mathrm{gd}^{-1}\phi,$$

$$x, = t - c \sin \phi, = c(\mathrm{gd}^{-1} \phi - \sin \phi), \quad y = c \cos \phi.$$

This is a convenient single-parameter form, which gives all

---

* This curve is used in Schiele's anti-friction pivot (Minchin's Statics, Vol. 1, p. 242); and in the theory of the skew circular arch, the horizontal projection of the joints being a tractory. (See " Arch," Encyclopædia Britannica.) The equation $\phi = \mathrm{gd}\ t/c$ furnishes a convenient method of plotting the curve.

values of x, y as ϕ increases from o to $\frac{1}{2}\pi$. The value of s, expressed in the same form, is found from the relation

$$ds = ST' = dt \sin \phi = c \tan \phi\, d\phi, \quad s = c \log_e \sec \phi.$$

At the point A, $\phi = 0$, $x = 0$, $s = 0$, $t = 0$, $y = c$. The Cartesian equation, obtained by eliminating ϕ, is

$$\frac{x}{c} = \mathrm{gd}^{-1}\left(\cos^{-1}\frac{y}{c}\right) - \sin\left(\cos^{-1}\frac{y}{c}\right) = \cosh^{-1}\frac{c}{y} - \sqrt{1 - \frac{y^2}{c^2}}.$$

If u be put for t/c, and be taken as independent variable, $\phi = \mathrm{gd}\, u$, $x/c = u - \tanh u$, $y/c = \mathrm{sech}\, u$, $s/c = \log \cosh u$.

Prob. 100. Given $t = 2c$, show that $\phi = 75° \, 35'$, $s = 1.3249c$, $y = .2658c$, $x = 1.0360c$. At what point is $t = c$?

Prob. 101. Show that the evolute of the tractory is the catenary. (See Prob. 92.)

Prob. 102. Find the radius of curvature of the tractory in terms of ϕ; and derive the intrinsic equation of the involute.

ART. 35. THE LOXODROME.

On the surface of a sphere a curve starts from the equator in a given direction and cuts all the meridians at the same angle. To find its equation in latitude-and-longitude coordinates:

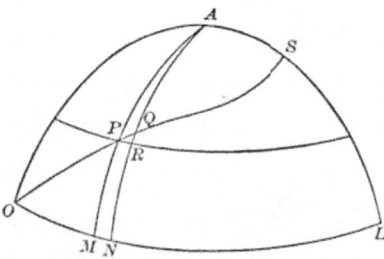

Let the loxodrome cross two consecutive meridians AM, AN in the points P, Q; let PR be a parallel of latitude; let $OM = x$, $MP = y$, $MN = dx$, $RQ = dy$, all in radian measure; and let the angle $MOP = RPQ = \alpha$; then

$$\tan \alpha = RQ/PR, \quad \text{but} \quad PR = MN \cos MP,*$$

hence $dx \tan \alpha = dy \sec y$, and $x \tan \alpha = \mathrm{gd}^{-1} y$, there being no integration-constant since y vanishes with x; thus the required equation is

$$y = \mathrm{gd}\,(x \tan \alpha).$$

* Jones, Trigonometry (Ithaca, 1890), p. 185.

To find the length of the arc OP: Integrate the equation

$$ds = \acute{a}y \csc \alpha, \quad \text{whence } s = y \csc \alpha.$$

To illustrate numerically, suppose a ship sails northeast, from a point on the equator, until her difference of longitude is 45°, find her latitude and distance:

Here $\tan \alpha = 1$, and $y = \text{gd } x = \text{gd } \frac{1}{4}\pi = \text{gd } (.7854) = .7152$ radians: $s = y \sqrt{2} = 1.0114$ radii. The latitude in degrees is 40.980.

If the ship set out from latitude y_1, the formula must be modified as follows: Integrating the above differential equation between the limits (x_1, y_1) and (x_2, y_2) gives

$$(x_2 - x_1) \tan \alpha = \text{gd}^{-1}y_2 - \text{gd}^{-1}y_1;$$

hence $\text{gd}^{-1}y_2 = \text{gd}^{-1}y_1 + (x_2 - x_1) \tan \alpha$, from which the final latitude can be found when the initial latitude and the difference of longitude are given. The distance sailed is equal to $(y_2 - y_1) \csc \alpha$ radii, a radius being $60 \times 180/\pi$ nautical miles.

Mercator's Chart.—In this projection the meridians are parallel straight lines, and the loxodrome becomes the straight line $y' = x \tan \alpha$, hence the relations between the coordinates of corresponding points on the plane and sphere are $x' = x$, $y' = \text{gd}^{-1}y$. Thus the latitude y is magnified into $\text{gd}^{-1}y$, which is tabulated under the name of "meridional part for latitude y"; the values of y and of y' being given in minutes. A chart constructed accurately from the tables can be used to furnish graphical solutions of problems like the one proposed above.

Prob. 103. Find the distance on a rhumb line between the points (30° N, 20° E) and (30° S, 40° E).

ART. 36. COMBINED FLEXURE AND TENSION.

A beam that is built-in at one end carries a load P at the other, and is also subjected to a horizontal tensile force Q applied at the same point; to find the equation of the curve assumed by its neutral surface: Let x, y be any point of the elastic curve, referred to the free end as origin, then the bending moment for this point is $Qy - Px$.   Hence, with the usual notation of the theory of flexure,*

$$EI\frac{d^2y}{dx^2} = Qy - Px, \quad \frac{d^2y}{dx^2} = n^2(y - mx), \quad \left[m = \frac{P}{Q}, \quad n^2 = \frac{Q}{EI} \right.$$

which, on putting $y - mx = u$, and $d^2y/dx^2 = d^2u/dx^2$, becomes

$$\frac{d^2u}{dx^2} = n^2u,$$

whence $u = A \cosh nx + B \sinh nx,$

that is, $y = mx + A \cosh nx + B \sinh nx.$

The arbitrary constants A, B are to be determined by the terminal conditions.   At the free end $x = 0, y = 0$; hence A must be zero, and

$$y = mx + B \sinh nx,$$
$$\frac{dy}{dx} = m + nB \cosh nx;$$

but at the fixed end, $x = l$, and $dy/dx = 0$, hence

$$B = -m/n \cosh nl,$$

and accordingly

$$y = mx - \frac{m \sinh nx}{n \cosh nl}.$$

To obtain the deflection of the loaded end, find the ordinate of the fixed end by putting $x = l$, giving

$$\text{deflection} = m(l - \frac{1}{n} \tanh nl).$$

Prob. 104. Compute the deflection of a cast-iron beam, 2×2 inches section, and 6 feet span, built-in at one end and carrying a load of 100 pounds at the other end, the beam being subjected to a horizontal tension of 8000 pounds.   [In this case $I = 4/3$, $E = 15 \times 10^6$, $Q = 8000$, $P = 100$; hence $n = 1/50$, $m = 1/80$, deflection $= \frac{1}{80}(72 - 50 \tanh 1.44) = \frac{1}{80}(72 - 44.69) = .341$ inches.]

* Merriman, Mechanics of Materials (New York, 1895), pp. 70–77, 267–269.

Prob. 105. If the load be uniformly distributed over the beam, say w per linear unit, prove that the differential equation is

$$EI\frac{d^2y}{dx^2} = Qy - \tfrac{1}{2}wx^2, \quad \text{or} \quad \frac{d^2y}{dx^2} = n^2(y - mx^2),$$

and that the solution is $y = A \cosh nx + B \sinh nx + mx^2 + \dfrac{2m}{n^2}$.

Show also how to determine the arbitrary constants.

ART. 37. ALTERNATING CURRENTS.*

In the general problem treated the cable or wire is regarded as having resistance, distributed capacity, self-induction, and leakage ; although some of these may be zero in special cases. The line will also be considered to feed into a receiver circuit of any description; and the general solution will include the particular cases in which the receiving end is either grounded or insulated. The electromotive force may, without loss of generality, be taken as a simple harmonic function of the time, because any periodic function can be expressed in a Fourier series of simple harmonics.† The E.M.F. and the current, which may differ in phase by any angle, will be supposed to have given values at the terminals of the receiver circuit ; and the problem then is to determine the E.M.F. and current that must be kept up at the generator terminals ; and also to express the values of these quantities at any intermediate point, distant x from the receiving end ; the four line-constants being supposed known, viz.:

$R =$ resistance, in ohms per mile,

$L =$ coefficient of self-induction, in henrys per mile,

$C =$ capacity, in farads per mile,

$G =$ coefficient of leakage, in mhos per mile.‡

It is shown in standard works § that if any simple harmonic

---

* See references in foot-note Art. 27.   † Chapter V, Art. 8.

‡ Kennelly denotes these constants by r, l, c, g. Steinmetz writes s for ωL, κ for ωC, θ for G, and he uses C for current.

§ Thomson and Tait, Natural Philosophy, Vol. I. p. 40; Rayleigh, Theory of Sound, Vol. I. p. 20; Bedell and Crehore, Alternating Currents, p. 214.

function $a \sin (\omega t + \theta)$ be represented by a vector of length a and angle θ, then two simple harmonics of the same period $2\pi/\omega$, but having different values of the phase-angle θ, can be combined by adding their representative vectors. Now the E.M.F. and the current at any point of the circuit, distant x from the receiving end, are of the form

$$e = e_1 \sin (\omega t + \theta), \quad i = i_1 \sin (\omega t + \theta'), \tag{64}$$

in which the maximum values e_1, i_1, and the phase-angles θ, θ', are all functions of x. These simple harmonics will be represented by the vectors $e_1/\underline{\theta}$, $i_1/\underline{\theta'}$; whose numerical measures are the complexes $e_1 (\cos \theta + j \sin \theta)$*, $i_1 (\cos \theta' + j \sin \theta')$, which will be denoted by $\bar{e}$, $\bar{i}$. The relations between $\bar{e}$ and $\bar{i}$ may be obtained from the ordinary equations †

$$\frac{di}{dx} = Ge + C\frac{de}{dt}, \quad \frac{de}{dx} = Ri + L\frac{di}{dt}; \tag{65}$$

for, since $de/dt = \omega e_1 \cos (\omega t + \theta) = \omega e_1 \sin (\omega t + \theta + \tfrac{1}{2}\pi)$, then de/dt will be represented by the vector $\omega e_1 /\underline{\theta + \tfrac{1}{2}\pi}$; and di/dx by the sum of the two vectors $Ge_1 /\underline{\theta}$, $C\omega e_1 /\underline{\theta + \tfrac{1}{2}\pi}$; whose numerical measures are the complexes $G\bar{e}$, $j\omega C\bar{e}$; and similarly for de/dx in the second equation; thus the relations between the complexes $\bar{e}$, $\bar{i}$ are

$$\frac{d\bar{i}}{dx} = (G + j\omega C)\bar{e}, \quad \frac{d\bar{e}}{dx} = (R + j\omega L)\bar{i}. \tag{66‡}$$

---

* In electrical theory the symbol j is used, instead of i, for $\sqrt{-1}$.

† Bedell and Crehore, Alternating Currents, p. 181. The sign of dx is changed, because x is measured from the receiving end. The coefficient of leakage, G, is usually taken zero, but is here retained for generality and symmetry.

‡ These relations have the advantage of not involving the time. Steinmetz derives them from first principles without using the variable t. For instance, he regards $R + j\omega L$ as a generalized resistance-coefficient, which, when applied to i, gives an E.M.F., part of which is in phase with i, and part in quadrature with i. Kennelly calls $R + j\omega L$ the conductor impedance; and $G + j\omega C$ the dielectric admittance; the reciprocal of which is the dielectric impedance.

Differentiating and substituting give

$$\frac{d^2\bar{e}}{dx^2} = (R + j\omega L)(G + j\omega C)\bar{e}, \\
\frac{d^2\bar{\imath}}{dx^2} = (R + j\omega L)(G + j\omega C)\bar{\imath}, \Bigg\} \qquad (67)$$

and thus $\bar{e}$, $\bar{\imath}$ are similar functions of x, to be distinguished only by their terminal values.

It is now convenient to define two constants m, m_1 by the equations *

$$m^2 = (R + j\omega L)(G + j\omega C), \ m_1^2 = (R + j\omega L)/(G + j\omega C); \ (68)$$

and the differential equations may then be written

$$\frac{d^2\bar{e}}{dx^2} = m^2\bar{e}, \quad \frac{d^2\bar{\imath}}{dx^2} = m^2\bar{\imath}, \qquad (69)$$

the solutions of which are †

$$\bar{e} = A \cosh mx + B \sinh mx, \quad \bar{\imath} = A' \cosh mx + B' \sinh mx,$$

wherein only two of the four constants are arbitrary; for substituting in either of the equations (66), and equating coefficients, give

$$(G + j\omega C) A = mB', \quad (G + j\omega C) B = mA',$$

whence $\qquad B' = A/m_1, \quad A' = B/m_1.$

Next let the assigned terminal values of $\bar{e}$, $\bar{\imath}$, at the receiver, be denoted by $\bar{E}$, $\bar{I}$; then putting $x = 0$ gives $\bar{E} = A$, $\bar{I} = A'$, whence $B = m_1\bar{I}$, $B' = \bar{E}/m_1$; and thus the general solution is

$$\bar{e} = \bar{E} \cosh mx + m_1\bar{I} \sinh mx, \\
\bar{\imath} = \bar{I} \cosh mx + \frac{1}{m_1}\bar{E} \sinh mx. \Bigg\} \qquad (70)$$

* The complex constants m, m_1, are written z, y by Kennelly; and the variable length x is written L_2. Steinmetz writes v for m.

† See Art. 14, Probs. 28–30; and Art. 27, foot-note.

If desired, these expressions could be thrown into the ordinary complex form $X + jY$, $X' + jY'$, by putting for the letters their complex values, and applying the addition-theorems for the hyperbolic sine and cosine. The quantities X, Y, X', Y' would then be expressed as functions of x; and the representative vectors of e, i, would be e_1 / θ, i_1 / θ', where $e_1^2 = X^2 + Y^2$, $i_1^2 = X'^2 + Y'^2$, $\tan \theta = Y/X$, $\tan \theta' = Y'/X'$.

For purposes of numerical computation, however, the formulas (70) are the most convenient, when either a chart,* or a table,† of cosh u, sinh u, is available, for complex values of u.

Prob. 106.‡ Given the four line-constants: $R = 2$ ohms per mile, $L = 20$ millihenrys per mile, $C = 1/2$ microfarad per mile, $G = 0$; and given ω, the angular velocity of E.M.F. to be 2000 radians per second; then

$\omega L = 40$ ohms, conductor reactance per mile;

$R + j\omega L = 2 + 40j$ ohms, conductor impedance per mile;

$\omega C = .001$ mho, dielectric susceptance per mile;

$G + j\omega C = .001j$ mho, dielectric admittance per mile;

$(G + j\omega C)^{-1} = -1000j$ ohms, dielectric impedance per mile;

$m^2 = (R + j\omega L)(G + j\omega C) = .04 + .002j$, which is the measure of $.04005 / 177° 8'$; therefore

$m = $ measure of $.2001 / 88° 34' = .0050 + .2000j$, an abstract coefficient per mile, of dimensions $[\text{length}]^{-1}$,

$mm_1 = m/(G + j\omega C) = 200 - 5j$ ohms per mile.

Next let the assigned terminals conditions at the receiver be: $I = 0$ (line insulated); and $E = 1000$ volts, whose phase may be taken as the standard (or zero) phase; then at any distance x, by (70),

$$\bar{e} = E \cosh mx, \qquad \bar{i} = \frac{E}{m_1} \sinh mx,$$

in which mx is an abstract complex.

Suppose it is required to find the E.M.F. and current that must be kept up at a generator 100 miles away; then

---

* Art. 30, foot-note.　　　　　　　† See Table II.

‡ The data for this example are taken from Kennelly's article.

$\bar{e} = 1000 \cosh (.5 + 20j), \quad \bar{\imath} = 200(40 - j)^{-1} \sinh (.5 + 20j),$
but, by Prob. 89, $\cosh (.5 + 20j) = \cosh (.5 + 20j - 6\pi j)$

$$= \cosh (.5 + 1.15j) = .4600 + .4750j$$

obtained from Table II, by interpolation between $\cosh (.5 + 1.1j)$
and $\cosh (.5 + 1.2j)$; hence

$$\bar{e} = 460 + 475j = e_1(\cos \theta + j \sin \theta),$$

where $\log \tan \theta = \log 475 - \log 460 = .0139$, $\theta = 45° 55'$, and
$e_1 = 460 \sec \theta = 625.9$ volts, the required E.M.F.

Similarly $\sinh (.5 + 20j) = \sinh (.5 + 1.15j) = .2126 + 1.0280j$,
and hence

$$\bar{\imath} = \frac{200}{1601}(100 + j)(.2126 + 1.028j) = \frac{1}{1601}(4046 + 2060j)$$
$$= i_1(\cos \theta' + j \sin \theta'),$$

where $\log \tan \theta' = 9.70684$, $\theta' = 26° 59'$, $i_1 = 4046 \sec \theta'/1601 = 2.77$
amperes, the phase and magnitude of required current.

Next let it be required to find e at $x = 8$; then

$$\bar{e} = 1000 \cosh (.04 + 1.6j) = 1000j \sinh (.04 + .03j),$$

by subtracting $\frac{1}{2}\pi j$, and applying Prob. 89.   Interpolation between $\sinh (0 + 0j)$ and $\sinh (0 + .1j)$ gives

$$\sinh (0 + .03j) = 00000 + .02995j.$$

Similarly $\sinh (.1 + .03j) = .10004 + .03004j.$

Interpolation between the last two gives

$$\sinh (.04 + .03j) = .04002 + .02999j.$$

Hence $\bar{e} = j(40.02 + 29.99j) = -29.99 + 40.02j = e_1(\cos \theta + j \sin \theta)$,
where

$\log \tan \theta = .12530$, $\theta = 126° 51'$, $e_1 = -29.99 \sec 126° 51' = 50.01$
volts.

Again, let it be required to find e at $x = 16$; here

$$\bar{e} = 1000 \cosh (.08 + 3.2j) = -1000 \cosh (.08 + .06j),$$

but $\cosh (0 + .06j) = .9970 + 0j$, $\cosh (.1 + .06j) = 1.0020 + .006j$;

hence $\cosh (.08 + .06j) = 1.0010 + .0048j$,

and $\bar{e} = -1001 + 4.8j = e_1(\cos \theta + j \sin \theta),$

where $\theta = 180° 17'$, $e_1 = 1001$ volts.   Thus at a distance of about
16 miles the E.M.F. is the same as at the receiver, but in opposite

phase. Since $\bar{e}$ is proportional to cosh $(.005 + .2j)x$, the value of x for which the phase is exactly $180°$ is $\pi/.2 = 15.7$. Similarly the phase of the E.M.F. at $x = 7.85$ is $90°$. There is agreement in phase at any two points whose distance apart is 31.4 miles.

In conclusion take the more general terminal conditions in which the line feeds into a receiver circuit, and suppose the current is to be kept at 50 amperes, in a phase $40°$ in advance of the electromotive force; then $\bar{I} = 50(\cos 40° + j \sin 40°) = 38.30 + 32.14 j$; and substituting the constants in (70) gives

$$\bar{e} = 1000 \cosh (.005 + .2j)x + (7821 + 6236j) \sinh (.005 + .2j)x$$
$$= 460 + 475j - 4748 + 9366j = -4288 + 9841j = e_1(\cos \theta + j \sin \theta),$$

where $\theta = 113° 33'$, $e_1 = 10730$ volts, the E.M.F. at sending end. This is 17 times what was required when the other end was insulated.

Prob. 107. If the receiving end be grounded, that is if $E = 0$; and if a current of 10 amperes be caused to flow to ground; find the E.M.F. and current to be kept up at the generator. Also compute these quantities, and their phases, at the distances 7.85, 15.7, 31.42, 94.25 miles from the receiver.

Prob. 108. If self-induction and capacity be zero, and the receiving end be insulated, show that the graph of the electromotive force is a catenary.

Prob. 109. Neglecting leakage and capacity, prove that the solution of equations (66) is $\bar{i} = \bar{I}$, $\bar{e} = \bar{E} + (R + j\omega L)\bar{I}x$.

Prob. 110. If x be measured from the sending end, show how equations (65), (66) are to be modified; and prove that

$$\bar{e} = \bar{E}_0 \cosh mx - m_1\bar{I}_0 \sinh mx, \quad \bar{i} = \bar{I}_0 \cosh mx - \frac{1}{m_1}\bar{E}_0 \sinh mx,$$

where $\bar{E}_0$, $\bar{I}_0$ refer to the sending end.

ART. 38.   MISCELLANEOUS APPLICATIONS.

1. The length of the arc of the logarithmic curve $y = a^x$ is $s = \frac{1}{2}(\cosh u + \log \tanh \frac{1}{2}u)$, in which $M = 1/\log a$, $\sinh u = y/M$.

2. The length of arc of the spiral of Archimedes $r = a\theta$ is $s = \frac{1}{2}a(\sinh 2u + 2u)$, where $\sinh u = \theta$.

3. In the hyperbola $x^2/a^2 - y^2/b^2 = 1$ the radius of curvature is $\rho = (a^2 \sinh^2 u + b^2 \cosh^2 u)^{\frac{3}{2}}/ab$; in which u is the measure of the sector AOP, i.e. $\cosh u = x/a$, $\sinh u = y/b$.

4. In an oblate spheroid, the superficial area of the zone

between the equator and a parallel plane at a distance y is $S = \pi b^2(\sinh 2u + 2u)/2e$, wherein b is the axial radius, e the eccentricity, $u = ey/p$, and p parameter of generating ellipse.

5. The length of the arc of the parabola $y^2 = 2px$, measured from the vertex of the curve, is $l = \frac{1}{2}p(\sinh 2u + 2u)$, in which $\sinh u = y/p = \tan \phi$, where ϕ is the inclination of the terminal tangent to the initial one.

6. The centre of gravity of this arc is given by

$$3l\bar{x} = p^2(\cosh^3 u - 1), \quad 64l\bar{y} = p^2(\sinh 4u - 4u);$$

and the surface of a paraboloid of revolution is $S = 2\pi \bar{y}l$.

7. The moment of inertia of the same arc about its terminal ordinate is $I = \mu[xl(x - 2\bar{x}) + \frac{1}{64}p^3N]$, where μ is the mass of unit length, and

$$N = u - \frac{1}{4} \sinh 2u - \frac{1}{4} \sinh 4u + \frac{1}{12} \sinh 6u.$$

8. The centre of gravity of the arc of a catenary measured from the lowest point is given by

$$4l\bar{y} = c^2(\sinh 2u + 2u), \; lx = c^2(u \sinh u - \cosh u + 1),$$

in which $u = x/c$; and the moment of inertia of this arc about its terminal abscissa is

$$I = \mu c^3(\frac{1}{12} \sinh 3u + \frac{3}{4} \sinh u - u \cosh u).$$

9. Applications to the vibrations of bars are given in Rayleigh, Theory of Sound, Vol. I, art. 170; to the torsion of prisms in Love, Elasticity, pp. 166–74; to the flow of heat and electricity in Byerly, Fourier Series, pp. 75–81; to wave motion in fluids in Rayleigh, Vol. I, Appendix, p. 477, and in Bassett, Hydrodynamics, arts. 120, 384; to the theory of potential in Byerly p. 135, and in Maxwell, Electricity, arts. 172–4; to Non-Euclidian geometry and many other subjects in Günther, Hyperbelfunktionen, Chaps. V and VI. Several numerical examples are worked out in Laisant, Essai sur les fonctions hyperboliques.

ART. 39. EXPLANATION OF TABLES.

In Table I the numerical values of the hyperbolic functions sinh u, cosh u, tanh u are tabulated for values of u increasing from 0 to 4 at intervals of .02. When u exceeds 4, Table IV may be used.

Table II gives hyperbolic functions of complex arguments, in which

$$\cosh (x \pm iy) = a \pm ib, \quad \sinh (x \pm iy) = c \pm id,$$

and the values of a, b, c, d are tabulated for values of x and of y ranging separately from 0 to 1.5 at intervals of .1. When interpolation is necessary it may be performed in three stages.   For example, to find $\cosh (.82 + 1.34i)$: First find $\cosh (.82 + 1.3i)$, by keeping y at 1.3 and interpolating between the entries under $x = .8$ and $x = .9$; next find $\cosh (.82 + 1.4i)$, by keeping y at 1.4 and interpolating between the entries under $x = .8$ and $x = .9$, as before; then by interpolation between $\cosh (.82 + 1.3i)$ and $\cosh (.82 + 1.4i)$ find $\cosh(.82 + 1.34i)$, in which x is kept at .82.   The table is available for all values of y, however great, by means of the formulas

$$\sinh (x + 2i\pi) = \sinh x, \quad \cosh (x + 2i\pi) = \cosh x, \text{ etc.}$$

It does not apply when x is greater than 1.5, but this case seldom occurs in practice.   This table can also be used as a complex table of circular functions, for

$$\cos (y \pm ix) = a \mp ib, \quad \sin (y \pm ix) = d \pm ic\,;$$

and, moreover, the exponential function is given by

$$\exp (\pm x \pm iy) = a \pm c \pm i(b \pm d),$$

in which the signs of c and d are to be taken the same as the sign of x, and the sign of i on the right is to be the product of the signs of x and of i on the left.

Table III gives the values of $v = \operatorname{gd} u$, and of the gudermanian angle $\theta = 180° v/\pi$, as u changes from 0 to 1 at intervals of .02, from 1 to 2 at intervals of .05, and from 2 to 4 at intervals of .1.

In Table IV are given the values of gd u, log sinh u, log cosh u, as u increases from 4 to 6 at intervals of .1, from 6 to 7 at intervals of .2, and from 7 to 9 at intervals of .5.

In the rare cases in which more extensive tables are necessary, reference may be made to the tables* of Gudermann, Glaisher, and Geipel and Kilgour. In the first the Gudermanian angle (written k) is taken as the independent variable, and increases from 0 to 100 grades at intervals of .01, the corresponding value of u (written Lk) being tabulated. In the usual case, in which the table is entered with the value of u, it gives by interpolation the value of the gudermanian angle, whose circular functions would then give the hyperbolic functions of u. When u is large, this angle is so nearly right that interpolation is not reliable. To remedy this inconvenience Gudermann's second table gives directly log sinh u, log cosh u, log tanh u, to nine figures, for values of u varying by .001 from 2 to 5, and by .01 from 5 to 12.

Glaisher has tabulated the values of e^x and e^{-x}, to nine significant figures, as x varies by .001 from 0 to .1, by .01 from 0 to 2, by .1 from 0 to 10, and by 1 from 0 to 500. From these the values of cosh x, sinh x are easily obtained.

Geipel and Kilgour's handbook gives the values of cosh x, sinh x, to seven figures, as x varies by .01 from 0 to 4.

There are also extensive tables by Forti, Gronau, Vassal, Callet, and Hoüel; and there are four-place tables in Byerly's Fourier Series, and in Wheeler's Trigonometry.

In the following tables a dash over a final digit indicates that the number has been increased.

* Gudermann in Crelle's Journal, vols. 6–9, 1831–2 (published separately under the title Theorie der hyperbolischen Functionen, Berlin, 1833). Glaisher in Cambridge Phil. Trans., vol. 13, 1881. Geipel and Kilgour's Electrical Handbook.

TABLE I.—HYPERBOLIC FUNCTIONS.

u.	sinh u.	cosh u.	tanh u.	u.	sinh u.	cosh u.	tanh u.
.00	.0000	1.0000	.0000	1.00	1.1752	1.5431	.7616
02	0200	1.0020	0200	1.02	1.2063	1.5669	7699
04	0400	1.0008	0400	1.04	1.2379	1.5913	7779
06	0600	1.0018	0599	1.06	1.2700	1.6164	7857
08	0801	1.0032	0798	1.08	1.3025	1.6421	7932
.10	.1002	1.0050	.0997	1.10	1.3356	1.6685	.8005
12	1203	1.0072	1194	1.12	1.3693	1.6956	8076
14	1405	1.0098	1391	1.14	1.4035	1.7233	8144
16	1607	1.0128	1586	1.16	1.4382	1.7517	8210
18	1810	1.0162	1781	1.18	1.4735	1.7808	8275
.20	.2013	1.0201	.1974	1.20	1.5095	1.8107	.8337
22	2218	1.0243	2165	1.22	1.5460	1.8412	8397
24	2423	1.0289	2355	1.24	1.5831	1.8725	8455
26	2629	1.0340	2543	1.26	1.6209	1.9045	8511
28	2837	1.0395	2729	1.28	1.6593	1.9373	8565
.30	.3045	1.0453	.2913	1.30	1.6984	1.9709	.8617
32	3255	1.0516	3095	1.32	1.7381	2.0053	8668
34	3466	1.0584	3275	1.34	1.7786	2.0404	8717
36	3678	1.0655	3452	1.36	1.8198	2.0764	8764
38	3892	1.0731	3627	1.38	1.8617	2.1132	8810
.40	.4108	1.0811	.3799	1.40	1.9043	2.1509	.8854
42	4325	1.0895	3969	1.42	1.9477	2.1894	8896
44	4543	1.0984	4136	1.44	1.9919	2.2288	8937
46	4764	1.1077	4301	1.46	2.0369	2.2691	8977
48	4986	1.1174	4462	1.48	2.0827	2.3103	9015
.50	.5211	1.1276	.4621	1.50	2.1293	2.3524	.9051
52	5438	1.1383	4777	1.52	2.1768	2.3955	9087
54	5666	1.1494	4930	1.54	2.2251	2.4395	9121
56	5897	1.1609	5080	1.56	2.2743	2.4845	9154
58	6131	1.1730	5227	1.58	2.3245	2.5305	9186
.60	.6367	1.1855	.5370	1.60	2.3756	2.5775	.9217
62	6605	1.1984	5511	1.62	2.4276	2.6255	9246
64	6846	1.2119	5649	1.64	2.4806	2.6746	9275
66	7090	1.2258	5784	1.66	2.5346	2.7247	9302
68	7336	1.2402	5915	1.68	2.5896	2.7760	9329
.70	.7586	1.2552	.6044	1.70	2.6456	2.8283	.9354
72	7838	1.2706	6169	1.72	2.7027	2.8818	9379
74	8094	1.2865	6291	1.74	2.7609	2.9364	9402
76	8353	1.3030	6411	1.76	2.8202	2.9922	9425
78	8615	1.3199	6527	1.78	2.8806	3.0492	9447
.80	.8881	1.3374	.6640	1.80	2.9425	3.1075	.9468
82	9150	1.3555	6751	1.82	3.0049	3.1669	9488
84	9423	1.3740	6858	1.84	3.0689	3.2277	9508
86	9700	1.3932	6963	1.86	3.1340	3.2897	9527
88	9981	1.4128	7064	1.88	3.2005	3.3530	9545
.90	1.0265	1.4331	.7163	1.90	3.2682	3.4177	.9562
92	1.0554	1.4539	7259	1.92	3.3372	3.4838	9579
94	1.0847	1.4753	7352	1.94	3.4075	3.5512	9595
96	1.1144	1.4973	7443	1.96	3.4792	3.6201	9611
98	1.1446	1.5199	7531	1.98	3.5523	3.6904	9626

TABLE I. HYPERBOLIC FUNCTIONS.

u.	sinh u.	cosh u.	tanh u.	u.	sinh u.	cosh u.	tanh u.
2.00	3.6269	3.7622	.9640	3.00	10.0179	10.0677	.99505
2.02	3.7028	3.8355	9654	3.02	10.2212	10.2700	99524
2.04	3.7803	3.9103	9667	3.04	10.4287	10.4765	99543
2.06	3.8593	3.9867	9680	3.06	10.6403	10.6872	99561
2.08	3.9398	4.0647	9693	3.08	10.8562	10.9022	99578
2.10	4.0219	4.1443	.9705	3.10	11.0765	11.1215	.99594
2.12	4.1056	4.2256	9716	3.12	11.3011	11.3453	99610
2.14	4.1909	4.3085	9727	3.14	11.5303	11.5736	99626
2.16	4.2779	4.3932	9737	3.16	11.7641	11.8065	99640
2.18	4.3666	4.4797	9748	3.18	12.0026	12.0442	99654
2.20	4.4571	4.5679	.9757	3.20	12.2459	12.2866	.99668
2.22	4.5494	4.6580	9767	3.22	12.4941	12.5340	99681
2.24	4.6434	4.7499	9776	3.24	12.7473	12.7864	99693
2.26	4.7394	4.8437	9785	3.26	13.0056	13.0440	99705
2.28	4.8372	4.9395	9793	3.28	13.2691	13.3067	99717
2.30	4.9370	5.0372	.9801	3.30	13.5379	13.5748	.99728
2.32	5.0387	5.1370	9809	3.32	13.8121	13.8483	99738
2.34	5.1425	5.2388	9816	3.34	14.0918	14.1273	99749
2.36	5.2483	5.3427	9823	3.36	14.3772	14.4120	99758
2.38	5.3562	5.4487	9830	3.38	14.6684	14.7024	99768
2.40	5.4662	5.5569	.9837	3.40	14.9654	14.9987	.99777
2.42	5.5785	5.6674	9843	3.42	15.2684	15.3011	99786
2.44	5.6929	5.7801	9849	3.44	15.5774	15.6095	99794
2.46	5.8097	5.8951	9855	3.46	15.8928	15.9242	99802
2.48	5.9288	6.0125	9861	3.48	16.2144	16.2453	99810
2.50	6.0502	6.1323	.9866	3.50	16.5426	16.5728	.99817
2.52	6.1741	6.2545	9871	3.52	16.8774	16.9070	99824
2.54	6.3004	6.3793	9876	3.54	17.2190	17.2480	99831
2.56	6.4293	6.5066	9881	3.56	17.5674	17.5958	99838
2.58	6.5607	6.6364	9886	3.58	17.9228	17.9507	99844
2.60	6.6947	6.7690	.9890	3.60	18.2854	18.3128	.99850
2.62	6.8315	6.9043	9895	3.62	18.6554	18.6822	99856
2.64	6.9709	7.0423	9899	3.64	19.0328	19.0590	99862
2.66	7.1132	7.1832	9903	3.66	19.4178	19.4435	99867
2.68	7.2583	7.3268	9906	3.68	19.8106	19.8358	99872
2.70	7.4063	7.4735	.9910	3.70	20.2113	20.2360	.99877
2.72	7.5572	7.6231	9914	3.72	20.6201	20.6443	99882
2.74	7.7112	7.7758	9917	3.74	21.0371	21.0609	99887
2.76	7.8683	7.9316	9920	3.76	21.4626	21.4859	99891
2.78	8.0285	8.0905	9923	3.78	21.8966	21.9194	99896
2.80	8.1919	8.2527	.9926	3.80	22.3394	22.3618	.99900
2.82	8.3586	8.4182	9929	3.82	22.7911	22.8131	99904
2.84	8.5287	8.5871	9932	3.84	23.2520	23.2735	99907
2.86	8.7021	8.7594	9935	3.86	23.7221	23.7432	99911
2.88	8.8791	8.9352	9937	3.88	24.2018	24.2224	99915
2.90	9.0596	9.1146	.9940	3.90	24.6911	24.7113	.99918
2.92	9.2437	9.2976	9942	3.92	25.1903	25.2101	99921
2.94	9.4315	9.4844	9944	3.94	25.6996	25.7190	99924
2.96	9.6231	9.6749	9947	3.96	26.2191	26.2382	99927
2.98	9.8185	9.8693	9949	3.98	26.7492	26.7679	99930

TABLE II.　VALUES OF COSH $(x + iy)$ AND SINH $(x + iy)$.

y	$x = 0$				$x = .1$			
	a	b	c	d	a	b	c	d
0	1.0000	0000	0000	.0000	1.0050	.00000	.10017̃	.0000
.1	0.9950	"	"	0998	1.0000̃	01000	09967	1000
.2	0.9801̃	"	"	1987̃	0.9850	01990̃	09817	1997̃
.3	0.9553	"	"	2955	0.9601	02960	09570̃	2970
.4	.9211̃	"	"	.3894	.9257̃	.03901	.09226	.3914
.5	8776	"	"	4794	8820̃	04802	08791̃	4818
.6	8253	"	"	5646	8295̃	05656	08267	5675̃
.7	7648	"	"	6442	7687̃	06453	07661	6474
.8	.6967	"	"	.7174̃	.7002̃	.07186̃	.06979̃	.7200
.9	6216	"	"	7833	6247̃	07847	06227̃	7872̃
1.0	5403	"	"	8415̃	5430	08429	05412	8457̃
1.1	4536	"	"	8912	4559̃	08927	04544	8957̃
1.2	.3624̃	"	"	.9320	.3645̃	.09336	.03630̃	0 9367̃
1.3	2675̃	"	"	9636̃	2688̃	09652̃	02680̃	0.9684
1.4	1700̃	"	"	9854	1708	09871	01703̃	0.9904̃
1.5	0707	"	"	9975̃	0711	09992̃	00709	1.0025
$\frac{1}{2}\pi$	0000	"	"	1.0000	0000	10017̃	00000	1.0050

y	$x = .4$				$x = .5$			
	a	b	c	d	a	b	c	d
0	1.0811̃	.0000	.4108̃	.0000	1.1276	.0000	.5211̃	.0000
.1	1.0756	0410	4087̃	1079	1.1220̃	0520	5185̃	1126
.2	1.0595	0816	4026̃	2148̃	1.1051̃	1025	5107	2240
.3	1.0328̃	1214	3924	3195	1.0773̃	1540	4978	3332
.4	.9957	.1600̃	.3783	.4210̃	1.0386̃	.2029	.4800̃	.4391
.5	9487	1969	3605̃	5183̃	0.9896	2498	4573	5406
.6	8922	2319	3390	6104	0.9306	2942	4301̃	6367
.7	8268	2646	3142̃	6964	0.8624	3357̃	3986̃	7264
.8	.7532̃	.2947	.2862̃	.7755	.7856	.3738̃	.3631̃	0.8089
.9	6720̃	3218	2553	8468	7009	4082̃	3239	0.8833
1.0	5841	3456	2219	9097̃	6093̃	4385	2815	0.9489
1.1	4904	3661̃	1863	9635	5115	4644	2364	1.0050
1.2	.3917	.3829̃	.1488	1.0076	.4086	.4857̃	.1888	1.0510
1.3	2892̃	3958̃	1099	1.0417	3016	5021	1394	1.0865
1.4	1838̃	4048	0698	1.0653	1917	5135	0886̃	1.1163
1.5	0765̃	4097	0291̃	1.0784̃	0798̃	5198	0369̃	1.1248
$\frac{1}{2}\pi$	0000	4108	0000	1.0811̃	0000	5211̃	0000	1.1276

TABLE II. VALUES OF COSH $(x + iy)$ AND SINH $(x + iy)$.

| | $x = .2.$ | | | | $x = .3$ | | | |
a	b	c	d	a	b	c	d	y
1.0201	.0000	.2013	.0000	1.0453	.0000	.3045	.0000	0
1.0150	0201	2003	1018	1.0401	0304	3030	1044	.1
0.9997	0400	1973	2027	1.0245	0605	2985	2077	.2
0.9745	0595	1923	3014	9987	0900	2909	3089	.3
.9395	.0784	.1854	.3972	.9628	.1186	.2805	.4071	.4
8952	0965	1767	4890	9174	1460	2672	5012	.5
8419	1137	1662	5760	8627	1719	2513	5903	.6
7802	1297	1540	6571	7995	1962	2329	6734	.7
.7107	.1444	.1403	.7318	.7283	.2184	.2122	.7498	.8
6341	1577	1252	7990	6498	2385	1893	8188	.9
5511	1694	1088	8584	5648	2562	1645	8796	1.0
4627	1795	0913	9091	4742	2714	1381	9316	1.1
.3696	.1877	.0730	0.9507	.3788	.2838	.1103	0.9743	1.2
2729	1940	0539	0.9829	2796	2934	0815	1.0072	1.3
1734	1984	0342	1.0052	1777	3001	0518	1.0301	1.4
0722	2008	0142	1.0175	0739	3038	0215	1.0427	1.5
0000	2013	0000	1.0201	0000	3045	0000	1.0453	$\frac{1}{2}\pi$

| | $x = .6$ | | | | $x = .7$ | | | |
a	b	c	d	a	b	c	d	y
1.1855	.0000	.6367	.0000	1.2552	.0000	.7586	.0000	0
1.1795	0636	6335	1183	1.2489	0757	7548	1253	.1
1.1618	1265	6240	2355	1.2301	1542	7435	2494	.2
1.1325	1881	6082	3503	1.1991	2242	7247	3709	.3
1.0918	.2479	.5864	.4617	1.1561	.2954	.6987	.4885	.4
1.0403	3052	5587	5684	1.1015	3637	6657	6018	.5
0.9784	3595	5255	6694	1.0359	4253	6261	7087	.6
0.9067	4101	4869	7637	0.9600	4887	5802	8086	.7
.8259	.4567	.4436	0.8504	.8745	.5442	.5285	0.9004	.8
7369	4987	3957	0.9286	7802	5942	4715	0.9832	.9
6405	5357	3440	0.9975	6782	6383	4099	1.0562	1.0
5377	5674	2888	1.0565	5693	6760	3441	1.1186	1.1
.4296	5934	.2307	1.1049	.4548	.7070	.2749	1.1699	1.2
3171	6135	1703	1.1422	3358	7309	2029	1.2094	1.3
2015	6274	1082	1.1682	2133	7475	1289	1.2369	1.4
0839	6351	0450	1.1825	0888	7567	0537	1.2520	1.5
0000	6367	0000	1.1855	0000	7586	0000	1.2552	$\frac{1}{2}\pi$

TABLE II. VALUES OF COSH $(x + iy)$ AND SINH $(x + iy)$.

y	x = .8				x = .9			
	a	b	c	d	a	b	c	d
0	1.3374	.0000	.8881	.0000	1.4331	.0000	1.0265	.0000
.1	1.3308	0887	8837	1335	1.4259	1025	1.0214	1431
.2	1.3108	1764	8704	2657	1.4045	2039	1.0061	2847
.3	1.2776	2625	8484	3952	1.3691	3034	0.9807	4235
.4	1.2319	3458	8180	.5208	1.3200	3997	.9455	.5581
.5	1.1737	4258	7794	6412	1.2577	4921	9008	6871
.6	1.1038	5015	7330	7552	1.1828	5796	8472	8092
.7	1.0229	5721	6793	8616	1.0961	6613	7851	9232
.8	.9318	.6371	.6188	0.9595	.9984	.7364	.7152	1.0280
.9	8314	6957	5521	1.0476	8908	8041	6381	1.1226
1.0	7226	7472	4798	1.1254	7743	8638	5546	1.2059
1.1	6067	7915	4028	1.1919	6500	9148	4656	1.2772
1.2	.4846	.8278	.3218	1.2465	.5193	0.9568	.3720	1.3357
1.3	3578	8557	2376	1.2887	3834	0.9891	2746	1.3809
1.4	2273	8752	1510	1.3180	2436	1.0124	1745	1.4122
1.5	0946	8859	0628	1.3341	1014	1.0239	0726	1.4295
½π	0000	.8881	0000	1.3374	0000	1.0265	0000	1.4331

y	x = 1.2				x = 1.3			
	a	b	c	d	a	b	c	d
0	1.8107	.0000	1.5095	.0000	1.9709	.0000	1.6984	.0000
.1	1.8016	1507	1.5019	1808	1.9611	1696	1.6899	1968
.2	1.7746	2999	1.4794	3598	1.9316	3374	1.6645	3916
.3	1.7298	4461	1.4420	5351	1.8829	5019	1.6225	5824
.4	1.6677	.5878	1.3903	0.7051	1.8153	.6614	1.5643	0.7675
.5	1.5890	7237	1.3247	0.8681	1.7296	8142	1.4905	0.9449
.6	1.4944	8523	1.2458	1.0224	1.6267	9590	1.4017	1.1131
.7	1.3849	9724	1.1545	1.1665	1.5074	1.0941	1.2990	1.2697
.8	1.2615	1.0828	1.0517	1.2989	1.3731	1.2183	1.1833	1.4139
.9	1.1255	1.1824	0.9383	1.4183	1.2251	1.3304	1.0557	1.5439
1.0	0.9783	1.2702	0.8156	1.5236	1.0649	1.4291	0.9176	1.6585
1.1	0.8213	1.3452	0.6847	1.6137	0.8940	1.5136	0.7704	1.7565
1.2	.6561	1.4069	.5470	1.6876	.7142	1.5830	.6154	1.8370
1.3	4844	1.4544	4038	1.7447	5272	1.6365	4543	1.8991
1.4	3078	1.4875	2566	1.7843	3350	1.6737	2887	1.9422
1.5	1281	1.5057	1068	1.8061	1394	1.6941	1201	1.9660
½π	0000	1.5095	0000	1.8107	0000	1.6984	0000	1.9709

TABLE II. VALUES OF COSH $(x + iy)$ AND SINH $(x + iy)$.

$x = 1.0$				$x = 1.1$				y
a	b	c	d	a	b	c	d	
1.5431	.0000	1.1752	.0000	1.6685	.0000	1.3356	.0000	0
1.5354	1173	1.1693	1541	1.6602	1333	1.3290	1666	.1
1.5123	2335	1.1518	3066	1.6353	2654	1.3090	3315	.2
1.4742	3473	1.1227	4560	1.5940	3946	1.2760	4931	.3
1.4213	.4576	1.0824	.6009	1.5368	.5201	1.2302	0 6498	.4
1.3542	5634	1.0314	7398	1.4643	6403	1.1721	0.7999	.5
1.2736	6636	0.9699	8718	1.3771	7542	1.1024	0.9421	.6
1.1802	7571	0.8988	9941	1.2762	8604	1.0216	1.0749	.7
1.0751	0.8430	.8188	1 1069	1.1625	0.9581	.9306	1 1969	.8
0.9592	0.9206	7305	1.2087	1.0372	1.0462	8302	1.3070	.9
0.8337	0.9889	6350	1.2985	0.9015	1.1239	7217	1.4040	1.0
0.6999	1.0473	5331	1.3752	0.7568	1.1903	6058	1.4870	1.1
.5592	1.0953	.4258	1.4382	.6046	1.2449	.4840	1.5551	1.2
4128	1.1324	3144	1.4869	4463	1.2870	3573	1.6077	1.3
2623	1.1581	1998	1.5213	2836	1.3162	2270	1.6442	1.4
1092	1.1723	0831	1.5392	1180	1.3323	0945	1.6643	1.5
0000	1.1752	0000	1 5431	0000	1.3356	0000	1.6685	$\frac{1}{2}\pi$

$x = 1.4$				$x = 1.5.$				y
a	b	c	d	a	b	c	d	
2.1509	.0000	1.9043	.0000	2.3524	.0000	2.1293	.0000	0
2.1401	1901	1.8948	2147	2 3413	2126	2.1187	2348	.1
2.1080	3783	1.8663	4273	2.3055	4230	2.0868	4674	.2
2.0548	5628	1.8192	6356	2 2473	6292	2.0342	6951	.3
1.9811	0.7416	1.7540	0.8376	2.1667	0.8292	1.9612	0.9161	.4
1.8876	0.9130	1.6712	1.0312	2.0644	1.0208	1.8686	1.1278	.5
1.7752	1.0753	1.5713	1.2145	1.9415	1.2023	1.7574	1.3283	.6
1.6451	1.2268	1.4565	1.3856	1.7992	1.3717	1.6286	1.5155	.7
1.4985	1.3661	1.3268	1.5430	1.6389	1.5275	1.4835	1.6875	.8
1.3370	1.4917	1.1838	1.6849	1.4623	1.6679	1.3236	1.8427	.9
1.1622	1.6024	1.0289	1.8099	1.2710	1.7917	1.1505	1.9795	1.0
0.9756	1.6971	0.8638	1.9168	1.0677	1.8976	0.9659	2.0965	1.1
.7794	1.7749	.6900	2.0047	.8524	1.9846	.7716	2.1925	1.2
5754	1.8349	5094	2.0725	6293	2.0517	5696	2.2667	1.3
3656	1.8766	3237	2.1196	3998	2.0983	3619	2.3182	1.4
1522	1.8996	1347	2.1455	1664	2.1239	1506	2.3465	1.5
.0000	1.9043	0000	2.1509	.0000	2.1293	.0000	2.3524	$\frac{1}{2}\pi$

TABLE III.

u	gd u	θ°	u	gd u	θ°	u	gd u	θ°
		°						°
.00	.0000	0.000	.60	.5669	32.483	1.50	1.1317	64.843
.02	0200	1.146	.62	5837	33.444	1.55	1.1525	66.034
.04	0400	2.291	.64	6003	34.395	1.60	1.1724	67.171
.06	0600	3.436	.66	6167	35.336	1.65	1.1913	68.257
.08	0799	0.579	.68	6329	36.265	1.70	1.2094	69.294
.10	.0998	5.720	.70	.6489	37.183	1.75	1.2267	70.284
.12	1197	6.859	.72	6648	38.091	1.80	1.2432	71.228
.14	1395	7.995	.74	6804	38.987	1.85	1.2589	72.128
.16	1593	9.128	.76	6958	39.872	1.90	1.2739	72.987
.18	1790	10.258	.78	7111	40.746	1.95	1.2881	73.805
.20	.1987	11.384	.80	.7261	41.608	2.00	1.3017	74.584
.22	2183	12.505	.82	7410	42.460	2.10	1.3271	76.037
.24	2377	13.621	.84	7557	43.299	2.20	1.3501	77.354
.26	2571	14.732	.86	7702	44.128	2.30	1.3710	78.549
.28	2764	15.837	.88	7844	44.944	2.40	1.3899	79.633
.30	.2956	16.937	.90	.7985	45.750	2.50	1.4070	80.615
.32	3147	18.030	.92	8123	46.544	2.60	1.4227	81.513
.34	3336	19.116	.94	8260	47.326	2.70	1.4366	82.310
.36	3525	20.195	.96	8394	48.097	2.80	1.4493	83.040
.38	3712	21.267	.98	8528	48.857	2.90	1.4609	83.707
.40	.3897	22.331	1.00	.8658	49.605	3.00	1.4713	84.301
.42	4082	23.386	1.05	8976	51.428	3.10	1.4808	84.841
.44	4264	24.434	1.10	9281	53.178	3.20	1.4894	85.336
.46	4446	25.473	1.15	9575	54.860	3.30	1.4971	85.775
.48	4626	26.503	1.20	9857	56.476	3.40	1.5041	86.177
.50	.4804	27.524	1.25	1.0127	58.026	3.50	1.5104	86.541
.52	4980	28.535	1.30	1.0387	59.511	3.60	1.5162	86.870
.54	5155	29.537	1.35	1.0635	60.933	3.70	1.5214	87 168
.56	5328	30.529	1.40	1.0873	62.295	3.80	1.5261	87.437
.58	5500	31.511	1.45	1.1100	63.598	3.90	1.5303	87.681

TABLE IV.

u	gd u	log sinh u	log cosh u	u	gd u	log sinh u	log cosh u
4.0	1.5342	1.4360	1.4363	5.5	1.5626	2.08758	2.08760
4.1	1.5377	1.4795	1.4797	5.6	1.5634	2.13101	2.13103
4.2	1.5408	1.5229	1.5231	5.7	1.5641	2.17444	2.17445
4.3	1.5437	1.5664	1.5665	5.8	1.5648	2.21787	2.21788
4.4	1.5462	1.6098	1.6099	5.9	1.5653	2.26130	2.26131
4.5	1.5486	1.6532	1.6533	6.0	1 5658	2.30473	2.30474
4.6	1.5507	1.6967	1.6968	6.2	1.5667	2.39159	2.39160
4.7	1 5526	1.7401	1.7402	6.4	1.5675	2.47845	2.47846
4.8	1.5543	1.7836	1.7836	6.6	1.5681	2.56531	2.56531
4.9	1.5559	1.8270	1.8270	6.8	1.5686	2.65217	2.65217
5.0	1 5573	1.8704	1.8705	7.0	1.5690	2.73903	2.73903
5.1	1.5586	1.9139	1.9139	7.5	1.5697	2.95618	3.95618
5.2	1.5598	1.9573	1.9573	8.0	1.5701	3.17333	3.17333
5.3	1.5608	2.0007	2.0007	8.5	1.5704	3.39047	3.39047
5.4	1.5618	2.0442	2.0442	9.0	1.5705	3.60762	3.60762
				∞	1.5708	∞	∞

CHAPTER V.

HARMONIC FUNCTIONS.

By WILLIAM E. BYERLY,

Professor of Mathematics in Harvard University.

ART. 1. HISTORY AND DESCRIPTION.

What is known as the Harmonic Analysis owed its origin and development to the study of concrete problems in various branches of Mathematical Physics, which however all involved the treatment of partial differential equations of the same general form.

The use of Trigonometric Series was first suggested by Daniel Bernouilli in 1753 in his researches on the musical vibrations of stretched elastic strings, although Bessel's Functions had been already (1732) employed by him and by Euler in dealing with the vibrations of a heavy string suspended from one end; and Zonal and Spherical Harmonics were introduced by Legendre and Laplace in 1782 in dealing with the attraction of solids of revolution.

The analysis was greatly advanced by Fourier in 1812–1824 in his remarkable work on the Conduction of Heat, and important additions have been made by Lamé (1839) and by a host of modern investigators.

The differential equations treated in the problems which have just been enumerated are

$$\frac{\partial^2 y}{\partial t^2} = a^2 \frac{\partial^2 y}{\partial x^2} \qquad (1)$$

for the transverse vibrations of a musical string;

$$\frac{\partial^2 y}{\partial t^2} = c^2\left(x\frac{\partial^2 y}{\partial x^2} + \frac{\partial y}{\partial x}\right) \tag{2}$$

for small transverse vibrations of a uniform heavy string suspended from one end;

$$\frac{\partial^2 V}{\partial x^2} + \frac{\partial^2 V}{\partial y^2} + \frac{\partial^2 V}{\partial z^2} = 0, \tag{3}$$

which is Laplace's equation; and

$$\frac{\partial u}{\partial t} = a^2\left(\frac{\partial^2 u}{\partial x^2} + \frac{\partial^2 u}{\partial y^2} + \frac{\partial^2 u}{\partial z^2}\right) \tag{4}$$

for the conduction of heat in a homogeneous solid.

Of these Laplace's equation (3), and (4) of which (3) is a special case, are by far the most important, and we shall concern ourselves mainly with them in this chapter. As to their interest to engineers and physicists we quote from an article in The Electrician of Jan. 26, 1894, by Professor John Perry:

"There is a well-known partial differential equation, which is the same in problems on heat-conduction, motion of fluids, the establishment of electrostatic or electromagnetic potential, certain motions of viscous fluid, certain kinds of strain and stress, currents in a conductor, vibrations of elastic solids, vibrations of flexible strings or elastic membranes, and innumerable other phenomena. The equation has always to be solved subject to certain boundary or limiting conditions, sometimes as to space and time, sometimes as to space alone, and we know that if we obtain any solution of a particular problem, then that is the true and only solution. Furthermore, if a solution, say, of a heat-conduction problem is obtained by any person, that answer is at once applicable to analogous problems in all the other departments of physics. Thus, if Lord Kelvin draws for us the lines of flow in a simple vortex, he has drawn for us the lines of magnetic force about a circular current; if Lord Rayleigh calculates for us the resistance of the mouth of an organ-pipe, he has also determined the end effect of a bar of iron which is magnetized; when Mr. Oliver Heaviside shows his matchless skill and familiarity with Bessel's functions in solving electro-magnetic problems, he is solving problems in heat-conductivity or the strains in prismatic shafts. How difficult it is to express exactly the distribution of strain in a twisted square shaft, for example, and yet how easy it is to understand thoroughly when one knows the perfect-fluid analogy! How easy, again, it is to imagine the electric current density everywhere in a conductor when transmitting alternating currents when we know Mr. Heaviside's viscous-fluid analogy, or even the heat-conduction analogy!

"Much has been written about the correlation of the physical sciences; but when we observe how a young man who has worked almost altogether at heat problems suddenly shows himself acquainted with the most difficult investigations in other departments of physics, we may say that the true correlation of the physical sciences lies in the equation of continuity

$$\frac{\partial u}{\partial t} = a^2 \left(\frac{\partial^2 u}{\partial x^2} + \frac{\partial^2 u}{\partial y^2} + \frac{\partial^2 u}{\partial z^2} \right)."$$

In the Theory of the Potential Function in the Attraction of Gravitation, and in Electrostatics and Electrodynamics,* V in Laplace's equation (3) is the value of the Potential Function, at any external point (x, y, z), due to any distribution of matter or of electricity; in the theory of the Conduction of Heat in a homogeneous solid † V is the temperature at any point in the solid after the stationary temperatures have been established, and in the theory of the irrotational flow of an incompressible fluid ‡ V is the Velocity Potential Function and (3) is known as the equation of continuity.

If we use spherical coördinates, (3) takes the form

$$\frac{1}{r^2} \left[r \frac{\partial^2 (rV)}{\partial r^2} + \frac{1}{\sin \theta} \frac{\partial \left(\sin \theta \frac{\partial V}{\partial \theta} \right)}{\partial \theta} + \frac{1}{\sin^2 \theta} \frac{\partial^2 V}{\partial \phi^2} \right] = 0; \quad (5)$$

* See Peirce's Newtonian Potential Function. Boston.
† See Fourier's Analytic Theory of Heat. London and New York, 1878 ; or Riemann's Partielle Differentialgleichungen. Brunswick.
‡ See Lamb's Hydrodynamics. London and New York, 1895.

and if we use cylindrical coördinates, the form

$$\frac{\partial^2 V}{\partial r^2} + \frac{1}{r}\frac{\partial V}{\partial r} + \frac{1}{r^2}\frac{\partial^2 V}{\partial \phi^2} + \frac{\partial^2 V}{\partial z^2} = 0. \tag{6}$$

In the theory of the Conduction of Heat in a homogeneous solid,* u in equation (4) is the temperature of any point (x, y, z) of the solid at any time t, and a^2 is a constant determined by experiment and depending on the conductivity and the thermal capacity of the solid.

ART. 2. HOMOGENEOUS LINEAR DIFFERENTIAL EQUATIONS.

The general solution of a differential equation is the equation expressing the most general relation between the primitive variables which is consistent with the given differential equation and which does not involve differentials or derivatives. A general solution will always contain arbitrary (i.e., undetermined) constants or arbitrary functions.

A particular solution of a differential equation is a relation between the primitive variables which is consistent with the given differential equation, but which is less general than the general solution, although included in it.

Theoretically, every particular solution can be obtained from the general solution by substituting in the general solution particular values for the arbitrary constants or particular functions for the arbitrary functions; but in practice it is often easy to obtain particular solutions directly from the differential equation when it would be difficult or impossible to obtain the general solution.

(a) If a problem requiring for its solution the solving of a differential equation is determinate, there must always be given in addition to the differential equation enough outside conditions for the determination of all the arbitrary constants or arbitrary functions that enter into the general solution of the equation; and in dealing with such a problem, if the differential equation can be readily solved the natural method of procedure is to obtain its general solution, and then to determine the constants or functions by the aid of the given conditions.

It often happens, however, that the general solution of the differential equation in question cannot be obtained, and then, since the problem, if determinate, will be solved, if by any means a solution of the equation can be found which will also satisfy the given outside conditions, it is worth while to try to get particular solutions and so to combine them as to form a result which shall satisfy the given conditions without ceasing to satisfy the differential equation.

(*b*) A differential equation is linear when it would be of the first degree if the dependent variable and all its derivatives were regarded as algebraic unknown quantities. If it is linear and contains no term which does not involve the dependent variable or one of its derivatives, it is said to be linear and homogeneous.

All the differential equations given in Art. 1 are linear and homogeneous.

(*c*) If a value of the dependent variable has been found which satisfies a given homogeneous, linear, differential equation, the product formed by multiplying this value by any constant will also be a value of the dependent variable which will satisfy the equation.

For if all the terms of the given equation are transposed to the first member, the substitution of the first-named value must reduce that member to zero; substituting the second value is equivalent to multiplying each term of the result of the first substitution by the same constant factor, which therefore may be taken out as a factor of the whole first member. The remaining factor being zero, the product is zero and the equation is satisfied.

(*d*) If several values of the dependent variable have been found each of which satisfies the given differential equation, their sum will satisfy the equation; for if the sum of the values in question is substituted in the equation, each term of the sum will give rise to a set of terms which must be equal to zero, and therefore the sum of these sets must be zero.

(*e*) It is generally possible to get by some simple device particular solutions of such differential equations as those we have collected in Art. 1. The object of this chapter is to find methods of so combining these particular solutions as to satisfy any given conditions which are consistent with the nature of the problem in question.

This often requires us to be able to develop any given function of the variables which enter into the expression of these conditions in terms of normal forms suited to the problem with which we happen to be dealing, and suggested by the form of particular solution that we are able to obtain for the differential equation.

These normal forms are frequently sines and cosines, but they are often much more complicated functions known as Legendre's Coefficients, or Zonal Harmonics; Laplace's Coefficients, or Spherical Harmonics; Bessel's Functions, or Cylindrical Harmonics; Lamé's Functions, or Ellipsoidal Harmonics; etc.

ART. 3. PROBLEM IN TRIGONOMETRIC SERIES.

As an illustration let us consider the following problem: A large iron plate π centimeters thick is heated throughout to a uniform temperature of 100 degrees centigrade; its faces are then suddenly cooled to the temperature zero and are kept at that temperature for 5 seconds. What will be the temperature of a point in the middle of the plate at the end of that time? Given $a^2 = 0.185$ in C.G.S. units.

Take the origin of coördinates in one face of the plate and the axis of X perpendicular to that face, and let u be the temperature of any point in the plate t seconds after the cooling begins.

We shall suppose the flow of heat to be directly across the plate so that at any given time all points in any plane parallel to the faces of the plate will have the same temperature. Then u depends upon a single space-coordinate x ; $\dfrac{\partial u}{\partial y} = 0$ and $\dfrac{\partial u}{\partial z} = 0$, and (4), Art. 1, reduces to

$$\frac{\partial u}{\partial t} = a^2 \frac{\partial^2 u}{\partial x^2}. \tag{1}$$

Obviously, $\quad u = 100°$ when $t = 0$, $\qquad\qquad$ (2)

$\qquad\qquad\quad u = \quad 0$ when $x = 0$, $\qquad\qquad$ (3)

and $\qquad\qquad u = \quad 0$ when $x = \pi$; $\qquad\qquad$ (4)

and we need to find a solution of (1) which satisfies the conditions (2), (3), and (4).

We shall begin by getting a particular solution of (1), and we shall use a device which always succeeds when the equation is linear and homogeneous and has constant coefficients.

Assume * $u = e^{\beta x + \gamma t}$, where β and γ are constants; substitute in (1) and divide through by $e^{\beta x + \gamma t}$ and we get $\gamma = a^2 \beta^2$; and if this condition is satisfied, $u = e^{\beta x + \gamma t}$ is a solution of (1).

$u = e^{\beta x + a^2 \beta^2 t}$ is then a solution of (1) no matter what the value of β.

We can modify the form of this solution with advantage. Let $\beta = \mu i$,† then $u = e^{-a^2 \mu^2 t} e^{\mu x i}$ is a solution of (1), as is also $u = e^{-a^2 \mu^2 t} e^{-\mu x i}$.

By (d), Art. 2,

$$u = e^{-a^2 \mu^2 t} \frac{\left(e^{\mu x i} + e^{-\mu x i}\right)}{2} = e^{-a^2 \mu^2 t} \cos \mu x \tag{5}$$

is a solution, as is also

$$u = e^{-a^2 \mu^2 t} \frac{\left(e^{\mu x i} - e^{-\mu x i}\right)}{2 i} = e^{-a^2 \mu^2 t} \sin \mu x ; \tag{6}$$

and μ is entirely arbitrary.

---

* This assumption must be regarded as purely tentative. It must be tested by substituting in the equation, and is justified if it leads to a solution.

† The letter i will be used to represent $\sqrt{-1}$.

By giving different values to μ we get different particular solutions of (1); let us try to so combine them as to satisfy our conditions while continuing to satisfy equation (1).

$u = e^{-a^2\mu^2 t} \sin \mu x$ is zero when $x = 0$ for all values of μ; it is zero when $x = \pi$ if μ is a whole number. If, then, we write u equal to a sum of terms of the form $A e^{-a^2 m^2 t} \sin mx$, where m is an integer, we shall have a solution of (1) (see (d), Art. 2) which satisfies (3) and (4).

Let this solution be

$$u = A_1 e^{-a^2 t} \sin x + A_2 e^{-4a^2 t} \sin 2x + A_3 e^{-9a^2 t} \sin 3x + \ldots, \quad (7)$$

$A_1, A_2, A_3, \ldots$ being undetermined constants.

When $t = 0$, (7) reduces to

$$u = A_1 \sin x + A_2 \sin 2x + A_3 \sin 3x + \ldots. \quad (8)$$

If now it is possible to develop unity into a series of the form (8) we have only to substitute the coefficients of that series each multiplied by 100 for $A_1, A_2, A_3, \ldots$ in (7) to have a solution satisfying (1) and all the equations of condition (2), (3), and (4).

We shall prove later (see Art. 6) that

$$1 = \frac{4}{\pi}\left[\sin x + \frac{1}{3} \sin 3x + \frac{1}{5} \sin 5x + \ldots \right]$$

for all values of x between 0 and π. Hence our solution is

$$u = \frac{400}{\pi}\left[e^{-a^2 t} \sin x + \frac{1}{3} e^{-9a^2 t} \sin 3x + \frac{1}{5} e^{-25 a^2 t} \sin 5x + \ldots \right] (9)$$

To get the answer of the numerical problem we have only to compute the value of u when $x = \frac{\pi}{2}$ and $t = 5$ seconds. As there is no object in going beyond tenths of a degree, four-place tables will more than suffice, and no term of (9) beyond the first will affect the result. Since $\sin \frac{\pi}{2} = 1$, we have to compute the numerical value of

$$\frac{400}{\pi}e^{-a^2t} \quad \text{where} \quad a^2 = 0.185 \quad \text{and} \quad t = 5.$$

$\log a^2 = 9.2672 - 10$ $\log 400 = 2.6061$

$\log t = 0.6990$ $\text{colog } \pi = 9.5059 - 10$

$\log a^2 t = 9.9662 - 10$ $\text{colog } e^{a^2t} = 9.5982 - 10$

$\log \log e = 9.6378 - 10$

$\log \log e^{a^2t} = 9.6040 - 10$ $\log u = 1.7102$

$\log e^{a^2t} = 0.4018$ $u = 51°.3.$

If the breadth of the plate had been c centimeters instead of π centimeters it is easy to see that we should have needed the development of unity in a series of the form

$$A_1 \sin \frac{\pi x}{c} + A_2 \sin \frac{2\pi x}{c} + A_3 \sin \frac{3\pi x}{c} + \dots.$$

Prob. 1. An iron slab 50 centimeters thick is heated to the temperature 100 degrees Centigrade throughout. The faces are then suddenly cooled to zero degrees, and are kept at that temperature for 10 minutes. Find the temperature of a point in the middle of the slab, and of a point 10 centimeters from a face at the end of that time. Assume that

$$1 = \frac{4}{\pi}\left(\sin \frac{\pi x}{c} + \frac{1}{3} \sin \frac{3\pi x}{c} + \frac{1}{5} \sin \frac{5\pi x}{c} + \dots\right) \text{ from } x = 0 \text{ to } x = c.$$

Ans. $84°.0$; $49°.4$.

ART. 4.   PROBLEM IN ZONAL HARMONICS.

As a second example let us consider the following problem: Two equal thin hemispherical shells of radius unity placed together to form a spherical surface are separated by a thin layer of air. A charge of statical electricity is placed upon one hemisphere and the other hemisphere is connected with the ground, the first hemisphere is then found to be at potential 1, the other hemisphere being of course at potential zero. At what potential is any point in the "field of force" due to the charge?

We shall use spherical coordinates and shall let V be the potential required. Then V must satisfy equation (5), Art. 1.

But since from the symmetry of the problem V is obviously independent of ϕ, if we take the diameter perpendicular to the plane separating the two conductors as our polar axis, $\dfrac{\partial^2 V}{\partial \phi^2}$ is zero, and our equation reduces to

$$\frac{r\partial^2(rV)}{\partial r^2} + \frac{1}{\sin\theta}\frac{\partial\left(\sin\theta\dfrac{\partial V}{\partial\theta}\right)}{\partial\theta} = 0. \tag{1}$$

V is given on the surface of our sphere, hence

$$V = f(\theta) \quad \text{when} \quad r = 1, \tag{2}$$

where $f(\theta) = 1$ if $0 < \theta < \dfrac{\pi}{2}$, and $f(\theta) = 0$ if $\dfrac{\pi}{2} < \theta < \pi$.

Equation (2) and the implied conditions that V is zero at an infinite distance and is nowhere infinite are our conditions.

To find particular solutions of (1) we shall use a method which is generally effective. Assume* that $V = R\Theta$ where R is a function of r but not of θ, and Θ is a function of θ but not of r. Substitute in (1) and reduce, and we get

$$\frac{1}{R}\frac{rd^2(rR)}{dr^2} = -\frac{1}{\Theta\sin\theta}\frac{d\left(\sin\theta\dfrac{d\Theta}{d\theta}\right)}{d\theta}. \tag{3}$$

Since the first member of (3) does not contain θ and the second does not contain r and the two members are identically equal, each must be equal to a constant. Let us call this constant, which is wholly undetermined, $m(m + 1)$; then

$$\frac{r}{R}\frac{d^2(rR)}{dr^2} = -\frac{1}{\Theta\sin\theta}\frac{d\left(\sin\theta\dfrac{d\Theta}{d\theta}\right)}{d\theta} = m(m + 1);$$

whence

$$r\frac{d^2(rR)}{dr^2} - m(m + 1)R = 0, \tag{4}$$

and

$$\frac{1}{\sin\theta}\frac{d\left(\sin\theta\dfrac{d\Theta}{d\theta}\right)}{d\theta} + m(m + 1)\Theta = 0. \tag{5}$$

* See the first foot-note on page 175.

Equation (4) can be expanded into

$$r^2\frac{d^2R}{dr^2} + 2r\frac{dR}{dr} - m(m+1)R = 0,$$

and can be solved by elementary methods. Its complete solution is

$$R = Ar^m + Br^{-m-1}. \tag{6}$$

Equation (5) can be simplified by changing the independent variable to x where $x = \cos\theta$. It becomes

$$\frac{d}{dx}\left[(1 - x^2)\frac{d\Theta}{dx}\right] + m(m+1)\Theta = 0, \tag{7}$$

an equation which has been much studied and which is known as Legendre's Equation.

We shall restrict m, which is wholly undetermined, to positive whole values, and we can then get particular solutions of (7) by the following device:

Assume* that Θ can be expressed as a sum or a series of terms involving whole powers of x multiplied by constant coefficients.

Let $\Theta = \Sigma a_n x^n$ and substitute in (7). We get

$$\Sigma[n(n-1)a_n x^{n-2} - n(n+1)a_n x^n + m(m+1)a_n x^n] = 0, \tag{8}$$

where the symbol Σ indicates that we are to form all the terms we can by taking successive whole numbers for n.

Since (8) must be true no matter what the value of x, the coefficient of any given power of x, as for instance x^k, must vanish. Hence

$$(k+2)(k+1)a_{k+2} - k(k+1)a_k + m(m+1)a_k = 0,$$

and

$$a_{k+2} = -\frac{m(m+1) - k(k+1)}{(k+1)(k+2)}a_k. \tag{9}$$

If now any set of coefficients satisfying the relation (9) be taken, $\Theta = \Sigma a_k x^k$ will be a solution of (7).

If $k = m$, then $a_{k+2} = 0$, $a_{k+4} = 0$, etc.

* See the first foot-note on page 175.

Since it will answer our purpose if we pick out the simplest set of coefficients that will obey the condition (9), we can take a set including a_m.

Let us rewrite (9) in the form

$$a_k = -\frac{(k+2)(k+1)}{(m-k)(m+k-1)}. \tag{10}$$

We get from (10), beginning with $k = m - 2$,

$$a_{m-2} = -\frac{m(m-1)}{2\cdot(2m-1)} m$$

$$a_{m-4} = \frac{m(m-1)(m-2)(m-3)}{2\cdot4\cdot(2m-1)(2m-3)} a_m,$$

$$a_{m-6} = -\frac{m(m-1)(m-2)(m-3)(m-4)(m-5)}{2\cdot4\cdot6\cdot(2m-1)(2m-3)(2m-5)} a_m, \text{ etc.}$$

If m is even we see that the set will end with a_0; if m is odd, with a_1.

$$\Theta = a_m \left[x^m - \frac{m(m-1)}{2\cdot(2m-1)} x^{m-2} \right.$$
$$\left. + \frac{m(m-1)(m-2)m-3)}{2\cdot4\cdot(2m-1)(2m-3)} x^{m-4} - \cdots \right],$$

where a_m is entirely arbitrary, is, then, a solution of (7). It is found convenient to take a_m equal to

$$\frac{(2m-1)(2m-3)\ldots 1}{m!},$$

and it will be shown later that with this value of a_m, $\Theta = 1$ when $x = 1$.

Θ is a function of x and contains no higher powers of x than x^m. It is usual to write it as $P_m(x)$.

We proceed to write out a few values of $P_m(x)$ from the formula

$$P_m(x) = \frac{(2m-1)(2m-3)\ldots 1}{m!} \left[x^m - \frac{m(m-1)}{2\cdot(2m-1)} x^{m-2} \right.$$
$$\left. + \frac{m(m-1)(m-2)(m-3)}{2\cdot4\cdot(2m-1)(2m-3)} x^{m-4} - \cdots \right] \tag{11}$$

We have:

$$\left.\begin{array}{ll}
P_0(x) = 1 & \text{or } P_0(\cos\theta) = 1, \\
P_1(x) = x & \text{or } P_1(\cos\theta) = \cos\theta, \\
P_2(x) = \tfrac{1}{2}(3x^2 - 1) & \text{or } P_2(\cos\theta) = \tfrac{1}{2}(3\cos^2\theta - 1), \\
P_3(x) = \tfrac{1}{2}(5x^3 - 3x) & \text{or } P_3(\cos\theta) = \tfrac{1}{2}(5\cos^3\theta - 3\cos\theta), \\
P_4(x) = \tfrac{1}{8}(35x^4 - 30x^2 + 3) & \text{or} \\
\quad P_4(\cos\theta) = \tfrac{1}{8}(35\cos^4\theta - 30\cos^2\theta + 3), \\
P_5(x) = \tfrac{1}{8}(63x^5 - 70x^3 + 15x) & \text{or} \\
\quad P_5(\cos\theta) = \tfrac{1}{8}(63\cos^5\theta - 70\cos^3\theta + 15\cos\theta).
\end{array}\right\} \quad (12)$$

We have obtained $\Theta = P_m(x)$ as a particular solution of (7), and $\Theta = P_m(\cos\theta)$ as a particular solution of (5). $P_m(x)$ or $P_m(\cos\theta)$ is a new function, known as a Legendre's Coefficient, or as a Surface Zonal Harmonic, and occurs as a normal form in many important problems.

$V = r^m P_m(\cos\theta)$ is a particular solution of (1), and $r^m P_m(\cos\theta)$ is sometimes called a Solid Zonal Harmonic.

$$V = A_0 P_0(\cos\theta) + A_1 r P_1(\cos\theta) + A_2 r^2 P_2(\cos\theta) \\ + A_3 r^3 P_3(\cos\theta) + \dots \quad (13)$$

satisfies (1), is not infinite at any point within the sphere, and reduces to

$$V = A_0 P_0(\cos\theta) + A_1 P_1(\cos\theta) + A_2 P_2(\cos\theta) \\ + A_3 P_3(\cos\theta) + \dots \quad (14)$$

when $r = 1$.

$$V = \frac{A_0 P_0(\cos\theta)}{r} + \frac{A_1 P_1(\cos\theta)}{r^2} + \frac{A_2 P_2(\cos\theta)}{r^3} \\ + \frac{A_3 P_3(\cos\theta)}{r^4} + \dots \quad (15)$$

satisfies (1), is not infinite at any point without the sphere, is equal to zero when $r = \infty$, and reduces to (14) when $r = 1$.

If then we can develop $f(\theta)$ [see eq. (2)] into a series of the form (14), we have only to put the coefficients of this series in place of the A_0, A_1, A_2, ... in (13) to get the value of V for a point within the sphere, and in (15) to get the value of V at a point without the sphere.

We shall see later (Art. 16, Prob. 22) that if $f(\theta) = 1$ for

$0 < \theta < \dfrac{\pi}{2}$ and $f(\theta) = 0$ for $\dfrac{\pi}{2} < \theta < \pi$,

$$f(\theta) = \frac{1}{2} + \frac{3}{4}P_1(\cos \theta) - \frac{7}{8} \cdot \frac{1}{2} \cdot P_3(\cos \theta)$$

$$+ \frac{11}{12} \cdot \frac{1 \cdot 3}{2 \cdot 4}P_5(\cos \theta) - \dots \quad (16)$$

Hence our required solution is

$$V = \frac{1}{2} + \frac{3}{4}rP_1(\cos \theta) - \frac{7}{8} \cdot \frac{1}{2} \cdot r^3 P_3(\cos \theta)$$

$$+ \frac{11}{12} \cdot \frac{1 \cdot 3}{2 \cdot 4} r^5 P_5(\cos \theta) - \dots \quad (17)$$

at an internal point; and

$$V = \frac{1}{2r} + \frac{3}{4} \frac{1}{r^2} P_1(\cos\theta) - \frac{7}{8} \cdot \frac{1}{2} \frac{1}{r^4} P_3(\cos \theta)$$

$$+ \frac{11}{12} \cdot \frac{1 \cdot 3}{2 \cdot 4} \frac{1}{r^6} P_5(\cos \theta) - \dots \quad (18)$$

at an external point.

If $r = \dfrac{1}{4}$ and $\theta = 0$, (17) reduces to

$$V = \frac{1}{2} + \frac{3}{4} \cdot \frac{1}{4} - \frac{7}{8} \cdot \frac{1}{2} \cdot \frac{1}{4^3} + \frac{11}{12} \cdot \frac{1 \cdot 3}{2 \cdot 4} \cdot \frac{1}{4^5} \dots, \text{ since } P_m(1) = 1.$$

To two decimal places $V = 0.68$, and the point $r = \dfrac{1}{4}$, $\theta = 0$ is at potential 0.68.

If $r = 5$ and $\theta = \dfrac{\pi}{4}$, (18) and Table I, at the end of this chapter, give

$$V = \frac{1}{2 \cdot 5} + \frac{3}{4} \cdot \frac{1}{5^2} \cdot 0.7071 + \frac{7}{8} \cdot \frac{1 \cdot 3}{2 \cdot 4} \cdot \frac{1}{5^4} \cdot 0.1768 + \dots = 0.12,$$

and the point $r = 5$, $\theta = \dfrac{\pi}{4}$ is at potential 0.12.

If the radius of the conductor is a instead of unity, we have only to replace r by $\dfrac{r}{a}$ in (17) and (18).

Prob. 2. One half the surface of a solid sphere 12 inches in diameter is kept at the temperature zero and the other half at 100 degrees centigrade until there is no longer any change of temperature at any point within the sphere. Required the temperature of the center; of any point in the diametral plane separating the hot and cold hemispheres; of points 2 inches from the center and in the axis of symmetry; and of points 3 inches from the center in a diameter inclined at an angle of 45° to the axis of symmetry.

Ans. 50°; 50°; 73°.9; 26°.1; 77°.1; 22°.9.

ART. 5.   PROBLEM IN BESSEL'S FUNCTIONS.

As a last example we shall take the following problem: The base and convex surface of a cylinder 2 feet in diameter and 2 feet high are kept at the temperature zero, and the upper base at 100 degrees centigrade. Find the temperature of a point in the axis one foot from the base, and of a point 6 inches from the axis and one foot from the base, after the permanent state of temperatures has been set up.

If we use cylindrical coördinates and take the origin in the base we shall have to solve equation (6), Art. 1; or, representing the temperature by u and observing that from the symmetry of the problem u is independent of ϕ,

$$\frac{\partial^2 u}{\partial r^2} + \frac{1}{r}\frac{\partial u}{\partial r} + \frac{\partial^2 u}{\partial z^2} = 0, \qquad (1)$$

subject to the conditions

$$u = 0 \quad \text{when} \quad z = 0, \qquad (2)$$
$$u = 0 \quad \text{``} \quad r = 1, \qquad (3)$$
$$u = 100 \quad \text{``} \quad z = 2. \qquad (4)$$

Assume $u = RZ$ where R is a function of r only and Z of z only; substitute in (1) and reduce.

We get $$\frac{1}{R}\frac{d^2 R}{dr^2} + \frac{1}{rR}\frac{dR}{dr} = -\frac{1}{Z}\frac{d^2 Z}{dz^2}. \qquad (5)$$

The first member of (5) does not contain z; therefore the second member cannot. The second member of (5) does not contain r; therefore the first member cannot. Hence each member of (5) is a constant, and we can write (5)

$$\frac{1}{R}\frac{d^2R}{dr^2} + \frac{1}{rR}\frac{dR}{dr} = -\frac{1}{Z}\frac{d^2Z}{dz^2} = -\mu^2, \tag{6}$$

when μ^2 is entirely undetermined.

Hence
$$\frac{d^2Z}{dz^2} - \mu^2 Z = 0, \tag{7}$$

and
$$\frac{d^2R}{dr^2} + \frac{1}{r}\frac{dR}{dr} + \mu^2 R = 0. \tag{8}$$

Equation (7) is easily solved, and its general solution is

$$Z = Ae^{\mu z} + Be^{-\mu z}, \qquad \text{or the equivalent form}$$
$$Z = C\cosh(\mu z) + D\sinh(\mu z). \tag{9}$$

We can reduce (8) slightly by letting $\mu r = x$, and it becomes

$$\frac{d^2R}{dx^2} + \frac{1}{x}\frac{dR}{dx} + R = 0. \tag{10}$$

Assume, as in Art. 4, that R can be expressed in terms of whole powers of x. Let $R = \Sigma a_n x^n$ and substitute in (10). We get

$$\Sigma[n(n-1)a_n x^{n-2} + na_n x^{n-2} + a_n x^n] = 0,$$

an equation which must be true, no matter what the value of x. The coefficient of any given power of x, as x^{k-2}, must, then, vanish, and

$$k(k-1)a_k + ka_k + a_{k-2} = 0,$$
or
$$k^2 a_k + a_{k-2} = 0,$$
whence we obtain
$$a_{k-2} = -k^2 a_k \tag{11}$$

as the only relation that need be satisfied by the coefficients in order that $R = \Sigma a_k x^k$ shall be a solution of (10).

If　　　　　$k = 0$,　$a_{k-2} = 0$,　$a_{k-4} = 0$,　etc.

We can, then, begin with $k = 0$ as the lowest subscript.

From (11) $a_k = -\dfrac{a_{k-2}}{k^2}.$

Then $a_2 = -\dfrac{a_0}{2^2},\quad a_4 = \dfrac{a_0}{2^2 . 4^2},\quad a_6 = -\dfrac{a_0}{2^2 . 4^2 . 6^2},$ etc.

Hence $R = a_0\left[1 - \dfrac{x^2}{2^2} + \dfrac{x^4}{2^2 . 4^2} - \dfrac{x^6}{2^2 . 4^2 . 6^2} + \cdots \right],$

where a_0 may be taken at pleasure, is a solution of (10), provided the series is convergent.

Take $a_0 = 1$, and then $R = J_0(x)$ where

$$J_0(x) = 1 - \frac{x^2}{2^2} + \frac{x^4}{2^2 . 4^2} - \frac{x^6}{2^2 . 4^2 . 6^2} + \frac{x^8}{2^2 . 4^2 . 6^2 . 8^2} - \cdots \quad (12)$$

is a solution of (10).

$J_0(x)$ is easily shown to be convergent for all values real or imaginary of x, it is a new and important form, and is called a Bessel's Function of the zero order, or a Cylindrical Harmonic.

Equation (10) was obtained from (8) by the substitution of $x = \mu r$; therefore

$$R = J_0(\mu r) = 1 - \frac{(\mu r)^2}{2^2} + \frac{(\mu r)^4}{2^2 . 4^2} - \frac{(\mu r)^6}{2^2 . 4^2 . 6^2} + \cdots$$

is a solution of (8), no matter what the value of μ; and $u = J_0(\mu r) \sinh(\mu z)$ and $u = J_0(\mu r) \cosh(\mu z)$ are solutions of (1). $u = J_0(\mu r) \sinh(\mu z)$ satisfies condition (2) whatever the value of μ. In order that it should satisfy condition (3) μ must be so taken that

$$J_0(\mu) = 0; \quad (13)$$

that is, μ must be a root of the transcendental equation (13).

It was shown by Fourier that $J_0(\mu) = 0$ has an infinite number of real positive roots, any one of which can be obtained to any required degree of approximation without serious difficulty. Let $\mu_1, \mu_2, \mu_3, \ldots$ be these roots; then

$$u = A_1 J_0(\mu_1 r) \sinh(\mu_1 z) + A_2 J_0(\mu_2 r) \sinh(\mu_2 z)$$
$$+ A_3 J_0(\mu_3 r) \sinh(\mu_3 z) + \cdots \quad (14)$$

is a solution of (1) which satisfies (2) and (3).

If now we can develop unity into a series of the form

$$1 = B_1 J_0(\mu_1 r) + B_2 J_0(\mu_2 r) + B_3 J_0(\mu_3 r) + \cdots,$$

$$u = 100\left[\frac{B_1 \sinh (\mu_1 z)}{\sinh (2\mu_1)} J_0(\mu_1 r) + \frac{B_2 \sinh (\mu_2 z)}{\sinh (2\mu_2)} J_0(\mu_2 r) + \cdots\right] \quad (15)$$

satisfies (1) and the conditions (2), (3), and (4).

We shall see later (Art. 21) that if $J_1(x) = -\dfrac{dJ_0(x)}{dx}$

$$1 = 2\left[\frac{J_0(\mu_1 r)}{\mu_1 J_1(\mu_1)} + \frac{J_0(\mu_2 r)}{\mu_2 J_1(\mu_2)} + \frac{J_0(\mu_3 r)}{\mu_3 J_1(\mu_3)} + \cdots\right] \quad (16)$$

for values of $r < 1$.

Hence

$$u = 200\left[\frac{J_0(\mu_1 r)}{\mu_1 J_1(\mu_1)} \frac{\sinh (\mu_1 z)}{\sinh (2\mu_1)} + \frac{J_0(\mu_2 r)}{\mu_2 J_1(\mu_2)} \frac{\sinh (\mu_2 z)}{\sinh (2\mu_2)} + \cdots\right] \quad (17)$$

is our required solution.

At the point $r = 0$, $z = 1$ (17) reduces to

$$u = 200\left[\frac{\sinh \mu_1}{\mu_1 J_1(\mu_1) \sinh (2\mu_1)} + \frac{\sinh \mu_2}{\mu_2 J_1(\mu_2) \sinh (2\mu_2)} + \cdots\right]$$

$$= 100\left[\frac{1}{\mu_1 J_1(\mu_1) \cosh \mu_1} + \frac{1}{\mu_2 J_1(\mu_2) \cosh \mu_2} + \cdots\right],$$

since $J_0(0) = 1$ and $\sinh (2x) = 2 \sinh x \cosh x$.

If we use a table of Hyperbolic functions* and Tables II and III, at the end of this chapter, the computation of the value of u is easy. We have

$\mu_1 = 2.405$	$\mu_2 = \quad 5.520$
$J_1(\mu_1) = 0.5190$	$J_1(\mu_2) = -0.3402$
colog $\mu_1 = 9.6189 - 10$	colog $\mu_2 = \quad 9.2581 \quad - 10$
" $J_1(\mu_1) = 0.2848$	" $J_1(\mu_2) = \quad 0.4683n$
" $\cosh \mu_1 = 9.2530 - 10$	" $\cosh \mu_2 = \quad 7.9037 \quad - 10$
$9.1567 - 10$	$7.6301n - 10$

* See Chapter IV, pp. 162, 163, for a four-place table on hyperbolic functions.

$$(\mu_1 J_1(\mu_1) \cosh \mu_1)^{-1} = \quad 0.1434$$
$$(\mu_2 J_1(\mu_2) \cosh \mu_2)^{-1} = - 0.0058$$

$$0.1376; \qquad u = 13°.8$$

At the point $r = \tfrac{1}{2}$, $z = 1$, (17), reduces to

$$u = 100 \left[\frac{J_0(\tfrac{1}{2}\mu_1)}{\mu_1 J_1(\mu_1) \cosh \mu_1} + \frac{J_0(\tfrac{1}{2}\mu_2)}{\mu_2 J_1(\mu_2) \cosh \mu_2} + \cdots \right].$$

$$J_0(\tfrac{1}{2}\mu_1) = 0.6698$$

$$\log J_0(\tfrac{1}{2}\mu_1) = 9.8259 \ - 10$$
$$\text{colog } \mu_1 J_1(\mu_1) \cosh \mu_1 = 9.1567 \ - 10$$

$$8.9826 \ - 10;$$

$$J_0(\tfrac{1}{2}\mu_2) = - 0.1678$$

$$\log J_0(\tfrac{1}{2}\mu_2) = \quad 9.2248n - 10$$
$$\text{colog } \mu_2 J_1(\mu_2) \cosh \mu_2 = \quad 7.6301n - 10$$

$$6.8549 \ - 10;$$

$$\frac{J_0(\tfrac{1}{2}\mu_1)}{\mu_1 J_1(\mu_1) \cosh \mu_1} = 0.0961$$

$$\frac{J_0(\tfrac{1}{2}\mu_2)}{\mu_2 J_1(\mu_2) \cosh \mu_2} = \frac{0.0007}{0.0968}; \qquad u = 9°.7$$

If the radius of the cylinder is a and the altitude b, we have only to replace μ by μa in (13); $2\mu_1$, $2\mu_2$, ... in the denominators of (15) and (17) by $\mu_1 b$, $\mu_2 b$, ...; and μ_1, μ_2, μ_3, ... in the denominators of (16) and (17) by $\mu_1 a$, $\mu_2 a$, $\mu_3 a$, ...

Prob. 3. One base and the convex surface of a cylinder 20 centimeters in diameter and 30 centimeters high are kept at zero temperature and the other base at 100 degrees Centigrade. Find the temperature of a point in the axis and 20 centimeters from the cold base, and of a point 5 centimeters from the axis and 20 centimeters from the cold base after the temperatures have ceased to change.

Ans. $13°.9$; $9°.6$.

ART. 6. THE SINE SERIES.

As we have seen in Art. 3, it is sometimes important to be able to express a given function of a variable, x, in terms of sines of multiples of x. The problem in its general form was first solved by Fourier in his "Théorie Analytique de la Chaleur" (1822), and its solution plays an important part in most branches of Mathematical Physics.

Let us endeavor to so develop a given function of x, $f(x)$, in terms of $\sin x$, $\sin 2x$, $\sin 3x$, etc., that the function and the series shall be equal for all values of x between o and π.

We can of course determine the coefficients $a_1, a_2, a_3, \ldots a_n$ so that the equation

$$f(x) = a_1 \sin x + a_2 \sin 2x + a_3 \sin 3x + \ldots + a_n \sin nx \quad (1)$$

shall hold good for any n arbitrarily chosen values of x between o and π; for we have only to substitute those values in turn in (1) to get n equations of the first degree, in which the n coefficients are the only unknown quantities.

For instance, we can take the n equidistant values Δx, $2\Delta x$, $3\Delta x$, $\ldots n\Delta x$, where $\Delta x = \dfrac{\pi}{n+1}$, and substitute them for x in (1). We get

$$
\left.
\begin{aligned}
f(\Delta x) &= a_1 \sin \Delta x + a_2 \sin 2\Delta x + a_3 \sin 3\Delta x + \ldots \\
&\quad + a_n \sin n\Delta x, \\
f(2\Delta x) &= a_1 \sin 2\Delta x + a_2 \sin 4\Delta x + a_3 \sin 6\Delta x + \ldots \\
&\quad + a_n \sin 2n\Delta x, \\
f(3\Delta x) &= a_1 \sin 3\Delta x + a_2 \sin 6\Delta x + a_3 \sin 9\Delta x + \ldots \\
&\quad + a_n \sin 3n\Delta x, \\
&\quad \vdots \qquad\quad \vdots \qquad\quad \vdots \\
f(n\Delta x) &= a_1 \sin n\Delta x + a_2 \sin 2n\Delta x + a_3 \sin 3n\Delta x + \ldots \\
&\quad + a_n \sin n^2\Delta x,
\end{aligned}
\right\} \quad (2)
$$

n equations of the first degree, to determine the n coefficients $a_1, a_2, a_3, \ldots a_n$.

Not only can equations (2) be solved in theory, but they can be actually solved in any given case by a very simple and ingenious method due to Lagrange,* and any coefficient a_m can be expressed in the form

$$a_m = \frac{2}{n+1} \sum_{\kappa=1}^{\kappa=n} f(\kappa \varDelta x) \sin (\kappa m \varDelta x). \tag{3}$$

If now n is indefinitely increased the values of x for which (1) holds good will come nearer and nearer to forming a continuous set; and the limiting value approached by a_m will probably be the corresponding coefficient in the series required to represent $f(x)$ for all values of x between zero and π.

Remembering that $(n+1)\varDelta x = \pi$, the limiting value in question is easily seen to be

$$a_m = \frac{2}{\pi} \int_0^\pi f(x) \sin mx \, dx. \tag{4}$$

This value can be obtained from equations (2) by the following device without first solving the equations:

Let us multiply each equation in (2) by the product of $\varDelta x$ and the coefficient of a_m in the equation in question, add the equations, and find the limiting form of the resulting equation as n increases indefinitely.

The coefficient of any a, a_κ in the resulting equation is

$$\sin \kappa \varDelta x \sin m \varDelta x . \varDelta x + \sin 2\kappa \varDelta x \sin 2m \varDelta x . \varDelta x + \ldots$$
$$+ \sin n\kappa \varDelta x \sin nm \varDelta x . \varDelta x.$$

Its limiting value, since $(n+1)\varDelta x = \pi$, is

$$\int_0^\pi \sin \kappa x \sin mx . dx;$$

but

$$\int_0^\pi \sin \kappa x \sin mx . dx = \tfrac{1}{2} \int_0^\pi [\cos (m-\kappa)x - \cos(m+\kappa)x] dx = 0$$

if m and κ are not equal.

* See Riemann's Partielle Differentialgleichungen, or Byerly's Fourier's Series and Spherical Harmonics.

The coefficient of a_n is

$$\Delta x(\sin^2 m\Delta x + \sin^2 2m\Delta x + \sin^2 3m\Delta x + \ldots + \sin^2 nm\Delta x).$$

Its limiting value is

$$\int_0^\pi \sin^2 mx \,.\, dx = \frac{\pi}{2}.$$

The first member is

$$f(\Delta x) \sin m\Delta x \,.\, \Delta x + f(2\Delta x) \sin 2m\Delta x \,.\, \Delta x + \ldots$$
$$+ f(n\Delta x) \sin mn\Delta x \,.\, \Delta x,$$

and its limiting value is

$$\int_0^\pi f(x) \sin mx \,.\, dx.$$

Hence the limiting form approached by the final equation as n is increased is

$$\int_0^\pi f(x) \sin mx \,.\, dx = \frac{\pi}{2} a_m.$$

Whence
$$a_m = \frac{2}{\pi} \int_0^\pi f(x) \sin mx \,.\, dx \tag{5}$$

as before.

This method is practically the same as multiplying the equation

$$f(x) = a_1 \sin x + a_2 \sin 2x + a_3 \sin 3x + \ldots \tag{6}$$

by $\sin mx \,.\, dx$ and integrating both members from zero to π.

It is important to realize that the considerations given in this article are in no sense a demonstration, but merely establish a probability.

An elaborate investigation * into the validity of the development, for which we have not space, entirely confirms the results formulated above, provided that between $x = 0$ and $x = \pi$ the

* See Art. 10 for a discussion of this question.

function is finite and single-valued, and has not an infinite number of discontinuities or of maxima or minima.

It is to be noted that the curve represented by $y = f(x)$ need not follow the same mathematical law throughout its length, but may be made up of portions of entirely different curves. For example, a broken line or a locus consisting of finite parts of several different and disconnected straight lines can be represented perfectly well by $y =$ a sine series.

As an example of the application of formula (5) let us take the development of unity.

Here
$$f(x) = 1.$$

$$a_m = \frac{2}{\pi} \int_0^\pi \sin mx \, . \, dx \, ;$$

$$\int \sin mx \, . \, dx = -\frac{\cos mx}{m}.$$

$$\int_0^\pi \sin mx \, . \, dx = \frac{1}{m}(1 - \cos m\pi) = \frac{1}{m}[1 - (-1)^m]$$

$$= 0 \text{ if } m \text{ is even}$$

$$= \frac{2}{m} \text{ if } m \text{ is odd.}$$

Hence $1 = \frac{4}{\pi}\left(\frac{\sin x}{1} + \frac{\sin 3x}{3} + \frac{\sin 5x}{5} + \frac{\sin 7x}{7} + \ldots\right).$ (7)

It is to be noticed that (7) gives at once a sine development for any constant c. It is,

$$c = \frac{4c}{\pi}\left(\frac{\sin x}{1} + \frac{\sin 3x}{3} + \frac{\sin 5x}{5} + \ldots\right). \tag{8}$$

Prob. 4. Show that for values of x between zero and π

(a) $\quad x = 2\left[\frac{\sin x}{1} - \frac{\sin 2x}{2} + \frac{\sin 3x}{3} - \frac{\sin 4x}{4} + \ldots\right],$

(b) $f(x) = \frac{4}{\pi}\left[\frac{\sin x}{1^2} - \frac{\sin 3x}{3^2} + \frac{\sin 5x}{5^2} - \frac{\sin 7x}{7^2} + \ldots\right]$

if $f(x) = x$ for $o < x < \dfrac{\pi}{2}$, and $f(x) = \pi - x$ for $\dfrac{\pi}{2} < x < \pi$.

(c) $f(x) =$

$$\dfrac{2}{\pi}\left[\dfrac{\sin x}{1} + \dfrac{2 \sin 2x}{2} + \dfrac{\sin 3x}{3} + \dfrac{\sin 5x}{5} + \dfrac{2 \sin 6x}{6} + \dfrac{\sin 7x}{7} + \ldots\right]$$

if $f(x) = 1$ for $o < x < \dfrac{\pi}{2}$, and $f(x) = o$ for $\dfrac{\pi}{2} < x < \pi$.

(d) $\sinh x =$

$$\dfrac{2 \sinh \pi}{\pi}\left[\dfrac{1}{2}\sin x - \dfrac{2}{5}\sin 2x + \dfrac{3}{10}\sin 3x - \dfrac{4}{17}\sin 4x + \ldots\right].$$

(e) $x^2 =$

$$\dfrac{2}{\pi}\left[\left(\dfrac{\pi^2}{1} - \dfrac{4}{1^3}\right)\sin x - \dfrac{\pi^2}{2}\sin 2x + \left(\dfrac{\pi^2}{3} - \dfrac{4}{3^3}\right)\sin 3x - \dfrac{\pi^2}{4}\sin 4x + \ldots\right].$$

ART. 7.   THE COSINE SERIES.

Let us now try to develop a given function of x in a series of cosines, using the method suggested by the last article.

Assume

$$f(x) = b_0 + b_1 \cos x + b_2 \cos 2x + b_3 \cos 3x + \ldots \qquad (1)$$

To determine any coefficient b_m multiply (1) by $\cos mx \cdot dx$ and integrate each term from o to π.

$$\int_0^\pi b_0 \cos mx \cdot dx = o.$$

$$\int_0^\pi b_k \cos kx \cos mx \cdot dx = o, \quad \text{if } m \text{ and } k \text{ are not equal.}$$

$$\int_0^\pi b_m \cos^2 mx \; dx = \dfrac{\pi}{2}b_m, \quad \text{if } m \text{ is not zero.}$$

Hence $b_m = \dfrac{2}{\pi}\displaystyle\int_0^\pi f(x) \cos mx \cdot dx,$ $\qquad (2)$

if m is not zero.

To get b_0 multiply (1) by dx and integrate from zero to π.

$$\int_0^\pi b_0 dx = b_0 \pi,$$

$$\int_0^\pi b_k \cos kx \cdot dx = 0.$$

Hence
$$b_0 = \frac{1}{\pi} \int_0^\pi f(x)dx, \tag{3}$$

which is just half the value that would be given by formula (2) if zero were substituted for m.

To save a separate formula (1) is usually written

$$f(x) = \tfrac{1}{2}b_0 + b_1 \cos x + b_2 \cos 2x + b_3 \cos 3x + \ldots, \tag{4}$$

and then the formula (2) will give b_0 as well as the other coefficients.

Prob. 5. Show that for values of x between 0 and π

(a)　$x = \dfrac{\pi}{2} - \dfrac{4}{\pi}\left(\dfrac{\cos x}{1^2} + \dfrac{\cos 3x}{3^2} + \dfrac{\cos 5x}{5^2} + \ldots\right);$

(b) $f(x) = \dfrac{\pi}{4} - \dfrac{8}{\pi}\left(\dfrac{\cos 2x}{2^2} + \dfrac{\cos 6x}{6^2} + \dfrac{\cos 10x}{10^2} + \ldots\right),$

if $f(x) = x$ for $0 < x < \dfrac{\pi}{2}$, and $f(x) = \pi - x$ for $\dfrac{\pi}{2} < x < \pi$;

(c) $f(x) = \dfrac{1}{2} + \dfrac{2}{\pi}\left(\dfrac{\cos x}{1} - \dfrac{\cos 3x}{3} + \dfrac{\cos 5x}{5} - \ldots\right),$

if $f(x) = 1$ for $0 < x < \dfrac{\pi}{2}$, and $f(x) = 0$ for $\dfrac{\pi}{2} < x < \pi$,

(d) $\sinh x = \dfrac{2}{\pi}\left[\dfrac{1}{2}(\cosh \pi - 1) - \dfrac{1}{2}(\cosh \pi + 1)\cos x\right.$

$\left. + \dfrac{1}{5}(\cosh \pi - 1)\cos 2x - \dfrac{1}{10}(\cosh \pi + 1)\cos 3x + \ldots\right];$

(e) $x^2 = \dfrac{\pi^2}{3} - 4\left(\dfrac{\cos x}{1^2} - \dfrac{\cos 2x}{2^2} + \dfrac{\cos 3x}{3^2} - \dfrac{\cos 4x}{4^2} + \ldots\right).$

ART. 8.   FOURIER'S SERIES.

Since a sine series is an odd function of x the development of an odd function of x in such a series must hold good from $x = -\pi$ to $x = \pi$, except perhaps for the value $x = 0$, where it is easily seen that the series is necessarily zero, no matter what the value of the function.   In like manner we see that if $f(x)$ is an even function of x its development in a cosine series must be valid from $x = -\pi$ to $x = \pi$.

Any function of x can be developed into a Trigonometric series to which it is equal for all values of x between $-\pi$ and π.

Let $f(x)$ be the given function of x.   It can be expressed as the sum of an even function of x and an odd function of x by the following device :

$$f(x) = \frac{f(x) + f(-x)}{2} + \frac{f(x) - f(-x)}{2} \tag{1}$$

identically; but $\dfrac{f(x) + f(-x)}{2}$ is not changed by reversing the sign of x and is therefore an even function of x; and when we reverse the sign of x, $\dfrac{f(x) - f(-x)}{2}$ is affected only to the extent of having its sign reversed, and is consequently an odd function of x.

Therefore for all values of x between $-\pi$ and π

$$\frac{f(x) + f(-x)}{2} = \frac{1}{2}b_0 + b_1 \cos x + b_2 \cos 2x + b_3 \cos 3x + \dots$$

where $$b_m = \frac{2}{\pi} \int_0^\pi \frac{f(x) + f(-x)}{2} \cos mx \,.\, dx;$$

and $$\frac{f(x) - f(-x)}{2} = a_1 \sin x + a_2 \sin 2x + a_3 \sin 3x + \dots$$

where $$a_m = \frac{2}{\pi} \int_0^\pi \frac{f(x) - f(-x)}{2} \sin mx \,.\, dx.$$

b_m and a_m can be simplified a little.

$$b_m = \frac{2}{\pi} \int_0^\pi \frac{f(x) + f(-x)}{2} \cos mx \cdot dx$$

$$= \frac{1}{\pi} \left[\int_0^\pi f(x) \cos mx \cdot dx + \int_0^\pi f(-x) \cos mx \cdot dx \right];$$

but if we replace x by $-x$, we get

$$\int_0^\pi f(-x) \cos mx \cdot dx = -\int_0^{-\pi} f(x) \cos mx \cdot dx = \int_{-\pi}^0 f(x) \cos mx \cdot dx,$$

and we have $b_m = \frac{1}{\pi} \int_{-\pi}^\pi f(x) \cos mx \cdot dx.$

In the same way we can reduce the value of a_m to

$$\frac{1}{\pi} \int_{-\pi}^\pi f(x) \sin mx \cdot dx.$$

Hence

$$f(x) = \frac{1}{2} b_0 + b_1 \cos x + b_2 \cos 2x + b_3 \cos 3x + \ldots$$

$$+ a_1 \sin x + a_2 \sin 2x + a_3 \sin 3x + \ldots, \quad (2)$$

where $b_m = \frac{1}{\pi} \int_{-\pi}^\pi f(x) \cos mx \cdot dx,$ (3)

and $a_m = \frac{1}{\pi} \int_{-\pi}^\pi f(x) \sin mx \cdot dx,$ (4)

and this development holds for all values of x between $-\pi$ and π.

The second member of (2) is known as a Fourier's Series.

The developments of Arts. 5 and 7 are special cases of development in Fourier's Series.

Prob 6. Show that for all values of x from $-\pi$ to π

$$e^x = \frac{2 \sinh \pi}{\pi} \left[\frac{1}{2} - \frac{1}{2} \cos x + \frac{1}{5} \cos 2x - \frac{1}{10} \cos 3x + \frac{1}{17} \cos 4x + \ldots \right]$$

$$+ \frac{2 \sinh \pi}{\pi} \left[\frac{1}{2} \sin x - \frac{2}{5} \sin 2x + \frac{3}{10} \sin 3x - \frac{4}{17} \sin 4x + \ldots \right].$$

Prob. 7. Show that formula (2), Art. 8, can be written

$$f(x) = \frac{1}{2} c_0 \cos \beta_0 + c_1 \cos (x - \beta_1) + c_2 \cos (2x - \beta_2)$$
$$+ c_3 \cos (3x - \beta_3) + \ldots,$$

where $c_m = (a_m^2 + b_m^2)^{\frac{1}{2}}$ and $\beta_m = \tan^{-1} \dfrac{a_m}{b_m}.$

Prob. 8. Show that formula (2), Art. 8, can be written

$$f(x) = \frac{1}{2} c_0 \sin \beta_0 + c_1 \sin (x + \beta_1) + c_2 \sin (2x + \beta_2)$$
$$+ c_3 \sin (3x + \beta_3) + \ldots,$$

where $c_m = (a_m^2 + b_m^2)^{\frac{1}{2}}$ and $\beta_m = \tan^{-1} \dfrac{b_m}{a_m}.$

ART. 9.   EXTENSION OF FOURIER'S SERIES.

In developing a function of x into a Trigonometric Series it is often inconvenient to be held within the narrow boundaries $x = -\pi$ and $x = \pi$. Let us see if we cannot widen them.

Let it be required to develop a function of x into a Trigonometric Series which shall be equal to $f(x)$ for all values of x between $x = -c$ and $x = c$.

Introduce a new variable

$$z = \frac{\pi}{c} x,$$

which is equal to $-\pi$ when $x = -c$, and to π when $x = c$.

$f(x) = f\left(\dfrac{c}{\pi} z\right)$ can be developed in terms of z by Art. 8, (2), (3), and (4). We have

$$f\left(\frac{c}{\pi} z\right) = \frac{1}{2} b_0 + b_1 \cos z + b_2 \cos 2z + b_3 \cos 3z + \ldots$$
$$+ a_1 \sin z + a_2 \sin 2z + a_3 \sin 3z + \ldots, \quad (\text{I})$$

where $b_m = \dfrac{1}{\pi} \displaystyle\int_{-\pi}^{\pi} f\left(\dfrac{c}{\pi} z\right) \cos mz \,.\, dz,$ (2)

and $$a_m = \frac{1}{\pi} \int_{-\pi}^{\pi} f\left(\frac{c}{\pi}z\right) \sin mz \cdot dz, \tag{3}$$

and (1) holds good from $z = -\pi$ to $z = \pi$.

Replace z by its value in terms of x and (1) becomes

$$f(x) = \frac{1}{2} b_0 + b_1 \cos \frac{\pi x}{c} + b_2 \cos \frac{2\pi x}{c} + b_3 \cos \frac{3\pi x}{c} + \ldots$$

$$+ a_1 \sin \frac{\pi x}{c} + a_2 \sin \frac{2\pi x}{c} + a_3 \sin \frac{3\pi x}{c} + \ldots; \tag{4}$$

and (2) and (3) can be transformed into

$$b_m = \frac{1}{c} \int_{-c}^{c} f(x) \cos \frac{m\pi x}{c} dx, \tag{5}$$

$$a_m = \frac{1}{c} \int_{-c}^{c} f(x) \sin \frac{m\pi x}{c} dx, \tag{6}$$

and (4) holds good from $x = -c$ to $x = c$.

In the formulas just obtained c may have as great a value as we please so that we can obtain a Trigonometric Series for $f(x)$ that will be equal to the given function through as great an interval as we may choose to take.

It can be shown that if this interval c is increased indefinitely the series will approach as its limiting form the double integral $\frac{1}{\pi} \int_{-\infty}^{\infty} f(\lambda) d\lambda \int_{0}^{\infty} \cos \alpha(\lambda - x) d\alpha$, which is known as a Fourier's Integral. So that

$$f(x) = \frac{1}{\pi} \int_{-\infty}^{+\infty} f(\lambda) d\lambda \int_{0}^{\infty} \cos \alpha(\lambda - x) d\alpha \tag{7}$$

for all values of x.

For the treatment of Fourier's Integral and for examples of its use in Mathematical Physics the student is referred to Riemann's Partielle Differentialgleichungen, to Schlömilch's Höhere Analysis, and to Byerly's Fourier's Series and Spherical Harmonics.

Prob. 9. Show that formula (4), Art. 9, can be written

$$f(x) = \frac{1}{2} c_0 \cos \beta_0 + c_1 \cos \left(\frac{\pi x}{c} - \beta_1 \right) + c_2 \cos \left(\frac{2 \pi x}{c} - \beta_2 \right)$$
$$+ c_3 \cos \left(\frac{3 \pi x}{c} - \beta_3 \right) + \ldots,$$

where $\qquad c_m = (a_m^2 + b_m^2)^{\frac{1}{2}} \quad$ and $\quad \beta_m = \tan^{-1} \dfrac{a_m}{b_m}.$

Prob. 10. Show that formula (4), Art. 9, can be written

$$f(x) = \frac{1}{2} c_0 \sin \beta_0 + c_1 \sin \left(\frac{\pi x}{c} + \beta_1 \right) + c_2 \sin \left(\frac{2 \pi x}{c} + \beta_2 \right)$$
$$+ c_3 \sin \left(\frac{3 \pi x}{c} + \beta_3 \right) + \ldots,$$

where $\qquad c_m = (a_m^2 + b_m^2)^{\frac{1}{2}} \quad$ and $\quad \beta_m = \tan^{-1} \dfrac{b_m}{a_m}.$

ART. 10.　DIRICHLET'S CONDITIONS.

In determining the coefficients of the Fourier's Series representing $f(x)$ we have virtually assumed, first, that a series of the required form and equal to $f(x)$ exists; and second, that it is *uniformly convergent*; and consequently we must regard the results obtained as only provisionally established.

It is, however, possible to prove rigorously that if $f(x)$ is finite and single-valued from $x = -\pi$ to $x = \pi$ and has not an infinite number of (finite) discontinuities, or of maxima or minima between $x = -\pi$ and $x = \pi$, the Fourier's Series of (2), Art. 8, and that Fourier's Series only, is equal to $f(x)$ for all values of x between $-\pi$ and π, excepting the values of x corresponding to the discontinuities of $f(x)$, and the values π and $-\pi$; and that if c is a value of x corresponding to a discontinuity of $f(x)$, the value of the series when $x = c$ is $\frac{1}{2} \underset{\epsilon = 0}{\operatorname{limit}} [f(c + \epsilon) + f(c - \epsilon)]$; and that when $x = \pi$ or $x = -\pi$ the value of the series is $\frac{1}{2}[f(\pi) + f(-\pi)]$.

This proof was first given by Dirichlet in 1829, and may be found in readable form in Riemann's Partielle Differential-gleichungen and in Picard's Traité d'Analyse, Vol. I.

A good deal of light is thrown on the peculiarities of trigonometric series by the attempt to construct approximately the curves corresponding to them.

If we construct $y = a_1 \sin x$ and $y = a_2 \sin 2x$ and add the ordinates of the points having the same abscissas, we shall obtain points on the curve

$$y = a_1 \sin x + a_2 \sin 2x.$$

If now we construct $y = a_3 \sin 3x$ and add the ordinates to those of $y = a_1 \sin x + a_2 \sin 2x$ we shall get the curve

$$y = a_1 \sin x + a_2 \sin 2x + a_3 \sin 3x.$$

By continuing this process we get successive approximations to

$$y = a_1 \sin x + a_2 \sin 2x + a_3 \sin 3x + a_4 \sin 4x + \cdots$$

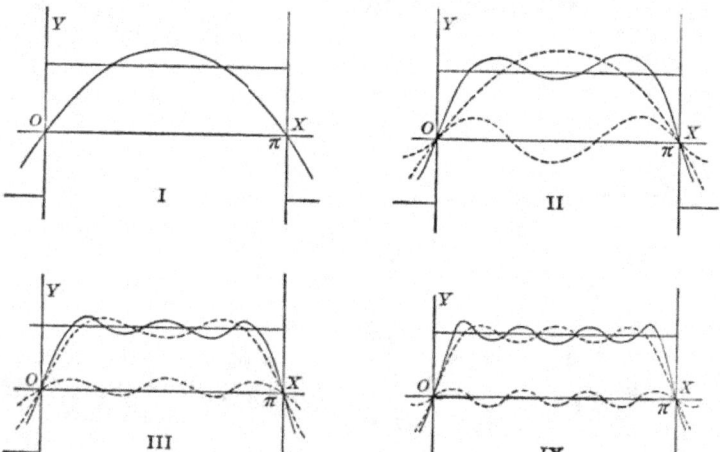

Let us apply this method to the series

$$y = \sin x + \tfrac{1}{3} \sin 3x + \tfrac{1}{5} \sin 5x + \cdots \qquad \text{(1) (See (7), Art. 6.)}$$

$y = 0$ when $x = 0$, $\dfrac{\pi}{4}$ from $x = 0$ to $x = \pi$, and 0 when $x = \pi$.

It must be borne in mind that our curve is periodic, having the period 2π, and is symmetrical with respect to the origin.

The preceding figures represent the first four approximation to this curve. In each figure the curve $y =$ the series, and the approximations in question are drawn in continuous lines, and the preceding approximation and the curve corresponding to the term to be added are drawn in dotted lines.

Prob. 11. Construct successive approximations to the series given in the examples at the end of Art. 6.

Prob. 12. Construct successive approximations to the Maclaurin's Series for sinh x, namely $x + \dfrac{x^3}{3!} + \dfrac{x^5}{5!} + \cdots$

ART. 11. APPLICATIONS OF TRIGONOMETRIC SERIES.

(a) Three edges of a rectangular plate of tinfoil are kept at potential zero, and the fourth at potential 1. At what potential is any point in the plate?

Here we have to solve Laplace's Equation (3), Art. 1, which, since the problem is two-dimensional, reduces to

$$\frac{\partial^2 V}{\partial x^2} + \frac{\partial^2 V}{\partial y^2} = 0, \tag{1}$$

subject to the conditions $V = 0$ when $x = 0$, (2)

$$V = 0 \quad \text{``} \quad x = a, \tag{3}$$
$$V = 0 \quad \text{``} \quad y = 0, \tag{4}$$
$$V = 1 \quad \text{``} \quad y = b. \tag{5}$$

Working as in Art. 3, we readily get $\sinh \beta y \sin \beta x$, $\sinh \beta y \cos \beta x$, $\cosh \beta y \sin \beta x$, and $\cosh \beta y \cos \beta x$ as particular values of V satisfying (1).

$V = \sinh \dfrac{m\pi y}{b} \sin \dfrac{m\pi x}{a}$ satisfies (1), (2), (3), and (4).

$$V = \frac{4}{\pi}\left[\frac{\sinh \dfrac{\pi y}{a}}{\sinh \dfrac{\pi b}{a}} \sin \frac{\pi x}{a} + \frac{1}{3}\frac{\sinh \dfrac{3\pi y}{a}}{\sinh \dfrac{3\pi b}{a}} \sin \frac{3\pi x}{a} + \cdots \right] \tag{6}$$

is the required solution, for it reduces to 1 when $y = b$. See (7), Art. 6.

(*b*) A harp-string is initially distorted into a given plane curve and then released ; find its motion.

The differential equation for the small transverse vibrations of a stretched elastic string is

$$\frac{\partial^2 y}{\partial t^2} = a^2 \frac{\partial^2 y}{\partial x^2},$$ (1)

as stated in Art. 1. Our conditions if we take one end of the string as origin are

$$y = 0 \text{ when } x = 0,$$ (2)
$$y = 0 \quad \text{``} \quad x = l,$$ (3)
$$\frac{\partial y}{\partial t} = 0 \quad \text{``} \quad t = 0,$$ (4)
$$y = fx \quad \text{``} \quad t = 0.$$ (5)

Using the method of Art. 3, we easily get as particular solutions of (1)

$$y = \sin \beta x \sin a\beta t, \qquad y = \sin \beta x \cos a\beta t,$$
$$y = \cos \beta x \sin a\beta t, \quad \text{and} \quad y = \cos \beta x \cos a\beta t.$$

$$y = \sin \frac{m\pi x}{l} \cos \frac{m\pi at}{l} \text{ satisfies } (1), (2), (3), \text{ and } (4).$$

$$y = \sum_{m=1}^{m=\infty} a_m \sin \frac{m\pi x}{l} \cos \frac{m\pi at}{l},$$ (6)

where $\qquad a_m = \frac{2}{l} \int_0^l f(x) \sin \frac{m\pi x}{l} . dx$ (7)

is our required solution ; for it reduces to $f(x)$ when $t = 0$. See Art. 9.

Prob. 13. Three edges of a square sheet of tinfoil are kept at potential zero, and the fourth at potential unity ; at what potential is the centre of the sheet? Ans. 0.25.

Prob. 14. Two opposite edges of a square sheet of tinfoil are kept at potential zero, and the other two at potential unity ; at what potential is the centre of the sheet ? Ans. 0.5.

Prob. 15. Two adjacent edges of a square sheet of tinfoil are kept at potential zero, and the other two at potential unity. At what potential is the centre of the sheet? Ans. o.5.

Prob. 16. Show that if a point whose distance from the end of a harp-string is $\frac{1}{n}$th the length of the string is drawn aside by the player's finger to a distance b from its position of equilibrium and then released, the form of the vibrating string at any instant is given by the equation

$$y = \frac{2bn^2}{(n-1)\pi^2} \sum_{m=1}^{m=\infty} \left(\frac{1}{m^2} \sin \frac{m\pi}{n} \sin \frac{m\pi x}{l} \cos \frac{m\pi at}{l} \right).$$

Show from this that all the harmonics of the fundamental note of the string which correspond to forms of vibration having nodes at the point drawn aside by the finger will be wanting in the complex note actually sounded.

Prob. 17.* An iron slab 10 centimeters thick is placed between and in contact with two other iron slabs each 10 centimeters thick. The temperature of the middle slab is at first 100 degrees Centigrade throughout, and of the outside slabs zero throughout. The outer faces of the outside slabs are kept at the temperature zero. Required the temperature of a point in the middle of the middle slab fifteen minutes after the slabs have been placed in contact. Given $a^2 = 0.185$ in C.G.S. units. Ans. $10°.3$.

Prob. 18.* Two iron slabs each 20 centimeters thick, one of which is at the temperature zero and the other at 100 degrees Centigrade throughout, are placed together face to face, and their outer faces are kept at the temperature zero. Find the temperature of a point in their common face and of points 10 centimeters from the common face fifteen minutes after the slabs have been put together.

Ans. $22°.8$; $15°.1$; $17°.2$.

Art. 12.† Properties of Zonal Harmonics.

In Art. 4, $z = P_m(x)$ was obtained as a particular solution of Legendre's Equation [(7), Art. 4] by the device of assuming that z could be expressed as a sum or a series of terms of the form $a_n x^n$ and then determining the coefficients. We

---

* See Art. 3.

† The student should review Art. 4 before beginning this article.

can, however, obtain a particular solution of Legendre's equation by an entirely different method.

The potential function for any point (x, y, z) due to a unit of mass concentrated at a given point (x_1, y_1, z_1) is

$$V = \frac{1}{\sqrt{(x - x_1)^2 + (y - y_1)^2 + (z - z_1)^2}}, \tag{1}$$

and this must be a particular solution of Laplace's Equation [(3), Art. 1], as is easily verified by direct substitution.

If we transform (1) to spherical coordinates we get

$$V = \frac{1}{\sqrt{r^2 - 2rr_1[\cos \theta \cos \theta_1 + \sin \theta \sin \theta_1 \cos (\phi - \phi_1)] + r_1^2}} \tag{2}$$

as a solution of Laplace's Equation in Spherical Coordinates [(5), Art. 1].

If the given point (x_1, y_1, z_1) is taken on the axis of X, as it must be in order that (2) may be independent of ϕ, $\theta_1 = 0$, and

$$V = \frac{1}{\sqrt{r^2 - 2rr_1 \cos \theta + r_1^2}} \tag{3}$$

is a solution of equation (1), Art. 4.

Equation (3) can be written

$$V = \frac{1}{r_1}\left(1 - 2\frac{r}{r_1} \cos \theta + \frac{r^2}{r_1^2}\right)^{-\frac{1}{2}}; \tag{4}$$

and if r is less than r_1 $\left(1 - 2\dfrac{r}{r_1} \cos \theta + \dfrac{r^2}{r_1^2}\right)^{-\frac{1}{2}}$ can be developed into a convergent power series. Let $\Sigma p_m \dfrac{r^m}{r_1^m}$ be this series, p_m being of course a function of θ. Then $V = \dfrac{1}{r_1}\Sigma p_m \dfrac{r^m}{r_1^m}$ is a solution of (1), Art. 4.

Substituting this value of V in the equation, and remembering that the result must be identically true, we get after a slight reduction

$$m(m + 1)p_m + \frac{1}{\sin \theta}\frac{d}{d\theta}\left[\sin \theta\frac{dp_m}{d\theta}\right] = 0;$$

but, as we have seen, the substitution of $x = \cos\theta$ reduces this to Legendre's equation [(7), Art. 4]. Hence we infer that the coefficient of the mth power of z in the development of $(1 - 2xz + z^2)^{-\frac{1}{2}}$ is a function of x that will satisfy Legendre's equation.

$$(1 - 2xz + z^2)^{-\frac{1}{2}} = [1 - z(2x - z)]^{-\frac{1}{2}},$$

and can be developed by the Binomial Theorem; the coefficient of z^m is easily picked out, and proves to be precisely the function of x which in Art. 4 we have represented by $P_m(x)$, and have called a Surface Zonal Harmonic.

We have, then,

$$(1 - 2xz + z^2)^{-\frac{1}{2}} = P_0(x) + P_1(x).z + P_2(x).z^2 + P_3(x).z^3 + \ldots \quad (5)$$

if the absolute value of z is less than 1.

If $x = 1$, (5) reduces to

$$(1 - 2z + z^2)^{-\frac{1}{2}} = P_0(1) + P_1(1).z + P_2(1).z^2 + P_3(1).z^3 + \ldots;$$

but $(1 - 2z + z^2)^{-\frac{1}{2}} = (1 - z)^{-1} = 1 + z + z^2 + z^3 + \ldots;$

hence $\qquad\qquad\qquad P_m(1) = 1. \qquad\qquad\qquad (6)$

Any Surface Zonal Harmonic may be obtained from the two of next lower orders by the aid of the formula

$$(n + 1)P_{n+1}(x) - (2n + 1)xP_n(x) + nP_{n-1}(x) = 0, \qquad (7)$$

which is easily obtained, and is convenient when the numerical value of x is given.

Differentiate (5) with respect to z, and we get

$$\frac{-(z - x)}{(1 - 2xz + z^2)^{\frac{3}{2}}} = P_1(x) + 2P_2(x).z + 3P_3(x).z^2 + \ldots,$$

whence

$$\frac{-(z - x)}{(1 - 2xz + z^2)^{\frac{1}{2}}} = (1 - 2xz + z^2)(P_1(x) + 2P_2(x).z + 3P_3(x).z^2 + \ldots),$$

or by (5)

$$(1 - 2xz + z^2)(P_1(x) + 2P_2(x).z + 3P_3(x).z^2 \cdots)$$
$$+ (z - x)(P_0(x) + P_1(x).z + P_2(x).z^2 + \cdots) = 0. \quad (8)$$

Now (8) is identically true, hence the coefficient of each power of z must vanish. Picking out the coefficient of z^n and writing it equal to zero, we have formula (7) above.

By the aid of (7) a table of Zonal Harmonics is easily computed since we have $P_0(x) = 1$, and $P_1(x) = x$. Such a table for $x = \cos\theta$ is given at the end of this chapter.

ART. 13. PROBLEMS IN ZONAL HARMONICS.

In any problem on Potential if V is independent of ϕ so that we can use the form of Laplace's Equation employed in Art. 4, and if the value of V on the axis of X is known, and can be expressed as $\Sigma a_m r^m$ or as $\sum \dfrac{b_m}{r^{m+1}}$, we can write out our required solution as

$$V = \Sigma a_m r^m P_m(\cos\theta) \quad \text{or} \quad V = \sum \frac{b_m P_m(\cos\theta)}{r^{m+1}};$$

for since $P_m(1) = 1$ each of these forms reduces to the proper value on the axis; and as we have seen in Art. 4 each of them satisfies the reduced form of Laplace's Equation.

As an example, let us suppose a statical charge of M units of electricity placed on a conductor in the form of a thin circular disk, and let it be required to find the value of the Potential Function at any point in the " field of force " due to the charge.

The surface density at a point of the plate at a distance r from its centre is

$$\sigma = \frac{M}{4a\pi \sqrt{a^2 - r^2}}$$

and all points of the conductor are at potential $\dfrac{\pi M}{2a}$. See Pierce's Newtonian Potential Function (§ 61).

The value of the potential function at a point in the axis of the plate at the distance x from the plate can be obtained without difficulty by a simple integration, and proves to be

$$V = \frac{M}{2a} \cos^{-1} \frac{x^2 - a^2}{x^2 + a^2}. \tag{1}$$

The second member of (1) is easily developed into a power series.

$$\frac{M}{2a} \cos^{-1} \frac{x^2 - a^2}{x^2 + a^2}$$

$$= \frac{M}{a}\left[\frac{\pi}{2} - \frac{x}{a} + \frac{x^3}{3a^3} - \frac{x^5}{5a^5} + \frac{x^7}{7a^7} - \dots\right] \text{ if } x < a \quad (2)$$

$$= \frac{M}{a}\left[\frac{a}{x} - \frac{a^3}{3x^3} + \frac{a^5}{5x^5} - \frac{a^7}{7x^7} + \dots\right] \text{ if } x > a. \quad (3)$$

Hence

$$V = \frac{M}{a}\left[\frac{\pi}{2} - \frac{r}{a}P_1(\cos\theta) + \frac{1}{3}\frac{r^3}{a^3}P_3(\cos\theta)\right.$$

$$\left. - \frac{1}{5}\frac{r^5}{a^5}P_5(\cos\theta) + \dots\right] \quad (4)$$

is our required solution if $r < a$ and $\theta < \dfrac{\pi}{2}$, as is

$$V = \frac{M}{a}\left[\frac{a}{r} - \frac{1}{3}\frac{a^3}{r^3}P_2(\cos\theta) + \frac{1}{5}\frac{a^5}{r^5}P_4(\cos\theta)\right.$$

$$\left. - \frac{1}{7}\frac{a^7}{r^7}P_6(\cos\theta) + \dots\right] \text{ if } r > a. \quad (5)$$

The series in (4) and (5) are convergent, since they may be obtained from the convergent series (2) and (3) by multiplying the terms by a set of quantities no one of which exceeds one in absolute value. For it will be shown in the next article that $P_m(\cos\theta)$ always lies between 1 and -1.

Prob. 19. Find the value of the Potential Function due to the attraction of a material circular ring of small cross-section.

The value on the axis of the ring can be obtained by a simple integration, and is $\dfrac{M}{\sqrt{c^2 + r^2}}$ if M is the mass and c the radius of the ring. At any point in space, if $r < c$

$$V = \frac{M}{c}\left[P_0(\cos\theta) - \frac{1}{2}\frac{r^2}{c^2}P_2(\cos\theta) + \frac{1\cdot 3}{2\cdot 4}\frac{r^4}{c^4}P_4(\cos\theta) - \dots\right],$$

and if $r > c$

$$V = \frac{M}{c}\left[\frac{c}{r}P_0(\cos\theta) - \frac{1}{2}\frac{c^3}{r^3}P_2(\cos\theta) + \frac{1\cdot3}{2\cdot4}\frac{c^5}{r^5}P_4(\cos\theta) - \dots\right].$$

Art. 14. Additional Forms.

(a) We have seen in Art. 12 that $P_m(x)$ is the coefficient of z^m in the development of $(1 - 2xz + z^2)^{-\frac{1}{2}}$ in a power series.

$$(1 - 2xz + z^2)^{-\frac{1}{2}} = [1 - z(e^{\theta i} + e^{-\theta i}) + z^2]^{-\frac{1}{2}}$$
$$= (1 - ze^{\theta i})^{-\frac{1}{2}}(1 - ze^{-\theta i})^{-\frac{1}{2}}.$$

If we develop $(1 - ze^{\theta i})^{-\frac{1}{2}}$ and $(1 - ze^{-\theta i})^{-\frac{1}{2}}$ by the Binomial Theorem their product will give a development for $(1 - 2xz + z^2)^{-\frac{1}{2}}$. The coefficient of z^m is easily picked out and reduced, and we get

$$P_m(\cos\theta) =$$
$$\frac{1\cdot3\cdot5\dots(2m-1)}{2\cdot4\cdot6\dots2m}\left[2\cos m\theta + 2\,\frac{1\cdot m}{1\cdot(2m-1)}\cos(m-2)\theta\right.$$
$$\left.+ 2\frac{1\cdot3\cdot m(m-1)}{1\cdot2\cdot(2m-1)(2m-3)}\cos(m-4)\theta + \dots\right] \qquad (1)$$

If m is odd the parenthesis in (1) ends with the term containing $\cos\theta$; if m is even, with the term containing $\cos 0$, but in the latter case the term in question will not be multiplied by the factor 2, which is common to all the other terms.

Since all the coefficients in the second member of (1) are positive, $P_m(\cos\theta)$ has its maximum value when $\theta = 0$, and its value then has already been shown in Art. 12 to be unity. Obviously, then, its minimum value cannot be less than -1.

(b) If we integrate the value of $P_m(x)$ given in (11), Art. 4, m times in succession with respect to x, the result will be found to differ from $\frac{1\cdot3\cdot5\dots(2m-1)}{(2m)!}(x^2-1)^m$ by terms involving lower powers of x than the mth.

Hence $\qquad P_m(x) = \frac{1}{2^m m!}\frac{d^m}{dx^m}(x^2-1)^m.$ $\qquad\qquad$ (2)

(*c*) Other forms for $P_m(x)$, which we give without demonstration, are

$$P_m(x) = \frac{(-1)^m}{m!} \frac{\partial^m}{\partial x^m} \frac{1}{\sqrt{x^2 + y^2 + z^2}}. \qquad (3)$$

$$P_m(x) = \frac{1}{\pi} \int_0^\pi [x + \sqrt{x^2 - 1} \cdot \cos \phi]^m d\phi. \qquad (4)$$

$$P_m(x) = \frac{1}{\pi} \int_0^\pi \frac{d\phi}{[x - \sqrt{x^2 - 1} \cdot \cos \phi]^{m+1}}. \qquad (5)$$

(4) and (5) can be verified without difficulty by expanding and integrating.

ART. 15. DEVELOPMENT IN TERMS OF ZONAL HARMONICS.

Whenever, as in Art. 4, we have the value of the Potential Function given on the surface of a sphere, and this value depends only on the distance from the extremity of a diameter, it becomes necessary to develop a function of θ into a series of the form

$$A_0 P_0(\cos \theta) + A_1 P_1(\cos \theta) + A_2 P_2(\cos \theta) + \ldots;$$

or, what amounts to the same thing, to develop a function of x into a series of the form

$$A_0 P_0(x) + A_1 P_1(x) + A_2 P_2(x) + \ldots.$$

The problem is entirely analogous to that of development in sine-series treated at length in Art. 6, and may be solved by the same method.

Assume $\quad f(x) = A_0 P_0(x) + A_1 P_1(x) + A_2 P_2(x) + \ldots \qquad (1)$

for $-1 < x < 1$. Multiply (1) by $P_m(x)dx$ and integrate from -1 to 1. We get

$$\int_{-1}^1 f(x) P_m(x) dx = \sum_{n=0}^{n=\infty} [A_n \int_{-1}^1 P_m(x) P_n(x) dx]. \qquad (2)$$

We shall show in the next article that

$$\int_{-1}^{1} P_m(x)P_n(x)dx = 0, \quad \text{unless } m = n,$$

and that
$$\int_{-1}^{1} [P_m(x)]^2 dx = \frac{2}{2m+1}.$$

Hence
$$A_m = \frac{2m+1}{2} \int_{-1}^{1} f(x)P_m(x)dx. \tag{3}$$

It is important to notice here, as in Art. 6, that the method we have used in obtaining A_m amounts essentially to determining A_m, so that the equation

$$f(x) = A_0 P_0(x) + A_1 P_1(x) + A_2 P_2(x) + \dots + A_n P_n(x)$$

shall hold good for $n + 1$ equidistant values of x between -1 and 1, and taking its limiting value as n is indefinitely increased.

ART. 16.   FORMULAS FOR DEVELOPMENT.

We have seen in Art. 4 that $z = P_m(x)$ is a solution of

Legendre's Equation $\dfrac{d}{dx}\left[(1 - x^2)\dfrac{d^2z}{dx}\right] + m(m+1)z = 0.$ (1)

Hence
$$\frac{d}{dx}\left[(1 - x^2)\frac{dP_m(x)}{dx}\right] + m(m+1)P_m(x) = 0, \tag{2}$$

and
$$\frac{d}{dx}\left[(1 - x^2)\frac{dP_n(x)}{dx}\right] + n(n+1)P_n(x) = 0. \tag{3}$$

Multiply (2) by $P_n(x)$ and (3) by $P_m(x)$, subtract, transpose, and integrate.   We have

$$[m(m+1) - n(n+1)]\int_{-1}^{1} P_m(x)P_n(x)dx$$

$$= \int_{-1}^{1} P_m(x)\frac{d}{dx}\left[(1 - x^2)\frac{dP_n(x)}{dx}\right]dx$$

$$- \int_{-1}^{1} P_n(x) \frac{d}{dx}\left[(1 - x^2)\frac{dP_m(x)}{dx} \right] dx \quad (4)$$

$$= \left[P_m(x)(1 - x^2)\frac{dP_n(x)}{dx} - P_n(x)(1 - x^2)\frac{dP_m(x)}{dx} \right]_{-1}^{1}$$

$$- \int_{-1}^{1} (1 - x^2)\frac{dP_n(x)}{dx} \frac{dP_m(x)}{dx} \cdot dx$$

$$+ \int_{-1}^{1} (1 - x^2)\frac{dP_m(x)}{dx} \frac{dP_n(x)}{dx} \cdot dx \quad (5)$$

by integration by parts,

= 0.

Hence
$$\int_{-1}^{1} P_m(x)P_n(x)dx = 0, \quad (6)$$

unless $m = n$.

If in (4) we integrate from x to 1 instead of from -1 to 1, we get an important formula.

$$\int_x^{1} P_m(x)P_n(x)dx = \frac{(1 - x^2)\left[P_n(x)\frac{dP_m(x)}{dx} - P_m(x)\frac{dP_n(x)}{dx} \right]}{m(m + 1) - n(n + 1)}, \quad (7)$$

and as a special case, since $P_0(x) = 1$.

$$\int_x^{1} P_m(x)dx = \frac{(1 - x^2)\frac{dP_m(x)}{dx}}{m(m + 1)}, \quad (8)$$

unless $m = 0$.

To get $\int_{-1}^{1}[P_m(x)]^2dx$ is not particularly difficult. By (2),

Art. 14,

$$\int_{-1}^{1} [P_m(x)]^2dx = \frac{1}{2^{2m}(m!)^2} \int_{-1}^{1} \frac{d^m(x^2 - 1)^m}{dx^m} \cdot \frac{d^m(x^2 - 1)^m}{dx^m} \cdot dx \quad (9)$$

By successive integrations by parts, noting that

$\frac{d^{m-\kappa}}{dx^{m-\kappa}}(x^2 - 1)^m$ contains $(x^2 - 1)^\kappa$ as a factor if $\kappa < m$, and that $\dfrac{d^{2m}(x^2-1)^m}{dx^{2m}} = (2m)!$ we get

$$\int_{-1}^{1}[P_m(x)]^2dx = \frac{(-1)^m(2m)!}{2^{2m}(m!)^2}\int_{-1}^{1}(x^2-1)^mdx. \tag{10}$$

$$\int_{-1}^{1}(x^2-1)^mdx = \int_{-1}^{1}(x-1)^m(x+1)^mdx$$

$$= -\frac{m}{m+1}\int_{-1}^{1}(x-1)^{m-1}(x+1)^{m+1}dx$$

$$= (-1)^m\frac{m!\,m!}{(2m)!}\int_{-1}^{1}(x+1)^{2m}dx = (-1)^m\frac{2^{2m+1}(m!)^2}{(2m+1)!}.$$

Hence $$\int_{-1}^{1}[P_m(x)]^2dx = \frac{2}{2m+1}. \tag{11}$$

Prob. 20. Show that $\int_{0}^{1}P_m(x)dx = 0$ if m is even and is not zero

$$= (-1)^{\frac{m-1}{2}}\frac{1}{m(m+1)}\cdot\frac{3\cdot5\cdot7\cdots m}{2\cdot4\cdot4\cdots(m-1)}\text{ if } m \text{ is odd.}$$

Prob. 21. Show that $\int_{0}^{1}[P_m(x)]^2dx = \frac{1}{2m+1}.$ Note that

$[P_m(x)]^2$ is an even function of x.

Prob. 22. Show that if $f(x) = 0$ from $x = -1$ to $x = 0$, and $f(x) = 1$ from $x = 0$ to $x = 1$,

$$f(x) = \frac{1}{2} + \frac{3}{4}P_1(x) - \frac{7}{8}\cdot\frac{1}{2}P_3(x) + \frac{11}{12}\cdot\frac{1\cdot3}{2\cdot4}P_5(x) - \dots$$

Prob. 23. Show that $F(\theta) = \overset{m=\infty}{\underset{m=0}{\Sigma}} B_mP_m(\cos\theta)$ where

$$B_m = \frac{2m+1}{2}\int_{0}^{\pi}F(\theta)P_m(\cos\theta)\sin\theta\,d\theta.$$

Prob. 24. Show that

$$\csc \theta = \frac{\pi}{2}\left[1 + 5\left(\frac{1}{2}\right)^2 P_2(\cos\theta) + 9\left(\frac{1\cdot 3}{2\cdot 4}\right)^2 P_4(\cos\theta) + \dots \right].$$

See (1), Art. 14.

Prob. 25. Show that

$$x^n = \frac{n!}{1\cdot 3\cdot 5 \cdots (2n+1)}\left[(2n+1)P_n(x) + (2n-3)\frac{2n+1}{2}P_{n-2}(x) \right.$$

$$\left. + (2n-7)\frac{(2n+1)(2n-1)}{2\cdot 4}P_{n-4}(x) + \dots \right].$$

Note that $\int_{-1}^{1} x^n P_n(x)dx = \frac{1}{2^m m!}\int_{-1}^{1} x^n \frac{d^m(x^2-1)^m}{dx^m}dx$, and use the

method of integration by parts freely.

Prob. 26. Show that if V is the value of the Potential Function at any point in a field of force, not imbedded in attracting or repelling matter; and if $V = f(\theta)$ when $r = a$,

$$V = \Sigma A_m \frac{r^m}{a^m} P_m(\cos\theta) \text{ if } r < a$$

and

$$V = \Sigma A_m \frac{a^{m+1}}{r^{m+1}} P_m(\cos\theta) \text{ if } r > a,$$

where

$$A_m = \frac{2m+1}{2}\int_0^{\pi} f(\theta)P_m(\cos\theta)\sin\theta d\theta.$$

Prob. 27. Show that if

$$V = c \text{ when } r = a; \quad V = c \text{ if } r < a, \text{ and } V = \frac{ca}{r} \text{ if } r > a.$$

ART. 17.   FORMULAS IN ZONAL HARMONICS.

The following formulas which we give without demonstration may be found useful for reference:

$$\frac{dP_n(x)}{dx} = (2n-1)P_{n-1}(x) + (2n-5)P_{n-3}(x) + (2n-9)P_{n-5}(x) + \dots (1)$$

$$\frac{dP_{n+1}(x)}{dx} - \frac{dP_{n-1}(x)}{dx} = (2n+1)P_n(x) \tag{2}$$

$$\int_x^1 P_n(x)dx = \frac{1}{2n+1}[P_{n-1}(x) - P_{n+1}(x)]. \tag{3}$$

ART. 18. SPHERICAL HARMONICS.

In problems in Potential where the value of V is given on the surface of a sphere, but is not independent of the angle ϕ, we have to solve Laplace's Equation in the form (5), Art. 1, and by a treatment analogous to that given in Art. 4 it can be proved that

$$V = r^m \cos n\phi \, \sin^n \theta \frac{d^n P_m(\mu)}{d\mu_n} \quad \text{and} \quad V = r^m \sin n\phi \, \sin^n \theta \frac{d^n P_m(\mu)}{d\mu^n},$$

where $\mu = \cos \theta$, are particular solutions of (5), Art. 1.

The factors multiplied by r^m in these values are known as Tesseral Harmonics. They are functions of ϕ and θ, and they play nearly the same part in unsymmetrical problems that the Zonal Harmonics play in those independent of ϕ.

$$Y_m(\mu, \phi) = A_0 P_m(\mu) + \sum_{m=1}^{n=m} (A_n \cos n\phi + B_n \sin n\phi)\sin^n \theta \frac{d^n P_m(\mu)}{d\mu^n}$$

is known as a Surface Spherical Harmonic of the mth degree,

and $\qquad V = r^m Y_m(\mu, \phi) \quad \text{and} \quad V = \frac{1}{r^{m+1}} Y_m(\mu, \phi)$

satisfy Laplace's Equation, (5), Art. 1.

The Tesseral and the Zonal Harmonics are special cases of the Spherical Harmonic, as is also a form $P_m(\cos \gamma)$ known as a Laplace's Coefficient or a Laplacian : γ standing for the angle between r and the radius vector r, of some fixed point.

For the properties and uses of Spherical Harmonics we refer the student to more extended treatises, namely, to Ferrer's Spherical Harmonics, to Heine's Kugelfunctionen, or to Byerly's Fourier's Series and Spherical Harmonics.

ART. 19.* BESSEL'S FUNCTIONS. PROPERTIES.

We have seen in Art. 5 that $z = J_0(x)$ where

$$J_0(x) = 1 - \frac{x^2}{2^2} + \frac{x^4}{2^2 \cdot 4^2} - \frac{x^6}{2^2 \cdot 4 \cdot 6^2} + \cdots \qquad (1)$$

* The student should review Art. 5 before reading this article.

is a solution of the equation

$$\frac{d^2z}{dx^2} + \frac{1}{x}\frac{dz}{dx} + z = 0;\tag{2}$$

and we have called $J_0(x)$ a Bessel's Function or Cylindrical Harmonic of the zero order.

$$J_1(x) = -\frac{dJ_0(x)}{dx} = \frac{x}{2}\left[1 - \frac{x^2}{2.4} + \frac{x^4}{2.4^2.6} - \frac{x^6}{2.4^2.6^2.8} + \cdots\right]\tag{3}$$

is called a Bessel's Function of the first order, and

$$z' = J_1(x)$$

is a solution of the equation

$$\frac{d^2z'}{dx^2} + \frac{1}{x}\frac{dz'}{dx} + \left(1 - \frac{1}{x^2}\right)z' = 0,\tag{4}$$

which is the result of differentiating (2) with respect to x.

A table giving values of $J_0(x)$ and $J_1(x)$ will be found at the end of this chapter.

If we write $J_0(x)$ for z in equation (2), then multiply through by xdx and integrate from zero to x, simplifying the resulting equation by integration by parts, we get

$$\frac{xdJ_0(x)}{dx} + \int_0^x xJ_0(x)dx = 0,$$

or, since $J_1(x) = -\dfrac{dJ_0(x)}{dx}$,

$$\int_0^x xJ_0(x)dx = xJ_1(x).\tag{5}$$

If we write $J_0(x)$ for z in equation (2), then multiply through by $x^2\dfrac{dJ_0(x)}{dx}$, and integrate from zero to x, simplifying by integration by parts, we get

$$\frac{x^2}{2}\left[\left(\frac{dJ_0(x)}{dx}\right)^2 + (J_0(x))^2\right] - \int_0^x x(J_0(x))^2dx = 0,$$

or

$$\int_0^x x(J_0(x))^2dx = \frac{x^2}{2}\left[(J_0(x))^2 + (J_1(x))^2\right].\tag{6}$$

If we replace x by μx in (2) it becomes

$$\frac{d^2z}{dx^2} + \frac{1}{x}\frac{dz}{dx} + \mu^2 z = 0 \qquad (7)$$

(See (8), Art. 5).   Hence $z = J_0(\mu x)$ is a solution of (7).

If we substitute in turn in (7) $J_0(\mu_\kappa x)$ and $J_0(\mu_\iota x)$ for z, multiply the first equation by $xJ_0(\mu_\iota x)$, the second by $xJ_0(\mu_\kappa x)$, subtract the second from the first, simplify by integration by parts, and reduce, we get

$$\int_0^a xJ_0(\mu_\kappa x)J_0(\mu_\iota x)dx$$

$$= \frac{1}{\mu_\kappa^2 - \mu_\iota^2}[\mu_\kappa aJ_0(\mu_\iota a)J_1(\mu_\kappa a) - \mu_\iota aJ_0(\mu_\kappa a)J_1(\mu_\iota a)]. \qquad (8)$$

Hence if μ_κ and μ_ι are different roots of $J_0(\mu a) = 0$, or of $J_1(\mu a) = 0$, or of $\mu a J_1(\mu a) - \lambda J_0(\mu a) = 0$,

$$\int_0^a xJ_0(\mu_\kappa x)J_0(\mu_\iota x)dx = 0. \qquad (9)$$

We give without demonstration the following formulas, which are sometimes useful :

$$J_0(x) = \frac{1}{\pi}\int_0^\pi \cos(x\cos\phi)d\phi. \qquad (10)$$

$$J_1(x) = \frac{x}{\pi}\int_0^\pi \sin^2\phi\cos(x\cos\phi)d\phi. \qquad (11)$$

They can be confirmed by developing $\cos(x\cos\phi)$, integrating, and comparing with (1) and (3).

ART. 20.   APPLICATIONS OF BESSEL'S FUNCTIONS.

(a) The problem of Art. 5 is a special case of the following : The convex surface and one base of a cylinder of radius a and length b are kept at the constant temperature zero, the temperature at each point of the other base is a given function of the distance of the point from the center of the base ; required the temperature of any point of the cylinder after the permanent temperatures have been established.

Here we have to solve Laplace's Equation in the form

$$\frac{\partial^2 u}{\partial r^2} + \frac{1}{r}\frac{\partial u}{\partial r} + \frac{\partial^2 u}{\partial z^2} = 0 \qquad (1)$$

(see Art. 5), subject to the conditions

$$u = 0 \text{ when } z = 0,$$

$$u = 0 \quad \text{``} \quad r = a,$$

$$u = f(r) \text{ `` } z = b.$$

Starting with the particular solution of (1),

$$u = \sinh(\mu z) J_0(\mu r), \qquad (2)$$

and proceeding as in Art. 5, we get, if $\mu_1, \mu_2, \mu_3, \ldots$ are roots of

$$J_0(\mu a) = 0, \qquad (3)$$

and $\quad f(r) = A_1 J_0(\mu_1 r) + A_2 J_0(\mu_2 r) + A_3 J_0(\mu_3 r) + \ldots, \quad (4)$

$$u = A_1 \frac{\sinh(\mu_1 z)}{\sinh(\mu_1 b)} J_0(\mu_1 r) + A_2 \frac{\sinh(\mu_2 z)}{\sinh(\mu_2 b)} J_0(\mu_2 r) + \ldots. \quad (5)$$

(b) If instead of keeping the convex surface of the cylinder at temperature zero we surround it by a jacket impervious to heat the equation of condition, $u = 0$ when $r = a$, will be re-placed by $\dfrac{\partial u}{\partial r} = 0$ when $r = a$, or if $u = \sinh(\mu z) J_0(\mu r)$ by

$$\frac{dJ_0(\mu r)}{dr} = 0 \quad \text{when } r = a,$$

that is, by $\qquad -\mu J_1(\mu a) = 0$

or $\qquad\qquad\qquad J_1(\mu a) = 0. \qquad (6)$

If now in (4) and (5) $\mu_1, \mu_2, \mu_3, \ldots$ are roots of (6), (5) will be the solution of our new problem.

(c) If instead of keeping the convex surface of the cylinder at the temperature zero we allow it to cool in air which is at the temperature zero, the condition $u = 0$ when $r = a$ will be replaced by $\dfrac{\partial u}{\partial r} + hu = 0$ when $r = a$, h being the coefficient of surface conductivity.

If $u = \sinh(\mu z)J_0(\mu r)$ this condition becomes

$$- \mu J_1(\mu r) + hJ_0(\mu r) = 0 \quad \text{when } r = a,$$

or
$$\mu aJ_1(\mu a) - ahJ_0(\mu a) = 0. \tag{7}$$

If now in (4) and (5) μ_1, μ_2, μ_3, ... are roots of (7), (5) will be the solution of our present problem.

It can be shown that

$$J_0(x) = 0, \tag{8}$$
$$J_1(x) = 0, \tag{9}$$

and
$$xJ_1(x) - \lambda J_0(x) = 0 \tag{10}$$

have each an infinite number of real positive roots.* The earlier roots of these equations can be obtained without serious difficulty from the table for $J_0(x)$ and $J_1(x)$ at the end of this chapter.

Art. 21. Development in Terms of Bessel's Functions.

We shall now obtain the developments called for in the last article.

Let
$$f(r) = A_1 J_0(\mu_1 r) + A_2 J_0(\mu_2 r) + A_3 J_0(\mu_3 r) + \ldots \tag{1}$$

μ_1, μ_2, μ_3, etc., being roots of $J_0(\mu a) = 0$, or of $J_1(\mu a) = 0$, or of

$$\mu aJ_1(\mu a) - \lambda J_0(\mu a) = 0.$$

To determine any coefficient A_k multiply (1) by $rJ_0(\mu_k r)dr$ and integrate from zero to a. The first member will become

$$\int_0^a rf(r)J_0(\mu_k r)dr.$$

Every term of the second member will vanish by (9), Art. 19, except the term

$$A_k \int_0^a r[J_0(\mu_k r)]^2 dr.$$

$$\int_0^a r[J_0(\mu_k r)]^2 dr = \frac{1}{\mu_k^2}\int_0^{\mu_k a} x[J_0(x)]^2 dx = \frac{a^2}{2}\left([J_0(\mu_k a)]^2 + [J_1(\mu_k a)]^2\right)$$

by (6), Art. 19.

* See Riemann's Partielle Differentialgleichungen, § 97.

Hence $A_k = \dfrac{2}{a^2([J_0(\mu_k a)]^2 + [J_1(\mu_k a)]^2)} \displaystyle\int_0^a rf(r)J_0(\mu_k r)dr.$ (2)

The development (1) holds good from $r = 0$ to $r = a$ (see Arts. 6 and 15).

If μ_1, μ_2, μ_3, etc., are roots of $J_0(\mu a) = 0$, (2) reduces to

$$A_k = \frac{2}{a^2[J_1(\mu_k a)]^2}\int_0^a rf(r)J_0(\mu_k r)dr. \tag{3}$$

If μ_1, μ_2, μ_3, etc., are roots of $J_1(\mu a) = 0$, (2) reduces to

$$A_k = \frac{2}{a^2[J_0(\mu_k a)]^2}\int_0^a rf(r)J_0(\mu_k r)dr. \tag{4}$$

If μ_1, μ_2, μ_3, etc., are roots of $\mu a J_1(\mu a) - \lambda J_0(\mu a) = 0$, (2) reduces to

$$A_k = \frac{2\mu_k^2}{(\lambda^2 + \mu_k^2 a^2)[J_0(\mu_k a)]^2}\int_0^a rf(r)J_0(\mu_k r)dr. \tag{5}$$

For the important case where $f(r) = 1$

$$\int_0^a rf(r)J_0(\mu_k r)dr = \int_0^a rJ_0(\mu_k r)dr = \frac{1}{\mu_k^2}\int_0^{\mu_k a} xJ_0(x)dx = \frac{a}{\mu_k}J_1(\mu_k a) \tag{6}$$

by (5), Art. 19; and (3) reduces to

$$A_k = \frac{2}{\mu_k a J_1(\mu_k a)}; \tag{7}$$

(4) reduces to

$$A_k = 0, \tag{8}$$

except for $k = 1$, when $\mu_k = 0$, and we have

$$A_1 = 1; \tag{9}$$

(5) reduces to $A_k = \dfrac{2\lambda}{(\lambda^2 + \mu_k^2 a^2)J_0(\mu_k a)}.$ (10)

Prob. 28. A cylinder of radius one meter and altitude one meter has its upper surface kept at the temperature 100°, and its base and convex surface at the temperature 15°, until the stationary temperatures are established. Find the temperature at points on the axis 25, 50, and 75 centimeters from the base, and also at a point 25 centimeters from the base and 50 centimeters from the axis.

Ans. 29°.6; 47°.6; 71°.2; 25°.8

Prob. 29. An iron cylinder one meter long and 20 centimeters in diameter has its convex surface covered with a so-called non-conducting cement one centimeter thick. One end and the convex surface of the cylinder thus coated are kept at the temperature zero, the other end at the temperature of 100 degrees. Given that the conductivity of iron is 0.185 and of cement 0.000162 in C. G. S. units.

Find to the nearest tenth of a degree the temperature of the middle point of the axis, and of the points of the axis 20 centimeters from each end after the temperatures have ceased to change.

Find also the temperature of a point on the surface midway between the ends, and of points of the surface 20 centimeters from each end. Find the temperatures of the three points of the axis, supposing the coating a perfect non-conductor, and again, supposing the coating absent. Neglect the curvature of the coating. Ans. $15°.4$; $40°.85$; $72°.8$; $15°.3$; $40°.7$; $72°.5$; $0°.0$; $0°.0$; $1°.3$.

Prob. 30. If the temperature at any point in an infinitely long cylinder of radius c is initially a function of the distance of the point from the axis, the temperature at any time must satisfy the equation $\dfrac{\partial u}{\partial t} = a^2 \left(\dfrac{\partial^2 u}{\partial r^2} + \dfrac{1}{r} \dfrac{\partial u}{\partial r} \right)$ (see Art. 1), since it is clearly independent of z and ϕ.

Show that

$$u = A_1 e^{-a^2\mu_1^2 t} J_0(\mu_1 r) + A_2 e^{-a^2\mu_2^2 t} J_0(\mu_2 r)$$
$$+ A_3 e^{-a^2\mu_3^2 t} J_0(\mu_3 r) + \cdots,$$

where, if the surface of the cylinder is kept at the temperature zero, μ_1, μ_2, μ_3, ... are roots of $J_0(\mu c) = 0$ and A_k is the value given in (3) with c written in place of a; if the surface of the cylinder is adiabatic μ_1, μ_2, μ_3, ... are roots of $J_1(\mu c) = 0$ and A_k is obtained from (4); and if heat escapes at the surface into air at the temperature zero μ_1, μ_2, μ_3, ... are roots of $\mu c J_1(\mu c) - \lambda J_0(\mu c) = 0$, and A_k is obtained from (5).

Prob. 31. If the cylinder described in problem 29 is very long and is initially at the temperature $100°$ throughout, and the convex surface is kept at the temperature $0°$, find the temperature of a point 5 centimeters from the axis 15 minutes after cooling has begun; first when the cylinder is coated, and second, when the coating is absent. Ans. $97°.2$; $0°.01$.

Prob. 32. A circular drumhead of radius a is initially slightly distorted into a given form which is a surface of revolution about the axis of the drum, and is then allowed to vibrate, and z is the ordinate of any point of the membrane at any time. Assuming that

z must satisfy the equation $\dfrac{\partial^2 z}{\partial t^2} = c^2\left(\dfrac{\partial^2 z}{\partial r^2} + \dfrac{1}{r}\dfrac{\partial z}{\partial r}\right)$, subject to the conditions $z = 0$ when $r = a$, $\dfrac{\partial z}{\partial t} = 0$ when $t = 0$, and $z = f(r)$ when $t = 0$, show that $z = A_1 J_0(\mu_1 r)\cos \mu_1 ct + A_2 J_0(\mu_2 r)\cos \mu_2 ct + \ldots$ where μ_1, μ_2, μ_3, ... are roots of $J_0(\mu a) = 0$ and A_k has the value given in (3).

Prob. 33. Show that if a drumhead be initially distorted as in problem 32 it will not in general give a musical note ; that it may be initially distorted so as to give a musical note ; that in this case the vibration will be a steady vibration ; that the periods of the various musical notes that can be given are proportional to the roots of $J_0(x) = 0$, and that the possible nodal lines for such vibrations are concentric circles whose radii are proportional to the roots of $J_0(x) = 0$.

ART. 22.   PROBLEMS IN BESSEL'S FUNCTIONS.

If in a problem on the stationary temperatures of a cylinder $u = 0$ when $z = 0$, $u = 0$ when $z = b$, and $u = f(z)$ when $r = a$, the problem is easily solved.   If in (2), Art. 20, and in the corresponding solution $z = \cosh(\mu z)J_0(\mu r)$ we replace μ by μi, we can readily obtain $z = \sin(\mu z)J_0(\mu r i)$ and $z = \cos(\mu z)J_0(\mu r i)$ as particular solutions of (1), Art. 20; and

$$J_0(xi) = 1 + \frac{x^2}{2} + \frac{x^4}{2^2 \cdot 4^2} + \frac{x^6}{2^2 \cdot 4^2 \cdot 6^2} + \cdots \qquad (1)$$

and is real.

$$f(z) = \sum_{k=1}^{k=\infty} A_k \sin\frac{k\pi z}{b}$$

where

$$A_k = \frac{2}{b}\int_0^b f(z)\sin\frac{k\pi z}{b}\,dz \qquad (2)$$

by Art. 9.

Hence

$$u = \sum_{k=1}^{k=\infty} A_k \sin\frac{k\pi z}{b}\frac{J_0\left(\dfrac{k\pi r i}{b}\right)}{J_0\left(\dfrac{k\pi a i}{b}\right)} \qquad (3)$$

is the required solution.

A table giving the values of $J_0(xi)$ will be found at the end of this chapter.

Prob. 34. A cylinder two feet long and two feet in diameter has its bases kept at the temperature zero and its convex surface at 100 degrees Centigrade until the internal temperatures have ceased to change. Find the temperature of a point on the axis half way between the bases, and of a point six inches from the axis, half way between the bases. Ans. 72.°1; 80°.1.

ART. 23. BESSEL'S FUNCTIONS OF HIGHER ORDER.

If we are dealing with Laplace's Equation in Cylindrical Coordinates and the problem is not symmetrical about an axis, functions of the form

$$J_n(x) = \frac{x^n}{2^n \Gamma(n+1)} \left[1 - \frac{x^2}{2^2(n+1)} + \frac{x^4}{2^4 . 2!(n+1)(n+2)} - \cdots \right]$$

play very much the same part as that played by $J_0(x)$ in the preceding articles. They are known as Bessel's Functions of the nth order. In problems concerning hollow cylinders much more complicated functions enter, known as Bessel's Functions of the second kind.

For a very brief discussion of these functions the reader is referred to Byerly's Fourier's Series and Spherical Harmonics; for a much more complete treatment to Gray and Matthews' admirable treatise on Bessel's Functions.

ART. 24. LAMÉ'S FUNCTIONS.

Complicated problems in Potential and in allied subjects are usually handled by the aid of various forms of curvilinear coördinates, and each form has its appropriate Harmonic Functions, which are usually extremely complicated. For instance, Lamé's Functions or Ellipsoidal Harmonics are used when solutions of Laplace's Equation in Ellipsoidal coordinates are required; Toroidal Harmonics when solutions of Laplace's Equation in Toroidal coordinates are needed.

For a brief introduction to the theory of these functions see Byerly's Fourier's Series and Spherical Harmonics.

TABLE I. SURFACE ZONAL HARMONICS.

θ	$P_1(\cos\theta)$	$P_2(\cos\theta)$	$P_3(\cos\theta)$	$P_4(\cos\theta)$	$P_5(\cos\theta)$	$P_6(\cos\theta)$	$P_7(\cos\theta)$
0°	1.0000	1.0000	1.0000	1.0000	1.0000	1.0000	1.0000
1	.9998	.9995	.9991	.9985	.9977	.9967	.9955
2	.9994	.9982	.9963	.9939	.9909	.9872	.9829
3	.9986	.9959	.9918	.9863	.9795	.9713	.9617
4	.9976	.9927	.9854	.9758	.9638	.9495	.9329
5	.9962	.9886	.9773	.9623	.9437	.9216	.8961
6	.9945	.9836	.9674	.9459	.9194	.8881	.8522
7	.9925	.9777	.9557	.9267	.8911	.8476	.7986
8	.9903	.9709	.9423	.9048	.8589	.8053	.7448
9	.9877	.9633	.9273	.8803	.8232	.7571	.6831
10	.9848	.9548	.9106	.8532	.7840	.7045	.6164
11	.9816	.9454	.8923	.8238	.7417	.6483	.5461
12	.9781	.9352	.8724	.7920	.6966	.5892	.4732
13	.9744	.9241	.8511	.7582	.6489	.5273	.3940
14	.9703	.9122	.8283	.7224	.5990	.4635	.3219
15	.9659	.8995	.8042	.6847	.5471	.3982	.2454
16	.9613	.8860	.7787	.6454	.4937	.3322	.1699
17	.9563	.8718	.7519	.6046	.4391	.2660	.0961
18	.9511	.8568	.7240	.5624	.3836	.2002	.0289
19	.9455	.8410	.6950	.5192	.3276	.1347	−.0443
20	.9397	.8245	.6649	.4750	.2715	.0719	−.1072
21	.9336	.8074	.6338	.4300	.2156	.0107	−.1662
22	.9272	.7895	.6019	.3845	.1602	−.0481	−.2201
23	.9205	.7710	.5692	.3386	.1057	−.1038	−.2681
24	.9135	.7518	.5357	.2926	.0525	−.1559	−.3095
25	.9063	.7321	.5016	.2465	.0009	−.2053	−.3463
26	.8988	.7117	.4670	.2007	−.0489	−.2478	−.3717
27	.8910	.6908	.4319	.1553	−.0964	−.2869	−.3921
28	.8829	.6694	.3964	.1105	−.1415	−.3211	−.4052
29	.8746	.6474	.3607	.0665	−.1839	−.3503	−.4114
30	.8660	.6250	.3248	.0234	−.2233	−.3740	−.4101
31	.8572	.6021	.2887	−.0185	−.2595	−.3924	−.4022
32	.8480	.5788	.2527	−.0591	−.2923	−.4052	−.3876
33	.8387	.5551	.2167	−.0982	−.3216	−.4126	−.3670
34	.8290	.5310	.1809	−.1357	−.3473	−.4148	−.3409
35	.8192	.5065	.1454	−.1714	−.3691	−.4115	−.3096
36	.8090	.4818	.1102	−.2052	−.3871	−.4031	−.2738
37	.7986	.4567	.0755	−.2370	−.4011	−.3898	−.2343
38	.7880	.4314	.0413	−.2666	−.4112	−.3719	−.1918
39	.7771	.4059	.0077	−.2940	−.4174	−.3497	−.1469
40	.7660	.3802	−.0252	−.3190	−.4197	−.3234	−.1003
41	.7547	.3544	−.0574	−.3416	−.4181	−.2938	−.0534
42	.7431	.3284	−.0887	−.3616	−.4128	−.2611	−.0065
43	.7314	.3023	−.1191	−.3791	−.4038	−.2255	.0398
44	.7193	.2762	−.1485	−.3940	−.3914	−.1878	.0846
45°	.7071	.2500	−.1768	−.4062	−.3757	−.1485	.1270

TABLE I. SURFACE ZONAL HARMONICS.

θ	$P_1(\cos\theta)$	$P_2\cos\theta$	$P_3(\cos\theta)$	$P_4(\cos\theta)$	$P_5(\cos\theta)$	$P_6(\cos\theta)$	$P_7(\cos\theta)$
45°	.7071	.2500	−.1768	−.4062	−.3757	−.1485	.1270
46	.6947	.2238	−.2040	−.4158	−.3568	−.1079	.1666
47	.6820	.1977	−.2300	−.4252	−.3350	−.0645	.2054
48	.6691	.1716	−.2547	−.4270	−.3105	−.0251	.2349
49	.6561	.1456	−.2781	−.4286	−.2836	.0161	.2627
50	.6428	.1198	−.3002	−.4275	−.2545	.0563	.2854
51	.6293	.0941	−.3209	−.4239	−.2235	.0954	.3031
52	.6157	.0686	−.3401	−.4178	−.1910	.1326	.3153
53	.6018	.0433	−.3578	−.4093	−.1571	.1677	.3221
54	.5878	.0182	−.3740	−.3984	−.1223	.2002	.3234
55	.5736	−.0065	−.3886	−.3852	−.0868	.2297	.3191
56	.5592	−.0310	−.4016	−.3698	−.0510	.2559	.3095
57	.5446	−.0551	−.4131	−.3524	−.0150	.2787	.2949
58	.5299	−.07-8	−.4229	−.3331	.0206	.2976	.2752
59	.5150	−.1021	−.4310	−.3119	.0557	.3125	.2511
60	.5000	−.1250	−.4375	−.2891	.0898	.3232	.2231
61	.4848	−.1474	−.4423	−.2647	.1229	.3298	.1916
62	.4695	−.1694	−.4455	−.2390	.1545	.3321	.1571
63	.4540	−.1908	−.4471	−.2121	.1844	.3302	.1203
64	.4384	−.2117	−.4470	−.1841	.2123	.3240	.0818
65	.4226	−.2321	−.4452	−.1552	.2381	.3138	.0422
66	.4067	−.2518	−.4419	−.1256	.2615	.2996	.0021
67	.3907	−.2710	−.4370	−.0955	.2824	.2819	−.0375
68	.3746	−.2896	−.4305	−.0650	.3005	.2605	−.0763
69	.3584	−.3074	−.4225	−.0344	.3158	.2361	−.1135
70	.3420	−.3245	−.4130	−.0038	.3281	.2089	−.1485
71	.3256	−.3410	−.4021	.0267	.3373	.1786	−.1811
72	.3090	−.3568	−.3898	.0568	.3434	.1472	−.2099
73	.2924	−.3718	−.3761	.0864	.3463	.1144	−.2347
74	.2756	−.3860	−.3611	.1153	.3461	.0795	−.2559
75	.2588	−.3995	−.3449	.1434	.3427	.0431	−.2730
76	.2419	−.4112	−.3275	.1705	.3362	.0076	−.2848
77	.2250	−.4241	−.3090	.1964	.3267	−.0284	−.2919
78	.2079	−.4352	−.2894	.2211	.3143	−.0644	−.2943
79	.1908	−.4454	−.2688	.2443	.2990	−.0989	−.2913
80	.1736	−.4548	−.2474	.2659	.2810	−.1321	−.2835
81	.1564	−.4633	−.2251	.2859	.2606	−.1635	−.2709
82	.1392	−.4709	−.2020	.3040	.2378	−.1926	−.2536
83	.1219	−.4777	−.1783	.3203	.2129	−.2193	−.2321
84	.1045	−.4836	−.1539	.3345	.1861	−.2431	−.2067
85	.0872	−.4886	−.1291	.3468	.1577	−.2638	−.1779
86	.0698	−.4927	−.1038	.3569	.1278	−.2811	−.1460
87	.0523	−.4959	−.0781	.3648	.0969	−.2947	−.1117
88	.0349	−.4982	−.0522	.3704	.0651	−.3045	−.0735
89	.0175	−.4995	−.0262	.3739	.0327	−.3105	−.0381
90°	.0000	−.5000	.0000	.3750	.0000	−.3125	.0000

TABLE II. BESSEL'S FUNCTIONS.

x	$J_0(x)$	$J_1(x)$	x	$J_0(x)$	$J_1(x)$	x	$J_0(x)$	$J_1(x)$
0.0	1.0000	0.0000	5.0	−.1776	−.3276	10.0	−.2459	.0435
0.1	.9975	.0499	5.1	−.1443	−.3371	10.1	−.2490	.0184
0.2	.9900	.0995	5.2	−.1103	−.3432	10.2	−.2496	.0066
0.3	.9776	.1483	5.3	−.0758	−.3460	10.3	−.2477	−.0313
0.4	.9604	.1960	5.4	−.0412	−.3453	10.4	−.2434	−.0555
0.5	.9385	.2423	5.5	−.0068	−.3414	10.5	−.2366	−.0789
0.6	.9120	.2867	5.6	.0270	−.3343	10.6	−.2276	−.1012
0.7	.8812	.3290	5.7	.0599	−.3241	10.7	−.2164	−.1224
0.8	.8463	.3688	5.8	.0917	−.3110	10.8	−.2032	−.1422
0.9	.8075	.4060	5.9	.1220	−.2951	10.9	−.1881	−.1604
1.0	.7652	.4401	6.0	.1506	−.2767	11.0	−.1712	−.1768
1.1	.7196	.4709	6.1	.1773	−.2559	11.1	−.1528	−.1913
1.2	.6711	.4983	6.2	.2017	−.2329	11.2	−.1330	−.2039
1.3	.6201	.5220	6.3	.2238	−.2081	11.3	−.1121	−.2143
1.4	.5669	.5419	6.4	.2433	−.1816	11.4	−.0902	−.2225
1.5	.5118	.5579	6.5	.2601	−.1538	11.5	−.0677	−.2284
1.6	.4554	.5699	6.6	.2740	−.1250	11.6	−.0446	−.2320
1.7	.3980	.5778	6.7	.2851	−.0953	11.7	−.0213	−.2333
1.8	.3400	.5815	6.8	.2931	−.0652	11.8	.0020	−.2323
1.9	.2818	.5812	6.9	.2981	−.0349	11.9	.0250	−.2290
2.0	.2239	.5767	7.0	.3001	−.0047	12.0	.0477	−.2234
2.1	.1666	.5683	7.1	.2991	.0252	12.1	.0697	−.2157
2.2	.1104	.5560	7.2	.2951	.0543	12.2	.0908	−.2060
2.3	.0555	.5399	7.3	.2882	.0826	12.3	.1108	−.1943
2.4	.0025	.5202	7.4	.2786	.1096	12.4	.1296	−.1807
2.5	−.0484	.4971	7.5	.2663	.1352	12.5	.1469	−.1655
2.6	−.0968	.4708	7.6	.2516	.1592	12.6	.1626	−.1487
2.7	−.1424	.4416	7.7	.2346	.1813	12.7	.1766	−.1307
2.8	−.1850	.4097	7.8	.2154	.2014	12.8	.1887	−.1114
2.9	−.2243	.3754	7.9	.1944	.2192	12.9	.1988	−.0912
3.0	−.2601	.3391	8.0	.1717	.2346	13.0	.2069	−.0703
3.1	−.2921	.3009	8.1	.1475	.2476	13.1	.2129	−.0489
3.2	−.3202	.2613	8.2	.1222	.2580	13.2	.2167	−.0271
3.3	−.3443	.2207	8.3	.0960	.2657	13.3	.2183	−.0052
3.4	−.3643	.1792	8.4	.0692	.2708	13.4	.2177	.0166
3.5	−.3801	.1374	8.5	.0419	.2731	13.5	.2150	.0380
3.6	−.3918	.0955	8.6	.0146	.2728	13.6	.2101	.0590
3.7	−.3992	.0538	8.7	−.0125	.2697	13.7	.2032	.0791
3.8	−.4026	.0128	8.8	−.0392	.2641	13.8	.1943	.0984
3.9	−.4018	−.0272	8.9	−.0653	.2559	13.9	.1836	.1166
4.0	−.3972	−.0660	9.0	−.0903	.2453	14.0	.1711	.1334
4.1	−.3887	−.1033	9.1	−.1142	.2324	14.1	.1570	.1488
4.2	−.3766	−.1386	9.2	−.1367	.2174	14.2	.1414	.1626
4.3	−.3610	−.1719	9.3	−.1577	.2004	14.3	.1245	.1747
4.4	−.3423	−.2028	9.4	−.1768	.1816	14.4	.1065	.1850
4.5	−.3205	−.2311	9.5	−.1939	.1613	14.5	.0875	.1934
4.6	−.2961	−.2566	9.6	−.2090	.1395	14.6	.0679	.1999
4.7	−.2693	−.2791	9.7	−.2218	.1166	14.7	.0476	.2043
4.8	−.2404	−.2985	9.8	−.2323	.0928	14.8	.0271	.2066
4.9	−.2097	−.3147	9.9	−.2403	.0684	14.9	.0064	.2069
5.0	−.1776	−.3276	10.0	−.2459	.0435	15.0	−.0142	.2051

TABLE III.—ROOTS OF BESSEL'S FUNCTIONS.

n	x_n for $J_0(x_n) = 0$	x_n for $J_1(x_n) = 0$	n	x_n for $J_0(x_n) = 0$	x_n for $J_1(x_n) = 0$
1	2.4048	3.8317	6	18.0711	19.6159
2	5.5201	7.0156	7	21.2116	22.7601
3	8.6537	10.1735	8	24.3525	25.9037
4	11.7915	13.3237	9	27.4935	29.0468
5	14.9309	16.4706	10	30.6346	32.1897

TABLE IV.—VALUES OF $J_0(xi)$.

x	$J_0(xi)$	x	$J_0(xi)$	x	$J_0(xi)$
0.0	1.0000	2.0	2.2796	4.0	11.3019
0.1	1.0025	2.1	2.4463	4.1	12.3236
0.2	1.0100	2.3	2.6291	4.2	13.4425
0.3	1.0226	2.3	2.8296	4.3	14.6680
0.4	1.0404	2.4	3.0493	4.4	16.0104
0.5	1.0635	2.5	3.2898	4.5	17.4812
0.6	1.0920	2.6	3.5533	4.6	19.0926
0.7	1.1263	2.7	3.8417	4.7	20.8585
0.8	1.1665	2.8	4.1573	4.8	22.7937
0.9	1.2130	2.9	4.5027	4.9	24.9148
1.0	1.2661	3.0	4.8808	5.0	27.2399
1.1	1.3262	3.1	5.2945	5.1	29.7889
1.2	1.3937	3.2	5.7472	5.2	32.5836
1.3	1.4963	3.3	6.2426	5.3	35.6481
1.4	1.5534	3.4	6.7848	5.4	39.0088
1.5	1.6467	3.5	7.3782	5.5	42.6946
1.6	1.7500	3.6	8.0277	5.6	46.7376
1.7	1.8640	3.7	8.7386	5.7	51.1725
1.8	1.9896	3.8	9.5169	5.8	56.0381
1.9	2.1277	3.9	10.3690	5.9	61.3766

CHAPTER VI.

FUNCTIONS OF A COMPLEX VARIABLE.

By THOMAS S. FISKE,

Adjunct Professor of Mathematics in Columbia University.

ART. 1.   DEFINITION OF FUNCTION.

If two or more quantities are such that no one of them, when any values whatsoever are assigned to the others, suffers any restriction in regard to the values which it can assume the quantities are said to be " independent."

If one quantity is so related to another quantity or to several independent quantities, that whenever particular values are assigned to the latter, the former is required to take one or another of a system of completely determined values, the former is said to be a " function " of the latter.   The quantity or quantities upon the values of which the value of the function depends, are said to be the " independent variables " of the function.

A function is " one-valued " when to every set of values assigned to the independent variables there corresponds but one value of the function.   It is said to be " n-valued " when to every set of values of the independent variables n values of the function correspond.

The " Theory of Functions " has among its objects the study of the properties of functions, their classification according to their properties, the derivation of formulas which exhibit the relations of functions to one another or to their independent variables, and the determination whether or not functions exist satisfying assigned conditions.

ART. 2. REPRESENTATION OF COMPLEX VARIABLE.

A variable quantity is capable, in general, of assuming both real and imaginary values. In fact, unless it be otherwise specified, every quantity w is to be regarded as having the "complex" form $u + v \sqrt{-1}$, u and v being real. It is customary to denote $\sqrt{-1}$ by i, and to write the preceding quantity thus: $u + iv$. If v is zero, w is real; if u is zero, w is a "pure imaginary."

A quantity $z = x + iy$ is said to vary "continuously" when between every pair of values which it takes, $c_1 = a_1 + ib_1$ and $c_2 = a_2 + ib_2$, the value of z varies in such a manner that x and y pass through all real values intermediate to a_1 and a_2, b_1 and b_2, respectively.

It is usual to give to a variable quantity $z = x + iy$ a graphical representation by drawing in a plane a pair of rectangular axes and constructing a point whose abscissa and ordinate are respectively equal to x and y. To every value of z will correspond a point; and, conversely, to every point will correspond a value of z. The terms "point" and value, then, may be interchanged without confusion. When z varies continuously the graphical representation of its variation, or its "path," will be a continuous line. This graphical representation is of the highest importance. By means of it some of the most complicated propositions may be given an exceedingly condensed and concrete expression.

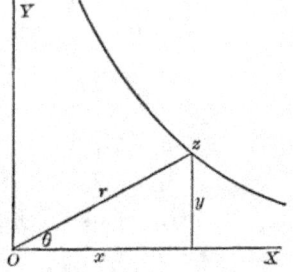

By putting $x = r \cos \theta$, $y = r \sin \theta$, where r is a positive real quantity, the point

$$z = r(\cos \theta + i \sin \theta)$$

is referred to polar coördinates. The quantity r is called the absolute value or "modulus" of z. It is often written $|z|$. θ is known as the "argument" of z.

A function is sometimes considered for only such values of each independent variable as are represented graphically by the points of a certain continuous line. In the study of functions of real variables, for example, the path of each variable is represented by a straight line, namely, the axis of real quantities, or $y = 0$.

ART. 3.   ABSOLUTE CONVERGENCE.

The representation of functions by means of infinite series is one of the most important branches of the theory of functions. In many problems, in fact, it is only by means of series that it is possible to determine functions satisfying the conditions assigned and to obtain the required numerical results. Frequent use will be made of the following theorem.

Theorem.—If the moduli of the terms of a series form a convergent series, the given series is convergent.

Let the given series be $W = w_0 + w_1 + \ldots + w_n + \ldots$ in which $w_0 = r_0 (\cos \theta_0 + i \sin \theta_0)$, $w_1 = r_1 (\cos \theta_1 + i \sin \theta_1) \ldots$ By hypothesis the series $R = r_0 + r_1 + \ldots + r_n + \ldots$ is convergent. Its terms being all positive, the sum of its first m terms constantly increases with m, but in such a manner as to approach a limit. The same will be true necessarily of any series formed by selecting terms from R. The sum of the first m terms of the series W is composed of two parts,

$$r_0 \cos \theta_0 + r_1 \cos \theta_1 \ldots + r_{m-1} \cos \theta_{m-1},$$
$$i(r_0 \sin \theta_0 + r_1 \sin \theta_1 + \ldots + r_{m-1} \sin \theta_{m-1}),$$

and each of these in turn may be divided into parts which have all their terms of the same sign. Every one of the four parts thus obtained approaches a limit as m is increased; for the terms of each part have the same sign, and cannot exceed, in absolute value, the corresponding terms of R. Hence, as m is increased, the sum of the first m terms of W approaches a limit; which was to be proved.

A series, the moduli of whose terms form a convergent series, is said to be " absolutely convergent."

Prob. 1. Show that the series $1 + z + z^2 + \ldots + z^n + \ldots$ is absolutely convergent, if $|z| < 1$.

ART. 4. ELEMENTARY FUNCTIONS.

In elementary mathematics the functions are usually considered for only real values of the independent variables. In the case of the algebraic functions, however, there is no difficulty in assuming that the independent variables are complex. The theory of elimination shows that every algebraic equation can be freed from radicals. Every algebraic function, therefore, is defined by an equation which may be put in a form wherein the second member is zero and the first member is rational and entire in the function and its independent variables.

Besides the algebraic functions, the functions most often occurring in elementary mathematics are the trigonometric and exponential functions and the functions inverse to them. The definitions, by which these functions are generally first introduced, have no significance in the case where the independent variables are complex. However, the following familiar series,

$$e^z = \exp z = 1 + z + \frac{z^2}{2} + \frac{z^3}{3!} + \frac{z^4}{4!} + \cdots,$$

$$\cos z = 1 - \frac{z^2}{2} + \frac{z^4}{4!} - \frac{z^6}{6!} + \cdots,$$

$$\sin z = z - \frac{z^3}{3!} + \frac{z^5}{5!} - \frac{z^7}{!} + \cdots$$

which have been established for the case where the variables are real, furnish most convenient general definitions for $\exp z$, $\cos z$, and $\sin z$, these series being absolutely convergent for every finite value of z. Defining the logarithmic function by the equation

$$e^{\log z} = \exp(\log z) = z,$$

it follows that

$$a^z = e^{z \log a} = \exp(z \log a).$$

The following equations also are to be regarded as equations of definition:

$$\tan z = \frac{\sin z}{\cos z}, \qquad \cot z = \frac{\cos z}{\sin z},$$

$$\sec z = \frac{1}{\cos z}, \qquad \operatorname{cosec} z = \frac{1}{\sin z}.$$

It may be shown that the formulas which are usually obtained on the supposition that the independent variables are real, and which express in that case properties of and relations between the preceding functions, still hold when the independent variables are complex.

Prob. 2. Show that $e^{m}e^{n} = e^{m+n}$, m and n being complex.

Prob. 3. Deduce $\cos z = \frac{1}{2}(e^{iz} + e^{-iz})$, $\sin z = \frac{1}{2i}(e^{iz} - e^{-iz})$.

Prob. 4. Deduce $\cos(z_1 + z_2) = \cos z_1 \cos z_2 - \sin z_1 \sin z_2$,
$$\sin(z_1 + z_2) = \cos z_1 \sin z_2 + \sin z_1 \cos z_2.$$

ART. 5. CONTINUITY OF FUNCTIONS.

Let a function of a single independent variable have a determinate value for a given value c of the independent variable. If, when the independent variable is made to approach c, whatever supposition be made as to the method of approach, the function approaches as a limit its determinate value at c, the function is said to be "continuous" at c.

This definition may be otherwise expressed as follows: A function of a single independent variable is continuous at the point c, when, being given any positive quantity ϵ, it is possible to construct a circle, with center at c and radius equal to a determinate quantity δ, so small that the modulus of the difference between the value of the function at the center and that at every other point within the circle is less than ϵ.

A function of several independent variables is said to be continuous for a particular set of values assigned to those variables, when it takes for that set of values a determinate value c, and for every new set of values, obtained by altering the variables by quantities of moduli less than some determinate positive quantity δ, the value of the function is altered by a quantity of modulus less than any previously chosen arbitrarily small positive quantity ϵ.

A function of one independent variable is said to be continuous in a given region of the plane upon which its independent variable is represented, if it is continuous at every point in that region.

From the principles of limits, it follows that if two functions are continuous at a given point, their sum, difference, and product are continuous at that point. As an immediate consequence, every rational entire function of z is continuous at every finite point; for every such function can be constructed from z and constant quantities by a finite number of additions, subtractions, and multiplications.

Let a function of a single independent variable be continuous at c, and let it take at that point the value t, different from zero. Suppose also that at any other point $c + \Delta c$ the function takes the value $t + \Delta t$. Then

$$\frac{1}{t + \Delta t} - \frac{1}{t} = - \frac{\Delta t}{t(t + \Delta t)}.$$

If it be assumed that $|\Delta t| < |t|$, the modulus of the preceding difference cannot exceed

$$\frac{|\Delta t|}{|t|(|t| - |\Delta t|)},$$

and will, therefore, be less than ϵ if

$$|\Delta t| < \frac{\epsilon |t|^2}{1 + \epsilon |t|}.$$

Hence if a function is continuous and different from zero at a point c, its reciprocal is also continuous at c. It follows at once that if two functions are both continuous at c, their ratio is continuous at c, unless the denominator reduces to zero at that point. But every rational function of z may be expressed as the ratio of two entire functions.   It is therefore continuous for all values of z except those for which its denominator vanishes.

Consider the function $\exp z$,

$$e^{z+\Delta z} - e^z = e^z(e^{\Delta z} - 1) = e^z\left(\Delta z + \frac{\Delta z^2}{2!} + \ldots\right).$$

Hence if $\qquad\qquad |\Delta z| < 1,$

$$|e^{z+\Delta z} - e^z| \lessgtr |e^z|\left(|\Delta z| + \frac{|\Delta z|^2}{2!} + \ldots\right) \lessgtr |e^z| \frac{|\Delta z|}{1 - |\Delta z|},$$

but the limit of the second member is zero when $|\Delta z|$ approaches zero.   Hence $\exp z$ is continuous for all finite values of z.

Prob. 5. Show that $\cos z$ and $\sin z$ are continuous for all finite values of z.

Prob. 6. Show that $\tan z$ is continuous in any circle described about the origin as a center with a radius less than $\frac{1}{2}\pi$.

ART. 6.   GRAPHICAL REPRESENTATION OF FUNCTIONS.

It was shown in Art. 2 that a plane suffices for the complete graphical representation of the values of an independent variable.   In the same way it is convenient to use a second plane to represent graphically the values of any one-valued function. For example, if $w = f(z)$ be such a function, to each point $x + iy$ of the independent variable will correspond a point $u + iv$ of the function. This point $u + iv$ is called the " image " of the point $x + iy$. If w is a continuous function of z, then every continuous curve in the z-plane will have an image in the w-plane, and this image will be also a continuous curve.

Consider the expression $u + iv = x^2 + y^2 + 2ixy$.   Here

$u = x^2 + y^2$ and $v = 2xy$. Since to every value of z correspond determinate values of x and y,
and consequently determinate values
of u and v, this expression falls under the general definition of a function of z.   It is evidently continuous.
Every straight line $x = t$ parallel to
the axis of y is converted by means
of it into a parabola $v^2 = 4t^2(u - t^2)$.

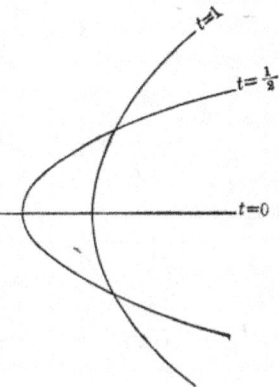

Prob. 7. Find the family of curves
into which the straight lines parallel to
the axis of y are converted by means of
the function $u + iv = x^2 - y^2 + 2ixy$.   Show that no two curves
of this family intersect.

ART. 7.   DERIVATIVES.

Let $w = f(z)$ be a given function of z.   If h is an "infinitesimal," that is, a variable having zero as its limit, and if the
expression

$$\frac{f(z + h) - f(z)}{h}$$

has a finite determinate limit, remaining the same under all
possible suppositions as to the way in which h approaches zero,
this limit is said to be the "derivative" of the function $f(z)$ at
the point z.   In this case $w = f(z)$ is said to be "monogenic"
at z.   The derivative is written $f'(z)$ or $\dfrac{dw}{dz}$.   A function is said
to be monogenic in a region of the plane of the independent
variable if it is monogenic at every point of that region.

Consider now the circumstances under which a function
$w = u + iv$ may have a derivative at the point $z = x + iy$.
If z be given a real increment, x is changed into $x + \Delta x$, while
y is unaltered, so that $\Delta z = \Delta x$; and

$$\frac{\Delta w}{\Delta z} = \frac{\Delta u}{\Delta x} + i\frac{\Delta v}{\Delta x}.$$

If, on the other hand, z is given a purely imaginary increment, $\Delta z = i\Delta y$, and

$$\frac{\Delta w}{\Delta z} = \frac{\Delta u}{i\Delta y} + \frac{\Delta v}{\Delta y}.$$

If the second members of these equations approach deter-
minate limits as Δx and Δy approach zero, and if these limits
are equal,

$$\frac{\partial u}{\partial x} + i\frac{\partial v}{\partial x} = -i\frac{\partial u}{\partial y} + \frac{\partial v}{\partial y}.$$

Hence, equating real and imaginary parts,

$$\frac{\partial u}{\partial x} = \frac{\partial v}{\partial y}, \qquad \frac{\partial v}{\partial x} = -\frac{\partial u}{\partial y},$$

which are necessary conditions for the existence of a derivative.

It can be shown that these conditions are also sufficient.
For let the increment of the independent variable be entirely
arbitrary, no supposition being made as to the relative magnitudes of its real and imaginary parts.   Then the differential of
the function, that is, that part of the increment of the function
which remains after subtracting the terms of order higher than
the first, is

$$du + idv = \left(\frac{\partial u}{\partial x} + i\frac{\partial v}{\partial x}\right)dx + \left(\frac{\partial u}{\partial y} + i\frac{\partial v}{\partial y}\right)dy.$$

Hence
$$\frac{du + idv}{dx + idy} = \frac{\left(\frac{\partial u}{\partial x} + i\frac{\partial v}{\partial x}\right) + \left(\frac{\partial u}{\partial y} + i\frac{\partial v}{\partial y}\right)\frac{dy}{dx}}{1 + i\frac{dy}{dx}},$$

which, by virtue of the conditions written above, is equal to
either member of the equation

$$\frac{\partial u}{\partial x} + i\frac{\partial v}{\partial x} = -i\frac{\partial u}{\partial y} + \frac{\partial v}{\partial y}.$$

The value thus obtained is independent of $\frac{dy}{dx}$, or, what is the
same thing, of the direction of approach to the point z.   The

existence of a derivative of the function w depends, therefore, only on the existence of partial derivatives $\dfrac{\partial u}{\partial x}, \; \dfrac{\partial v}{\partial x}, \; \dfrac{\partial u}{\partial y}, \; \dfrac{\partial v}{\partial y}$ satisfying the specified equations of conditions.

The same equations of condition express the fact that $w = u + iv$, supposed to be an analytical expression involving x and y, involves z as a whole, that is, may be constructed from z by some series of operations, not introducing x or y except in the combination $x + iy$. In other words, they indicate that x and y may both be eliminated from $w = \phi(x, y)$ by means of the equation $z = x + iy$. This property might have been used to define monogenic function, but such a definition would have had the disadvantage of assuming a priori that the function was capable of analytical expression in terms of the independent variable.

A monogenic function is necessarily continuous; that is, the existence of a derivative involves continuity. For, if

$$\text{limit} \; \frac{f(z + h) - f(z)}{h} = f'(z),$$

it follows that

$$f(z + h) = f(z) + h\,[f'(z) + \eta],$$

where η approaches zero with h. Hence $f(z)$ is the limit of $f(z + h)$ when h approaches zero, or $f(z)$ is continuous at the point z.

The following pages relate almost exclusively to functions which are monogenic except for special isolated values of z. Functions which are discontinuous for every value of the independent variable, and functions which are continuous but admit no derivatives, have been little studied except in the case of real variables.*

---

* In this connection see G. Darboux, Sur les fonctions discontinues, Annales de l'École Normale, Series 2, Vol. 4 (1875), pp. 51–112. For a systematic treatment of functions of a real variable, see the German translation of Dini's treatise by Lüroth and Schepp, Leipzig, 1892.

ART. 8.   CONFORMAL REPRESENTATION.

Let z start from the point z_0 and trace two different paths forming a given angle at the point z_0, and let z_1 and z_2 be arbitrary points on the first and second paths respectively.   Then

$$z_1 - z_0 = r_1(\cos \theta_1 + i \sin \theta_1) = r_1 e^{i\theta_1},$$

where r_1 denotes the length of the straight line joining z_0 and

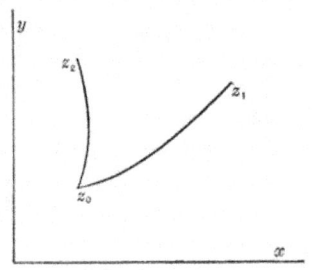
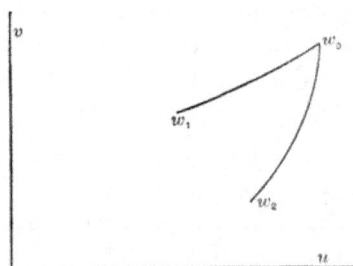

z_1, and θ_1 denotes the inclination of this line to the axis of reals.   In the same way, for the point z_2, there is an equation

$$z_2 - z_0 = r_2(\cos \theta_2 + i \sin \theta_2) = r_2 e^{i\theta_2}.$$

If now w is a one-valued monogenic function of z, in the region of the z-plane considered, to the points z_0, z_1, z_2 correspond points w_0, w_1, w_2; and for these points can be formed the equations

$$w_1 - w_0 = \rho_1 e^{i\phi_1}, \quad w_2 - w_0 = \rho_2 e^{i\phi_2}.$$

From the supposition that w is monogenic, it follows at once that, when z_1 and z_2 are assumed to approach z_0,

$$\text{limit } \frac{w_1 - w_0}{z_1 - z_0} = \text{limit } \frac{w_2 - w_0}{z_2 - z_0}.$$

If the members of this equation are not equal to zero, it may be put in the form

$$\text{limit } \frac{w_1 - w_0}{w_2 - w_0} = \text{limit } \frac{z_1 - z_0}{z_2 - z_0},$$

or

$$\text{limit } \frac{\rho_1}{\rho_2}e^{i(\phi_1 - \phi_2)} = \text{limit } \frac{r_1}{r_2}e^{i(\theta_1 - \theta_2)}.$$

Hence

$$\text{limit } (\phi_1 - \phi_2) = \text{limit } (\theta_1 - \theta_2) \; ;$$

and the images in the w-plane of the two paths traced by z form at w_0 an angle equal to that at z_0 in the z-plane. Accordingly, if z be supposed to trace any configuration whatever in a portion of the z-plane in which $\dfrac{dw}{dz}$ is determinate and not equal to zero, every angle in the image traced by w will be equal to the corresponding angle in the z-plane. If, for example, such a portion of the z-plane be divided into infinitesimal triangles, the corresponding portion of the z-plane will be divided in the same manner, and the corresponding triangles will be mutually equiangular. Such a copy upon a plane, or upon any surface, of a configuration in another surface is called a " conformal representation."

The modulus of the derivative $\left|\dfrac{dw}{dz}\right| = \text{limit } \left|\dfrac{\Delta w}{\Delta z}\right|$ is the " magnification." Its value, which, in general, changes from point to point, may be obtained from the relations

$$\left|\frac{dw}{dz}\right|^2 = \left(\frac{\partial u}{\partial x}\right)^2 + \left(\frac{\partial v}{\partial x}\right)^2 = \left(\frac{\partial u}{\partial y}\right)^2 + \left(\frac{\partial v}{\partial y}\right)^2$$

$$= \frac{\partial u}{\partial x}\frac{\partial v}{\partial y} - \frac{\partial u}{\partial y}\frac{\partial v}{\partial x}.$$

The theory of conformal representation has interesting applications to map drawing.*

* For the literature of the subject, see Forsyth, Theory of Functions, p. 500, and Holzmüller, Einführing in die Theorie der isogonalen Verwandschaften und der conformen Abbildungen, verbunden mit Anwendungen auf mathematische Physik.

ART. ⌣.   EXAMPLES OF CONFORMAL REPRESENTATION.

Case I.—Let $w = z + c$.   This function is formed from the independent variable by the addition of a constant.   Putting for w, z, and c, respectively, $u + iv$, $x + iy$ and $a + ib$, one obtains

$$u = x + a, \quad v = y + b.$$

Any configuration in the z-plane appears, therefore, in the w-plane unaltered in magnitude, and is situated with respect to the axes as if it had been moved parallel to the axis of reals through the distance a and parallel to the axis of imaginaries through the distance b.   The following diagrams represent the transformation of a network of squares by means of the relation $w = z + c$.

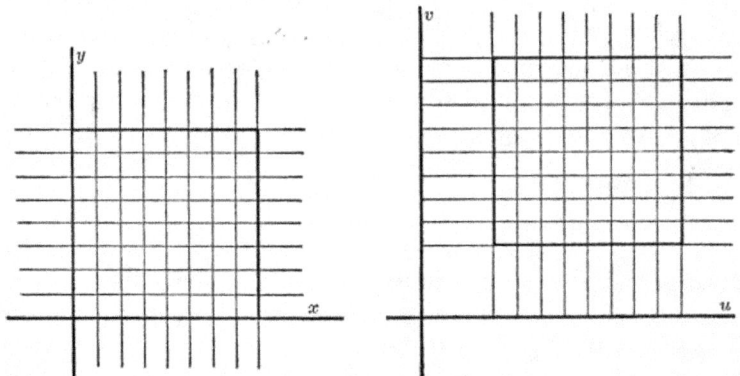

Case II. — Let $w = cz$.   Writing $w = \rho e^{i\phi}$, $z = r e^{i\theta}$, and $c = r_1 e^{i\theta_1}$, the following equations result :

$$\rho = r_1 r, \quad \phi = \theta_1 + \theta.$$

The origin transforms into the origin, all distances measured from the origin are multiplied by a constant quantity, and all straight lines passing through the origin are turned through a constant angle.   See the following diagrams.

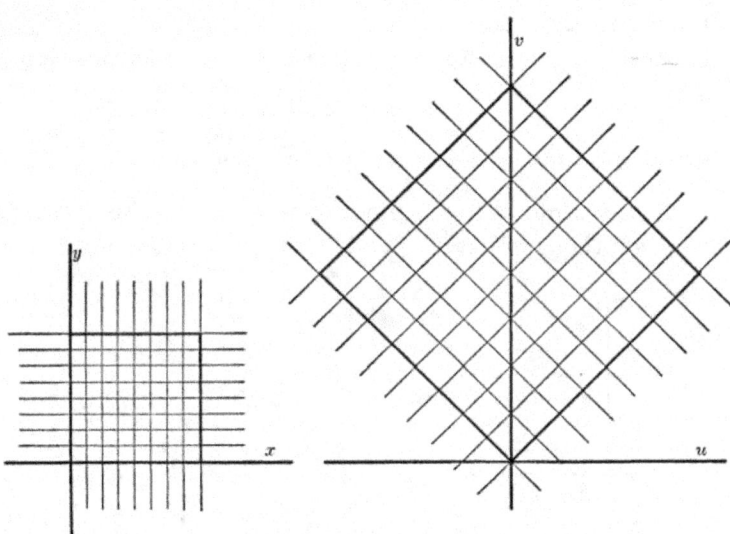

Case III.—Let $w = e^z$. Writing $z = x + iy$, the function becomes

$$w = e^x e^{iy} = e^x(\cos y + i \sin y).$$

Every straight line $x = t_1$, parallel to the axis of y is transformed into a circle $\rho = e^{t_1}$ described about the origin as a center, the axis of y becoming the unit circle. Points to the right of the axis of y fall without the unit circle, while points to the left of this axis fall within.   Every straight line $y = t_2$, parallel to the axis of x becomes a straight line $v/u = \tan t_2$, passing through the origin. The accompanying diagrams* exhibit in a simple manner the periodicity expressed by the equation

$$\exp (z + 2n\pi i) = \exp (z),$$

where n is any positive or negative integer.

To every point in the w-plane, excluding the origin, correspond an infinite number of points in the z-plane. These points are all situated on a straight line parallel to the axis of

* The figures of this and the following example are taken from Holzmüller's treatise.

y, and divide it into segments, each of length 2π. If z' be one of these points, the general value of the inverse function is

$$\log w = z' + 2ni\pi,$$

where n is any positive or negative integer.

If any straight line beginning at the origin be drawn in the w-plane, there will correspond in the z-plane an infinite number

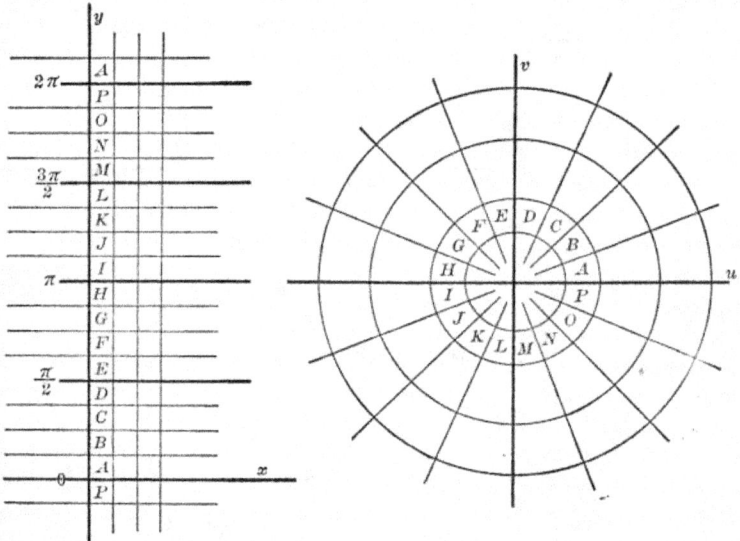

of straight lines parallel to the axis of x, dividing that plane into strips of equal width. To any curve in the w-plane which does not meet the line just drawn, will correspond in the z-plane an infinite number of curves, of which there will be one in each strip.

Case IV.—Let $w = \cos z$. Writing $w = u + iv$, $z = x + iy$, and employing as equations of definition $\cos(iy) = \cosh y$, $\sin(iy) = i \sinh y$, the given function takes the form

$$u + iv = \cos x \cosh y - i \sin x \sinh y.$$

Hence $u = \cos x \cosh y, \; v = -\sin x \sinh y.$

Any straight line, $x = t_1$, parallel to the axis of y, is transformed into one branch of a hyperbola,

$$\frac{u^2}{\cos^2 t_1} - \frac{v^2}{\sin^2 t_1} = 1,$$

having its foci at the points $+1$ and -1.   Any straight line, $y = t_2$, parallel to the axis of x, is transformed into an ellipse,

$$\frac{u^2}{\cosh^2 t_2} + \frac{v^2}{\sinh^2 t_2} = 1,$$

having its foci at the same points, any segment of the straight line equal in length to 2π corresponding to the entire curve taken once.   By means of these confocal conics, the w-plane is divided into curvilinear rectangles, the conformal representation breaking down only at the foci, where the condition that $\frac{du}{dz}$ should be different from zero is not fulfilled.   The periodicity of the function, expressed by the equation

$$\cos(z + 2\pi) = \cos z,$$

15	16	1	2	3	4	5	6	7	8	9	10	11	12	13	14	15	16	1	2
O	P	A	B	C	D	E	F	G	H	I	J	K	L	M	N	O	P	A	B
B	A	P	O	N	M	L	K	J	I	H	G	F	E	D	C	B	A	P	O
2	1	16	15	14	13	12	11	10	9	8	7	6	5	4	3	2	1	16	15

is exhibited graphically in the accompanying diagrams.

It is interesting to note in this example, as also in the preceding one, that the conformal representation introduces well-known systems of curvilinear coordinates, the cartesian coordinates, x, y of a point in the

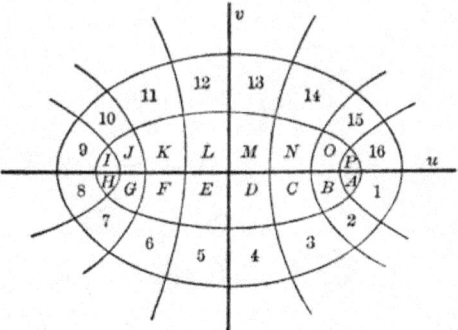

z-plane serving to determine its image in the w-plane as an intersection of orthogonal curves.

Case V.—Let $w = z^3$. Writing $w = u + iv$, $z = x + iy$, the relations

$$u = x^3 - 3xy^2, \quad v = 3x^2y - y^3$$

follow at once. If one of the variables x, y be eliminated from these two equations by means of the equation $lx + my + n = 0$, representing a straight line in the z-plane, equations are obtained representing a unicursal cubic in the w-plane.

By putting $w = \rho(\cos\phi + i\sin\phi)$, $z = r(\cos\theta + i\sin\theta)$, the relations $\rho = r^3$, $\phi = 3\theta$, are obtained. Hence the circle

$$r^2 - 2ar\cos\theta + a^2 = c^2$$

gives the curve

$$\rho^{\frac{2}{3}} - 2a\rho^{\frac{1}{3}}\cos\frac{\theta}{3} + a^2 = c^2,$$

which enwraps three times the point corresponding to the center. The accompanying figure represents this transformation, the straight line feg giving the curve feg.

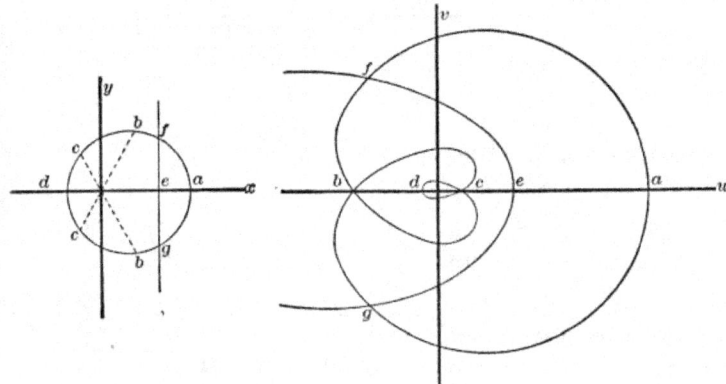

To each point in the w-plane, excluding the origin, at which $\dfrac{dw}{dz} = 0$ and the conformal representation is not maintained, there correspond three points in the z-plane, having for their

arguments $\dfrac{\phi}{3}, \dfrac{\phi + 2\pi}{3}, \dfrac{\phi + 4\pi}{3}$, respectively.   Any straight line

drawn from the origin in the w-plane will have, therefore, three images in the z-plane, viz., three straight lines diverging from the origin, and dividing the plane into three equal regions. Any continuous curve in the w-plane not meeting the line just drawn will be represented in the z-plane by three curves, of which one will be situated within each of these regions.   In the figure here given are exhibited the three conformal representations of a square formed in the w-plane by lines $u = t_1$, $u = t_2$, $v = t_1$, $v = t_2$, parallel to the axes.

If the relation between w and z be reversed, and z be taken as a function of w, z will be a three-valued function, its values giving rise to three branches which will remain distinct and continuous except when w becomes equal to zero.

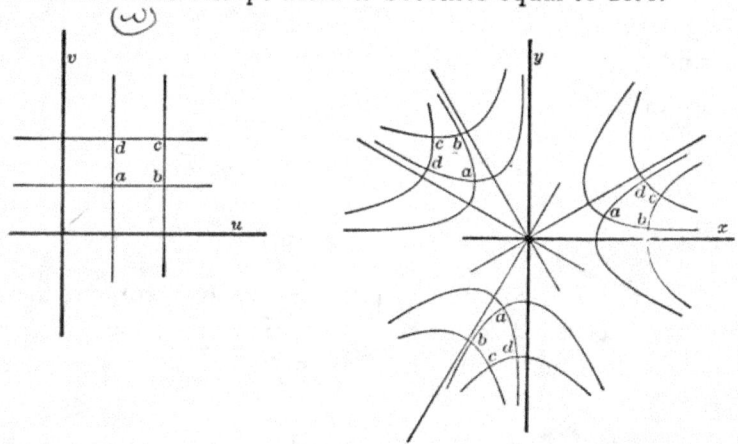

Prob. 8. If $w = z + \dfrac{1}{z}$, show that circles in the z-plane having

a common center at the origin transform into confocal ellipses.

Prob. 9. If $w = \dfrac{z - i}{z + i}$, show that the axis of reals in the z-plane

transforms into the circle $|w| = 1$, and the upper half of the z-plane into the interior of this circle.

ART. 10.   CONFORMAL REPRESENTATION OF A SPHERE.

Let OPO' be a sphere having its diameter OO' equal in length to unity. Construct tangent planes at
at O and O'. Draw in
the tangent plane at
O rectangular axes Ox
and Oy; and in the
other plane draw as
axes $O'u$, parallel to Ox
and measured in the
same direction, and $O'v$
parallel to Oy but measured in a contrary di-
rection. Join any point
z in the plane xOy to

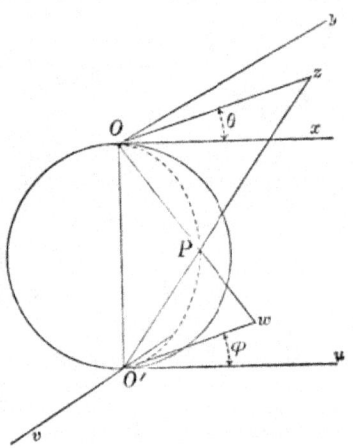

O' by a straight line, and let $O'z$ meet the sphere in P. Draw
OP and produce it to meet the plane $uO'v$ in w.

From the similar triangles $O'Oz$ and $OO'w$

$$\frac{Oz}{OO'} = \frac{OO'}{O'w}, \quad \text{or} \quad Oz \cdot O'w = \overline{OO'}^2 ;$$

that is,
$$|z| \cdot |w| = r\rho = 1.$$

To an observer standing on the sphere at O' rotation about
OO' from $O'u$ toward $O'v$ is positive, while to an observer
standing on the sphere at O such a rotation is negative.
Hence
$$\angle xOz = - \angle uO'w, \quad \text{or} \quad \theta = - \phi.$$

The following equation results :

$$wz = \rho r e^{i(\phi + \theta)} = 1.$$

The w- and z-planes are therefore conformal representations of one another. Any configuration in one plane can be
formed from its image in the other by an inversion with respect

to the origin as a center, combined with a reflection in the axis of reals. Such a transformation was termed by Cayley a " quasi-inversion." By it points at a great distance from the origin in one plane are brought near together in the immediate neighborhood of the origin in the other plane.

Since the line $O'Pz$ makes the same angle with the plane tangent to the sphere at P as with the plane xOy, any spherical angle having its vertex at P is projected into an equal angle at z. The sphere is thus seen to be related conformally to the plane xOy, and it must be also so related to the plane $uO'v$.

The representation of the sphere upon a tangent plane in the manner described above is termed a "stereographic projection." When to this representation is applied a logarithmic transformation, that is, one inverse to the transformation described in Case III of the preceding article, the so-called " Mercator's projection" is obtained.

ART. 11. CONJUGATE FUNCTIONS.

The real and imaginary parts of a monogenic function, $w = u + iv$, have been shown to satisfy the partial differential equations

$$\frac{\partial u}{\partial x} = \frac{\partial v}{\partial y}, \qquad \frac{\partial v}{\partial x} = -\frac{\partial u}{\partial y}.$$

At any point, therefore, where u and v admit second partial derivatives, one obtains

$$\frac{\partial^2 u}{\partial x^2} + \frac{\partial^2 u}{\partial y^2} = 0, \qquad \frac{\partial^2 v}{\partial x^2} + \frac{\partial^2 v}{\partial y^2} = 0;$$

that is, the functions u and v are solutions of Laplace's equation for two dimensions. Any two real solutions p and q of this equation, such that $p + iq$ is a monogenic function of $x + iy$, are called "conjugate functions."[*] Thus the examples of Art. 9 furnish the following pairs of conjugate functions:

---

[*] Maxwell, Electricity and Magnetism, 1873, vol. I, p. 227.

$x + a$, $y + b$; $r_1 r \cos(\theta_1 + \theta)$, $r_1 r \sin(\theta_1 + \theta)$; $e^x \cos y$, $e^x \sin y$; $\cos x \cosh y$, $- \sin x \sinh y$; $x^3 - 3xy^2$, $3x^2 y - y^3$. The second pair is expressed in polar coordinates, but may be transformed to cartesian coordinates by means of the relations

$$r = \sqrt{x^2 + y^2}, \quad \cos\theta = \frac{x}{\sqrt{x^2 + y^2}}, \quad \sin\theta = \frac{y}{\sqrt{x^2 + y^2}}.$$

If one of two conjugate functions be given, the other is thereby determined except for an additive constant. Let u, for example, be given. Then

$$dv = \frac{\partial v}{\partial x}dx + \frac{\partial v}{\partial y}dy$$
$$= -\frac{\partial u}{\partial y}dx + \frac{\partial u}{\partial x}dy,$$

and therefore the value of v is

$$\int \left(-\frac{\partial u}{\partial y}dx + \frac{\partial u}{\partial x}dy \right).$$

The equations $u = c_1$, $v = c_2$, obtained by assigning constant values to two conjugate functions, represent in the w-plane straight lines parallel to the coordinate axes. It follows that the curves which these equations define in the z-plane intersect at right angles. Consequently, by varying the quantities c_1 and c_2, two orthogonal systems of curves are obtained; and c_1 and c_2 may be taken as orthogonal curvilinear coordinates for the determination of position in the z-plane.

Prob. 10. Show that if p and q are conjugate functions of u and v, where u and v are conjugate functions of x and y, p and q will be conjugate functions of x and y.

Prob. 11. Show that if u and v are conjugate functions of x and y, x and y are conjugate functions of u and v.

ART. 12. APPLICATION TO FLUID MOTION.

Consider an incompressible fluid, in which it is assumed that every element can move only parallel to the z-plane, and has a velocity of which the components parallel to the coordi-

nate axes are functions of x and y alone.   The whole motion
of the fluid is known as soon as the motion in the z-plane is
ascertained.   When any curve in the z-plane is given, by the
"flux across the curve"* will be meant the volume of fluid
which in unit time crosses the right cylindrical surface having
the curve as base and included between the z-plane and a parallel plane at a unit distance.

The flux across any two curves joining the points z_0 and z
is the same, provided the curves enclose a region covered with
the moving fluid.   For, corresponding to the enclosed region,
there must be neither a gain nor a loss of matter.   Let z_0 be
fixed, and z be variable.   Let ψ denote the flux across any curve
z_0z, reckoned from left to right for an observer stationed at z_0
and looking along the curve toward z.   If l, m be the direction
cosines of the normal (drawn to the right) at any point of the
curve, and p, q be the components parallel to the axes of the
velocity of any moving element, the value of ψ will be

$$\psi = \int_{z_0}^{z} (lp + mq)ds,$$

where the path of integration is the curve joining z_0 and z.
The function ψ is a one-valued function of z in any region
within which every two curves joining z_0 to z enclose a region
covered with the moving fluid.

If z moves in such a manner that the value of ψ does not
vary, it will trace a curve such that no fluid crosses it, i.e., a
"stream-line."   The curves $\psi = $ const. are all stream-lines, and
ψ is called the "stream-function."   If p and q are continuous,
and if z be given infinitesimal increments parallel to x and y
respectively, one obtains

$$\frac{\partial \psi}{\partial x} = -q, \quad \frac{\partial \psi}{\partial y} = p.$$

If now the motion of the fluid be characterized, as is the

* Lamb's Hydrodynamics (1895), p. 69.

case in the so-called "irrotational" motion,* by the existence of a velocity-potential ϕ, so that

$$p = \frac{\partial \phi}{\partial x}, \quad q = \frac{\partial \phi}{\partial y},$$

the following equations result :

$$\frac{\partial \phi}{\partial x} = \frac{\partial \psi}{\partial y}, \quad \frac{\partial \psi}{\partial x} = -\frac{\partial \phi}{\partial y}.$$

Hence $\phi + i\psi$ is a monogenic function of $x + iy$. The curves $\phi = $ const., which are orthogonal to the stream-lines, are called the "equipotential curves."

Consider, as an example, the motion corresponding to the function† $w = z^3$. The equipotential curves are given by the equations

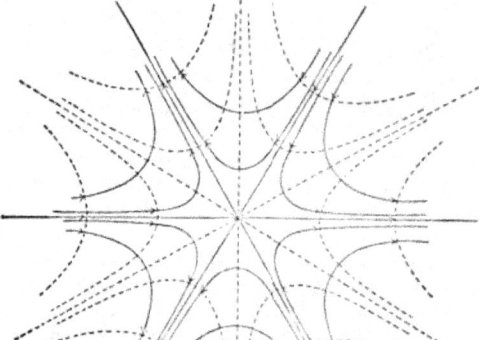

$$u = x^3 - 3xy^2 = \text{const.},$$

the stream-lines by the equations

$$v = 3x^2y - y^3 = \text{const.}$$

In the following figure the stream-lines are the heavy lines, while the equipotential curves are dotted.

The fluid moves in toward the origin, which is called a "cross-point," from three directions, and flows out again in three other directions. At the cross-point the fluid is at a standstill, since at that point the velocity, for which the general expression is

$$\sqrt{\left(\frac{\partial u}{\partial x}\right)^2 + \left(\frac{\partial u}{\partial y}\right)^2},$$

---

* In irrotational motion each element is subject to translation and pure strain, but not to rotation.

† F. Klein: Riemann's Theory of Algebraic Functions ; translated by Frances Hardcastle (1893), p. 3.

is equal to zero.   The stream-lines in the figure represent the
motion of the fluid in each of six different angles, as if the fluid
were confined between walls perpendicular to the z-plane.

It is of importance to note that if the function considered
be multiplied by i, the equipotential curves and stream-lines
are interchanged, since the function $\phi + i\psi$ then becomes
$-\psi + i\phi$.

An example of particular interest is

$$w = -\mu \log \frac{z-a}{z+a}.$$

Let $z - a = r_1 e^{i\theta_1}$, $z + a = r_2 e^{i\theta_2}$; then

$$u = -\mu \log \frac{r_1}{r_2}, \quad v = -\mu(\theta_1 - \theta_2).$$

The curves $u =$ const., $v =$ const. form two orthogonal systems of circles, either of which may be regarded as the stream-
lines, the other constituting the equipotential curves.

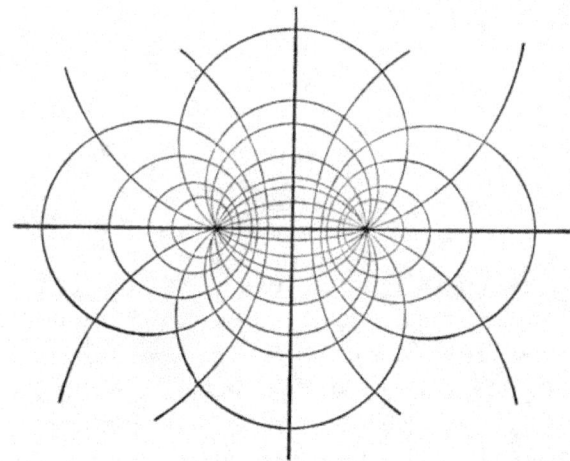

The velocities are everywhere, except at the points $\pm a$,
finite and determinate.   If the circles $r_1/r_2 =$ const. be taken
as the stream-lines, each of the points $\pm a$ is a " vortex-point."
If the circles $\theta_1 - \theta_2 =$ const. be taken as the stream-lines, one

of the points $\pm a$ is a "source," the other a "sink." In the latter case, besides the hydrodynamical interpretation, a very simple electrical illustration is afforded by attaching the poles of a battery to a conducting plate of indefinite extent at two fixed points of the plate.

As another example may be taken the relation $w = \cos z$. As has been shown, the curves $x = $ const. form a system of confocal hyperbolas, while the curves $y = $ const. form an orthogonal system of ellipses. Either system may be regarded as stream-lines. In one case the motion of the fluid would be such as would occur if a thin wall were constructed along the axis of reals, except between the foci, and the fluid should be impelled through the aperture thus formed. In the other case the fluid would circulate around a barrier placed on the axis of reals and included between the foci.

Besides their application to fluid-motion, conjugate functions have important applications in the theory of electricity and magnetism * and in elasticity.†

ART. 13.   CRITICAL POINTS.

Let w be any rational function of z. It can be written in the form

$$w = \frac{f(z)}{\phi(z)},$$

where $f(z)$ and $\phi(z)$ are entire and without common factors. This function is finite and admits an infinite number of successive derivatives for every finite value of z, except the roots of the equation $\phi(z) = 0$. Let a be such a root. Then the reciprocal of the given function is finite and admits an infinite nnmber of successive derivatives at the point a. Such a point

---

* J. J. Thomson, Recent Researches in Electricity and Magnetism (1893), p. 208.

† Love, Theory of Elasticity (1892), vol. 1, p. 331.

is called a "pole." Any rational function having a pole at a can be put by the method of partial fractions in the form

$$w = \frac{A_1}{z-a} + \ldots + \frac{A_k}{(z-a)^k} + \psi(z),$$

where $A_1, \ldots, A_k$ are constants, A_k being different from zero, and $\psi(z)$ is finite at the point a. The integer k is said to be the "order" of the pole, and the function is said to have for its value at a infinity of the kth order. In accordance with the definition of a derivative, w does not admit a derivative at a. From the character of the derivative in the immediate neighborhood of a, however, the derivative is sometimes said to become infinite at a.

The trigonometric function $\cot z$ has a pole of the first order at every point $z = m\pi$, m being zero or any integer positive or negative.

The function $w = \log(z-a)$ has for every finite value of z, except $z = a$, an infinite number of values. If $z - a$ is written in the form $Re^{i\Theta}$,

$$w = \log R + i(\Theta + 2m\pi),$$

where $\log R$ is real, and m is zero or any positive or negative integer. If z describes a straight line, beginning at a, Θ will remain fixed, but R will vary. The images in the z-plane will therefore be straight lines parallel to the axis of reals, dividing the plane into horizontal strips of width 2π. If now the z-plane is supposed to be divided along the straight line just drawn, and z varies along any continuous path, subject only to the restriction that it cannot cross this line of division, there will be a continuous curve as the image of the path of z in each strip of the w-plane. Each of these images is said to correspond to a "branch" of the function, or, expressed otherwise, the function is said to have a branch situated in each strip. The line of division in the z-plane, which serves to separate the branches from one another is called a "cut."

At the point $z = a$ no definite value is attached to the function. As z approaches that point the modulus of the real part of the function increases without limit, while the imaginary part is entirely indeterminate.

Let z_0 be an arbitrary point, distinct from a, and let

$$\log R_0 + i\Theta_0 + 2m\pi i$$

be any one of the corresponding values of the function. Suppose that z starts from z_0 and describes a closed path around the point a, the values of the function being taken so as to give a continuous variation. Upon returning to the point z_0 the value of the function will be

$$\log R_0 + i\Theta_0 + 2(m + 1)\pi i,$$

or $$\log R_0 + i\Theta_0 + 2(m - 1)\pi i,$$

according as the curve is described in a positive or negative direction. By repeating the curve a sufficient number of times it is evidently possible to pass from any value of the function at z_0 to any other. When a point is such that a z-path enclosing it may lead in this manner from one value of a function to another value, it is called a "branch-point." In the case of the function here considered, the point $z = a$ is called a "logarithmic branch-point," or a point of "logarithmic discontinuity."

The function $w = \log \dfrac{f(z)}{\phi(z)}$, where $f(z)$ and $\phi(z)$ are entire, has a point of logarithmic discontinuity at every point where either $f(z)$ or $\phi(z)$ is equal to zero. For, writing

$$f(z) = A(z - a_1)^{p_1}(z - a_2)^{p_2} \dots$$
$$\phi(z) = B(z - b_1)^{q_1}(z - b_2)^{q_2} \dots$$

the value of w may be written

$$w = \log \frac{A}{B} + \underset{m}{\Sigma} p_m \log (z - a_m) - \underset{n}{\Sigma} q_n \log (z - b_n).$$

Take now the function $w = e^{\frac{1}{z}}$. It has a single finite value for every value of z except $z = 0$. If z is supposed to approach zero, the limit of the value of the function is indeterminate.

For let $p + iq$ be perfectly arbitrary, and write

$$e^{p+iq} = c + id.$$

If now $a + ib$ is the reciprocal of $p + iq$, so that

$$a = \frac{p}{p^2 + q^2}, \qquad b = \frac{-q}{p^2 + q^2},$$

the preceding equation may be written

$$e^{\frac{1}{a+ib}} = c + id.$$

But whatever the value of the integer m, $q + 2m\pi$ may be substituted for q without altering the value of $c + id$, and hence both a and b may be made less than any assignable quantity. The given function $e^{\frac{1}{z}}$ therefore takes the value $c + id$ at points $a + ib$ indefinitely near to the origin. A point such that, when z approaches it, a function elsewhere one-valued tends toward an indeterminate limiting value is called an "essential singularity."

Prob. 12. Show that for the function $e^{\frac{1}{z-a}}$ $z = a$ is an essential singularity.

Prob. 13. The function $e^{-\frac{1}{z^2}}$ considered as a function of a real variable is continuous for every finite value of z, and the same is true of each of its successive derivatives. Show that when it is regarded as a function of a complex variable, $z = 0$ is an essential singularity.

In order to illustrate still another class of special points take the function

$$w = \sqrt{(z - a_1)(z - a_2) \ldots (z - a_n)}.$$

This function has at every finite point, except $a_1, a_2, \ldots, a_n$, two distinct values differing in sign.   At these points, however, it takes but a single value, zero.   From each of the points $a_1, a_2, \ldots, a_n$ let a straight line of indefinite extent be drawn in such a manner that no one of them intersects any other, and suppose the z-plane to be divided, or cut, along each of these lines.   Along any continuous path in the z-plane thus divided the values of the function form two distinct branches.

For, writing

$$z - a_1 = r_1 e^{i\theta_1}, \quad z - a_2 = r_2 e^{i\theta_2}, \quad \ldots, \quad z - a_n = r_n e^{i\theta_n},$$

the function takes the form

$$w = \sqrt{r_1 r_2 \ldots r_n} \; e^{i\frac{\theta_1 + \theta_2 + \ldots + \theta_n}{2}}.$$

No closed path in the divided plane will enclose any of the points $a_1, a_2, \ldots, a_n$, and the quantities $\theta_1, \theta_2, \ldots, \theta_n$, after continuous variation along such a path, must resume at the initial point their original values.   No such path, therefore, can lead from one value of the function at any point to a new value of the function at the same point.   If, however, the cuts are disregarded and z traces in a positive direction, a closed curve including an odd number of the points $a_1, a_2, \ldots, a_n$, and not intersecting itself, then an odd number of the quantities $\theta_1, \theta_2, \ldots, \theta_n$ are each increased by 2π; and the value of the function is altered by a factor $e^{(2k+1)\pi i}$, and so changed in sign.   In the same way any closed path described about one of these points, and enwrapping it an odd number of times, leads from one value of the function to the other.   On the other hand, a simple closed path enclosing an even number of these points, or a closed path which encloses but one of the points, enwrapping it an even number of times, leads back to the initial value of the function.   It fol-

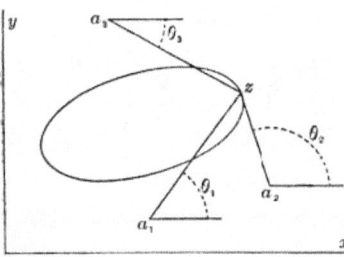

lows that each of the points $a_1, a_2, \ldots, a_n$ is a branch-point. Any point in the z-plane, closed paths about which lead from one to another of a set of different values of a function, the number of values in the set being finite, is called an "algebraic branch-point."

As a further illustration, consider the function

$$w = z^{\frac{1}{2}} + (z - a)^{\frac{1}{3}},$$

which is a root of the equation of the sixth degree,

$$w^6 - 3zw^4 - 2(z - a)w^3 + 3z^2w^2 - 6z(z - a)w + (z-a)^2 - z^3 = 0.$$

The function has at every point, except $z = 0$ and $z = a$, six distinct values. Six branches are thereby formed which can be completely separated from one another by making cuts from the points $z = 0$ and $z = a$ to infinity. Putting ω for the cube root of unity, these six branches can be written

$$w_1 = z^{1/2} + (z - a)^{1/3}, \quad w_2 = - z^{1/2} + (z - a)^{1/3},$$
$$w_3 = z^{1/2} + \omega (z - a)^{1/3}, \quad w_4 = - z^{1/2} + \omega(z - a)^{1/3},$$
$$w_5 = z^{1/2} + \omega^2(z - a)^{1/3}, \quad w_6 = - z^{1/2} + \omega^2(z - a)^{1/3}.$$

The branches w_1 and w_2, w_3 and w_4, w_5 and w_6 are interchanged by a small closed circuit described about $z = 0$, while a small circuit described about $z = a$ permutes cyclically the branches $w_1, w_3, w_5,$ and also the branches w_2, w_4, w_6.

All of the special points examined above, poles, points of logarithmic discontinuity, essential singularities, and branch-points, are called critical points. In fact, a function, or a branch of a function, is said to have a "critical point" at each point where it fails to have a continuous derivative,* or about which as a center it is impossible to describe a circle of determinate radius within which the function, or branch, is one-valued.

Any point not a critical point is called an "ordinary point."

---

* Continuity and. therefore, finiteness of the function are implied in the existence of a derivative.

An ordinary point at which a function reduces to zero is called a " zero " of the function.

If in a certain region of the z-plane there are no critical points for a given function, the function is said to be " synec. tic" or " holomorphic " in that region. If in a certain region the only critical points are poles, the function is said to be " meromorphic " in that region. Under similar conditions a branch of a function is also described as holomorphic or meromorphic.

Prob. 14. When w and z are connected by the relation $w - g = (z - h)^t$ show that if z describes a circle about h as a center, w describes a circle about g as a center, an angle in the z-plane having its vertex at h is transformed into an angle in the w-plane t times as great and having its vertex at g, and that $z = h$ is a branch-point of w except when t is an integer.

Art. 14.   Point at Infinity.

In determining the limiting value of a function when the modulus of the independent variable z is increased indefinitely, it is usual to introduce a new independent variable z' by the relation $z = 1/z'$, and consider the function at the point $z' = 0$. This is equivalent to passing from the z-plane to another plane, the z'-plane, related to the former by the geometrical construction described in Art. 10. It is often very convenient, however, to go further and to substitute for the z-plane the surface of the sphere of unit diameter touching the z-plane at the origin. No difficulty is thus introduced since, as explained in the article just cited, any configuration in the z-plane obtains a conformal representation upon the sphere; and the advantage is gained that the entire surface upon which the variation of the independent variable is studied is of finite extent. The point of the sphere diametrically opposite to its point of contact with the z-plane coincides with the point written above as $z' = 0$. It is called the point at infinity, $z = \infty$, since a point on the sphere approaches it at the same time that its image in the z-plane recedes indefinitely from the origin.

The point at infinity may be either an ordinary or a critical point. For the function $e^{\frac{1}{z}}$, for example, it is an ordinary point, since $e^{\frac{1}{z}} = e^{z'}$. For a rational entire function of the nth degree it is a pole of order n. Consider it for the function $\sqrt{(z - a_1)(z - a_2)\ldots(z - a_n)}$, discussed in the preceding article. Let a circle of great radius be described in the z-plane inclosing all the branch-points $a_1, a_2, \ldots, a_n$. Its conformal representation on the sphere will be a small closed curve surrounding the point $z = \infty$. This point must, therefore, be regarded as a branch-point or not, according as the function changes value or not when the curve surrounding it is described, that is according as n, the number of finite branch-points, is odd or even. When the point at infinity is taken into account, then, the total number of branch-points of this function is always even. The character of the point $z = \infty$ for this function can be determined directly, by changing z into $1/z'$ and considering the point $z' = 0$.

Prob. 15. Show that $z = \infty$ is an ordinary point for $\dfrac{\phi(z)}{\psi(z)}$, where $\phi(z)$ and $\psi(z)$ are rational and entire if the degree of $\phi(z)$ does not exceed that of $\psi(z)$.

ART. 15. INTEGRAL OF A FUNCTION.

Let $w = f(z)$ be a continuous function of a complex variable z, and suppose z to describe a continuous path L from the point z_0 to the point Z. Let a series of points $z_1, z_2, \ldots, z_n$ be taken on L, and let $t_0, t_1, \ldots, t_n$ be points arbitrarily chosen on the arcs $z_0 z_1, z_1 z_2, \ldots, z_n Z$ respectively. Form the sum

$$S = (z_1 - z_0)f(t_0) + (z_2 - z_1)f(t_1) + \ldots + (Z - z_n)f(t_n).$$

If now the number of points $z_1, \ldots, z_n$ be increased indefinitely in such a manner that the length * of each of the arcs

---

* It is assumed in regard to every path of integration that the idea of length may be associated with the portion of it included between any two of its points, or, what is the same thing, that the path is rectifiable. This condition is evidently satisfied if the current coordinates x and y can be expressed in terms of

$z_0z_1, z_1z_2, \ldots, z_nZ$ approaches zero as a limit, the sum S approaches a finite limit which is independent of the choice of the points z_1, $z_2, \ldots, z_n$ and $t_0, t_1, \ldots, t_n$.

For take any other sum

$$S' = (z_1' - z_0)f(t_0') + \\ (z_2' - z_1')f(t_1') + \cdots$$

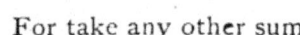

formed in a similar manner. Suppose for the sake of greater definiteness that the points $z_1, \ldots$ and $z_1', \ldots$ follow one another on the line L in the order

$$z_1, z_1', z_2', z_2, z_3, z_3', \ldots,$$

and form a third sum

$$S'' = (z_1 - z_0)f(\tau_0) + (z_1' - z_1)f(\tau_1) + (z_2' - z_1')f(\tau_2) \\ + (z_2 - z_2')f(\tau_3) + \cdots,$$

in which both series of points occur. It may be shown that as the number of points in each of the series $z_1, \ldots$ and $z_1', \ldots$ is increased, the differences $S'' - S$ and $S'' - S'$ both approach zero, from which it follows that the difference $S - S'$ has a limit equal to zero. For example, the difference $S'' - S$ has the value

$$(z_1 - z_0)[f(\tau_0) - f(t_0)] + (z_1' - z_1)[f(\tau_1) - f(t_1)] \\ + (z_2' - z_1')[f(\tau_2) - f(t_1)] + \cdots$$

If M denotes the upper extreme of the quantities

$$|f(\tau_0) - f(t_0)|, \qquad |f(\tau_1) - f(t_1)|, \qquad |f(\tau_2) - f(t_1)|, \cdots$$

the modulus of $S'' - S$ will be less than

$$M[|z_1 - z_0| + |z_1' - z_1| + |z_2' - z_1'| + \cdots].$$

any parameter t so that $\dfrac{dx}{dt}$ and $\dfrac{dy}{dt}$ are continuous. For then the integral

$\displaystyle\int \sqrt{dx^2 + dy^2}$ is finite. See, in this connection, Jordan, Cours d'Analyse, 2d Edition, Vol. I., p. 100.

But $|z_1 - z_0|$ is equal to the chord of the arc $z_0 z_1$, and must therefore be less than or equal to this arc, and a similar result holds for each of the quantities $|z_1' - z_1|, |z_2' - z_1'|, \ldots$ Hence

$$|S'' - S| \lessgtr Ml,$$

where l denotes the length of the path of integration. When the number of points of division on the line L is increased, the differences

$$f(\tau_0) - f(t_0), \qquad f(\tau_1) - f(t_1), \qquad f(\tau_2) - f(t_2), \ldots$$

decrease indefinitely, for $f(z)$ is continuous. M acccordingly decreases indefinitely and the difference $S'' - S$ approaches zero.

The limit, the existence of which has just been demonstrated, is called the integral of $f(z)$ along the path L. It is written $\int_L f(z)dz$. The definition here given is similar to that given for the integral of a function of a real variable. It is unnecessary to specify the path of integration when the independent variable is restricted to real values, since in that case it must be the portion of the axis of reals included between the limits of integration.

The following well-known principles, applicable to the case of a real independent variable, may be readily extended to the general case :

1. The modulus of the integral cannot exceed the length of the path of integration multiplied by the upper extreme of the modulus of the function along that path.

2. The independent variable may be altered by any equation of transformation, but L', the path of integration in the transformed integral, must be such that it is described by the new variable while z describes L.

3. If $F(z)$ is any one-valued function having everywhere $f(z)$ for its derivative, the equation

$$\int_L f(z)dz = F(Z) - F(z_0)$$

must be true.

To prove the third principle, write $F(Z) - F(z_0)$ in the form

$$F(Z) - F(z_n) + F(z_n) - F(z_{n-1}) + \ldots + F(z_2) - F(z_1) + F(z_1) - F(z_0).$$

Since the derivative of $F(z)$ is $f(z)$,

$$F(z_{m+1}) - F(z_m) = [f(z_m) + \eta_m](z_{m+1} - z_m),$$

where η_m has zero for its limit when z_{m+1} is made to approach z_m. Hence

$$F(Z) - F(z_0) = \text{limit } \Sigma f(z_m)(z_{m+1} - z_m) + \text{limit } \Sigma \eta_m(z_{m+1} - z_m);$$

or, since the second term of the right-hand member is equal to zero,

$$F(Z) - F(z_0) = \int_L f(z)dz.$$

If no function $F(z)$ fulfilling the preceding conditions is known, the value of the integral requires further investigation.

Consider as an example the integral $\int \frac{dz}{z^2}$ taken from the point $z = -1$ to the point $z = 1$, the path of integration being the upper half of the circumference of a unit circle described about the origin as a center. Writing $z = \exp(i\theta)$, z will describe the required path while θ varies from π to o.

The equations $\frac{1}{z^2} = e^{-2i\theta}$, $dz = ie^{i\theta}d\theta$,

$$\frac{dz}{z^2} = ie^{-i\theta}d\theta = i\cos\theta\, d\theta + \sin\theta\, d\theta = id(\sin\theta) - d(\cos\theta),$$

follow at once. Hence for the path specified

$$\int_{-1}^{+1} \frac{dz}{z^2} = i\int_{\pi}^{0} d(\sin\theta) - \int_{\pi}^{0} d(\cos\theta) = -2.$$

The application of the direct and more familiar method gives the same result:

$$\int_{-1}^{+1} \frac{dz}{z^2} = \left[-\frac{1}{z}\right]_{z=1} - \left[-\frac{1}{z}\right]_{z=-1} = -2.$$

For a path along the axis of reals between the limits of integration this result is unintelligible. The discontinuity of the differential, $\dfrac{dz}{z^3}$, at the point $z = 0$, prevents the consideration of such a path; and that the result should be negative when the differential is at every point of the path positive has no significance. The introduction of the complex variable furnishes a perfectly satisfactory explanation of the result.

Prob. 16. Show that the integral of $\dfrac{dz}{z}$ along any semi-circumference described about the origin as a center is equal to πi.

ART. 16.   REDUCTION OF COMPLEX INTEGRALS TO REAL.

The integral

$$\int_L f(z)\,dz$$

may be written in the form

$$\int_L (u + iv)(dx + idy),$$

or, separating the real and imaginary terms,

$$\int_L (u\,dx - v\,dy) + i\int_L (v\,dx + u\,dy).$$

Hence the calculation of the integral may be reduced to the calculation of two real curvilinear integrals.

The equations

$$\frac{\partial u}{\partial x} = \frac{\partial v}{\partial y}, \quad \frac{\partial u}{\partial y} = -\frac{\partial v}{\partial x},$$

which express the condition that $u + iv$ should be monogenic, express also that

$$u\,dx - v\,dy, \quad v\,dx + u\,dy$$

are the exact differentials of two real functions of the variables x, y. Consider the case where these functions are one-valued.

Denoting them by $P(x, y)$ and $Q(x, y)$ respectively, the integral may be written

$$[P(X, Y) - P(x_0, y_0)] + i[Q(X, Y) - Q(x_0, y_0)],$$

(x_0, y_0) and (X, Y) being the initial and terminal points respectively of the path of integration.

ART. 17.   CAUCHY'S THEOREM.

Cauchy's Theorem furnishes the necessary and sufficient conditions that a one-valued function $f(z)$, having a continuous derivative $f'(z)$, should yield a one-valued integral, that is, an integral the value of which, when the lower limit is fixed, depends simply on the upper limit, and not on the path of integration. It will be more convenient, before considering Cauchy's Theorem, to demonstrate the following lemma :

Lemma.—Let A be a portion of the z-plane, having a boundary S which consists of a closed curve not intersecting itself, or of several closed curves not intersecting themselves or one another.   Denote by λ the inclination to the axis of x of the exterior normal at any point of the boundary,* that is, the normal drawn to the right as the boundary is described in a positive direction.   If at every point of the region A, including its boundary S, a function W of the real variables x and y is one-valued and continuous and has continuous partial derivatives $\dfrac{\partial W}{\partial x}, \dfrac{\partial W}{\partial y}$, the relations

$$\int_S W dy = \int \int_A \frac{\partial W}{\partial x} dx\, dy, \tag{1}$$

$$\int_S W dx = - \int \int_A \frac{\partial W}{\partial y} dx\, dy \tag{2}$$

exist, the integrals in the first members being taken along the

* It is assumed that the boundary has a determinate tangent at every point. If the boundary of a given region is not of this sort, the theorem holds for any interior curve of which this assumption is true.

boundary in the positive direction, and those in the second
members being taken over the enclosed area.

If any straight line parallel to the axis of x be traced in
the direction of increasing values of x, at each point where
it passes into the area A,
cos λ is negative, and therefore in the first member of
(1) $dy = \cos \lambda \, ds$ is negative.
At each point where this
straight line passes out of
the area A, cos λ, and therefore dy, in the first member
of equation (1), is positive.
Hence in the first member
of equation (1) the differentials Wdy corresponding

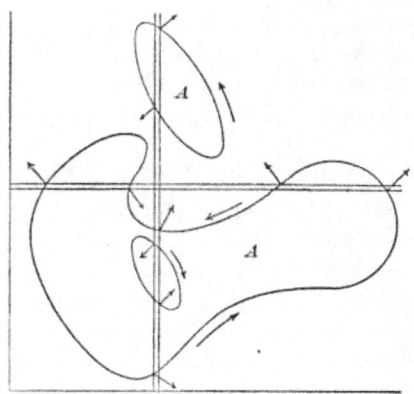

to a given value of y, and taken in the order of increasing
values of x, have signs which, compared with those of the
corresponding values of W, first differ, then agree, and so
on alternately. In order now to compare the integral in the
first member of equation (1) with the integral in the second
member, it is necessary to take dy as essentially positive.
The sum of the differentials in the first member, corresponding to a fixed value of y, must therefore be written in the
form

$$dy(- W_1 + W_2 - W_3 + W_4 - \ldots),$$

where W_1, W_2, . . . are the corresponding values of W taken in
the order of increasing values of x. But performing now in
the second member of equation (1) an integration with respect
to x, the same result is obtained, so that the two members of
equation (1) become identical, and the equation is verified.

To obtain equation (2) the same method is used. It is
necessary in this case to observe that if a line parallel to the
axis of y is traced in the direction of increasing values of y, at
each point where it enters A, dx in the integral of the first

member must be taken as positive; and at each point where this line passes out of A, dx in that integral must be taken as negative.

By means of the preceding lemma, Cauchy's Theorem is easily proved. This theorem may be stated as follows:

Theorem.—If, on the boundary of and within a given region A, a one-valued function $w = f(z)$ is monogenic, and its derivative $f'(z)$ is continuous,* the integral $\int_S f(z)dz$ taken along the boundary S is equal to zero.

For writing the integral in the form

$$\int_S wdz = \int_S (udx - vdy) + i \int_S udy + vdx),$$

the preceding lemma gives

$$\int_S (udx - vdy) = -\int\int_A \left(\frac{\partial u}{\partial y} + \frac{\partial v}{\partial x}\right)dx\,dy,$$

$$\int_S (udy + vdx) = \int\int_A \left(\frac{\partial u}{\partial x} - \frac{\partial v}{\partial y}\right)dx\,dy;$$

but since at every point of A

$$\frac{\partial u}{\partial y} + \frac{\partial v}{\partial x} = 0, \qquad \frac{\partial u}{\partial x} - \frac{\partial v}{\partial y} = 0,$$

the given integral reduces to zero.

ART. 18.   APPLICATION OF CAUCHY'S THEOREM.

From Cauchy's Theorem it follows that, if two different paths L_1 and L_2 lead from the point z_0 to the point Z, and if along these paths and in the region inclosed between them a given function $f(z)$ has no critical points, the integrals of the function taken along these two paths are equal. For two such paths taken together, one described directly, the other reversed, constitute a closed curve, and the integral taken along

---

* Otherwise expressed, the one-valued function $f(z)$ has no critical points on the boundary of or within A, or $f(z)$ is holomorphic in A.

it is equal to zero. But, since reversing the direction of the path of integration is equivalent to changing the sign of the integral, the equation

$$\int_{L_1} f(z)dz - \int_{L_2} f(z)dz = 0$$

is obtained.

The result just established may be stated in the following theorem:

Theorem I.—If a function is holomorphic in any simply connected region bounded by a continuous closed curve, the integral of the function, from a fixed lower limit in that region to any point contained therein, is independent of the path of integration, and is a one-valued function of its upper limit.

A region whose boundary is composed of disconnected curves is not necessarily characterized by the property stated in the theorem. Take, for example, the function

$$w = \sqrt{(z - a_1)(z - a_2) \ldots (z - a_n)},$$

and suppose that $0 < |a_1| < |a_2| < \ldots < |a_n|$. With the origin as a center, construct a system of concentric circles C_1, $C_2, \ldots, C_n$, C_1 passing through a_1, C_2 through a_2, and so on. Denote by S_0 the region inclosed within the first circle C_1, by S_1 that inclosed between C_1 and C_2, and so on, the portion of the plane exterior to the last circle C_n being denoted by S_n. At an initial point z_0 interior to one of these regions, assign to w one of the two values possible, and consider the branch of w resulting from a continuous variation. Then however z may vary within any such region, this branch of w will be a monogenic function, and its derivative will be continuous. Having regard to the branch-points $a_1, a_2, \ldots, a_n$, it is evident that in the regions $S_0, S_2, \ldots$ it will be one-valued, and in the regions $S_1, S_3, \ldots$, it will be two-valued. Thus in the former regions $S_0, S_2, \ldots$, the branch fulfils all the conditions required by the theorem above. The theorem is applicable, however, only to S_0, for in any other region two paths may be drawn joining the same two points, and such that the branch is not one-valued throughout the enclosed portion of the z-plane.

Theorem II.—If $f(z)$ is holomorphic in any simply connected region S bounded by a continuous closed curve, the integral $\int f(z)dz$, taken from a fixed lower limit z_0 in that region to any point Z contained therein, is a holomorphic function of its upper limit.

Let L be any path from z_0 to Z. When the upper limit is at the point $Z + dZ$, L followed by a straight line from Z to $Z + dZ$ can be taken as the path of integration. Hence

$$\int_{z_0}^{Z+dZ} f(z)dz - \int_{z_0}^{Z} f(z)dz = \int_{Z}^{Z+dZ} f(z)dz$$

$$= f(Z)\int_{Z}^{Z+dZ} dz + \int_{Z}^{Z+dZ} [f(z) - f(Z)]dz.$$

The first term is equal to $f(Z)dZ$. The modulus of second term is equal to or less than $M|dZ|$, where M is the upper extreme of $|f(z) - f(Z)|$ along the line joining Z to $Z + aZ$. But since $f(z)$ is continuous, the limit of M when $Z + dZ$ approaches Z is zero. Hence

$$\int_{z_0}^{Z+dZ} f(z)dz - \int_{z_0}^{Z} f(z)dz = [f(Z) + \eta]dZ,$$

where η approaches zero with dZ. The integral therefore has $f(Z)$ for a derivative, and is holomorphic in S.

In the case of a region bounded by several disconnected closed curves, of which one is exterior to all the others, Cauchy's Theorem may be stated in the following form:

Theorem III.—Let a function $f(z)$ be holomorphic in a region A bounded by a closed curve C and one or more closed curves $C_1, C_2, \ldots$ interior to C. The integral of $f(z)$ taken along C will be equal to the sum of its integrals taken in the same direction along the curves $C_1, C_2, \ldots$

For the integral of $f(z)$ taken in a positive direction completely around the boundary of A is equal to zero. But the curves $C_1, C_2, \ldots$ are then described in the direction opposite to that in which C is described.   Hence if all the curves are described in the same direction, the result may be written

$$\int_C f(z)dz = \int_{C_1} f(z)dz + \int_{C_2} f(z)dz + \dots$$

If there is but one interior curve, so that the region A is included between two curves C and C_1, the integral taken along every closed curve containing C_1 but interior to C has the same value, viz., the common value corresponding to the paths C and C_1.

ART. 19.   THEOREMS ON CURVILINEAR INTEGRALS.

Theorem I.—If $f(z)$ be continuous in a given region except at the point a, the integral $\int f(z)dz$, taken around a small circle c, having its center at a, will approach zero as a limit simultaneously with the radius r of the circle c, provided only

$$\lim (z - a)f(z) = 0 \quad \text{when} \quad z = a.$$

For let the upper extreme of the modulus of $(z - a)f(z)$ on the circle c be denoted by M.   Then at every point of c,

$$\text{mod } f(z) \lessgtr \frac{M}{|z - a|} \lessgtr \frac{M}{r},$$

and consequently

$$\text{mod} \int_c f(z)dz \lessgtr \frac{M}{r} \int ds \lessgtr 2\pi M.$$

Theorem II.—The integral $\int \frac{dz}{(z - a)^n}$, taken around any closed curve C containing the point a, is equal to zero, except when $n = 1$.   When $n = 1$, this integral is equal to $2\pi i$.

For the value of the integral will be the same if any circle described about a as a center be taken as the path of integration.   Let then $z - a = re^{i\theta}$, where r is a constant and θ varies from 0 to 2π.   The integral becomes

$$\frac{i}{r^{n-1}} \int_0^{2\pi} e^{-(n-1)i\theta} \, d\theta$$

which reduces to zero except when $n = 1$. If $n = 1$, its value is $2\pi i$, whence

$$\int \frac{dz}{z - a} = 2\pi i.$$

Theorem III.—If $f(z)$ is a function holomorphic in a given region S, C a closed curve the interior of which is wholly within S, and a a point situated within C, then

$$\int_C \frac{f(z)}{z - a} dz = 2\pi i f(a).$$

For describing about a as a center a small circle c of radius r, the equation

$$\int_c \frac{f(z)}{z - a} dp = \int_c \frac{f(z)}{z - a} dz$$

is obtained. But at every point of c,

$$f(z) = f(a) + \eta,$$

where, by choosing r sufficiently small, the modulus of η may be made less than any fixed positive quantity. Hence

$$\int_c \frac{f(z)}{z - a} dz = \int_c \frac{f(a)}{z - a} dz + \int_c \frac{\eta}{z - a} dz,$$

but by the preceding theorems the first term of the right-hand member is equal to $2\pi i f(a)$, and the second term is equal to zero.

If the equation of the theorem just established be differentiated with respect to a, the following important formulas, expressing the successive derivatives of a holomorphic function at a given point, are obtained:

$$\int_C \frac{f(z)}{(z - a)^2} dz = 2\pi i f'(a),$$

$$1 \cdot 2 \int_C \frac{f(z)}{(z - a)^3} dz = 2\pi i f''(a),$$

$$\cdot \quad \cdot \quad \cdot \quad \cdot \quad \cdot \quad \cdot \quad \cdot \quad \cdot$$

$$1 \cdot 2 \ldots n \int_C \frac{f(z)}{(z - a)^{n+1}} dz = 2\pi i f^{(n)}(a).$$

The integrals in the first members of these equations are all finite and determinate for every position of a within the curve C. Therefore any function holomorphic in a given region admits an infinite number of successive derivatives at every interior point. Each of these derivatives being monogenic must be continuous. Hence the following:

Theorem IV.—If $f(z)$ is holomorphic within a given region, there exists an infinite number of successive derivatives of $f(z)$, which are all holomorphic within the same region.

Denote by r the shortest distance from the point a to the curve C. Then at every point of this curve $|z - a| \gtreqless r$. Let M be the upper extreme of the modulus $f(z)$ on C, and l the length of C. Then

$$\mod \int_C \frac{f(z)}{(z-a)^{n+1}} dz \lesseqgtr \int_C \frac{M}{r^{n+1}} ds \lesseqgtr \frac{Ml}{r^{n+1}},$$

and consequently $\mod f^{(n)}(a) \lesseqgtr \frac{1 \cdot 2 \ldots n}{2\pi} \cdot \frac{Ml}{r^{n+1}}.$

In particular, if C is a circle having a for its center,

$$\mod f^{(n)}(a) \lesseqgtr \frac{1 \cdot 2 \ldots n \cdot M}{r^n}.$$

ART. 20. TAYLOR'S SERIES.

Theorem.—Let $f(z)$ be holomorphic in a region S, and let C be any circle situated in the interior of S. If a be the center and $a + t$ any other point interior to C,

$$f(a + t) = f(a) + tf'(a) + \frac{t^2}{1 \cdot 2} f''(a) + \ldots$$

$$+ \frac{t^n}{1 \cdot 2 \ldots n} f^{(n)}(a) + \ldots$$

From the preceding article, denoting a variable point on C by ζ,

$$f(a + t) = \frac{1}{2\pi i} \int_C \frac{f(\zeta)}{\zeta - a - t} d\zeta$$

$$= \frac{1}{2\pi i} \int_C \frac{f(\zeta) d\zeta}{\zeta - a}\left[1 + \frac{t}{\zeta - a} + \cdots + \frac{t^n}{(\zeta-a)^n} + \frac{t^{n+1}}{(\zeta-a)^n(\zeta-a-t)}\right]$$

$$= f(a) + tf'(a) + \frac{t^2}{1 \cdot 2}f''(a) + \cdots + \frac{t^n}{1 \cdot 2 \ldots n}f^{(n)}(a) + R,$$

where

$$R = \frac{1}{2\pi i} \int_C \frac{t^{n+1}f(\zeta)}{(\zeta - a)^{n+1}(\zeta - a - t)} d\zeta.$$

By taking n sufficiently great the modulus of R may be made less than any given positive quantity. Let M be the upper extreme of the modulus of $f(z)$ on the circle C, ρ the modulus of t, and r the modulus of $\zeta - a$ or radius of C. Then

$$|R| \leqq \frac{1}{2\pi} \int_C M \frac{\rho^{n+1}}{r^{n+1}(r - \rho)} ds \leqq \frac{Mr}{r - \rho}\left(\frac{\rho}{r}\right)^{n+1},$$

which, since $\rho < r$, has zero for its limit when $n = \infty$.

Writing now z for $a + t$, Taylor's Series becomes

$$f(z) = f(a) + (z-a)f'(a) + \frac{(z-a)^2}{1 \cdot 2}f''(a) + \cdots + \frac{(z-a)^n}{1 \cdot 2 \ldots n}f^{(n)}(a) + \cdots$$

The series is convergent and the equality is maintained for every point z included within a circle described about a as a center with a radius less than the distance from a to the nearest critical point of $f(z)$.

When a is equal to zero, Taylor's Series takes the form

$$f(z) = f(0) + zf'(0) + \frac{z^2}{1 \cdot 2}f''(0) + \cdots + \frac{z^n}{1 \cdot 2 \ldots n}f^{(n)}(0) + \cdots,$$

expressing $f(z)$ in terms of powers of z. This form is known as Maclaurin's Series.

ART. 21. LAURENT'S SERIES.

Theorem.—Let S, a portion of the z-plane bounded by two concentric circles C_1 and C_2, be situated in the interior of the region E, in which a given function $f(z)$ is holomorphic. If a be the common center of the two circles, and $a + t$ a point interior to S, $f(a + t)$ can be expressed in a convergent double series of the form ·

$$f(a + t) = \sum_{m=-\infty}^{m=\infty} A_m t^m.$$

With $a + t$ as a center construct a circle c sufficiently small to be contained within the region S. If then C_1 be the greater of the two given circles, it follows from Article 18 that

$$\frac{1}{2\pi i}\int_{C_1}\frac{f(\zeta)d\zeta}{\zeta - a - t} = \frac{1}{2\pi i}\int_{C_2}\frac{f(\zeta)d\zeta}{\zeta - a - t} + \frac{1}{2\pi i}\int_{c}\frac{f(\zeta)d\zeta}{\zeta - a - t}.$$

But from Article 19,

$$\frac{1}{2\pi i}\int_{c}\frac{f(\zeta)d\zeta}{\zeta - a - t} = f(a + t),$$

whence

$$f(a + t) = \frac{1}{2\pi i}\int_{C_1}\frac{f(\zeta)d\zeta}{\zeta - a - t} - \frac{1}{2\pi i}\int_{C_2}\frac{f(\zeta)d\zeta}{\zeta - a - t}.$$

The two integrals of the right-hand member may be written :

$$\frac{1}{2\pi i}\int_{C_1}\frac{f(\zeta)d\zeta}{\zeta - a}\left[1 + \frac{t}{\zeta - a} + \cdots + \frac{t^n}{(\zeta - a)^n}\right] + R_1,$$

$$-\frac{1}{2\pi i}\int_{C_2}f(\zeta)d\zeta\left[\frac{1}{t} + \frac{\zeta - a}{t^2} + \cdots + \frac{(\zeta - a)^n}{t^{n+1}}\right] + R_2,$$

where

$$R_1 = \frac{1}{2\pi i}\int_{C_1}\frac{t^{n+1}f(\zeta)d\zeta}{(\zeta - a)^{n+1}(\zeta - a - t)},$$

$$R_2 = \frac{1}{2\pi i}\int_{C_2}\frac{(\zeta - a)^{n+1}f(\zeta)d\zeta}{t^{n+1}(\zeta - a - t)}.$$

But $|t| < |\zeta - a|$ at every point of C_1, and $|t| > |\zeta - a|$ at every point of C_2, so that R_1 and R_2 both have zero for a limit

when $n = \infty$. The value of $f(a + t)$ can therefore be expressed
in the form

$$f(a + t) = A_0 + A_1 t + A_2 t^2 + A_3 t^3 + \cdots$$
$$+ \frac{A_{-1}}{t} + \frac{A_{-2}}{t^2} + \frac{A_{-3}}{t^3} + \cdots$$

Since in the region S the function $f(z)/(z - a)^{m+1}$ is holomorphic for both positive and negative values of m, A_m may be
written

$$A_m = \frac{1}{2i\pi} \int \frac{f(\zeta)}{(\zeta - a)^{m+1}} d\zeta,$$

where C is any circle concentric with C_1 and C_2 and included
between them.

The series thus obtained is convergent at every point $a + t$
contained within the region S. It is important to notice, however, that when the positive and negative powers of t are considered separately, the two resulting series have different
regions of convergence. The series containing the positive
powers of t converges over the whole interior of the circle C_1;
while the series of negative powers of t converges at every
point exterior to the circle C_2. The region S can be regarded,
therefore, as resulting from an overlapping of two other
regions in which different parts of Laurent's Series converge.

Writing z for $a + t$, Laurent's Series takes the form

$$f(z) = A_0 + A_1(z - a) + A_2(z - a)^2 + \cdots$$
$$+ A_{-1}(z - a)^{-1} + A_{-2}(z - a)^{-2} + \cdots$$

Consider as a special numerical example the fraction

$$\frac{1}{(z - 1)(z - 2)(z - 3)} = \frac{1}{2(z - 1)} - \frac{1}{z - 2} + \frac{1}{2(z - 3)}$$

If $|z| < 1$, all three terms of the second member, when
developed in powers of z, give only positive powers. If
$1 < |z| < 2$, the first term of the second member gives a series
of negative descending powers, but the others give the same
series as before. If $2 < |z| < 3$, the first and second terms
both give negative powers. If $|z| > 3$, all three terms give

negative powers, and the development of the given fraction can contain no positive powers. Thus a system of concentric annular regions is obtained in each of which the given fraction is expressed by a convergent power-series. Laurent's Series gives analogous results for every function which is holomorphic except at isolated points of the z-plane.

ART. 22. FOURIER'S SERIES.

Let $w = f(z)$ be holomorphic in a region S_0, and let it be periodic, having a period equal to ω, so that $f(z + n\omega) = f(z)$, where n is any positive or negative integer. Denote by S_n the region obtained from S_0 by the addition of $n\omega$ to z; and suppose that the regions $\ldots, S_{-n}, \ldots, S_{-1}, S_0, S_1, \ldots, S_n, \ldots$ meet or overlap in such a manner as to form a continuous strip S, in which, of course, the function w will be holomorphic. Draw two parallel straight lines, inclined to the axis of reals at an angle equal to the argument of ω, and contained within the strip S. The band T included between these parallels will be wholly interior to S.

By means of the transformation $z' = e^{\frac{2\pi i z}{\omega}}$ the band T in the z-plane becomes in the z'-plane a ring T' bounded by two concentric circles described about the origin as a center, z and $z + n\omega$ falling at the same point z'. Since w is holomorphic in a region including T, and

$$\frac{dw}{dz'} = \frac{dw}{dz}\frac{dz}{dz'} = \frac{\omega}{2\pi i} e^{-\frac{2\pi i z}{\omega}} \frac{dw}{dz},$$

w regarded as a function of z' will be holomorphic in T'. Hence, by Laurent's Theorem,

$$w = \sum_{m = -\infty}^{m = \infty} A_m z'^m,$$

the quantity a in the general formula of the preceding article being in this case equal to zero. Substituting for z' its value, the preceding equation becomes

$$w = \sum_{m = -\infty}^{m = \infty} A_m e^{\frac{2m\pi i z}{\omega}},$$

where

$$A_m = \frac{1}{2\pi i}\int_c \frac{wdz'}{z'^{m+1}} = \frac{1}{\omega}\int_z^{z+\omega} e^{-\frac{2m\pi iz}{\omega}} wdz.$$

In the latter integral the path is rectilinear. Denoting its independent variable by ζ for the purpose of avoiding confusion, the value of w becomes

$$f(z) = \frac{1}{\omega}\sum_{m=-\infty}^{m=\infty}\int_\zeta^{\zeta+\omega} e^{\frac{2m\pi i}{\omega}(z-\zeta)} f(\zeta)d\zeta$$

$$= \frac{1}{\omega}\int_\zeta^{\zeta+\omega} f(\zeta)d\zeta + \frac{2}{\omega}\sum_{m=1}^{m=\infty}\int_\zeta^{\zeta+\omega}\cos\frac{2m\pi}{\omega}(z-\zeta)f(\zeta)d\zeta$$

$$= \frac{1}{\omega}\int_\zeta^{\zeta+\omega} f(\zeta)d\zeta + \frac{2}{\omega}\sum_{m=1}^{m=\infty}\cos\frac{2m\pi z}{\omega}\int_\zeta^{\zeta+\omega}\cos\frac{2m\pi\zeta}{\omega}f(\zeta)d\zeta$$

$$+ \frac{2}{\omega}\sum_{m=1}^{m=\infty}\sin\frac{2m\pi z}{\omega}\int_\zeta^{\zeta+\omega}\sin\frac{2m\pi\zeta}{\omega}f(\zeta)d\zeta.$$

ART. 23.   UNIFORM CONVERGENCE.

Let the series $W = w_0 + w_1 + w_2 + \ldots + w_n + \ldots$, each term of which is a function of z, be convergent at every point of a given region S. Denote by W_n the sum of the first n terms of W. If it is possible, whatever the value of the positive quantity ϵ, to determine an integer ν, such that whenever $n > \nu$

$$|W - W_n| < \epsilon$$

at every point of S, the series W is said to be uniformly convergent in the region S.

Uniformly convergent series can in many respects be treated in exactly the same manner as sums containing a finite number of terms.

Theorem I.—A uniformly convergent series, the terms of which are continuous functions of z, is itself a continuous function of z.

For at any point z, W may be written in the form

$W = W_n + R$; and at a neighboring point z', $W' = W_n' + R'$. Hence

$$W - W' = W_n - W_n' + R - R',$$

and $\qquad | W - W' | \lesseqgtr | W_n - W_n' | + | R | + | R' |.$

But by choosing n sufficiently great, $| R |$ and $| R' |$ may both be made less than any given positive quantity $\dfrac{\epsilon}{3}$. Having chosen n thus, W_n becomes the sum of a finite number of continuous functions. It is then continuous, and, by making $| z' - z |$ less than a suitable quantity δ, $| W - W_n' |$ may be made less than $\dfrac{\epsilon}{3}$. But, under these suppositions,

$$| W - W' | < \epsilon.$$

W is, therefore, continuous at the point z.

Theorem II.—If all the terms of a uniformly convergent series

$$W = w_0 + w_1 + \ldots + w_n + \ldots$$

are continuous, the integral of the series, for any path L situated in the region of uniform convergence, is the sum of the integrals of its terms:

$$\int_L W dz = \int_L w_0 dz + \int_L w_1 dz + \ldots + \int_L w_n dz + \ldots$$

For, writing $W = W_n + R$, it is possible to choose n so that, however small ϵ may be, $| R | < \epsilon$ at every point of L. If n be so chosen,

$$\int_L W dz = \int_L W_n dz + \int_L R dz.$$

But, by Article 15, denoting by l the length of the path L,

$$\text{mod} \int_L R dz < \epsilon l,$$

which, when $n = \infty$, has zero for its limit. Hence

$$\int_L W dz = \lim_{n = \infty} \int_L W_n dz.$$

Theorem III.—If the series $W = w_0 + w_1 + \ldots + w_n + \ldots$ is convergent, and the series

$$W' = \frac{dw_0}{dz} + \frac{dw_1}{dz} + \ldots + \frac{dw_n}{dz} + \ldots$$

is uniformly convergent in a region S, and if further the terms of W' are continuous in S, W' will be the derivative of W.

For, integrating W' from a to z along a path L contained in S,

$$\int_L W'dz = w_0(z) - w_0(a) + \ldots + w_n(z) - w_n(a) + \ldots$$
$$= W(z) - W(a).$$

But the derivative of the first member is W', which must also be the derivative of the second member, and therefore of W.

An immediate consequence of the preceding theorems is the following :

Theorem IV.—If the terms of the convergent series

$$W = w_0 + w_1 + \ldots + w_n + \ldots$$

are holomorphic in a given region S, contained in the region of convergence, and if the series

$$W' = \frac{dw_0}{dz} + \frac{dw_1}{dz} + \ldots + \frac{dw_n}{dz} + \ldots$$

is uniformly convergent, W will be holomorphic in the region S, and will have W' for its derivative.

To illustrate by an example that uniformity of convergence is essential to the preceding theorems, take the series

$$W = \frac{1}{1+z} + \sum_{1}^{\infty} \frac{z^n(1-z)}{(1+z^n)(1+z^{n+1})}.$$

At the point $z = 1$ each term is continuous, and the series is convergent, having the value $1/2$. The series is, however, discontinuous at $z = 1$. For, writing it in the form

$$W = \frac{1}{1+z} + \left(\frac{1}{1+z^2} - \frac{1}{1+z}\right) + \left(\frac{1}{1+z^3} - \frac{1}{1+z^2}\right) + \ldots.$$

the sum of the first n terms is

$$W_n = \frac{1}{1 + z^n}.$$

But W is the limit of W_n when $n = \infty$, and is therefore unity at every point z for which $|z| < 1$, and zero at every point for which $|z| > 1$.

If now this series be considered for the points within and upon a circle described about the origin as a center with an assigned radius less than unity, the remainder after n terms, or $1 - W_n = \frac{z^n}{1 + z^n}$ can, by a suitable choice of n, be made less in absolute value than any given quantity. In such a region, then, the series converges uniformly, and, by Theorem I, can have no point of discontinuity. A similar result holds for the region exterior to any circle described about the origin as a center with an assigned radius greater than unity.

By means of Theorem II given above it can be shown that Laurent's Series is unique. For, assuming the notation used in the determination of the series, the series is uniformly convergent in the region included between any two given circles concentric with C_1 and C_2, both being interior to C_1 and exterior to C_2.

Suppose, now, that two such series are possible :

$$f(a + t) = \sum_{m = -\infty}^{m = \infty} A_m t^m = \sum_{m = -\infty}^{m = \infty} A_m' t^m.$$

Divide by t^{n+1}, and integrate along any circle described about a as a center and included in the region of uniform convergence. The integral $\int t^{m-n-1} dt$ for such a path is zero, except when $m = n$; the integral $\int t^{-1} dt = 2i\pi$.

Hence for such a path,

$$\int \frac{f(a + t) dt}{t^{n+1}} = 2i\pi A_n = 2i\pi A_n';$$

from which it follows that $A_n = A_n'$, and the two series are identical.

ART. 24.   ONE-VALUED FUNCTIONS WITH CRITICAL POINTS.

Theorem I.—A function holomorphic in a region S and not equal to a constant, can take the same value only at isolated points of S.

For in the neighborhood of any point a interior to S, by Taylor's theorem,

$$f(z) - f(a) = (z - a)f'(a) + \frac{(z-a)^2}{1 \cdot 2} f''(a) + \ldots$$

Unless $f(z)$ is constant over the entire circle of convergence of this series, the derivatives $f'(a)$, $f''(a)$, ... cannot all be equal to zero.   Let $f^{(n)}(a)$ be the first which is not equal to zero.   Then

$$f(z) - f(a) = (z-a)^n \left[\frac{f^{(n)}(a)}{1 \cdot 2 \ldots n} + \frac{f^{(n+1)}(a)}{1 \cdot 2 \ldots (n+1)}(z-a) + \ldots \right]$$

If $|z - a|$ be given a finite value sufficiently small, the modulus of the first term of the series within the brackets will exceed the sum of the moduli of all the other terms, and the same result will hold for every still smaller value of $|z - a|$. For values of z, then, distant from a by less than a certain finite amount, $f(z) - f(a)$ is different from zero.

If, on the other hand, the function is constant over the entire circle, described about a as a center, within which Taylor's series converges, it will be possible, by giving in succession new positions to the point a, to show that the value of the function is constant over the whole region S.

Theorem II.—Two functions which are both holomorphic in a given region S and are equal to each other for a system of points which are not isolated from one another, are equal to each other at every point of S.

For let $f(z)$ and $\phi(z)$ be two such functions.   By the preceding theorem, the difference $f(z) - \phi(z)$ must be equal to zero at every point of S.

Theorem III.—A function which is holomorphic in every part of the z-plane, even at infinity, is constant.

For, a being any given point, whatever the value of z,

$$f(z) = f(a) + (z - a)f'(a) + \ldots + \frac{(z - a)}{1 \cdot 2 \ldots n} f^{(n)}(a) + \ldots$$

But by Article 20, r being the radius of any arbitrary circle having its center at a, and M being the upper extreme of the modulus of $f(z)$ on the circumference of this circle,

$$\text{mod } f^{(n)}(a) \underset{<}{=} \frac{1 \cdot 2 \ldots n M}{r^n}.$$

But M is always finite, and r may be made indefinitely great. Hence $f^{(n)}(a) = 0$ for all values of n, and

$$f(z) = f(a).$$

Theorem IV.—If a function $f(z)$, holomorphic in a region S, is equal to zero at the point a situated within S, the function can be expressed in the form

$$f(z) = (z - a)^m \phi(z),$$

where m is a positive integer, and $\phi(z)$ is holomorphic in S and different from zero at a.

For in the neighborhood of the point a, by Taylor's Theorem,

$$f(z) = f(a) + (z - a)f'(a) + \ldots$$

Let $f^{(m)}(a)$ be the first of the successive derivatives at a which is not equal to zero. Then

$$f(z) = (z-a)^m \left[\frac{f^{(m)}(a)}{1 \cdot 2 \ldots m} + \frac{f^{(m+1)}(a)}{1 \cdot 2 \ldots (m+1)} (z-a) + \ldots \right],$$

which is the required form. The point a is a zero of $f(z)$, and m is its order.

Theorem V.—If the point a is a critical point of a given function $f(z)$, but is interior to a region S, in which the reciprocal of $f(z)$ is holomorphic, the function can be expressed in the form

$$f(z) = \frac{\chi(z)}{(z - a)^m},$$

where m is a positive integer, and $\chi(z)$ is holomorphic in the neighborhood of a.

For by the preceding theorem

$$\frac{1}{f(z)} = (z - a)^m \phi(z),$$

where $\phi(z)$ is holomorphic and not equal to zero at $z = a$. Hence

$$f(z) = \frac{1}{(z - a)^m} \cdot \frac{1}{\phi(z)} = \frac{\chi(z)}{(z - a)^m}.$$

Further, since in a region of finite extent including the point a

$$\chi(z) = A_0 + A_1(z - a) + \dots,$$
$$f(z) = \frac{A_0}{(z - a)^m} + \dots + \frac{A_{m-1}}{z - a} + \psi(z),$$

a being an ordinary point for $\psi(z)$.

The point a is a pole of $f(z)$ and m is its order.

Theorem VI.—A function, not constant in value, and having no finite critical points except poles, must take values arbitrarily near to every assignable value.

For suppose that $f(z)$ is such a function, but that it takes no value for which the modulus of $f(z) - A$ is less than a given positive quantity ϵ. Then the function

$$\frac{1}{f(z) - A}$$

will be holomorphic in every part of the z-plane, which, by Theorem III, is impossible unless $f(z)$ is a constant.

Theorem VII.—A function $f(z)$, having no critical point except a pole at infinity, is a rational entire function of z.

For the only critical point of $f\left(\frac{1}{z}\right)$ is a pole at the origin. Hence

$$f\left(\frac{1}{z}\right) = \frac{A_m}{z^m} + \dots + \frac{A_1}{z} + \phi(z),$$

where $\phi(z)$ is holomorphic over the entire plane, including the point at infinity. $\phi(z)$ is consequently equal to a constant A_0. The given function therefore can be written in the form

$$f(z) = A_m z^m + \dots + A_1 z + A_0.$$

Theorem VIII.—A function $f(z)$ whose only critical points are poles is a rational function of z.

The poles must be at determinate distances from one another; otherwise the reciprocal of $f(z)$ would be equal to zero for points not isolated from one another. The number of poles cannot increase indefinitely as $|z|$ is increased; for then the reciprocal of $f\left(\dfrac{1}{z}\right)$ would be an infinite number of zeros indefinitely near to the origin. The total number of poles is therefore finite. Let $a, b, \ldots$ denote them. In the neighborhood of a the function can be expressed in the form

$$\frac{A_m}{(z-a)^m} + \cdots + \frac{A_1}{z-a} + \phi(z),$$

a being an ordinary point for $\phi(z)$. In the neighborhood of b, $\phi(z)$ can be expressed in the form

$$\frac{B_n}{(z-b)^n} + \cdots + \frac{B_1}{z-b} + \psi(z),$$

a and b being both ordinary points for $\psi(z)$. Proceeding in this way the given function will be expressed as the sum of a finite number of rational fractions and a term which can have no critical point except a pole at infinity. This term is a rational entire function.

Theorem IX.—If the function $f(z)$ has no zeros and no critical points for finite values of z, it can be expressed in the form $f(z) = e^{g(z)}$, where $g(z)$ is holomorphic in every finite region of the z-plane.

For $\dfrac{f'(z)}{f(z)}$ can have no critical points except at infinity, since in every finite region of the z-plane $f(z)$ and $f'(z)$ are holomorphic and $f(z)$ is different from zero. Hence, choosing an arbitrary lower limit z_0, the integral

$$\int_{z_0} \frac{f'(z)}{f(z)} = h(z)$$

is holomorphic in every finite region. The function $f(z)$ consequently must take the form

$$f(z) = f(z_0)e^{h(z)} = e^{g(z)},$$

where
$$g(z) = h(z) + \log f(z_0).$$

Theorem X.—If two functions $f(z)$ and $\phi(z)$ have no critical points in the finite portion of the z-plane except poles, and if these poles are identical in position and in order for the two functions, and their zeros are also identical in position and order, there must exist a relation of the form

$$f(z) = \phi(z)e^{g(z)},$$

where $g(z)$ is holomorphic in every finite region of the z-plane.

For the ratio of the two functions has no zeros and no critical points in the finite portion of the z-plane.

ART. 25. RESIDUES.

If a one-valued function has an isolated critical point a, it is expressible by Laurent's series in the region comprised between any two concentric circles described about a with radii less than the distance from a to the nearest critical point. Hence in the neighborhood of a

$$f(z) = A_0 + A_1(z - a) + A_2(z - a)^2 + \ldots$$
$$+ B_1(z - a)^{-1} + B_2(z - a)^{-2} + \ldots$$

The coefficient of $(z - a)^{-1}$ in this expansion is called the "residue" of $f(z)$ at the point a.

If any closed curve C including the point a be drawn in the region of convergence of this series, and $f(z)$ be integrated along C in a positive direction, the result will be

$$\int_C (fz)dz = 2\pi i B_1.$$

The following may be regarded as an extension of Cauchy's theorem :

Theorem I.—If in a region S the only critical points of the one-valued function $f(z)$ are the interior points $a, a', \ldots$, the integral $\int f(z)dz$ taken around its boundary C in a positive direction is equal to

$$\int_C f(z)dz = 2\pi i(B + B' + \ldots),$$

where $B, B', \ldots$ are the residues of $f(z)$ at the critical points. For the integral taken along C is equal to the sum of the integrals whose paths are mutually exterior small circles described about the points $a, a', \ldots$

The following theorems are immediate consequences of the preceding :

Theorem II.—If in a region having a given boundary C the only critical points of the one-valued function $f(z)$ are poles interior to C, an equation

$$\int_C \frac{f'(z)}{f(z)}dz = 2i\pi(M - N)$$

exists, M denoting the number of zeros and N the number of poles within C, each such point being taken a number of times equal to its order.

For in the neighborhood of the point a

$$f(z) = (z - a)^m \phi(z)$$

where $\phi(z)$ is finite and different from zero at a, and m is a positive integer if a is a zero, a negative integer if a is a pole. Hence

$$\frac{f'(z)}{f(z)} = \frac{m}{z - a} + \frac{\phi'(z)}{\phi(z)}.$$

The integrand, therefore, has a pole at every zero and pole of $f(z)$, and its residue is the order, taken positively for a zero, and negatively for a pole.

Theorem III.—Every algebraic equation of degree n has n roots.

For let $f(z)$ represent the first member of the equation $z^n + a_1 z^{n-1} + \ldots + a_n = 0$. Since $f(z)$ has no poles in the finite part of the z-plane, the number of roots contained within any closed curve C will be given by the integral

$$\frac{1}{2\pi i}\int_C \frac{f'(z)}{f(z)}\,dz.$$

But taking for C a circle described about the origin as a center with a very great radius, this integral is

$$\frac{1}{2\pi i}\int_C \frac{nz^{n-1}+(n-1)a_1 z^{n-2}+\cdots}{z^n+a_1 z^{n-1}+\cdots}\,dz = \frac{1}{2\pi i}\int_C \frac{n\,dz}{z}(1+\epsilon)$$

where ϵ has zero for a limit when $|z|=\infty$. Hence the limit of the preceding integral, as $|z|$ is increased, is n.

Prob. 17. Show that if $z=\infty$ is an ordinary point of $f(z)$, that is, if $f(z)$ is expressible for very great value of z by a series containing only negative powers of z, the integral of $f(z)$ around an infinitely great circle is equal to $2\pi i$ into the coefficient of $\frac{1}{z}$. This coefficient is called the residue for $z=\infty$.

Prob. 18. Show that the sum of all the residues of $f(z)$, of the preceding problem, including the residue at infinity, is equal to zero.

Prob. 19. If $\frac{\phi(z)}{\psi(z)}$ is a rational function of which the numerator is of degree lower by 2 than the denominator, and if the zeros $a_1, a_2, \ldots, a_n$ of the denominator are of the first order, show that

$$\sum_1^n \frac{\phi(a_\nu)}{\psi'(a_\nu)}=0.$$

ART. 26.   INTEGRAL OF A ONE-VALUED FUNCTION.

It was shown in Article 18 that, if a function $f(z)$ is holomorphic in a given region S, its integral taken from a fixed lower limit contained in S to a variable upper limit z is a one-valued function of z within S. If $F(z)$ is a function which takes a determinate value $F(z_0)$ at $z=z_0$ and is one-valued while z remains within S, having at every point $f(z)$ for its derivative, the integral of $f(z)$ from z_0 to z is equal to $F(z)-F(z_0)$. If $F_1(z)$ is another function fulfilling these conditions, so that the integral of $f(z)$ can be written also in the form $F_1(z) - F_1(z_0)$, the functions $F(z)$ and $F_1(z)$ differ only by a constant term ; for

$$F_1(z) = F(z) + [F_1(z_0) - F(z_0)].$$

Suppose now that $f(z)$ is still one-valued in S, but that it has isolated critical points $a_1, a_2, \ldots$ interior to S. Any two paths from z_0 to z, which inclose between them a region containing none of the points $a_1, a_2, \ldots$, will give integrals identical in value. Let the two paths L_1, L include between them a single critical point a_κ; and consider the integrals along these two paths. The integral along L_1 will be equal to the integral along the composite path $L_1 L^{-1} L$, where the exponent $-$ 1 indicates that the corresponding path is reversed ; for the integral along $L^{-1} L$ is equal to zero. But $L_1 L^{-1}$ is a closed curve, or " loop," including the critical point a_κ, and, assuming that it is described in a positive direction about a_κ, the integral along it is equal to $2\pi i B_\kappa$, where B_κ is the residue of $f(z)$ at a_κ. Hence

$$\int_{L_1} f(z)dz = 2\pi i B_\kappa + \int_L f(z)dz.$$

If now the two paths L_1, L from z_0 to z include between them several critical points $a_\kappa, a_\lambda, a_\mu, \ldots$, draw intermediate paths $L_2, \ldots, L_m$, so that the region between any two consecutive paths contains only one critical point. The integral along L_1 will be equal to the integral along the composite path $L_1 L_2^{-1} L_2 \ldots L_m^{-1} L_m L^{-1} L$, since the integrals corresponding to $L_2^{-1} L_2, \ldots, L_m^{-1} L_m$, $L^{-1} L$ are all equal to zero. But $L_1 L_2^{-1}$, $L_2 L_3^{-1}, \ldots, L_m L^{-1}$ are all closed paths or loops, each including a single critical point, so that, assuming that each is described in a positive direction and that $B_\kappa, B_\lambda, B_\mu, \ldots$ denote the residues of $f(z)$ at the critical points,

$$\int_{L_1} f(z)dz = 2\pi i(B_\kappa + B_\lambda + B_\mu + \ldots) + \int_L f(z)dz.$$

It has been assumed in the preceding that neither of the paths L_1, L intersects itself. In the case where a path, for example L_1, intersects itself in several points $c_1, c_2, \ldots$, it is possible to consider L, as made up of a path L_1' not intersecting itself, together with a series of loops attached to L_1' at the points $c_1, c_2, \ldots$ Each of these loops encloses a single critical point a_κ and, if described in a positive direction, adds to the integral a term $2\pi i B_\kappa$. Each such loop described in a negative direction adds a term of the form $-2\pi i B_\kappa$. It is evident that the form of each loop and the point at which it is attached to L_1' may be altered arbitrarily without altering the value of the integral, provided no critical point be introduced into or removed from the loop. In fact all the loops may be regarded as attached to L_1' at z_0.

It can be proved by similar reasoning that the most general path that can be drawn from z_0 to z will be equivalent, so far as the value of the integral is concerned, to any given path L preceded by a series of loops, each of which includes a single critical point and is described in either a positive or negative direction. The value of the integral is therefore of the form

$$\int_L f(z)dz + 2\pi i(m_1 B_1 + m_2 B_2 + \ldots),$$

where $m_1, m_2, \ldots$ are any integers positive or negative.

As an example consider the integral $\int_{z_0}^{z} \frac{dz}{z - a}$. The only critical point is $z = a$. Any path whatsoever from z_0 to z is equivalent to a determinate path, for example, a rectilinear path, preceded by a loop containing a and described a certain number of times in a positive or negative direction. If w denote the integral for a selected path, the general value of the integral will be $w + 2n\pi i$. If now a straight line be drawn joining z_0 to a, and if along its prolongation from a to infinity the z-plane be cut or divided, the integral in the z-plane thus divided is one-valued. But, with the variation of z thus restricted, any branch of the function $\log (z - a)$ is one-valued. Select that branch, for example, which reduces to zero when $z = a + 1$. It takes a determinate value for $z = z_0$, and its derivative for every value of z is $\dfrac{1}{z-a}$. Hence, denoting it by Log $(z-a)$,

$$\int_{z_0}^{z}\frac{dz}{z-a} = \mathrm{Log}\,(z-a) - \mathrm{Log}\,(z_0-a) = \mathrm{Log}\,\frac{z-a}{z_0-a}.$$

For a path not restricted in any way, the value of the integral is

$$\int_{z_0}^{z}\frac{dz}{z-a} = \mathrm{Log}\,\frac{z-a}{z_0-a} \pm 2n\pi i = \log\frac{z-a}{z_0-a}.$$

Prob. 20. If $\dfrac{\phi(z)}{\psi(z)}$ is a rational function of z of which the numerator is of degree lower by 2 than the denominator, and if the zeros $a_1, a_2, \ldots, a_n$ of the denominator be of the first order, show that

$$\int_{z_0}^{z}\frac{dz}{z-a} = \sum_{1}^{n}\frac{\phi(a_\nu)}{\psi'(a_\nu)}\log\frac{z-a_\nu}{z_0-a_\nu},$$

where $\displaystyle\sum_{1}^{n}\phi(a_\nu)/\psi'(a_\nu) = 0.$ (See Prob. 18, Art. 25.)

ART. 27. WEIERSTRASS'S THEOREM.

Any rational entire function of z, having its zeros at the points $a_1, a_2, \ldots, a_m$, can be put in the form

$$A(z-a_1)^{n_1}(z-a_2)^{n_2}\ldots(z-a_m)^{n_m},$$

where A is a constant and $n_1, n_2, \ldots, n_m$ are positive integers. More generally, any function which has no critical point in the finite portion of the z-plane and has the points $a_1, \ldots, a_m$ as its zeros, is of the form

$$e^{g(z)}(z-a_1)^{n_1}\ldots(z-a_m)^{n_m},$$

where $g(z)$ is holomorphic in every finite region.

The extension of this result to the case where a function without finite critical points has an infinite number of zeros is due to Weierstrass. It is effected by means of the following theorem :

Theorem.—Given an infinite number of isolated points $a_1,$

$a_2, \ldots, a_n, \ldots$, a function can be constructed holomorphic except at infinity and equal to zero at each of the given points only.*

For the given points can be taken so that

$$|a_1| \lesseqqgtr |a_2| \lesseqqgtr \ldots |a_n| \lesseqqgtr \ldots,$$

$|a_n|$ increasing indefinitely with n. Consider the infinite product

$$\phi(z) = \prod_1^\infty \left(1 - \frac{z}{a_n}\right) e^{P_n(z)},$$

where $P_n(z)$ denotes the rational entire function

$$P_n(z) = \frac{z}{a_n} + \ldots + \frac{z^n}{na_n^n}.$$

Any factor may be written in the form

$$\left(1 - \frac{z}{a_n}\right) e^{P_n(z)} = e^{\log\left(1 - \frac{z}{a_n}\right) + P_n(z)}.$$

But since

$$\log\left(1 - \frac{z}{a_n}\right) = -\int_0^z \frac{dz}{a_n - z} = -\frac{z}{a_n} - \ldots - \frac{z^n}{na_n^n} - \int_0^z \frac{z^n dz}{a_n^n(a_n - z)},$$

the path of integration being arbitrary except that it avoids the points $a_1, a_2, \ldots$, the product may be expressed as

$$\prod_1^\infty e^{\psi_n(z)}, \text{ in which } \psi_n(z) = -\int_0^z \frac{z^n dz}{a_n^n(a_n - z)}.$$

In any given finite region of the z-plane it will be possible to assume that $|z| \lesseqqgtr \rho < a_m$, since $|a_n|$ increases indefinitely with n. Divide the product into two parts,

$$\prod_1^{m-1}\left(1 - \frac{z}{a_n}\right) e^{P_n(z)} \cdot \prod_m^\infty e^{\psi_n(z)}.$$

The second part is equal to

$$e^{\sum_m^\infty \psi_n(z)}.$$

* The following proof is taken from Jordan, Cours d'Analyse, 2d edition, Vol. II.

Consider the series $\overset{\infty}{\underset{m}{\Sigma}}\psi_n(z)$ and $\overset{\infty}{\underset{m}{\Sigma}}\psi_n'(z)$, each term of the second being the derivative of the corresponding term of the first. In the given region

$$|\psi_n'(z)| = \left| -\frac{z^n}{a_n{}^n(a_n - z)} \right| < \frac{\rho^n}{|a_m|^n(|a_m| - \rho)}.$$

Each term of $\overset{\infty}{\underset{m}{\Sigma}}\psi_n'(z)$ is accordingly less in absolute value than the corresponding term of a convergent geometrical progression independent of z. The series $\overset{\infty}{\underset{m}{\Sigma}}\psi_n'(z)$, therefore, converges uniformly. The series $\overset{\infty}{\underset{m}{\Sigma}}\psi_n(z)$ also converges, since

$$|\psi_n(z)| = \text{mod} \int_0^z \psi_n'(z)dz \leqq \frac{\rho^n l}{|a_m|^n(|a_m| - \rho)},$$

where l denotes the length of the path of integration.

By Theorem IV, of Article 23, the series $\overset{\infty}{\underset{m}{\Sigma}}\psi_n(z)$ represents in the given region a holomorphic function. The exponential

$$e^{\overset{\infty}{\underset{m}{\Sigma}}\psi_n(z)}$$

also must be holomorphic. The other part of the product

$$\overset{n-1}{\underset{1}{\Pi}}\left(1 - \frac{z}{a_n}\right)e^{P_n(z)}$$

containing only a finite number of factors is everywhere holomorphic, vanishing at all of the points $a_1, a_2, \ldots$, which are situated within the given finite region. But this region may be extended arbitrarily. The product therefore fulfils the required conditions.

In the preceding demonstration it was tacitly assumed that none of the given points $a_1, a_2, \ldots$ was situated at the origin. To introduce a zero at the origin it is necessary merely to multiply the result by a power of z.

The most general function without finite critical points

having its only zeros at the given points $a_1, a_2, \ldots, a_n \ldots$, can
be expressed in the form

$$f(z) = e^{g(z)} \prod_{1}^{\infty} \left(1 - \frac{z}{a_n}\right) e^{P_n(z)},$$

where $g(z)$ is holomorphic except at infinity; for the ratio of
any two functions satisfying the required conditions is neither
infinite nor zero at any finite point.

By means of Weierstrass's theorem it is possible to express
any function, $F(z)$, whose only finite critical points are poles as
the ratio of two functions holomorphic except at infinity. For,
construct a function $\psi(z)$ having the poles of $F(z)$ as its zeros.
The product $F(z) \cdot \psi(z) = \phi(z)$ will have no finite critical point.
The given function can, therefore, be written

$$F(z) = \frac{\phi(z)}{\psi(z)},$$

which is the required form.

In applying Weierstrass's theorem to particular examples,
it will rarely be found necessary to include in the polynomials
$P_n(z)$ so many terms as were employed in the demonstration
given above. It is quite sufficient, of course, to choose these
polynomials in any way which will make the product converge
for finite values of z to a holomorphic function. Factors of the
form

$$\left(1 - \frac{z}{a_n}\right) e^{P_n(z)},$$

where $P_n(z)$ is chosen in such a manner, are called "primary
factors."

As an application of Weierstrass's Theorem take the resolution of $\sin z$ into primary factors. The zeros of $\sin z$ are o
$\pm \pi, \pm 2\pi, \ldots, \pm n\pi, \ldots$. Consider factors of the form

$$\left(1 - \frac{z}{n\pi}\right) e^{\frac{z}{n\pi}}$$

so that $P_n(z)$ contains only one term $\dfrac{z}{n\pi}$, and

$$\psi_n(z) = -\int_0^z \frac{z\, dz}{n\pi(n\pi - z)}.$$

The series $\sum_m \psi_n'(z)$ will converge uniformly in any region at every point of which $|z| \leqq \rho < m\pi$; for, since

$$|\psi_n'(z)| = \left| -\frac{z}{n\pi(n\pi - z)} \right| \leqq \frac{\rho}{n^2\pi^2\left(1 - \frac{\rho}{|n\pi|}\right)} \leqq \frac{\rho}{n^2\pi^2\left(1 - \frac{\rho}{m\pi}\right)},$$

each term is less in absolute value than the corresponding term of the series

$$\frac{\rho}{\pi^2\left(1 - \frac{\rho}{m\pi}\right)} \sum_m^{\infty} \frac{1}{n^2}.$$

A similar result holds for the series $\sum_{-m}^{-\infty} \psi_n'(z)$. The two series

$$\sum_m^{\infty} \psi_n(z), \quad \sum_{-m}^{-\infty} \psi_n(z)$$

are also convergent; for $|\psi_n(z)|$ cannot exceed the upper extreme of $|\psi_n'(z)|$ multiplied by l, the length of the path of integration from the origin to the point z. These series accordingly represent holomorphic functions in the region for which $|z| \leqq \rho$. Hence the expression required is

$$\sin z = e^{g(z)} \prod_{-\infty}^{+\infty} \left(1 - \frac{z}{n\pi}\right)e^{\frac{z}{n\pi}}.$$

It will be shown in the course of the next article that $e^{g(z)} = 1$.

Prob. 21. If ω_1 and ω_2 be two quantities not having a real ratio, the doubly infinite series of which the general term is $\frac{1}{(m\omega_1 + n\omega_2)^p}$ is absolutely convergent if $p > 2$. Hence show that the product

$$\sigma(z) = z\prod\left(1 - \frac{z}{\omega}\right)e^{\frac{z}{\omega} + \frac{z^2}{2\omega^2}},$$

where $\omega = m\omega_1 + n\omega_2$, defines a holomorphic function in any finite region of the z-plane. This function is Weierstrass's sigma function, and is the basis of his system of elliptic functions.

ART. 28.   MITTAG-LEFFLER'S THEOREM.

Any one-valued function $f(z)$ with isolated critical points $a_1, a_2, \ldots$ can be represented in the neighborhood of one of these points by Laurent's series; viz.:

$$f(z) = A_0 + A_1(z - a_n) + A_2(z - a_n)^2 + \cdots$$
$$+ B_1(z - a_n)^{-1} + B_2(z - a_n)^{-2} + \cdots$$

Hence
$$f(z)^n = \phi(z) + G_n\left(\frac{1}{z - a_n}\right),$$

where $\phi(z)$ is holomorphic in a region containing the point a_n, and $G\left(\frac{1}{z - a_n}\right)$ is holomorphic over the whole plane excluding the point a_n. If a_n is a pole of $f(z)$, $G_n\left(\frac{1}{z - a_n}\right)$ contains a finite number of terms; otherwise it is an infinite series. If the number of critical points is finite, and the function $G_n\left(\frac{1}{z - a_n}\right)$ is formed at each such point, by subtracting the sum of these functions from $f(z)$ a remainder will be obtained which has no critical point in the finite part of the plane. This remainder can be expressed as a series of ascending powers $G(z)$ converging for every finite value of z. The function $f(z)$ can therefore be written in the form

$$f(z) = G(z) + \Sigma G_n\left(\frac{1}{z - a_n}\right),$$

analogous to the expression of a rational function by means of partial fractions.

The extension of this result to the case where the number of critical points is infinite is due to Mittag-Leffler. Let $a_1, a_2, \ldots, a_n, \ldots$ be the critical points of the one-valued function $f(z)$, and suppose that

$$|a_1| \lessgtr |a_2| \lessgtr \ldots |a_n| \lessgtr \ldots,$$

$|a_n|$ increasing without limit when n is increased indefinitely. Let, further, $G_n\left(\frac{1}{z - a_n}\right)$ be the series of negative powers of

$z - a_n$ contained in the expansion of $f(z)$ according to Laurent's Series in the neighborhood of a_n.

The function $G_n\left(\dfrac{1}{z - a_n}\right)$, having no critical point except at a_n, may be developed by Maclaurin's series in the form

$$G_n\left(\frac{1}{z - a_n}\right) = A_0^{(n)} + A_1^{(n)}z + \ldots + A_\nu^{(n)}z + \ldots,$$

and the series will converge uniformly within a circle described about the origin as a center with any determinate radius $\rho_n < |a_n|$. Within the same circle Maclaurin's series, applied to $G_n'\left(\dfrac{1}{z - a_n}\right)$, the derivative with respect to z of $G_n\left(\dfrac{1}{z - a_n}\right)$, converges uniformly. Hence, for any point within the circle $|z| = \rho_n$,

$$G_n\left(\frac{1}{z - a_n}\right) = F_n(z) + R, \quad G_n'\left(\frac{1}{z - a_n}\right) = F_n'(z) + R',$$

$F_n(z)$ representing the first $\nu + 1$ terms of the development of $G_n\left(\dfrac{1}{z - a_n}\right)$ by Maclaurin's theorem, $F_n'(z)$ its derivative, and R, R' remainders which by a suitable choice of ν may be made less in absolute value than any given quantity.

Choose the positive quantities $E_1, E_2, \ldots, E_n, \ldots$ so that the series $E_1 + E_2 + \ldots + E_n + \ldots$ is convergent. Choose also in connection with each of the points $a_1, a_2, \ldots, a_n, \ldots$, an integer ν such that

$$\mathrm{mod}\left[G_1\left(\frac{1}{z - a_1}\right) - F_1(z)\right] < E_1, \quad \mathrm{mod}\left[G_1'\left(\frac{1}{z - a_1}\right) - F_1'(z)\right] < E_1,$$

if $|z| \lessgtr \rho_1 < |a_1|$;

$$\mathrm{mod}\left[G_2\left(\frac{1}{z - a_2}\right) - F_2(z)\right] < E_2, \quad \mathrm{mod}\left[G_2'\left(\frac{1}{z - a_2}\right) - F_2'(z)\right] < E_2,$$

if $|z| \lessgtr \rho_2 < |a_2|$; and, in general,

$$\mathrm{mod}\left[G_n\left(\frac{1}{z - a_n}\right) - F_n(z)\right] < E_n, \quad \mathrm{mod}\left[G_n'\left(\frac{1}{z - a_n}\right) - F_n'(z)\right] < E_n,$$

if $|z| \lessgtr \rho_n < |a_n|$.

Consider now the series

$$\sum_1^\infty \left[G_n\left(\frac{1}{z-a_n}\right) - F_n(z) \right], \quad \sum_1^\infty \left[G_n'\left(\frac{1}{z-a_n}\right) - F'(z) \right]$$

in any finite region of the plane, the points $a_1, a_2, \ldots, a_n, \ldots$ being excluded. Since $|a_n|$ increases indefinitely with n, it is possible, in any finite region of the z-plane, to assume that $|z| \lessgtr \rho_m < |a_m|$. Separate from each of these two series its first $m - 1$ terms. These terms will have in each case a finite sum. The remaining terms of either series taken in order will be less in absolute value than $E_m, E_{m+1}, \ldots$ respectively, $|z|$ being less than each of the quantities $\rho_m, \rho_{m+1}, \ldots$. Accordingly, each of the series

$$\sum_1^\infty \left[G_n\left(\frac{1}{z-a_n}\right) - F_n(z) \right], \quad \sum_1^\infty \left[G_n'\left(\frac{1}{z-a_n}\right) - F_n'(z) \right]$$

is absolutely convergent for every value of z except $a_1, a_2, \ldots, a_n, \ldots$. It is evident, further, that in any given finite region, from which the points $a_1, a_2, \ldots, a_n, \ldots$ are excluded, the two series converge uniformly. In such a region any term of either series is holomorphic; and, therefore, by Theorem IV of Article 23, the first of these series defines a holomorphic function.

The point a_n is an ordinary point for the difference

$$f(z) - \left[G_n\left(\frac{1}{z-a_n}\right) - F_n(z) \right] = \left[f(z) - G_n\left(\frac{1}{z-a_n}\right) \right] + F_n(z),$$

since in its neighborhood this difference may be developed as a convergent series containing only positive powers of $z - a_n$. In the same way each of the points $a_1, a_2, \ldots, a_n, \ldots$ is an ordinary point for the function

$$f(z) - \sum_1^\infty \left[G_n\left(\frac{1}{z-a_n}\right) - F_n(z) \right].$$

This function, therefore, can have no critical point except at infinity. and must be expressible as a series $G(z)$ containing only positive powers of z and converging uniformly in any finite region of the z-plane. Hence the function $f(z)$ may be put in the form

$$f(z) = G(z) + \sum_{1}^{\infty} \left[G_n \left(\frac{1}{z - a_{n}} \right) - F_n(z) \right],$$

in which the character of each critical point is exhibited.

As an application of Mittag-Leffler's theorem consider cot z. Its critical points are $z = 0, \pm \pi, \pm 2\pi, \ldots$. In the neighborhood of $z = 0$, $\cot z - \dfrac{1}{z}$ is holomorphic; and in the neighborhood of $z = n\pi$, n being any positive or negative integer, $\cot z - \dfrac{1}{z - n\pi}$ is holomorphic. The series

$$\sum_{m}^{+\infty} \frac{1}{z - n\pi},$$

in which m is an arbitrary positive integer, is not convergent for finite values of z, even when $|z| < m$. The series

$$\sum_{m}^{+\infty} \left[\frac{1}{z - n\pi} + \frac{1}{n\pi} \right] = \sum_{m}^{+\infty} \frac{z}{n\pi(z - n\pi)} = \sum_{m}^{+\infty} \frac{-z}{n^2 \pi^2 \left(1 - \dfrac{z}{n\pi} \right)}$$

is, however, absolutely convergent at every point for which $|z| < m$. For the modulus of any term is equal to

$$\frac{|z|}{n^2 \pi^2 \left| 1 - \dfrac{z}{n\pi} \right|} \overset{=}{<} \frac{|z|}{n^2 \pi^2 \left(1 - \dfrac{|z|}{n\pi} \right)},$$

and, therefore, less than the corresponding term in the series

$$\frac{|z|}{\pi^2 \left(1 - \dfrac{|z|}{m\pi} \right)} \sum_{m}^{\infty} \frac{1}{n^2}.$$

A similar result holds for the series

$$\sum_{m}^{\infty} \left[\frac{1}{z + n\pi} - \frac{1}{n\pi} \right].$$

It is easy to see now that the reasoning employed in the demonstration of Mittag-Leffler's theorem may be applied to show that the series

$$\frac{1}{z} + \sum_{-\infty}^{+\infty} \left[\frac{1}{z - n\pi} + \frac{1}{n\pi} \right],$$

where the summation does not include $n = 0$, defines a function holomorphic in any finite region of the z-plane, the points $0, \pm \pi, \pm 2\pi, \ldots$ being excluded. The difference

$$\cot z - \frac{1}{z} - \sum_{-\infty}^{+\infty} \left[\frac{1}{z - n\pi} + \frac{1}{n\pi} \right]$$

can have no critical point except at infinity. It must, therefore, be expressible as a series $G(z)$ of positive powers of z, having an infinite circle of convergence. Hence

$$\cot z = G(z) + \frac{1}{z} + \sum_{-\infty}^{+\infty} \left[\frac{1}{z - n\pi} + \frac{1}{n\pi} \right].$$

The next step is to determine $G(z)$. It is to be observed that, if $G(z)$ is a constant, its value must be zero, since $\cot(-z) = -\cot z$. If $G(z)$ is not a constant, differentiation of the preceding expression for $\cot z$ gives

$$-\frac{1}{\sin^2 z} = G'(z) - \frac{1}{z^2} - \sum_{-\infty}^{+\infty} \frac{1}{(z - n\pi)^2}.$$

It follows, by changing z into $z + \pi$, that

$$G'(z + \pi) = G'(z).$$

Hence $G'(z)$ is periodic, having a period equal to π; and as the point z traces a line parallel to the axis of reals, $G'(z)$ passes again and again through the same range of values. But $G'(z)$, being the derivative of $G(z)$, is holomorphic for every finite value of z. It can, therefore, become infinite, if at all, only when the imaginary part of z is infinite. If z be written in the form $x + iy$, the value of $G'(z)$ may be expressed as

$$G'(z) = \frac{1}{(x + iy)^2} + \sum_{-\infty}^{+\infty} \frac{1}{(x + iy - n\pi)^2} - \left(\frac{2ie^{iy}(\cos x + i \sin x)}{(\cos 2x + i \sin 2x) - e^{2iy}} \right)^2.$$

When $y = \pm \infty$ the first and last terms of the second member vanish. In regard to the series it can be proved that, for any given region is which y is finite and different from zero, an integer ν can be found such that the sum of the moduli of those terms for which $|n| > \nu$ is less in absolute value than any previously assigned quantity ϵ. As $|y|$ is increased the modulus of each of these terms is diminished. The modulus of their sum, therefore, cannot exceed ϵ when $y = \pm \infty$. But when $y = \pm \infty$ the sum of any finite number of terms of the series is zero. Hence the limit of the whole series is zero. $G'(z)$, therefore, never becomes infinite. Hence, by Theorem III, Article 24, it is constant, and is equal to zero. It follows that $G(z)$ is equal to zero.

The expression for $\cot z$ is accordingly

$$\cot z = \frac{1}{z} + \sum_{-\infty}^{+\infty} \left[\frac{1}{z - n\pi} + \frac{1}{n\pi} \right].$$

The logarithmic derivative of the product expression for $\sin z$, given in the preceding article as an example of Weierstrass's theorem, is

$$\cot z = g'(z) + \frac{1}{z} + \sum_{-\infty}^{+\infty} \left[\frac{1}{z - n\pi} + \frac{1}{n\pi} \right].$$

Hence $g(z)$ in that expression is a constant. Making $z = 0$, its value is seen to be unity.

Prob. 22. From the expression for $\cot z$ deduce the equation

$$\operatorname{cosec}^2 z = \sum_{-\infty}^{+\infty} \frac{1}{(z - n\pi)^2},$$

where the summation does not exclude $n = 0$.

Prob. 23. Show that the doubly infinite series

$$\wp(z) = \frac{1}{z^2} + \sum \left[\frac{1}{(z - \omega)^2} - \frac{1}{\omega^2} \right],$$

where $\omega = m\omega_1 + n\omega_2$, defines a function whose only finite critical points are $z = \omega$. This function is Weierstrass's $\wp$-function. (Compare Problem 21.)

Prob. 24. Prove that

$$\wp(z) = - \frac{d^2}{dz^2} \log \sigma(z).$$

Prob. 25. Prove that $\wp'(z) = -2\sum \dfrac{1}{(z-\omega)^3}$, where the summation does not exclude $\omega = 0$.

ART. 29.   CRITICAL LINES AND REGIONS.

The functions whose properties have been considered in the preceding articles have been assumed to have only isolated critical points.   That an infinite number of critical points may be grouped together in the neighborhood of a single finite point is evident, however, from the consideration of such examples as

$$w = \cot\frac{1}{z}, \qquad w = e^{\operatorname{cosec}\frac{1}{z-a}}.$$

In the former an infinite number of poles are grouped in the neighborhood of the origin.   In the latter an infinite number of essential singularities are situated in the vicinity of the point $z = a$.

It is easy to illustrate by an example the occurrence of lines and regions of discontinuity.   Take the series *

$$\theta(z) = \frac{1}{1-z} + \frac{z}{z^2-1} + \frac{z^2}{z^4-1} + \frac{z^4}{z^8-1} + \cdots.$$

The sum of its first n terms is

$$-\frac{1}{z^{2^{n-1}}-1},$$

which converges to unity if $|z| < 1$, and to zero if $|z| > 1$. Hence the circle $|z| = 1$ is a line of discontinuity for this series.

Consider now any two regions S_1 and S_2, the former situated within, the latter without, the unit circle.   Let $\phi(z)$ and $\psi(z)$ be two arbitrary functions both completely defined in these regions.   The expression

$$\phi(z)\theta(z) + \psi(z)[1 - \theta(z)]$$

---

* This series is due to J. Tannery.   See Weierstrass, Abhandlungen aus der Functionenlehre (1886), p. 102.

will be equal to $\phi(z)$ in S_1 and $\psi(z)$ in S_2. In regions completely separated from one another by a critical line, the same literal expression may thus represent entirely independent functions.

For a single continuous region, however, in the interior of which exist only isolated critical points, the character of the function in one part determines its character in every other part. Let S be such a region, and assume that its boundary is a critical line. In the neighborhood of any interior point a, not a critical point, the given function is expressible as a power series, viz.:

$$f(z) = f(a) + (z-a)f'(a) + \ldots + \frac{(z-a)^n}{1 \cdot 2 \ldots n} f^{(n)}(a) + \ldots$$

This series will converge uniformly over a circle described about a as a center with any determinate radius less than the distance from a to the nearest critical point. It serves for the calculation of $f(z)$ and all its successive derivatives at any point b interior to this circle. From the preceding power series, accordingly, can be obtained another

$$f(z) = f(b) + (z-b)f'(b) + \ldots + \frac{(z-b)^n}{1 \cdot 2 \ldots n} f^{(n)}(b) + \ldots,$$

representing the $f(z)$ within a circle described about b as a center. In general, the point b can be so chosen that a portion of this new circle will lie without the circle of convergence of the former power series. At any new point c within the circle whose center is b, the value of the function and all its successive derivatives can be calculated; and so, as before, a power series can be obtained convergent in a circle described about c as a center and, in general, including points not contained in either of the preceding circles. By continuing in this manner it will be possible, starting from a given point a with the expression of $f(z)$ in ascending powers, to obtain an expression of the same character at any other point k which can be connected with a by a continuous line everywhere at a finite distance from the nearest critical point. It follows that the character of the function everywhere within S can be determined completely from its expression in ascending power series in the neighborhood of a single interior point.

It will be impossible by the process just explained to derive any information in regard to the function at points exterior to S. The example given above, furthermore, shows that a complete definition of $f(z)$ within S may carry with it the definition of an entirely independent function without S.

As an example of a function having a critical region consider the function defined by the series

$$1 + 2z + 2z^4 + 2z^9 + \ldots,$$

which represents a function without critical points in the interior of the circle $|z| = 1$. For points on or without this circle the series is divergent; and, further, it is impossible to obtain from it an expression converging when $|z| \gtreqless 1$. The function thus defined, consequently, exists only in the region interior to the unit circle. By changing z into $1/z$ a series

$$1 + \frac{2}{z} + \frac{2}{z^4} + \frac{2}{z^9} + \ldots$$

is obtained, representing a function which has no existence in the interior of the unity circle. Functions in connection with which such regions arise are called "lacunary functions." *

ART. 30. FUNCTIONS HAVING n VALUES.

Let the function $w = f(z)$ take at the point z_0 of a given region S a value $w^{(0)}$. Suppose that along any continuous path, beginning at z_0, and subject only to the conditions that it shall remain in the interior of S and shall not pass through certain isolated points $a_1, a_2, \ldots, w$ is continuous and has a continuous derivative. If it is impossible, when z traces such a path, to return to the point z_0 so as to obtain there a value of w different from $w^{(0)}$, w is one-valued in the region S. On the other

* Poincaré, American Journal of Mathematics, Vol. XIV; Harkness and Morley, Theory of Functions (1893), p. 119

hand, certain paths may lead back to z_0 with new values of w.

Suppose that at each point of S, except $a_1, a_2, \ldots, w$ has n different values, and that starting from such a point z_0 and tracing any continuous curve not passing through $a_1, a_2, \ldots,$ the several values of w give rise to n branches $w_1, w_2, \ldots, w_n$, each of which is characterized by a continuous derivative. In the neighborhood of a_k any one of the points $a_1, a_2, \ldots$ these branches are said to be distinct or not, according as small closed curves described about this point lead from each value of w back to the same value again, or cause some of the branches to interchange values. In the latter case the point is a branch point.

About any branch point a_k as a center describe a small circle ; and suppose that, starting from any point of it with the value w_α corresponding to a certain branch, the values $w_\beta, w_\gamma \ldots$ are obtained by successive revolutions about a_k, the original value being reproduced after p revolutions. Introduce now a new independent variable z' such that

$$z' = (z - a_k)^{\frac{1}{p}}.$$

It can be shown that when z makes one revolution about a_k, z' makes only one pth part of a revolution about the origin of the z'-plane, and that to a complete revolution of z' about the origin of the z'-plane correspond p revolutions of z about a_k. Considering then the branch w_α as a function of z', the origin cannot be a branch point, for whenever z' describes a small circle about it, the value w_α is reproduced. The branch w_α must accordingly be expressible by Laurent's series in the form

$$w_\alpha = \sum_{-\infty}^{+\infty} A_m z'^m,$$

or, substituting for z' its value,

$$w_\alpha = A_0 + A_1 (z - a_k)^{\frac{1}{p}} + A_2 (z - a_k)^{\frac{2}{p}} + \ldots$$
$$+ A_{-1}(z - a_k)^{-\frac{1}{p}} + A_{-2}(z - a_k)^{-\frac{2}{p}} + \ldots$$

This expression makes plain the relation between the different branches of a function in the neighborhood of a branch point. When the development of a branch in the neighborhood of one of its branch points gives rise to only a finite number of terms containing negative powers, the branch point is called a " polar branch point."

Consider the functions

$$P_1 = w_1 + w_2 + \ldots + w_n ,$$
$$P_2 = w_1 w_2 + w_1 w_3 = \ldots + w_{n-1} w_n ,$$
$$\cdot \quad \cdot \quad \cdot \quad \cdot \quad \cdot \quad \cdot \quad \cdot \quad \cdot$$
$$P_n = w_1 w_2 \ldots w_n.$$

Each of these functions is unchanged in value when several or all of the quantities $w_1, w_2, \ldots, w_n$ are interchanged, and is consequently a one-valued function of z within S. Hence w must satisfy an equation of the nth degree,

$$w^n + P_1 w^{n-1} + P_2 w^{n-2} + \ldots + P^n = 0,$$

the coefficients of which are one-valued functions of z having only isolated critical points within S. When the entire z-plane can be taken as the region S, and those branch points at which the branches do not all remain finite are polar branch points, the only other critical points being poles for one or more branches, the functions $P_1, P_2, \ldots, P_n$ are rational functions of z. In this case w is an algebraic function of z.

CHAPTER VII.

DIFFERENTIAL EQUATIONS.

By W. WOOLSEY JOHNSON,
Professor of Mathematics in the U. S. Naval Academy.

ART. 1. EQUATIONS OF FIRST ORDER AND DEGREE.

In the Integral Calculus, supposing y to denote an unknown function of the independent variable x, the derivative of y with respect to x is given in the form of a function of x, and it is required to find the value of y as a function of x.   In other words, given an equation of the form

$$\frac{dy}{dx} = f(x), \quad \text{or} \quad dy = f(x)dx, \tag{1}$$

of which the general solution is written in the form

$$y = \int f(x)dx, \tag{2}$$

it is the object of the Integral Calculus to reduce the expression in the second member of equation (2) to the form of a known function of x.   When such reduction is not possible, the equation serves to define a new function of x.

In the extension of the processes of integration of which the following pages give a sketch the given expression for the derivative may involve not only x, but the unknown function y; or, to write the equation in a form analogous to equation (1), it may be

$$Mdx + Ndy = 0, \tag{3}$$

in which M and N are functions of x and y.   This equation is in fact the general form of the differential equation of the first order and degree; either variable being taken as the independent variable, it gives the first derivative of the other variable in terms of x and y. So also the solution is not necessarily an expression of either variable as a function of the other, but is generally a relation between x and y which makes either an implicit function of the other.

When we recognize the left member of equation (3) as an "exact differential," that is, the differential of some function of x and y, the solution is obvious. For example, given the equation

$$x\,dy + y\,dx = 0, \tag{4}$$

the solution $$xy = C, \tag{5}$$

where C is an arbitrary constant, is obtained by "direct integration." When a particular value is attributed to C, the result is a "particular integral;" thus $y = x^{-1}$ is a particular integral of equation (4), while the more general relation expressed by equation (5) is known as the "complete integral."

In general, the given expression $M\,dx + N\,dy$ is not an exact differential, and it is necessary to find some less direct method of solution.

The most obvious method of solving a differential equation of the first order and degree is, when practicable, to "separate the variables," so that the coefficient of dx shall contain x only, and that of dy, y only. For example, given the equation

$$(1 - y)\,dx + (1 + x)\,dy = 0. \tag{6}$$

the variables are separated by dividing by $(1 + x)(1 - y)$.

Thus $$\frac{dx}{1 + x} + \frac{dy}{1 - y} = 0.$$

Each term is now directly integrable, and hence

$$\log(1 + x) - \log(1 - y) = c.$$

The solution here presents itself in a transcendental form, but it is readily reduced to an algebraic form. For, taking the exponential of each member, we find

$$\frac{1 + x}{1 - y} = e^c = C, \quad \text{whence} \quad 1 + x = C(1 - y), \tag{7}$$

where C is put for the constant e^c.

To verify the result in this form we notice that differentiation gives $dx = - Cdy$, and substituting in equation (6) we find

$$- C(\mathbf{1} - y) + \mathbf{1} + x = 0,$$

which is true by equation (7).

Prob. 1. Solve the equation $dy + y \tan x \, dx = 0$.

(Ans. $y = C \cos x$.)

Prob. 2. Solve $\dfrac{dy}{dx} + b^2 y^2 = a^2$. $\left(\text{Ans. } \dfrac{by + a}{by - a} = ce^{2abx}.\right)$

Prob. 3. Solve $\dfrac{dy}{dx} = \dfrac{y^2 + 1}{x^2 + 1}$. $\left(\text{Ans. } y = \dfrac{x + c}{1 - cx}.\right)$

Prob. 4. Helmholtz's equation for the strength of an electric current C at the time t is

$$C = \frac{E}{R} - \frac{l}{R} \frac{dC}{dt},$$

where E, R, and L are given constants. Find the value of C, determining the constant of integration by the condition that its initial value shall be zero.

ART. 2. GEOMETRICAL REPRESENTATION.

The meaning of a differential equation may be graphically illustrated by supposing simultaneous values of x and y to be the rectangular coordinates of a variable point. It is convenient to put p for the value of the ratio $dy : dx$. Then P being the moving point (x, y) and ϕ denoting the inclination of its path to the axis of x, we have

$$p = \frac{dy}{dx} = \tan \phi.$$

The given differential equation of the first order is a relation between p, x, and y, and, being of the first degree with respect to p, determines in general a single value of p for any assumed values of x and y. Suppose in the first place that, in addition to the differential equation, we were given one pair of simultaneous values of x and y, that is, one position of the point P. Now let P start from this fixed initial point and begin to move in either direction along the straight line whose inclination is determined by the value of p corresponding to the initial values of x and y. We thus have a moving point satisfying the given differential equation. As the point P moves the values of x and y vary, and we must suppose the direction of its motion to vary in such a way that the simultaneous values of x, y, and p continue to satisfy the differential equation. In that case, the path of the moving point is said to satisfy the differential equation. The point P may return to its initial position, thus describing a closed curve, or it may pass to infinity in each direction from the initial point describing an infinite branch of a curve.* The ordinary cartesian equation of the path of P is a particular integral of the differential equation.

If no pair of associated values of x and y be known, P may be assumed to start from any initial point, so that there is an unlimited number of curves representing particular integrals of the equation. These form a "system of curves," and the complete integral is the equation of the system in the usual form of a relation between x, y, and an arbitrary "parameter." This parameter is of course the constant of integration. It is constant for any one curve of the system, and different values of it determine different members of the system of curves, or different particular integrals.

As an illustration, let us take equation (4) of Art. 1, which may be written

$$\frac{dy}{dx} = -\frac{y}{x}.$$

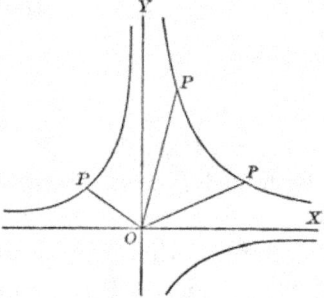

Denoting by θ the inclination to the axis of x of the line joining P with the origin, the equation is equivalent to $\tan \phi = - \tan \theta$, and therefore expresses that P moves in a direction inclined equally with OP to either axis, but on the other

* When the form of the functions M and N is unrestricted, there is no reason why either of these cases should exist, but they commonly occur among such differential equations as admit of solution.

side. Starting from any position in the plane, the point P thus moving must describe a branch of an hyperbola having the two axes as its asymptotes; accordingly, the complete integral $xy = C$ is the equation of the system consisting of these hyperbolas.

Prob. 5. Write the differential equation which requires P to move in a direction always perpendicular to OP, and thence derive the equation of the system of curves described.

$$\left(\text{Ans. } \frac{dy}{dx} = -\frac{x}{y}; \; x^2 + y^2 = C. \right)$$

Prob. 6. What is the system described when ϕ is the complement of θ? $\left(\text{Ans. } x^2 - y^2 = C. \right)$

Prob. 7. If $\phi = 2\theta$, show geometrically that the system described consists of circles, and find the differential equation.

$$\left(\text{Ans. } 2xy\,dx = (x^2 - y^2)\,dy. \right)$$

ART. 3. PRIMITIVE OF A DIFFERENTIAL EQUATION.

Let us now suppose an ordinary relation between x and y, which may be represented by a curve, to be given. By differentiation we may obtain an equation of which the given equation is of course a solution or particular integral. But by combining this with the given equation any number of differential equations of which the given equation is a solution may be found. For example, from

$$y^2 = m(x - a) \tag{1}$$

we obtain directly

$$2y\,dy = m\,dx, \tag{2}$$

of which equation (1) is an integral; again, dividing (2) by (1) we have

$$\frac{2\,dy}{y} = \frac{dx}{x - a}, \tag{3}$$

and of this equation also (1) is an integral.

If in equation (1) m be regarded as an arbitrary parameter, it is the equation of a system of parabolas having a common axis and vertex. The differential equation (3), which does not contain m, is satisfied by every member of this system of curves.

Hence equation (1) thus regarded is the complete integral of equation (3), as will be found by solving the equation in which the variables are already separated.

Now equation (3) is obviously the only differential equation independent of m which could be derived from (1) and (2), since it is the result of eliminating m. It is therefore the "differential equation of the system;" and in this point of view the integral equation (1) is said to be its "primitive."

Again, if in equation (1) a be regarded as the arbitrary constant, it is the equation of a system of equal parabolas having a common axis. Now equation (2) which does not contain a is satisfied by every member of this system of curves; hence it is the differential equation of the system, and its primitive is equation (1) with a regarded as the arbitrary constant.

Thus, a primitive is an equation containing as well as x and y an arbitrary constant, which we may denote by C, and the corresponding differential equation is a relation between x, y, and p, which is found by differentiation, and elimination of C if necessary. This is therefore also a method of verifying the complete integral of a given differential equation. For example, in verifying the complete integral (7) in Art. 1 we obtain by differentiation $1 = - Cp$. If we use this to eliminate C from equation (7) the result is equation (6); whereas the process before employed was equivalent to eliminating p from equation (6), thereby reproducing equation (7).

Prob. 8. Write the equation of the system of circles in Prob. 7, Art. 2, and derive the differential equation from it as a primitive.

Prob. 9. Write the equation of the system of circles passing through the points (o, b) and $(o, - b)$, and derive from it the differential equation of the system.

ART. 4. EXACT DIFFERENTIAL EQUATIONS.

In Art. 1 the case is mentioned in which $Mdx + Ndy$ is an "exact differential," that is, the differential of a function of x and y. Let u denote this function; then

$$du = Mdx + Ndy, \tag{1}$$

and in the notation of partial derivatives

$$M = \frac{\partial u}{\partial x}, \qquad N = \frac{\partial u}{\partial y}.$$

Then, since by a theorem of partial derivatives $\dfrac{\partial^2 u}{\partial y \partial x} = \dfrac{\partial^2 u}{\partial x \partial y}$,

$$\frac{\partial M}{\partial y} = \frac{\partial N}{\partial x}. \qquad (2)$$

This condition must therefore be fulfilled by M and N in order that equation (1) may be possible. When it is fulfilled $Mdx + Ndy = 0$ is said to be an "exact differential equation," and its complete integral is

$$u = C. \qquad (3)$$

For example, given the equation

$$x(x + 2y)dx + (x^2 - y^2)dy = 0,$$

$M = x(x + 2y)$, $N = x^2 - y^2$, $\dfrac{\partial M}{\partial y} = 2x$, and $\dfrac{\partial N}{\partial x} = 2x$; the

condition (2) is fulfilled, and the equation is exact. To find the function u, we may integrate Mdx, treating y as a constant; thus,

$$\tfrac{1}{8}x^3 + x^2 y = Y,$$

in which the constant of integration Y may be a function of y. The result of differentiating this is

$$x^2 dx + 2xy\, dx + x^2 dy = dY,$$

which should be identical with the given equation; therefore, $dY = y^2\, dy$, whence $Y = \tfrac{1}{3}y^3 + C$, and substituting, the complete integral may be written

$$x^3 + 3x^2 y = y^3 + C.$$

The result is more readily obtained if we notice that all terms containing x and dx only, or y and dy only, are exact differentials; hence it is only necessary to examine the terms containing both x and y. In the present case, these are $2xy\, dx + x^2 dy$, which obviously form the differential of $x^2 y$; whence, integrating and multiplying by 3, we obtain the result above.

The complete integral of any equation, in whatever way it

was found, can be put in the form $u = C$, by solving for C. Hence an exact differential equation $du = 0$ can be obtained, which must be equivalent to the given equation

$$Mdx + Ndy = 0, \qquad (4)$$

here supposed not to be exact. The exact equation $du = 0$ must therefore be of the form

$$\mu(Mdx + Ndy) = 0, \qquad (5)$$

where μ is a factor containing at least one of the variables x and y. Such a factor is called an "integrating factor" of the given equation. For example, the result of differentiating equation (7), Art. 1, when put in the form $u = C$, is

$$\frac{(1 - y)dx + (1 + x)dy}{(1 - y)^2} = 0,$$

so that $(1 - y^2)^{-2}$ is an integrating factor of equation (6). It is to be noticed that the factor by which we separated the variables, namely, $(1 - y)^{-1}(1 - x)^{-1}$, is also an integrating factor.

It follows that if an integrating factor can be discovered, the given differential equation can at once be solved.* Such a factor is sometimes suggested by the form of the equation. Thus, given $\qquad (y - x)dy + ydx = 0$,

the terms $ydx - xdy$, which contain both x and y, are not exact, but become so when divided by either x^2 or y^2; and because the remaining term contains y only, y^{-2} is an integrating factor of the whole expression. The resulting integral is

$$\log y + \frac{x}{y} = C.$$

Prob. 10. Show from the integral equation in Prob. 9, Art. 3. that x^{-2} is an integrating factor of the differential equation.

Prob. 11. Solve the equation $x(x^2 + 3y^2)dx + y(y^2 + 3x^2)dy = 0$.
$\qquad\qquad\qquad$ (Ans. $x^4 + 6x^2y^2 + y^4 = c$.)

---

* Since μM and μN in the exact equation (5) must satisfy the condition (2), we have a partial differential equation for μ; but as a general method of finding μ this simply comes back to the solution of the original equation.

Prob. 12. Solve the equation $y\,dy + x\,dx + \dfrac{x\,dy - y\,dx}{x^2 + y^2}$.

$$\left(\text{Ans.} \ \ \frac{x^2 + y^2}{2} + \tan^{-1}\frac{y}{x} = c.\right)$$

Prob. 13. If $u = c$ is a form of the complete integral and μ the corresponding integrating factor, show that $\mu f(u)$ is the general expression for the integrating factors.

Prob. 14. Show that the expression $x^\alpha y^\beta (my\,dx + nx\,dy)$ has the integrating factor $x^{km-1-\alpha}y^{kn-1-\beta}$; and by means of such a factor solve the equation $y(y^2 + 2x^4)dx + x(x^4 - 2y^2)dy = 0$.

$$\left(\text{Ans.} \ \ 2x^4y - y^4 = cx^2.\right)$$

Prob. 15. Solve $(x^2 + y^2)dx - 2xy\,dy = 0$. $\left(\text{Ans.} \ \ x^2 - y^2 = cx.\right)$

ART. 5.   HOMOGENEOUS EQUATION.

The differential equation $Mdx + Ndy = 0$ is said to be homogeneous when M and N are homogeneous functions of x and y of the same degree; or, what is the same thing, when $\dfrac{dy}{dx}$ is expressible as a function of $\dfrac{y}{x}$. If in such an equation the variables are changed from x and y to x and v, where

$$v = \frac{y}{x}; \quad \text{whence} \ \ y = xv \ \ \text{and} \ \ dy = xdv + vdx,$$

the variables x and v will be separable. For example, the equation

$$(x - 2y)dx + ydy = 0$$

is homogeneous; making the substitutions indicated and dividing by x,

$$(1 - 2v)dx + v(xdv + vdx) = 0,$$

whence

$$\frac{dx}{x} + \frac{vdv}{(v - 1)^2} = 0.$$

Integrating, $\log x + \log(v - 1) - \dfrac{1}{v - 1} = C;$

and restoring y,

$$\log(y - x) - \frac{x}{y - x} = C.$$

The equation $Mdx + Ndy = 0$ can always be solved when

M and N are functions of the first degree, that is, when it is of the form

$$(ax + by + c)dx + (a'x + b'y + c')dy = 0.$$

For, assuming $x = x' + h$, $y = y' + k$, it becomes

$$(ax' + b'y' + ah + bk + c)dx' + (a'x' + b'y' + a'h + b'k + c')dy' = 0,$$

which, by properly determining h and k, becomes

$$(ax' + by')dx' + (a'x' + b'y')dy',$$

a homogeneous equation.

This method fails when $a : b = a' : b'$, that is, when the equation takes the form

$$(ax + by + c)dx + [m(ax + by) + c']dy = 0;$$

but in this case if we put $z = ax + by$, and eliminate y, it will be found that the variables x and z can be separated.

Prob. 16. Show that a homogeneous differential equation represents a system of similar and similarly situated curves, the origin being the center of similitude, and hence that the complete integral may be written in a form homogeneous in x, y, and c.

Prob. 17. Solve $x\,dy - y\,dx - \sqrt{(x^2 + y^2)}\,dx = 0$.
(Ans. $x^2 = c^2 - 2cy$.)

Prob. 18. Solve $(3y - 7x + 7)dx + (7y - 3x + 3)dy = 0$.
(Ans. $(y - x + 1)^2(y + x - 1)^5 = c$.)

Prob. 19. Solve $(x^2 + y^2)dx - 2xy\,dy = 0$. (Ans. $x^2 - y^2 = cx$.)

Prob. 20. Solve $(1 + xy)y\,dx + (1 - xy)x\,dy = 0$ by introducing the new variable $z = xy$.
(Ans. $x = Cye^{\frac{1}{xy}}$.)

Prob. 21. Solve $\dfrac{dy}{dx} = ax + by + c$. (Ans. $abx + b^2y + a + bc = Ce^{bx}$.)

Art. 6.　The Linear Equation.

A differential equation is said to be "linear" when (one of the variables, say x, being regarded as independent,) it is of the first degree with respect to y, and its derivatives. The linear equation of the first order may therefore be written in the form

$$\frac{dy}{dx} + Py = Q. \tag{1}$$

where P and Q are functions of x only. Since the second member is a function of x, an integrating factor of the first member will be an integrating factor of the equation provided it contains x only.   To find such a factor, we solve the equation

$$\frac{dy}{dx} + Py = 0, \tag{2}$$

which is done by separating the variables; thus, $\dfrac{dy}{y} = -\,Pdx$;

whence $\log y = c - \int Pdx$ or

$$y = Ce^{-\int Pdx}. \tag{3}$$

Putting this equation in the form $u = c$, the corresponding exact equation is

$$e^{\int Pdx}(dy + Pydx) = 0,$$

whence $e^{\int Pdx}$ is the integrating factor required.   Using this factor, the general solution of equation (1) is

$$e^{\int Pdx}y = \int e^{\int Pdx}Qdx + C. \tag{4}$$

In a given example the integrating factor should of course be simplified in form if possible.   Thus

$$(1 + x^2)dy = (m + xy)dx$$

is a linear equation for y; reduced to the form (1), it is

$$\frac{dy}{dx} - \frac{x}{1+x^2}y = \frac{m}{1 + x^2},$$

from which

$$\int Pdx = -\int \frac{x\,dx}{1 + x^2} = -\frac{1}{2}\log (1 + x^2).$$

The integrating factor is, therefore,

$$e^{\int Pdx} = \frac{1}{\sqrt{(1 + x^2)}};$$

whence the exact equation is

$$\frac{dy}{\sqrt{(1 + x^2)}} - \frac{xy\,dx}{(1 + x^2)^{\frac{3}{2}}} = \frac{mdx}{(1 + x^2)^{\frac{3}{2}}}.$$

Integrating, there is found

$$\frac{y}{\sqrt{(1+x^2)}} = \frac{mx}{\sqrt{(1+x^2)}} + C,$$

or

$$y = mx + C\sqrt{(1+x^2)}.$$

An equation is sometimes obviously linear, not for y, but for some function of y. For example, the equation

$$\frac{dy}{dx} + \tan y = x \sec y$$

when multiplied by $\cos y$ takes a form linear for $\sin y$; the integrating factor is e^x, and the complete integral

$$\sin y = x - 1 + ce^{-x}.$$

In particular, the equation $\frac{dy}{dx} + Py = Qy^n$, which is known as "the extension of the linear equation," is readily put in a form linear for y^{1-n}.

Prob. 22. Solve $x^2\frac{dy}{dx} + (1-2x)y = x^2$.   (Ans. $y = x^2(1 + ce^{\frac{1}{x}})$.)

Prob. 23. Solve $\cos x \frac{dy}{dx} + y - 1 + \sin x = 0$.

(Ans. $y(\sec x + \tan x) = x + c$.)

Prob. 24. Solve $\frac{dy}{dx}\cos x + y \sin x = 1$.

(Ans. $y = \sin x + c \cos x$.)

Prob. 25. Solve $\frac{dy}{dx} = x^3y^3 - xy$.   (Ans. $\frac{1}{y^2} = x^2 + 1 + ce^{x^2}$.)

Prob. 26. Solve $\frac{dy}{dx} = \frac{1}{xy + x^2y^3}$.   (Ans. $\frac{1}{x} = 2 - y^2 + ce^{-\frac{1}{2}y^2}$.)

ART. 7. FIRST ORDER AND SECOND DEGREE.

If the given differential equation of the first order, or relation between x, y, and p, is a quadratic for p, the first step in the solution is usually to solve for p. The resulting value of p will generally involve an irrational function of x and y; so that an equation expressing such a value of p, like some of those solved in the preceding pages, is not properly to be regarded as an equation of the first degree. In the exceptional case when the expression whose root is to be extracted is a perfect square, the equation is decomposable into two equations properly of the first degree. For example, the equation

$$xy(1 + 4p^2) = 2p(x^2 + y^2)$$

when solved for p gives $2p = \dfrac{y}{x}$, or $2p = \dfrac{x}{y}$; it may therefore be written in the form

$$(2px - y)(2py - x) = 0,$$

and is satisfied by putting either

$$\frac{dy}{dx} = \frac{y}{2x}, \quad \text{or} \quad \frac{dy}{dx} = \frac{x}{2y}.$$

The integrals of these equations are

$$y^2 = cx \quad \text{and} \quad 2y^2 - x^2 = C,$$

and these form two entirely distinct solutions of the given equation.

As an illustration of the general case, let us take the equation

$$xp^2 = y^2, \quad \text{or} \quad \frac{dy}{dx} = \pm \frac{\sqrt{y}}{\sqrt{x}}. \tag{1}$$

Separating the variables and integrating,

$$\sqrt{x} \pm \sqrt{y} = \pm \sqrt{c}, \tag{2}$$

and this equation rationalized becomes

$$(x - y)^2 - 2c(x + y) + c^2 = 0. \tag{3}$$

There is thus a single complete integral containing one arbitrary constant and representing a single system of curves; namely, in this case, a system of parabolas touching each axis at the same distance c from the origin. The separate equations given in the form (2) are merely branches of the same parabola.

Recurring now to the geometrical interpretation of a differential equation, as given in Art. 2, it was stated that an equation of the first degree determines, in general, for assumed values of x and y, that is, at a selected point in the plane, a single value of p. The equation was, of course, then supposed rational in x and y.* The only exceptions occur at points for which the value of p takes the indeterminate form; that is, the equation being $Mdx + Ndy = 0$, at points (if any exist) for which $M = 0$ and $N = 0$. It follows that, except at such points, no two curves of the system representing the complete integral intersect, while through such points an unlimited number of the curves may pass, forming a "pencil of curves." †

On the other hand, in the case of an equation of the second degree, there will in general be two values of p for any given point. Thus from equation (1) above we find for the point $(4, 1)$, $p = \pm \frac{1}{2}$; there are therefore two directions in which a point starting from the position $(4, 1)$ may move while satisfying the differential equation. The curves thus described represent two of the particular integrals. If the same values of x and y be substituted in the complete integral (3), the result is a quadratic for c, giving $c = 9$ and $c = 1$, and these determine the two particular integral curves, $\sqrt{x} + \sqrt{y} = 3$, and $\sqrt{x} - \sqrt{y} = 1$.

In like manner the general equation of the second degree, which may be written in the form

$$Lp^2 + Mp + N = 0,$$

where L, M, and N are one-valued functions of x and y, represents a system of curves of which two intersect in any given point for which p is found to have two real values. For these points, therefore, the complete integral should generally give two real values of c. Accordingly we may assume, as the standard form of its equation,

$$Pc^2 + Qc + R = 0,$$

* In fact p was supposed to be a one-valued function of x and y; thus, $p = \sin^{-1}x$ would not in this connection be regarded as an equation of the first degree.

† In Prob. 6, Art. 3, the integral equation represents the pencil of circles passing through the points $(0, b)$ and $(0, -b)$; accordingly p in the differential equation is indeterminate at these points. In some cases, however, such a point is merely a node of one particular integral. Thus in the illustration given in Art. 2, p is indeterminate at the origin, and this point is a node of the only particular integral, $xy = 0$, which passes through it.

where P, Q, and R are also one-valued functions of x and y. If there are points which make p imaginary in the differential equation, they will also make c imaginary in the integral.

Prob. 27. Solve the equation $p^2 + y^2 = 1$ and reduce the integral to the standard form.

$$\text{(Ans. } (y + \cos x)c^2 - 2c \sin x + y - \cos x = 0.)$$

Prob. 28. Solve $yp^2 + 2xp - y = 0$, and show that the intersecting curves at any given point cut at right angles.

Prob. 29. Solve $(x^2 + 1)p^2 = 1$.   (Ans. $c^2e^{2y} - 2cxe^y = 1.$)

ART. 8. SINGULAR SOLUTIONS.

A differential equation not of the first degree sometimes admits of what is called a " singular solution ; " that is to say, a solution which is not included in the complete integral.   For suppose that the system of curves representing the complete integral has an envelope.   Every point A of this envelope is a point of contact with a particular curve of the complete integral system ; therefore a point moving in the envelope when passing through A has the same values of x, y, and p as if it were moving through A in the particular integral curve. Hence such a point satisfies the differential equation and will continue to satisfy it as long as it moves in the envelope. The equation of the envelope is therefore a solution of the equation.

As an illustration, let us take the system of straight lines whose equation is

$$y = cx + \frac{a}{c}, \tag{1}$$

where c is the arbitrary parameter.   The differential equation derived from this primitive is

$$y = px + \frac{a}{p}, \tag{2}$$

of which therefore (1) is the complete integral.

Now the lines represented by equation (1), for different values of c, are the tangents to the parabola

$$y^2 = 4ax. \tag{3}$$

A point moving in this parabola has the same value of p as if it

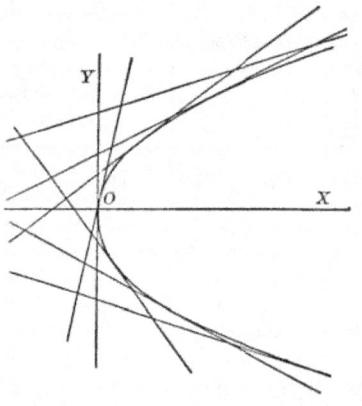

were moving in one of the tangents, and accordingly equation (3) will be found to satisfy the differential equation (2).

It will be noticed that for any point on the convex side of the parabola there are two real values of p; for a point on the other side the values of p are imaginary, and for a point on the curve they are equal. Thus its equation (3) expresses the relation between x and y which must exist in order that (2) regarded as a quadratic for p may have equal roots, as will be seen on solving that equation.

In general, writing the differential equation in the form

$$Lp^2 + Mp + N = 0, \qquad (4)$$

the condition of equal roots is

$$M^2 - 4LN = 0. \qquad (5)$$

The first member of this equation, which is the "discriminant" of equation (4), frequently admits of separation into factors rational in x and y. Hence, if there be a singular solution, its equation will be found by putting the discriminant of the differential equation, or one of its factors, equal to zero.

It does not follow that every such equation represents a solution of the differential equation. It can only be inferred that it is a locus of points for which the two values of p become equal. Now suppose that two distinct particular integral curves touch each other. At the point of contact, the two values of p, usually distinct, become equal. The locus of such points is called a "tac-locus." Its equation plainly satisfies the discriminant, but does not satisfy the differential equation. An illustration is afforded by the equation

$$y^2(p^2 + 1) = a^2,$$

of which the complete integral is $y^2 + (x - c)^2 = a^2$, and the discriminant, see equation (5), is $y^2(y^2 - a^2) = 0$.

This is satisfied by $y = a$, $y = -a$, and $y = 0$, the first two of which satisfy the differential equation, while $z = 0$ does not. The complete integral represents in this case all circles of radius a with center on the axis of x. Two of these circles touch at every point of the axis of x, which is thus a tac-locus, while $y = a$ and $y = -a$ constitute the envelope.

The discriminant is the quantity which appears under the radical sign when the general equation (4) is solved for p, and therefore it changes sign as we cross the envelope. But the values of p remain real as we cross the tac-locus, so that the discriminant cannot change sign. Accordingly the factor which indicates a tac-locus appears with an even exponent (as y^2 in the example above), whereas the factor indicating the singular solution appears as a simple factor, or with an odd exponent.

A simple factor of the discriminant, or one with an odd exponent, gives in fact always the boundary between a region of the plane in which p is real and one in which p is imaginary; nevertheless it may not give a singular solution. For the two arcs of particular integral curves which intersect in a point on the real side of the boundary may, as the point is brought up to the boundary, become tangent to each other, but not to the boundary curve. In that case, since they cannot cross the boundary, they become branches of the same particular integral forming a cusp. A boundary curve of this character is called a "cusp-locus"; the value of p for a point moving in it is of course different from the equal values of p at the cusp, and therefore its equation does not satisfy the differential equation.*

Prob. 30. To what curve is the line $y = mx + a\sqrt{(1 - m^2)}$ always tangent?   (Ans. $y^2 - x^2 = a^2$.)

Prob. 31. Show that the discriminant of a decomposable differ-

---

* Since there is no reason why the values of p referred to should be identical, we conclude that the equation $Lp^2 + Mp + N = 0$ has not in general a singular solution, its discriminant representing a cusp-locus except when a certain condition is fulfilled.

ential equation cannot be negative.   Interpret the result of equating it to zero in the illustrative example at the beginning of Art. 7.

Prob. 32. Show that the singular solutions of a homogeneous differential equation represent straight lines passing through the origin.

Prob. 33. Solve the equation $xp^2 - 2yp + ax = 0$.

(Ans. $x^2 - 2cy + ac^2 = 0$; singular solution $y^2 = ax^2$.)

Prob. 34. Show that the equation $p^2 + 2xp - y = 0$ has no singular solution, but has a cusp-locus, and that the tangent at every cusp passes through the origin.

ART. 9.   SINGULAR SOLUTION FROM COMPLETE INTEGRAL.

When the complete integral of a differential equation of the second degree has been found in the standard form

$$Pc^2 + Qc + R = 0 \qquad (1)$$

(see the end of Art. 7), the substitution of special values of x and y in the functions P, Q, and R gives a quadratic for c whose roots determine the two particular curves of the system which pass through a given point.   If there is a singular solution, that is, if the system of curves has an envelope, the two curves which usually intersect become identical when the given point is moved up to the envelope.   Every point on the envelope therefore satisfies the condition of equal roots for equation (1), which is

$$Q^2 - 4PR = 0 ; \qquad (2)$$

and, reasoning exactly as in Art. 8, we infer that the equation of the singular solution will be found by equating to zero the discriminant of the equation in c or one of its factors.   Thus the discriminant of equation (1), Art. 8, or "c-discriminant," is the same as the "p-discriminant," namely, $y^2 - 4ax$, which equated to zero is the equation of the envelope of the system of straight lines.

But, as in the case of the p-discriminant, it must not be inferred that every factor gives a singular solution.   For example, suppose a squared factor appears in the c-discriminant. The locus on which this factor vanishes is not a curve in crossing which c and p become imaginary.   At any point of it there will be two distinct values of p, corresponding to arcs of particular integral curves passing through that point; but, since there is but one value of c, these arcs belong to the same particular integral, hence the point is a double point or node. The locus is therefore called a "node-locus." The factor representing it does not appear in the p-discriminant, just as that representing a tac-locus does not appear in the c-discriminant.

Again, at any point of a cusp-locus, as shown at the end of Art. 8, the two branches of particular integrals become arcs of the same particular integral; the values of c become equal, so that a cusp-locus also makes the c-discriminant vanish.

The conclusions established above obviously apply also to equations of a degree higher than the second. In the case of the c-equation the general method of obtaining the condition for equal roots, which is to eliminate c between the original and the derived equation, is the same as the process of finding the envelope or "locus of the ultimate intersections" of a system of curves in which c is the arbitrary parameter.

Now suppose the system of curves to have for all values of c * a double point, it is obvious that among the intersections of two neighboring curves there are two in the neighborhood of the nodes, and that ultimately they coincide with the node, which accounts for the node-locus appearing twice in the discriminant or locus of ultimate intersections. In like manner,

---

* It is noticed in the second foot-note to Art. 7 that for an equation of the first degree p takes the indeterminate form, not only at a point through which all curves of the system pass (where the value of c would also be found indeterminate), but at a node of a particular integral. So also when the equation is of the nth degree, if there is a node for a particular value of c, the n values of c at the point (which is not on a node-locus where two values of c are equal) determine $n + 1$ arcs of particular integrals passing through the point; and therefore there are $n + 1$ distinct values of p at the point, which can only happen when p takes the indeterminate form, that is to say, when all the coefficients of the p-equation (which is of the nth degree) vanish. See Cayley on Singular Solutions in the Messenger of Mathematics, New Series, Vol. II, p. 10 (Collected Mathematical Works, Vol. VIII, p. 529). The present theory of Singular Solutions was established by Cayley in this paper and its continuation, Vol. VI, p. 23. See also a paper by Dr. Glaisher, Vol. XII, p. 1.

if there is a cusp for all values of c, there are three intersections of neighboring curves (all of which may be real) which ultimately coincide with the cusp; therefore a cusp-locus will appear as a cubed factor in the discriminant.*

Prob. 35. Show that the singular solutions of a homogeneous equation must be straight lines passing through the origin.

Prob. 36. Solve $3p^2y^2 - 2xyp + 4y^2 - x^2 = 0$, and show that there is a singular solution and a tac-locus.

Prob. 37. Solve $yp^2 + 2xp - y = 0$, and show that there is an imaginary singular solution.   (Ans. $y^2 = 2cx + c^2$.)

Prob. 38. Show that the equation $(1 - x^2)p^2 = 1 - y^2$ represents a system of conics touching the four sides of a square.

Prob. 39. Solve $yp^2 - 4xp + y = 0$; examine and interpret both discriminants.   (Ans. $c^2 + 2cx(3y^2 - 8x^2) - 3x^2y^4 + y^6 = 0$.)

ART. 10.   SOLUTION BY DIFFERENTIATION.

The result of differentiating a given differential equation of the first order is an equation of the second order, that is, it contains the derivative $\frac{d^2y}{dx^2}$; but, if it does not contain y explicitly, it may be regarded as an equation of the first order for the variables x and p. If the integral of such an equation can be obtained it will be a relation between x, p, and a constant of integration c, by means of which p can be eliminated from the original equation, thus giving the relation between x, y, and c which constitutes the complete integral. For example, the equation

$$\frac{dy}{dx} + 2xy = x^2 + y^2, \tag{1}$$

---

* The discriminant of $Pc^2 + Qc + R = 0$ represents in general an envelope, no further condition requiring to be fulfilled as in the case of the discriminant of $Lp^2 + Mp + N = 0$. Compare the foot-note to Art. 8. Therefore where there is an integral of this form there is generally a singular solution, although $Lp^2 + Mp + N = 0$ has not in general a singular solution. We conclude, therefore, that this equation (in which L, M, and N are one-valued functions of x and y) has not in general an integral of the above form in which P, Q, and R are one-valued functions of x and y. Cayley, Messenger of Mathematics, New Series, Vol. VI, p. 23.

when solved for y, becomes

$$y = x + \sqrt{p};$$ (2)

whence by differentiation

$$p = 1 + \frac{1}{2\sqrt{p}}\frac{dp}{dx}.$$ (3)

The variables can be separated in this equation, and its integral is

$$\sqrt{p} = \frac{C + e^{2x}}{C - e^{2x}}.$$

Substituting in equation (2), we find

$$y = x + \frac{C + e^{2x}}{C - e^{2x}},$$

which is the complete integral of equation (1).

This method sometimes succeeds with equations of a higher degree when the solution with respect to p is impossible or leads to a form which cannot be integrated. A differential equation between p and one of the two variables will be obtained by direct integration when only one of the variables is explicitly present in the equation, and also when the equation is of the first degree with respect to x and y. In the latter case after dividing by the coefficient of y, the result of differentiation will be a linear equation for x as a function of p, so that an expression for x in terms of p can be found, and then by substitution in the given equation an expression for y in terms of p. Hence, in this case, any number of simultaneous values of x and y can be found, although the elimination of p may be impracticable.

In particular, a homogeneous equation which cannot be solved for p may be soluble for the ratio $y : x$, so as to assume the form $y = x\phi(p)$. The result of differentiation is

$$p = \phi(p) + \phi'(p)\frac{dp}{dx},$$

in which the variables x and p can be separated.

Another special case is of the form

$$y = px + f(p),$$ (1)

which is known as Clairaut's equation. The result of differentiation is

$$p = p + x\frac{dp}{dx} + f'(p)\frac{dp}{dx},$$

which implies either

$$x + f'(p) = 0, \quad \text{or} \quad \frac{dp}{dx} = 0.$$

The elimination of p from equation (1) by means of the first of these equations* gives a solution containing no arbitrary constant, that is, a singular solution. The second is a differential equation for p; its integral is $p = c$, which in equation (1) gives the complete integral

$$y = cx + f(c). \tag{2}$$

This complete integral represents a system of straight lines, the singular solution representing the curve to which they are all tangent. An example has already been given in Art. 8.

A differential equation is sometimes reducible to Clairaut's form by means of a more or less obvious transformation of the variables. It may be noticed in particular that an equation of the form

$$y = nxp + \phi(x, p)$$

is sometimes so reducible by transformation to the independent variable z, where $x = z^n$; and an equation of the form

$$y = nxp + \phi(y, p),$$

by transformation to the new dependent variable $v = y^n$. A double transformation of the form indicated may succeed when the last term is a function of both x and y as well as of p.

Prob. 40. Solve the equation $3y = 2p^3 + 3p^2$; find a singular solution and a cusp-locus. (Ans. $(x + y + c - \frac{1}{3})^3 = \frac{4}{9}(x+c)^2$.)

Prob. 41. Solve $2y = xp + \dfrac{a}{p}$, and find a cusp-locus.

(Ans. $a^2c^2 - 12acxy + 8cy^3 - 12x^2y^2 + 16ax^3 = 0$.)

---

* The equation is in fact the same that arises in the general method for the condition of equal roots. See Art. 9.

Prob. 42. Solve $(x^2 - a^2)p^2 - 2xyp + y^2 - a^2 = 0$.

(Ans. The circle $x^2 + y^2 = 0$, and its tangents.)

Prob. 43. Solve $y = -xp + x^4p^2$.

(Ans. $c^2x + c - xy = 0$, and $1 + 4x^3y = 0$.)

Prob. 44. Solve $p^3 - 4xyp + 8y^2 = 0$.

(Ans. $y = c(x - c)^2$; $27y = 4x^3$ and $y = 0$ are singular solutions; $y = 0$ is also a particular integral.)

Prob. 45. Solve $x^2(y - px) = yp^2$.　　　(Ans. $y^2 = cx^2 + c^2$.)

ART. 11. GEOMETRIC APPLICATIONS; TRAJECTORIES.

Every property of a curve which involves the direction of its tangents admits of statement in the form of a differential equation. The solution of such an equation therefore determines the curve having the given property. Thus, let it be required to determine the curve in which the angle between the radius vector and the tangent is n times the vectorial angle. Using the expression for the trigonometric tangent of that angle, the expression of the property in polar coordinates is

$$\frac{r d\theta}{dr} = \tan n\theta.$$

Separating the variables and integrating, the complete integral is

$$r^n = c^n \sin n\theta.$$

The mode in which the constant of integration enters here shows that the property in question is shared by all the members of a system of similar curves.

The solution of a question of this nature will thus in general be a system of curves, the complete integral of a differential equation, but it may be a singular solution. Thus, if we express the property that the sum of the intercepts on the axes made by the tangent to a curve is equal to the constant a, the straight lines making such intercepts will themselves constitute the complete integral system, and the curve required is the singular solution, which, in accordance with Art. 8, is the envelope of these lines. The result in this case will be found to be the parabola $\sqrt{x} + \sqrt{y} = \sqrt{a}$.

An important application is the determination of the "orthogonal trajectories" of a given system of curves, that is to say, the curves which cut at right angles every curve of the given system. The differential equation of the trajectory is readily derived from that of the given system; for at every point of the trajectory the value of p is the negative reciprocal of its value in the given differential equation. We have therefore only to substitute $-p^{-1}$ for p to obtain the differential equation of the trajectory. For example, let it be required to determine the orthogonal trajectories of the system of parabolas

$$y^2 = 4ax$$

having a common axis and vertex. The differential equation of the system found by eliminating a is

$$2\,x dy = y\,dx.$$

Putting $-\dfrac{dx}{dy}$ in place of $\dfrac{dy}{dx}$, the differential equation of the system of trajectories is

$$2x\,dx + y\,dy = 0,$$

whence, integrating,

$$2x^2 + y^2 = c^2.$$

The trajectories are therefore a system of similar ellipses with axes coinciding with the coordinate axes.

Prob. 46. Show that when the differential equation of a system is of the second degree, its discriminant and that of its trajectory system will be identical; but if it represents a singular solution in one system, it will constitute a cusp locus of the other.

Prob. 47. Determine the curve whose subtangent is constant and equal to a.   (Ans. $ce^x = y^a$.)

Prob. 48. Show that the orthogonal trajectories of the curves $r^n = c^n \sin n\theta$ are the same system turned through the angle $\dfrac{\pi}{2n}$ about the pole. Examine the cases $n = 1$, $n = 2$, and $n = \frac{1}{2}$.

Prob. 49. Show that the orthogonal trajectories of a system of circles passing through two given points is another system of circles having a common radical axis.

Prob. 50. Determine the curve such that the area inclosed by any two ordinates, the curve and the axis of x, is equal to the product of the arc and the constant line a. Interpret the singular solution.

$$\text{(Ans. The catenary } y = \tfrac{1}{2}a(e^{\frac{x}{a}} - e^{-\frac{x}{a}}).)$$

Prob. 51. Show that a system of confocal conics is self-orthogonal.

ART. 12.   SIMULTANEOUS DIFFERENTIAL EQUATIONS.

A system of n equations between $n + 1$ variables and their differentials is a "determinate" differential system, because it serves to determine the n ratios of the differentials; so that, taking any one of the variables as independent, the others vary in a determinate manner, and may be regarded as functions of the single independent variable.   Denoting the variables by x, y, z, etc., the system may be written in the symmetrical form

$$\frac{dx}{X} = \frac{dy}{Y} = \frac{dz}{Z} = \dots,$$

where X, Y, $Z \dots$ may be any functions of the variables.

If any one of the several equations involving two differentials contains only the two corresponding variables, it is an ordinary differential equation; and its integral, giving a relation between these two variables, may enable us by elimination to obtain another equation containing two variables only, and so on until n integral equations have been obtained. Given, for example, the system

$$\frac{dx}{x} = \frac{dy}{z} = \frac{dz}{y}. \tag{1}$$

The relation between dy and dz above contains the variables y and z only, and its integral is

$$y^2 - z^2 = a. \tag{2}$$

Employing this to eliminate z from the relation between dx and dy it becomes

$$\frac{dx}{x} = \frac{dy}{\sqrt{(y^2 + a)}},$$

of which the integral is

$$y + \sqrt{(y^2 + a)} = bx. \tag{3}$$

The integral equations (2) and (3), involving two constants of integration, constitute the complete solution. It is in like manner obvious that the complete solution of a system of n equations should contain n arbitrary constants.

Confining ourselves now to the case of three variables, an extension of the geometrical interpretation given in Art. 2 presents itself. Let x, y, and z be rectangular coordinates of P referred to three planes. Then, if P starts from any given position A, the given system of equations, determining the ratios $dx : dy : dz$, determines the direction in space in which P moves. As P moves, the ratios of the differentials (as determined by the given equations) will vary, and if we suppose P to move in such a way as to continue to satisfy the differential equations, it will describe in general a curve of double curvature which will represent a particular solution. The complete solution is represented by the system of lines which may be thus obtained by varying the position of the initial point A. This system is a " doubly infinite " one ; for the two relations between x, y, and z which define it analytically must contain two arbitrary parameters, by properly determining which we can make the line pass through any assumed initial point.*

Each of the relations between x, y and z, or integral equations, represents by itself a surface, the intersection of the two surfaces being a particular line of the doubly infinite system. An equation like (2) in the example above, which contains only one of the constants of integration, is called an *integral* of the differential system, in contradistinction to an " integral equa-

---

* It is assumed in the explanation that X, Y, and Z are one-valued functions of x, y, and z. There is then but one direction in which P can move when passing a given point, and the system is a non-intersecting system of lines. But if this is not the case, as for example when one of the equations giving the ratio of the differentials is of higher degree, the lines may form an intersecting system, and there would be a theory of singular solutions, into which we do not here enter.

tion " like (3), which contains both constants.   An integral represents a surface which contains a singly infinite system of lines representing particular solutions selected from the doubly infinite system.   Thus equation (2) above gives a surface on which lie all those lines for which a has a given value, while b may have any value whatever ; in other words, a surface which passes through an infinite number of the particular solution lines.

The integral of the system which corresponds to the constant b might be found by eliminating a between equations (2) and (3).   It might also be derived directly from equation (1) ; thus we may write

$$\frac{dx}{x} = \frac{dy}{z} = \frac{dz}{y} = \frac{dy + dz}{y + z} = \frac{du}{u},$$

in which a new variable $u = y + z$ is introduced.   The relation between dx and du now contains but two variables, and its integral,

$$y + z = bx, \tag{4}$$

is the required integral of the system ; and this, together with the integral (2), presents the solution of equations (1) in its standard form.   The form of the two integrals shows that in this case the doubly infinite system of lines consists of hyperbolas, namely, the sections of the system of hyperbolic cylinders represented by (2) made by the system of planes represented by (4).

A system of equations of which the members possess a certain symmetry may sometimes be solved in the following manner.   Since

$$\frac{dx}{X} = \frac{dy}{Y} = \frac{dz}{Z} = \frac{\lambda dx + \mu dy + \nu dz}{\lambda X + \mu Y + \nu Z},$$

if we take multipliers λ, μ, ν such that

$$\lambda X + \mu Y + \nu Z = 0,$$

we shall have $\lambda dx + \mu dy + \nu dz = 0.$

If the expression in the first member is an exact differential, direct integration gives an integral of the given system. For example, let the given equations be

$$\frac{dx}{mz - ny} = \frac{dy}{nx - lz} = \frac{dz}{ly - mx};$$

l, m and n form such a set of multipliers, and so also do x, y and z. Hence we have

$$l\,dx + m\,dy + n\,dz = 0,$$

and also

$$x\,dx + y\,dy + z\,dz = 0.$$

Each of these is an exact equation, and their integrals

$$lx + my + nz = a$$

and

$$x^2 + y^2 + z^2 = b^2$$

constitute the complete solution. The doubly infinite system of lines consists in this case of circles which have a common axis, namely, the line passing through the origin and whose direction cosines are proportional to l, m, and n.

Prob. 52. Solve the equations $\dfrac{dx}{x^2 - y^2 - z^2} = \dfrac{dy}{2xy} = \dfrac{dz}{2xz}$, and interpret the result geometrically. (Ans. $y = az$, $x^2 + y^2 + z^2 = bz$.)

Prob. 53. Solve $\dfrac{dx}{y + z} = \dfrac{dy}{z + x} = \dfrac{dz}{x + y}$.

$$\left(\text{Ans. } \sqrt{(x + y + z)} = \frac{a}{z - y} = \frac{b}{x - z}.\right)$$

Prob. 54. Solve $\dfrac{dx}{(b - c)yz} = \dfrac{dy}{(c - a)zx} = \dfrac{dz}{(a - b)xy}$.

(Ans. $x^2 + y^2 + z^2 = A$, $ax^4 + by^2 + cz^2 = B$.}

ART. 13.   EQUATIONS OF THE SECOND ORDER.

A relation between two variables and the successive derivatives of one of them with respect to the other as independent variable is called a differential equation of the order indicated by the highest derivative that occurs. For example,

$$(1 + x^2)\frac{d^2y}{dx^2} + x\frac{dy}{dx} + mx = 0$$

is an equation of the second order, in which x is the independent variable.   Denoting as heretofore the first derivative by p, this equation may be written

$$(1 + x^2)\frac{dp}{dx} + xp + mx = 0, \tag{1}$$

and this, in connection with

$$\frac{dy}{dx} = p, \tag{2}$$

which defines p, forms a pair of equations of the first order, connecting the variables $x, y,$ and p.   Thus any equation of the second order is equivalent to a pair of simultaneous equations of the first order.

When, as in this example, the given equation does not contain y explicitly, the first of the pair of equations involves only the two variables x and p; and it is further to be noticed that, when the derivatives occur only in the first degree, it is a linear equation for p.   Integrating equation (1) as such, we find

$$p = -m + \frac{c_1}{\sqrt{(1 + x^2)}}; \tag{3}$$

and then using this value of p in equation (2), its integral is

$$y = c_2 - mx + c_1 \log [x + \sqrt{(1 + x^2)}], \tag{4}$$

in which, as in every case of two simultaneous equations of the first order, we have introduced two constants of integration.

An equation of the first order is readily obtained also when the independent variable is not explicitly contained in the equation.   The general equation of rectilinear motion in dynamics affords an illustration.   This equation is $\frac{d^2s}{dt^2} = f(s)$, where s denotes the distance measured from a fixed center of force upon the line of motion.   It may be written $\frac{dv}{dt} = f(s)$, in connection with $\frac{ds}{dt} = v$, which defines the velocity.   Eliminating dt from these equations, we have $v\,dv = f(s)ds$, whose integral is $\frac{1}{2}v^2 = \int f(s)ds + c$, the "equation of energy" for the unit mass.   The substitution of the value found for v in the second equation gives an equation from which t is found in terms of s by direct integration.

The result of the first integration, such as equation (3) above, is called a "first integral" of the given equation of the second order; it contains one constant of integration, and its complete integral, which contains a second constant, is also the "complete integral" of the given equation.

A differential equation of the second order is "exact" when, all its terms being transposed to the first member, that member is the derivative with respect to x of an expression of the first order, that is, a function of x, y and p. It is obvious that the terms containing the second derivative, in such an exact differential, arise solely from the differentiation of the terms containing p in the function of x, y and p. For example, let it be required to ascertain whether

$$(1 - x^2)\frac{d^2y}{dx^2} - x\frac{dy}{dx} + y = 0 \qquad (5)$$

is an exact equation. The terms in question are $(1 - x^2)\frac{dp}{dx}$, which can arise only from the differentiation of $(1 - x^2)p$. Now subtract from the given expression the complete derivative of $(1 - x^2)p$, which is

$$(1 - x^2)\frac{d^2y}{dx^2} - 2x\frac{dy}{dx};$$

the remainder is $x\frac{dy}{dx} + y$, which is an exact derivative, namely, that of xy. Hence the given expression is an exact differential, and

$$(1 - x^2)\frac{dy}{dx} + xy = c_1 \qquad (6)$$

is the first integral of the given equation. Solving this linear equation for y, we find the complete integral

$$y = c_1 x + c_2 \sqrt{(1 - x^2)}. \qquad (7)$$

Prob. 55. Solve $(1 - x^2)\frac{d^2y}{dx^2} - x\frac{dy}{dx} = 2$.

(Ans. $y = (\sin^{-1} x)^2 + c_1 \sin^{-1} x + c_2$.)

Prob. 56. Solve $\dfrac{d^2y}{dx^2} = \dfrac{2y}{x^2}$. $\left(\text{Ans. } y = \dfrac{c_1}{x} + c_2 x^2.\right)$

Prob. 57. Solve $\dfrac{d^2y}{dx^2} = a^2x - b^2y$.

$$(\text{Ans. } a^2x - b^2y = A \sin bx + B \cos bx.)$$

Prob. 58. Solve $y\dfrac{d^2y}{dx^2} + \left(\dfrac{dy}{dx}\right)^2 = 1$. $(\text{Ans. } y^2 = x^2 + c_1 x + c_2.)$

ART. 14. THE TWO FIRST INTEGRALS.

We have seen in the preceding article that the complete integral of an equation of the second order is a relation between x, y and two constants c_1 and c_2. Conversely, any relation between x, y and two arbitrary constants may be regarded as a primitive, from which a differential equation free from both arbitrary constants can be obtained. The process consists in first obtaining, as in Art. 3, a differential equation of the first order independent of one of the constants, say c_2, that is, a relation between x, y, p and c_1, and then in like manner eliminating c_1 from the derivative of this equation. The result is the equation of the second order or relation between x, y, p and q (q denoting the second derivative), of which the original equation is the complete primitive, the equation of the first order being the first integral in which c_1 is the constant of integration. It is obvious that we can, in like manner, obtain from the primitive a relation between x, y, p and c_2, which will also be a first integral of the differential equation. Thus, to a given form of the primitive or complete integral there corresponds two first integrals.

Geometrically the complete integral represents a doubly infinite system of curves, obtained by varying the values of c_1 and of c_2 independently. If we regard c_1 as fixed and c_2 as arbitrary, we select from that system a certain singly infinite system; the first integral containing c_1 is the differential equation of this system, which, as explained in Art. 2, is a relation between the coordinates of a moving point and the direction of its motion common to all the curves of the system. But the equation of the second order expresses a property involving curvature as well as direction of path, and this property being independent of c_1 is common to all the systems corresponding to different values of c_1, that is, to the entire doubly infinite system. A moving point, satisfying this equation, may have any position and move in any direction, provided its path has the proper curvature as determined by the value of q derived from the equation, when the selected values of x, y and p have been substituted therein.*

For example, equation (7) of the preceding article represents an ellipse having its center at the origin and touching the lines $x = \pm 1$, as in the diagram ; c_1 is the ordinate of the point of contact with $x = 1$, and c_2 that of the point in which the ellipse cuts the axis of y. If we regard c_1 as fixed and c_2 as arbitrary, the equation represents the system of ellipses touching the two lines at fixed points, and equation (6) is the

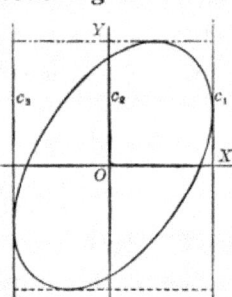

differential equation of this system. In like manner, if c_2 is fixed and c_1 arbitrary, equation (7) represents a system of ellipses cutting the axis of y in fixed points and touching the lines $x = \pm 1$. The corresponding differential equation will be found to be

$$(y - xp)\sqrt{(1 - x^2)} = c_2.$$

Finally, the equation of the second order, independent of c_1 and c_2 [(5) of the preceding article] is the equation of the doubly infinite system of conics † with center at the origin, and touching the fixed lines $x = \pm 1$.

---

* If the equation is of the second or higher degree in q, the condition for equal roots is a relation between x, y and p, which may be found to satisfy the given equation. If it does, it represents a system of singular solutions; each of the curves of this system, at each of its points, not only touches but osculates with a particular integral curve. It is to be remembered that a singular solution of a first integral is not generally a solution of the given differential equation; for it represents a curve which simply touches but does not osculate a set of curves belonging to the doubly infinite system.

† Including hyperbolas corresponding to imaginary values of c_2.

But, starting from the differential equation of second order, we may find other first integrals than those above which correspond to c_1 and c_2. For instance, if equation (5) be multiplied by p, it becomes

$$(1 - x^2)p\frac{dp}{dx} - xp^2 + yp = 0,$$

which is also an exact equation, giving the first integral

$$(1 - x^2)p^2 + y^2 = c_3^2,$$

in which c_3 is a new constant of integration.

Whenever two first integrals have thus been found independently, the elimination of p between them gives the complete integral without further integration.* Thus the result of eliminating p between this last equation and the first integral containing c_1 [equation (6), Art. 13] is

$$y^2 - 2c_1xy + c_1^2x^2 = c_3^2 - c_1^2,$$

which is therefore another form of the complete integral. It is obvious from the first integral above that c_3 is the maximum value of y, so that it is the differential equation of the system of ellipse inscribed in the rectangle drawn in the diagram. A comparison of the two forms of the complete integral shows that the relation between the constants is $c_3^2 = c_1^2 + c_2^2$.

If a first integral be solved for the constant, that is, put in the form $\phi(x, y, p) = c$, the constant will disappear on differentiation, and the result will be the given equation of second order multiplied, in general, by an integrating factor. We can thus find any number of integrating factors of an equation already solved, and these may suggest the integrating factors of more general equations, as illustrated in Prob. 59 below.

---

* The principle of this method has already been applied in Art. 10 to the solution of certain equations of the first order; the process consisted of forming the equation of the second order of which the given equation is a first integral (but with a particular value of the constant), then finding another first integral and deriving the complete integral by elimination of p.

Prob. 59. Solve the equation $\dfrac{d^2y}{dx^2} + a^2y = 0$ in the form

$$y = A \cos ax + B \sin ax;$$

and show that the corresponding integrating factors are also integrating factors of the equation

$$\frac{d^2y}{dx^2} + a^2y = X,$$

where X is any function of x; and thence derive the integral of this equation.

(Ans. $ay = \sin ax \displaystyle\int \cos ax . Xdx - \cos ax \displaystyle\int \sin ax . Xdx$).

Prob. 60. Find the rectangular and also the polar differential equation of all circles passing through the origin.

$\left(\text{Ans. } (x^2 + y^2)\dfrac{d^2y}{dx^2} = 2\left[1 + \left(\dfrac{dy}{dx}\right)^2\right]\left(x\dfrac{dy}{dx} - y\right), \text{ and } r + \dfrac{d^2r}{d\theta^2} = 0.\right)$

ART. 15. LINEAR EQUATIONS.

A linear differential equation of any order is an equation of the first degree with respect to the dependent variable y and each of its derivatives, that is, an equation of the form

$$P_0\frac{d^ny}{dx^n} + P_1\frac{d^{n-1}y}{dx^{n-1}} + \ldots + P_n y = X, \qquad (1)$$

where the coefficients $P_0, \ldots P_n$ and the second member X are functions of the independent variable only.

The solution of a linear equation is always supposed to be in the form $y = f(x)$; and if y_1 is a function which satisfies the equation, it is customary to speak of the function y_1, rather than of the equation $y = y_1$, as an "integral" of the linear equation. The general solution of the linear equation of the first order has been given in Art. 6. For orders higher than the first the general expression for the integrals cannot be effected by means of the ordinary functional symbols and the integral sign, as was done for the first order in Art. 6.

The solution of equation (1) depends upon that of

$$P_0\frac{d^ny}{dx^n} + P_1\frac{d^{n-1}y}{dx^{n-1}} + \ldots + P_n y = 0. \qquad (2)$$

The complete integral of this equation will contain n arbitrary constants, and the mode in which these enter the expression for y is readily inferred from the form of the equation. For let y_1 be an integral, and c_1 an arbitrary constant; the result of putting $y = c_1y_1$ in equation (2) is c_1 times the result of putting $y = y_1$; that is, it is zero; therefore c_1y_1 is an integral. So too, if y_2 is an integral, c_2y_2 is an integral; and obviously also $c_1y_1 + c_2y_2$ is an integral. Thus, if n distinct integrals y_1, $y_2, \ldots y_n$ can be found,

$$y = c_1y_1 + c_2y_2 + \ldots + c_ny_n \tag{3}$$

will satisfy the equation, and, containing, as it does, the proper number of constants, will be the complete integral.

Consider now equation (1); let Y be a particular integral of it, and denote by u the second member of equation (3), which is the complete integral when $X = 0$. If

$$y = Y + u \tag{4}$$

be substituted in equation (1), the result will be the sum of the results of putting $y = Y$ and of putting $y = u$; the first of these results will be X, because Y is an integral of equation (1), and the second will be zero because u is an integral of equation (2). Hence equation (4) expresses an integral of (2); and since it contains the n arbitrary constants of equation (3), it is the complete integral of equation (1). With reference to this equation Y is called "the particular integral," and u is called "the complementary function." The particular integral contains no arbitrary constant, and any two particular integrals may differ by any multiple of a term belonging to the complementary function.

If one term of the complementary function of a linear equation of the second order be known, the complete solution can be found. For let y_1 be the known term; then, if $y = y_1v$ be substituted in the first member, the coefficient of v in the result will be the same as if v were a constant: it will therefore be zero, and v being absent, the result will be a linear equation of the first order for v', the first derivative of v. Under the same circumstances the order of any linear equation can in like manner be reduced by unity.

A very simple relation exists between the coefficients of an exact linear equation. Taking, for example, the equation of the second order, and indicating derivatives by accents, if

$$P_0 y'' + P_1 y' + P_2 y = X$$

is exact, the first term of the integral will be $P_0 y'$ Subtracting the derivative of this from the first member, the remainder is $(P_1 - P_0') y' + P_2 y$. The second term of the integral must therefore be $(P_1 - P_0') y$; subtracting the derivative of this expression, the remainder, $(P_2 - P_1' + P_0'') y$, must vanish. Hence $P_2 - P_1' + P_0'' = 0$ is the criterion for the exactness of the given equation. A similar result obviously extends to equations of higher orders.

Prob. 61. Solve $x \dfrac{d^2 y}{dx} - (3 + x) \dfrac{dy}{dx} + 3y = 0$, noticing that e^x is

an integral.   (Ans. $y = c_1 e^x + c_2 (x^3 + 3x^2 + 6x + 6)$.)

Prob. 62. Solve $(x^2 - x) \dfrac{d^2 y}{dx^2} + 2(2x + 1) \dfrac{dy}{dx} + 2y = 0$.

   (Ans. $(x - 1)^5 y = c_1 (x^4 - 6x^2 + 2x - \tfrac{1}{3} - 4x^2 \log x) + c_2 x^3$.)

Prob. 63. Solve $\dfrac{d^3 y}{d\theta^3} + \cos \theta \dfrac{d^2 y}{d\theta^2} - 2 \sin \theta \dfrac{dy}{d\theta} - y \cos \theta = \sin 2\theta$.

$\left(\text{Ans. } y = e^{-\sin \theta} \displaystyle\int e^{\sin \theta} (c_1 \theta + c_2) d\theta + c_3 e^{-\sin \theta} - \dfrac{\sin \theta - 1}{2}. \right)$

ART. 16. LINEAR EQUATIONS WITH CONSTANT COEFFICIENTS.

The linear equation with constant coefficients and second member zero may be written in the form

$$A_0 D^n y + A_1 D^{n-1} y + \ldots + A_n y = \tag{1}$$

in which D stands for the operator $\dfrac{d}{dx}$, D^2 for $\dfrac{d^2}{dx^2}$, etc., so that D^n indicates that the operator is to be applied n times. Then, since $De^{mx} = me^{mx}$, $D^2 e^{mx} = m^2 e^{mx}$, etc., it is evident that if

$y = e^{mx}$ be substituted in equation (1), the result after rejecting the factor e^{mx} will be

$$A_0 m^n + A_1 m^{n-1} + \ldots + A_n = 0. \tag{2}$$

Hence, if m satisfies equation (2), e^{mx} is an integral of equation (1); and if $m_1, m_2, \ldots m_n$ are n distinct roots of equation (2), the complete integral of equation (1) will be

$$y = c_1 e^{m_1 x} + c_2 e^{m_2 x} + \ldots + c_n e^{m_n x}. \tag{3}$$

For example, if the given equation is

$$\frac{d^2 y}{dx^2} - \frac{dy}{dx} - 2y = 0,$$

the equation to determine m is

$$m^2 - m - 2 = 0,$$

of which the roots are $m_1 = 2, m_2 = -1$; therefore the integral is

$$y = c_1 e^{2x} + c_2 e^{-x}.$$

The general equation (1) may be written in the symbolic form $f(D) . y = 0$, in which f denotes a rational integral function. Then equation (2) is $f(m) = 0$, and, just as this last equation is equivalent to

$$(m - m_1)(m - m_2) \ldots (m - m_n) = 0, \tag{4}$$

so the symbolic equation $f(D) . y = 0$ may be written

$$(D - m_1)(D - m_2) \ldots (D - m_n)y = 0. \tag{5}$$

This form of the equation shows that it is satisfied by each of the quantities which satisfy the separate equations

$$(D - m_1)y = 0, \quad (D - m_2)y = 0 \ldots (D - m_n)y = 0; \tag{6}$$

that is to say, by the separate terms of the complete integral.

If two of the roots of equation (2) are equal, say to m_1, two of the equations (6) become identical, and to obtain the full number of integrals we must find two terms corresponding to the equation

$$(D - m_1)^2 y = 0; \tag{7}$$

in other words, the complete integral of this equation of which $y_1 = e^{m_1 x}$ is known to be one integral. For this purpose we put, as explained in the preceding article, $y = y_1 v$. By differentiation, $Dy = D e^{m_1 x} v = e^{m_1 x}(m_1 v + Dv)$; therefore

$$(D - m_1)e^{m_1 x} v = e^{m_1 x} Dv. \tag{8}$$

In like manner we find

$$(D - m_1)^2 e^{m_1 x} v = e^{m_1 x} D^2 v. \tag{9}$$

Thus equation (7) is transformed to $D^2 v = 0$, of which the complete integral is $v = c_1 x + c_2$; hence that of equation (7) is

$$y = e^{m_1 x}(c_1 x + c_2). \tag{10}$$

These are therefore the two terms corresponding to the squared factor $(D - m_1)^2$ in $f(D)y = 0$.

It is evident that, in a similar manner, the three terms corresponding to a case of three equal roots can be shown to be $e^{m_1 x}(c_1 x^2 + c_2 x + c_3)$, and so on.

The pair of terms corresponding to a pair of imaginary roots, say $m_1 = \alpha + i\beta$, $m_2 = \alpha - i\beta$, take the imaginary form

$$c_1 e^{(\alpha + i\beta)x} + c_2 e^{(\alpha - i\beta)x} = e^{\alpha x}(c_1 e^{i\beta x} + c_2 e^{-i\beta x}).$$

Separating the real and imaginary parts of $e^{i\beta x}$ and $e^{-i\beta x}$, and changing the constants, the expression becomes

$$e^{\alpha x}(A \cos \beta x + B \sin \beta x). \tag{11}$$

For a multiple pair of imaginary roots the constants A and B must be replaced by polynomials as above shown in the case of real roots.

When the second member of the equation with constant coefficients is a function of X, the particular integral can also be made to depend upon the solution of linear equations of the first order. In accordance with the symbolic notation introduced above, the solution of the equation

$$\frac{dy}{dx} - ay = X, \quad \text{or} \quad (D - a)y = X \tag{12}$$

is denoted by $y = (D - a)^{-1}X$, so that, solving equation (12), we have

$$\frac{1}{D - a}X = e^{ax}\int e^{-ax}X dx \tag{13}$$

as the value of the inverse symbol whose meaning is "that

function of x which is converted to X by the direct operation expressed by the symbol $D - a$." Taking the most convenient special value of the indefinite integral in equation (13), it gives the particular integral of equation (12). In like manner, the particular integral of $f(D)y = X$ is denoted by the inverse symbol $\frac{1}{f(D)}X$. Now, with the notation employed above, the symbolic fraction may be decomposed into partial fractions with constant numerators thus:

$$\frac{1}{f(D)}X = \frac{N_1}{D - m_1}X + \frac{N_2}{D - m_2}X + \ldots + \frac{N_n}{D - m_n}X, \text{*} \quad (14)$$

in which each term is to be evaluated by equation (13), and may be regarded (by virtue of the constant involved in the indefinite integral) as containing one term of the complementary function. For example, the complete solution of the equation

$$\frac{d^2y}{dx^2} - \frac{dy}{dx} - 2y = X$$

is thus found to be

$$y = \tfrac{1}{3}e^{2x}\int e^{-2x}Xdx - \tfrac{1}{3}e^{-x}\int e^x Xdx.$$

When X is a power of x the particular integral may be found as follows, more expeditiously than by the evaluation of the integrals in the general solution. For example, if $X = x^2$ the particular integral in this example may be evaluated by development of the inverse symbol, thus:

$$y = \frac{1}{D^2 - D - 2}x^2 = -\frac{1}{2}\frac{1}{1 + \tfrac{1}{2}(D - D^2)}x^2$$
$$= -\tfrac{1}{2}[1 - \tfrac{1}{2}(D - D^2) + \tfrac{1}{4}(D - D^2)^2 - \ldots]x^2$$
$$= -\tfrac{1}{2}[1 - \tfrac{1}{2}D + \tfrac{3}{4}D^2 - \ldots]x^2 = -\tfrac{1}{2}x^2 + \tfrac{1}{2}x - \tfrac{3}{4}.$$

---

* The validity of this equation depends upon the fact that the operations expressed in the second member of
$$f(D) = (D - m_1)(D - m_2) + \ldots + (D - m_n)$$
are commutative, hence the process of verification is the same as if the equation were an algebraic identity. This general solution was published by Boole in the Cambridge Math. Journal, First Series, vol. II, p. 114. It had, however, been previously published by Lobatto, Théorie des Charactéristiques, Amsterdam, 1837.

The form of the operand shows that, in this case, it is only necessary to carry the development as far as the term containing D^2.

For other symbolic methods applicable to special forms of X we must refer to the standard treatises on this subject.

Prob. 64. Solve $4\dfrac{d^2y}{dx^2} - 3\dfrac{dy}{dx} + y = 0$.

$$\text{(Ans. } y = e^{\frac{1}{4}x}(Ax + B) + ce^{-x}.)$$

Prob. 65. Show that $\dfrac{1}{f(D)}e^{ax} = \dfrac{1}{f(a)}e^{ax}$

and that $\dfrac{1}{f(D^2)}\sin(ax + \beta) = \dfrac{1}{f(-a^2)}\sin(ax + \beta)$.

Prob. 66. Solve $(D^2 + 1)y = e^x + \sin 2x + \sin x$.   (Compare Prob. 59, Art. 14.)

$$\text{(Ans. } y = A \sin x + B \cos x + \tfrac{1}{2}e^x - \tfrac{1}{3}\sin 2x - \tfrac{1}{2}x \cos x.)$$

ART. 17. HOMOGENEOUS LINEAR EQUATIONS.

The linear differential equation

$$A_0 x^n \frac{d^n y}{dx^n} + A_1 x^{n-1} \frac{d^{n-1}y}{dx^{n-1}} + \ldots + A_n y = 0, \qquad (1)$$

in which A_0, A_1, etc., are constants, is called the "homogeneous linear equation." It bears the same relation to x^m that the equation with constant coefficients does to e^{mx}. Thus, if $y = x^m$ be substituted in this equation, the factor x^m will divide out from the result, giving an equation for determining m, and the n roots of this equation will in general determine the n terms of the complete integral. For example, if in the equation

$$x^2 \frac{d^2 y}{dx^2} + 2x \frac{dy}{dx} - 2y = 0$$

we put $y = x^m$, the result is $m(m - 1) + 2m - 2 = 0$, or $(m - 1)(m + 2) = 0$.

The roots of this equation are $m_1 = 1$ and $m_2 = -2$. Hence
$$y = c_1 x + c_2 x^{-2}$$
is the complete integral.

Equation (1) might in fact have been reduced to the form with constant coefficients by changing the independent variable to θ, where $x = e^\theta$, or $\theta = \log\ x$. We may therefore at once infer from the results established in the preceding article that the terms corresponding to a pair of equal roots are of the form

$$(c_1 + c_2 \log\ x)x^m, \tag{2}$$

and also that the terms corresponding to a pair of imaginary roots, $\alpha \pm i\beta$, are

$$x^a[A \cos\ (\beta \log\ x) + B \sin\ (\beta \log\ x)]. \tag{3}$$

The analogy between the two classes of linear equations considered in this and the preceding article is more clearly seen when a single symbol $\vartheta = xD$ is used for the operation of taking the derivative and then multiplying by x, so that $\vartheta x^m = mx^m$. It is to be noticed that the operation x^2D^2 is not the same as ϑ^2 or $xDxD$, because the operations of taking the derivative and multiplying by a variable are not "commutative," that is, their order is not indifferent. We have, on the contrary, $x^2D^2 = \vartheta(\vartheta - 1)$; then the equation given above, which is

$$(x^2D^2 + 2xD - 2)y = 0,$$

becomes

$$[\vartheta(\vartheta - 1) + 2\vartheta - 2]\,y = 0, \quad \text{or} \quad (\vartheta - 1)(\vartheta + 2)y = 0,$$

the function of ϑ produced being the same as the function of m which is equated to 0 in finding the values of m.

A linear equation of which the first member is homogeneous and the second member a function of x may be reduced to the form

$$f(\vartheta) \cdot y = X. \tag{4}$$

The particular integral may, as in the preceding article (see eq. (14)), be separated into parts each of which depends upon the solution of a linear equation of the first order. Thus, solving the equation

$$x\frac{dy}{dx} - ay = X, \quad \text{or} \quad (\vartheta - a)y = X, \tag{5}$$

we find

$$\frac{1}{\vartheta - a}X = x^a \int x^{-a-1} X dx. \tag{6}$$

The more expeditious method which may be employed

when X is a power of x is illustrated in the following example :

Given $x^2\dfrac{d^3y}{dx^3} - 2\dfrac{dy}{dx} = x^2$. The first member becomes homogeneous when multiplied by x, and the reduced equation is $(\vartheta^3 - 3\vartheta^2)y = x^3$.

The roots of $f(\vartheta) = 0$ are 3 and the double root zero, hence the complementary function is $c_1x^3 + c_2 + c_3 \log x$. Since in general $f(\vartheta)x^r = f(r)x^r$, we infer that in operating upon x^3 we may put $\vartheta = 3$. This gives for the particular integral

$$\frac{1}{\vartheta - 3}\frac{1}{\vartheta^2}x^3 = \frac{1}{9}\frac{1}{\vartheta - 3}x^3,$$

but fails with respect to the factor $\vartheta - 3$.* We therefore now fall back upon equation (6), which gives

$$\frac{1}{\vartheta - 3}x^3 = x^3\int x^{-1}dx = x^3 \log x.$$

The complete integral therefore is

$$y = c_1x^3 + c_2 + c_3 \log x + \tfrac{1}{9}x^3 \log x.$$

Prob. 67. Solve $2x^2\dfrac{d^2y}{dx^2} + 3x\dfrac{dy}{dx} - 3y = x^2$.

(Ans. $y = c_1x + c_2x^{-\frac{3}{2}} + \tfrac{1}{4}x^2$.)

Prob. 68. Solve $(x^3D^3 + 3xD^2 + D)y = \dfrac{1}{x}$.

(Ans. $y = c_1 + c_2 \log x + c_3(\log x)^2 + \tfrac{1}{6}(\log x)^3$.)

ART. 18. SOLUTIONS IN INFINITE SERIES.

We proceed in this article to illustrate the method by which the integrals of a linear equation whose coefficients are algebraic functions of x may be developed in series whose terms are powers of x. For this purpose let us take the equation

$$x^2\frac{d^2y}{dx^2} + x\frac{dy}{dx} + (x^2 - n^2)y = 0, \tag{1}$$

---

* The failure occurs because x^3 is a term of the complementary function having an indeterminate coefficient; accordingly the new term is of the same form as the second term necessary when 3 is a double root, but of course with a determinate coefficient.

which is known as "Bessel's Equation," and serves to define the "Besselian Functions."

If in the first member of this equation we substitute for y the single term Ax^m the result is

$$A(m^2 - n^2)x^m + Ax^{m+2}, \qquad (2)$$

the first term coming from the homogeneous terms of the equation and the second from the term x^2y which is of higher degree. If this last term did not exist the equation would be satisfied by the assumed value of y, if m were determined so as to make the first term vanish, that is, in this case, by Ax^n or Bx^{-n}. Now these are the first terms of two series each of which satisfies the equation. For, if we add to the value of y a term containing x^{m+2}, thus $y = A_0x^m + A_1x^{m+2}$, the new term will give rise, in the result of substitution, to terms containing x^{m+2} and x^{m+4} respectively, and it will be possible so to take A_1 that the entire coefficient of x^{m+2} shall vanish. In like manner the proper determination of a third term makes the coefficient of x^{m+4} in the result of substitution vanish, and so on. We therefore at once assume

$$y = \sum_{0}^{\infty} A_r x^{m+2r} = A_0 x^m + A_1 x^{m+2} + A_2 x^{m+4} + \ldots, \qquad (3)$$

in which r has all integral values from 0 to ∞. Substituting in equation (1)

$$\sum_{r=0}^{\infty} [\{(m + 2r)^2 - n^2\} A_r x^{m+2r} + A_r x^{m+2(r+1)}] = 0. \qquad (4)$$

The coefficient of each power of x in this equation must separately vanish; hence, taking the coefficient of x^{m+2r}, we have

$$[(m + 2r)^2 - n^2]A_r + A_{r-1} = 0. \qquad (5)$$

When $r = 0$, this reduces to $m^2 - n^2 = 0$, which determines the values of m, and for other values of r it gives

$$A_r = - \frac{1}{(m + 2r + n)(m + 2r - n)} A_{r-1}, \qquad (6)$$

the relation between any two successive coefficients.

For the first value of m, namely n, this relation becomes

$$A_r = - \frac{1}{2^2(n + r)r} A_{r-1};$$

whence, determining the successive coefficients in equation (3), the first integral of the equation is

$$A_0 y_1 = A_0 x^n \left[1 - \frac{1}{n+1} \frac{x^2}{2^2} + \frac{1}{(n+1)(n+2)} \frac{x^4}{2^4 \cdot 2!} - \cdots \right]. \quad (7)$$

In like manner, the other integral is found to be

$$B_0 y_2 = B_0 x^{-n} \left[1 + \frac{1}{n-1} \frac{x^2}{2^2} + \frac{1}{(n-1)(n-2)} \frac{x^4}{2^4 \cdot 2!} + \cdots \right], \quad (8)$$

and the complete integral is $y = A_0 y_1 + B_0 y_2$.*

This example illustrates a special case which may arise in this form of solution. If n is a positive integer, the second series will contain infinite coefficients. For example, if $n = 2$, the third coefficient, or B_2, is infinite, unless we take $B_0 = 0$, in which case B_2 is indeterminate and we have a repetition of the solution y_1. This will always occur when the same powers of x occur in the two series, including, of course, the case in which m has equal roots. For the mode of obtaining a new integral in such cases the complete treatises must be referred to.†

It will be noticed that the simplicity of the relation between consecutive coefficients in this example is due to the fact that equation (1) contained but two groups of terms producing different powers of x, when Ax^m is substituted for y as in expression (2). The group containing the second derivative necessarily gives rise to a coefficient of the second degree in m, and from it we obtained two values of m. Moreover, because the other group was of a degree higher by two units, the assumed series was an ascending one, proceeding by powers of x^2.

---

* The Besselian function of the nth order usually denoted by J_n is the value of y_1 above, divided by $2^n n!$ if n is a positive integer, or generally by $2^n \Gamma(n+1)$. For a complete discussion of these functions see Lommel's Studien über die Bessel'schen Functionen, Leipzig, 1868; Todhunter's Treatise on Laplace's, Lamé's and Bessel's Functions, London, 1875, etc.

† A solution of the kind referred to contains as one term the product of the regular solution and $\log x$, and is sometimes called a "logarithmic solution." See also American Journal of Mathematics, Vol. XI, p. 37. In the case of Bessel's equation, the logarithmic solution is the "Besselian Function of the second kind."

In the following example,

$$\frac{d^2y}{dx^2} + a\frac{dy}{dx} - 2\frac{y}{x^2} = 0,$$ (9)

there are also two such groups of terms, and their difference of degree shows that the series must ascend by simple powers. We assume therefore at once

$$y = \sum_0^\infty A_r x^{m+r}.$$ (10)

The result of substitution is

$$\sum_0^\infty [\{(m+r)(m+r-1)-2\}A_r x^{m+r-2} + a(m+r)A_r x^{m+r-1}] = 0. \quad (11,$$

Equating to zero the coefficient of x^{m+r-2},

$$(m+r+1)(m+r-2)A_r + a(m+r-1)A^{r-1} = 0, \quad (12)$$

which, when $r = 0$, gives

$$(m+1)(m-2)A_0 = 0,$$ (13)

and when $r > 0$,

$$A_r = -a\frac{m+r-1}{(m+r+1)(m+r-2)}A_{r-1}.$$ (14)

The roots of equation (13) are $m = 2$ and $m = -1$; taking $m = 2$, the relation (14) becomes

$$A_r = -a\frac{r+1}{(r+3)r}A_{r-1},$$

whence the first integral is

$$A_0 y_1 = A_0 x^2 \left[1 - \frac{2}{4}ax + \frac{3}{4\cdot5}a^2x^2 - \frac{4}{4\cdot5\cdot6}a^3x^3 + \dots \right]. \quad (15)$$

Taking the second value $m = -1$, equation (14) gives

$$B_r = -a\frac{r-2}{r(r-3)}B_{r-1},$$

whence $B_1 = -\frac{a}{2}B_0$, and $B_2 = 0$*; therefore the second integral is the finite expression

$$B_0 y_2 = B_0 x^{-1}\left[1 - \frac{1}{2}ax \right] = B_0 \left[\frac{1}{x} - \frac{a}{2} \right].$$ (16)

---

* B_3 would take the indeterminate form, and if we suppose it to have a finite value, the rest of the series is equivalent to $B_3 y_1$, reproducing the first integral.

When the coefficient of the term of highest degree in the result of substitution, such as equation (11), contains m, it is possible to obtain a solution in descending powers of x. In this case, m occurring only in the first degree, but one such solution can be found; it would be identical with the finite integral (16). In the general case there will be two such solutions, and they will be convergent for values of x greater than unity, while the ascending series will converge for values less than unity.*

When the second member of the equation is a power of x, the particular integral can be determined in the form of a series in a similar manner. For example, suppose the second member of equation (9) to have been $x^{\frac{1}{2}}$. Then, making the substitution as before, we have the same relation between consecutive coefficients; but when $r = 0$, instead of equation (13) we have

$$(m + 1)(m - 2)A_{0}x^{m-2} = x^{\frac{1}{2}}$$

to determine the initial term of the series. This gives $m = 2\frac{1}{2}$ and $A_{0} = \frac{4}{7}$; hence, putting $m = \frac{5}{2}$ in equation (14), we find for the particular integral †

$$y = \frac{4}{7}x^{\frac{5}{2}}\left[1 - \frac{2 \cdot 5}{9 \cdot 3}ax + \frac{2^{2} \cdot 5 \cdot 7}{9 \cdot 11 \cdot 3 \cdot 5}a^{2}x^{2} - \cdots \right].$$

A linear equation remains linear for two important classes of transformations; first, when the independent variable is changed to any function of x, and second, when for y we put $vf(x)$. As an example of the latter, let $y = e^{-ax}v$ be substituted in equation (9) above. After rejecting the factor e^{-ax}, the result is

$$\frac{d^{2}v}{dx^{2}} - a\frac{dv}{dx} - \frac{2v}{x^{2}} = 0.$$

Since this differs from the given equation only in the sign

---

* When there are two groups of terms, the integrals are expressible in terms of Gauss's " Hypergeometric Series."

† If the second member is a term of the complementary function (for example, in this case, if it is any integral power of x), the particular integral will take the logarithmic form referred to in the foot-note on p. 346.

of a, we infer from equation (16) that it has the finite integral $v = \dfrac{1}{x} + \dfrac{a}{2}$. Hence the complete integral of equation (9) can be written in the form

$$xy = c_1(2 - ax) + c_2 e^{-ax}(2 + ax).$$

Prob. 69. Integrate in series the equation $\dfrac{d^2y}{dx^2} + xy = 0$.

$$\left(\text{Ans. } y = A\left(1 - \frac{1}{3!}x^3 + \frac{1 \cdot 4}{6!}x^6 - \ldots\right) + B\left(x - \frac{2}{4!}x^4 + \frac{2 \cdot 5}{7!}x^7 - \ldots\right).\right)$$

Prob. 70. Integrate in series $x^2\dfrac{d^2y}{dx^2} + x^2\dfrac{dy}{dx} + (x - 2)y = 0$.

Prob. 71. Derive for the equation of Prob. 70 the integral $y_2 = e^{-x}(x^{-1} + 1 + \frac{1}{2}x)$, and find its relation to those found above.

ART. 19.   SYSTEMS OF DIFFERENTIAL EQUATIONS.

It is shown in Art. 12 that a determinate system of n differential equations of the first order connecting $n + 1$ variables has for its complete solution as many integral equations connecting the variables and also involving n constants of integration. The result of eliminating $n - 1$ variables would be a single relation between the remaining two variables containing in general the n constants. But the elimination may also be effected in the differential system, the result being in general an equation of the nth order of which the equation just mentioned is the complete integral. For example, if there were two equations of the first order connecting the variables x and y with the independent variable t, by differentiating each we should have four equations from which to eliminate one variable, say y, and its two derivatives * with respect to t, leaving a single equation of the second order between x and t.

It is easy to see that the same conclusions hold if some of the given equations are of higher order, except that the order of the result will be correspondingly higher, its index being in

---

* In general, there would be n^2 equations from which to eliminate $n - 1$ variables and n derivatives of each, that is, $(n - 1)(n + 1) = n^2 - 1$ quantities leaving a single equation of the nth order.

general the sum of the indices of the orders of the given equations. The method is particularly applicable to linear equations with constant coefficients, since we have a general method
of solution for the final result. Using the symbolic notation,
the differentiations are performed simply by multiplying by
the symbol D, and therefore the whole elimination is of exactly
the same form as if the equations were algebraic. For example, the system

$$2\frac{d^2y}{dt^2} - \frac{dx}{dt} - 4y = 2t, \qquad 4\frac{dx}{dt} + 2\frac{dy}{dt} - 3x = 0,$$

when written symbolically, is

$$(2D^2 - 4)y - Dx = 2t, \qquad 2Dy + (4D - 3)x = 0;$$

whence, eliminating x,

$$\begin{vmatrix} 2D^2 - 4 & -D \\ 2D & 4D-3 \end{vmatrix} y = \begin{vmatrix} 2t & -D \\ 0 & 4D-3 \end{vmatrix},$$

which reduces to

$$(D-1)^2(2D+3)y = 2 - \tfrac{9}{2}t.$$

Integrating,

$$y = (A + Bt)e^t + Ce^{-\frac{3}{2}t} - \tfrac{1}{2}t,$$

the particular integral being found by symbolic development,
as explained at the end of Art. 16.

The value of x found in like manner is

$$x = (A' + B't)e^t + C'e^{-\frac{3}{2}t} - \tfrac{1}{3}.$$

The complementary function, depending solely upon the determinant of the first members,* is necessarily of the same form
as that for y, but involves a new set of constants. The relations between the constants is found by substituting the
values of x and y in one of the given equations, and equating
to zero in the resulting identity the coefficients of the several
terms of the complementary function. In the present example we should thus find the value of x, in terms of A, B,
and C, to be

$$x = (6B - 2A - 2Bt)e^t - \tfrac{1}{3}Ce^{-\frac{3}{2}t} - \tfrac{1}{3}.$$

---

* The index of the degree in D of this determinant is that of the order of
the final equation ; it is not necessarily the sum of the indices of the orders of
the given equations, but cannot exceed this sum.

In general, the solution of a system of differential equations depends upon our ability to combine them in such a way as to form exact equations.   For example, from the dynamical system

$$\frac{d^2x}{dt^2} = X, \qquad \frac{d^2y}{dt^2} = Y, \qquad \frac{d^2z}{dt^2} = Z, \qquad (1)$$

where X, Y, Z are functions of x, y, and z, but not of t, we form the equation

$$\frac{dx}{dt}d\frac{dx}{dt} + \frac{dy}{dt}d\frac{dz}{dt} + \frac{dz}{dt}d\frac{dz}{dt} = Xdx + Ydy + Zdz.$$

The first member is an exact differential, and we know that for a conservative field of force the second member is also exact, that is, it is the differential of a function U of x, y, and z. The integral

$$\frac{1}{2}\left[\left(\frac{dx}{dt}\right)^2 + \left(\frac{dy}{dt}\right)^2 + \left(\frac{dz}{dt}\right)^2\right] = U + C \qquad (2)$$

is that first integral of the system (1) which is known as the equation of energy for the unit mass.

Just as in Art. 13 an equation of the second order was regarded as equivalent to two equations of the first order, so the system (1) in connection with the equation defining the resolved velocities forms a system of six equations of the first order, of which system equation (2) is an "integral" in the sense explained in Art. 12.

Prob. 72. Solve the equations $\dfrac{dx}{-my} = \dfrac{dy}{mx} = dt$ as a system linear in t.   (Ans. $x = A\cos mt + B\sin mt$, $y = A\sin mt - B\cos mt$.)

Prob. 73. Solve the system $\dfrac{dz}{dx} + n^2y = e^x$, $\dfrac{dy}{dx} + z = 0$.

(Ans. $y = Ae^{nx} + Be^{-nx} + \dfrac{e^x}{n^2 - \ldots}$, $z = -nAe^{nx} + nBe^{-nx} - \dfrac{e^x}{n^2 - 1}$.)

Prob. 74. Find for the system $\dfrac{d^2x}{dt^2} = x\phi(x, y)$, $\dfrac{d^2y}{dt^2} = y\phi(x, y)$ a first integral independent of the function ϕ.

$$\left(\text{Ans. } x\frac{dy}{dt} - y\frac{dx}{dt} = C.\right)$$

Prob. 75. The approximate equations for the horizontal motion of a pendulum, when the earth's rotation is taken into account, are

$$\frac{d^2x}{dt^2} - 2r\frac{dy}{dt} + \frac{gx}{l} = 0, \quad \frac{d^2y}{dt^2} + 2r\frac{dx}{dt} + \frac{gy}{l} = 0;$$

show that both x and y are of the form

$$A \cos n_1 t + B \sin n_1 t + C \cos n_2 t + D \sin n_2 t.$$

ART. 20. FIRST ORDER AND DEGREE WITH THREE VARIABLES.

The equation of the first order and degree between three variables x, y and z may be written

$$Pdx + Qdy + Rdz = 0, \tag{1}$$

where P, Q and R are functions of x, y and z. When this equation is exact, P, Q and R are the partial derivatives of some function u, of x, y and z; and we derive, as in Art. 4,

$$\frac{\partial P}{\partial y} = \frac{\partial Q}{\partial x}, \quad \frac{\partial Q}{\partial z} = \frac{\partial R}{\partial y}, \quad \frac{\partial R}{\partial x} = \frac{\partial P}{\partial z} \tag{2}$$

for the conditions of exactness. In the case of two variables, when the equation is not exact integrating factors always exist; but in this case. there is not always a factor μ such that μP, μQ and μR (put in place of P, Q, and R) will satisfy all three of the conditions (2). It is easily shown that for this purpose the relation

$$P\left(\frac{\partial Q}{\partial z} - \frac{\partial R}{\partial y}\right) + Q\left(\frac{\partial R}{\partial x} - \frac{\partial P}{\partial z}\right) + R\left(\frac{\partial P}{\partial y} - \frac{\partial Q}{\partial x}\right) = 0 \tag{3}$$

must exist between the given values of P, Q, and R. This is therefore the " condition of integrability " of equation (1).*

When this condition is fulfilled equation (1) may be integrated by first supposing one variable, say z, to be constant. Thus, integrating $Pdx + Qdy = 0$, and supposing the constant of integration C to be a function of z, we obtain the integral, so

---

* When there are more than three variables such a condition of integrability exists for each group of three variables, but these conditions are not all independent. Thus with four variables there are but three independent conditions.

far as it depends upon x and y. Finally, by comparing the total differential of this result with the given equation we determine dC in terms of z and dz, and thence by integration the value of C.

It may be noticed that when certain terms of an exact equation forms an exact differential, the remaining terms must also be exact. It follows that if one of the variables, say z can be completely separated from the other two (so that in equation (1) R becomes a function of z only and P and Q functions of x and y, but not of z) the terms $Pdx + Qdy$ must be thus rendered exact if the equation is integrable.* For example,

$$zydx - zxdy - y^2dz = 0.$$

is an integrable equation. Accordingly, dividing by y^2z, which we notice separates the variable z from x and y, puts it in the exact form

$$\frac{ydx - xdy}{y^2} - \frac{dz}{z} = 0,$$

of which the integral is $x = y \log cz$.

Regarding x, y and z as coordinates of a moving point, an integrable equation restricts the point to motion upon one of the surfaces belonging to the system of surfaces represented by the integral; in other words, the point (x, y, z) moves in an arbitrary curve drawn on such a surface. Let us now consider in what way equation (1) restricts the motion of a point when it is not integrable. The direction cosines of a moving point are proportional to dx, dy, and dz; hence, denoting them by l, m and n, the direction of motion of the point satisfying equation (1) must satisfy the condition

$$Pl + Qm + Rn = 0. \tag{4}$$

It is convenient to consider in this connection an auxiliary system of lines represented, as explained in Art. 12, by the simultaneous equations

$$\frac{dx}{P} = \frac{dy}{Q} = \frac{dz}{R}. \tag{5}$$

---

* In fact for this case the condition (3) reduces to its last term, which expresses the exactness of $Pdx + Qdy$.

The direction cosines of a point moving in one of the lines of this system are proportional to P, Q and R. Hence, denoting them by λ, μ, ν, equation (4) gives

$$\lambda l + \mu m + \nu n = 0 \tag{6}$$

for the relation between the directions of two moving points, whose paths intersect, subject respectively to equation (1) and to equations (5). The paths in question therefore intersect at right angles; therefore equation (1) simply restricts a point to move in a path which cuts orthogonally the lines of the auxiliary system.

Now, if there be a system of surfaces which cut the auxiliary lines orthogonally, the restriction just mentioned is completely expressed by the requirement that the line shall lie on one of these surfaces, the line being otherwise entirely arbitrary. This is the case in which equation (1) is integrable.*

On the other hand, when the equation is not integrable, the restriction can only be expressed by two equations involving an arbitrary function. Thus if we assume in advance one such relation, we know from Art. 12 that the given equation (1) together with the first derivative of the assumed relation forms a system admitting of solution in the form of two integrals. Both of these integrals will involve the assumed function. For any particular value of that function we have a system of lines satisfying equation (1), and the arbitrary character of the function makes the solution sufficiently general to include all lines which satisfy the equation.†

Prob. 76. Show that the equation
$$(mz - ny)dx + (nx - lz)dy + (ly - mx)dz = 0$$
is integrable, and infer from the integral the character of the auxil-

---

* It follows that, with respect to the system of lines represented by equations (5), equation (3) is the condition that the system shall admit of surfaces cutting them orthogonally. The lines of force in any field of conservative forces form such a system, the orthogonal surfaces being the equipotential surfaces.

† So too there is an arbitrary element about the path of a point when the single equation to which it is subject is integrable, but this enters only into *one* of the two equations necessary to define the path.

iary lines.   (Compare the illustrative example at the end of Art. 12.)

$$(\text{Ans. } nx - lz = C(ny - mz).)$$

Prob. 77. Solve $yz^2dx - z^2dy - e^xdz = 0$.   (Ans. $yz = e^x(1 + cz)$.)

Prob. 78. Find the equation which in connection with $y = f(x)$ forms the solution of $dz = aydx + bdy$.

Prob. 79. Show that a general solution of

$$ydx = (x - z)(dy - dz)$$

is given by the equations

$$y - z = \phi(x), \qquad y = (x - z)\phi'(x).$$

(This is an example of " Monge's Solution.")

ART. 21. PARTIAL DIFFERENTIAL EQUATIONS OF FIRST ORDER AND DEGREE.

Let x denote an unknown function of the two independent variables x and y, and let

$$p = \frac{\partial z}{\partial x}, \qquad q = \frac{\partial z}{\partial y}$$

denote its partial derivatives: a relation between one or both of these derivatives and the variables is called a " partial differential equation " of the first order.   A value of z in terms of x and y which with its derivatives satisfies the equation, or a relation between x, y and z which makes z implicitly such a function, is a " particular integral."   The most general equation of this kind is called the " general integral."

If only one of the derivatives, say p, occurs, the equation may be solved as an ordinary differential equation.   For if y is considered as a constant, p becomes the ordinary derivative of z with respect to x; therefore, if in the complete integral of the equation thus regarded we replace the constant of integration by an arbitrary function of y, we shall have a relation which includes all particular integrals and has the greatest possible generality.   It will be found that, in like manner, when both p and q are present, the general integral involves an arbitrary function.

We proceed to give Lagrange's solution of the equation of

the first order and degree, or "linear equation," which may be written in the form

$$Pp + Qq = R,\tag{1}$$

P, Q and R denoting functions of x, y and z. Let $u = a$, in which u is a function of x, y and z, and a, a constant, be an integral of equation (1). Taking derivatives with respect to x and y respectively, we have

$$\frac{\partial u}{\partial x} + \frac{\partial u}{\partial z}p = 0, \qquad \frac{\partial u}{\partial y} + \frac{\partial u}{\partial z}q = 0,$$

and substitution of the values of p and q in equation (1) gives the symmetrical relation

$$P\frac{\partial u}{\partial x} + Q\frac{\partial u}{\partial y} + R\frac{\partial u}{\partial z} = 0.\tag{2}$$

Consider now the system of simultaneous ordinary differential equations

$$\frac{dx}{P} = \frac{dy}{Q} = \frac{dz}{R}\tag{3}$$

Let $u = 0$ be one of the integrals (see Art. 12) of this system. Taking its total differential,

$$\frac{\partial u}{\partial x}dx + \frac{\partial u}{\partial y}dy + \frac{\partial u}{\partial z}dz = 0;$$

and since by equations (3) dx, dy and dz are proportional to P, Q and R, we obtain by substitution

$$\frac{\partial u}{\partial x}P + \frac{\partial u}{\partial y}Q + \frac{\partial u}{\partial z}R = 0,$$

which is identical with equation (2). It follows that every integral of the system (3) satisfies equation (1), and conversely, so that the general expression for the integrals of (3) will be the general integral of equation (1).

Now let $v = b$ be another integral of equations (3), so that v is also a function which satisfies equation (2). As explained in Art. 12, each of the equations $u = a$, $v = b$ is the equation of a surface passing through a singly infinite system of lines belonging to the doubly infinite system represented by equations (3). What we require is the general expression for any surface passing through lines of the system (and intersecting none of them). It is evident that $f(u, v) = f(a, b) = C$ is such an equation,* and accordingly $f(u, v)$, where f is an arbitrary function, will be found to satisfy equation (2). Therefore, to solve equation (1), we find two independent integrals $u = a$, $v = b$ of the auxiliary system (3), (sometimes called Lagrange's equations,) and then put

$$u = \phi(v), \tag{4}$$

an equation which is evidently equally general with $f(u, v) = 0$.

Conversely, it may be shown that any equation of the form (4), regarded as a primitive, gives rise to a definite partial differential equation of Lagrange's linear form. For, taking partial derivatives with respect to the independent variables x and y, we have

$$\frac{\partial u}{\partial x} + \frac{\partial u}{\partial z}p = \phi'(v)\left[\frac{\partial v}{\partial x} + \frac{\partial v}{\partial z}p\right],$$

$$\frac{\partial u}{\partial y} + \frac{\partial u}{\partial z}q = \phi'(v)\left[\frac{\partial v}{\partial y} + \frac{\partial v}{\partial z}q\right];$$

and eliminating $\phi'(v)$ from these equations, the term containing pq vanishes, giving the result

$$\begin{vmatrix} \frac{\partial u}{\partial y} & \frac{\partial u}{\partial z} \\ \frac{\partial v}{\partial y} & \frac{\partial v}{\partial z} \end{vmatrix} p + \begin{vmatrix} \frac{\partial u}{\partial z} & \frac{\partial u}{\partial x} \\ \frac{\partial v}{\partial z} & \frac{\partial v}{\partial x} \end{vmatrix} q = \begin{vmatrix} \frac{\partial u}{\partial x} & \frac{\partial u}{\partial y} \\ \frac{\partial v}{\partial x} & \frac{\partial v}{\partial y} \end{vmatrix}, \tag{5}$$

which is of the form $Pp + Qq = R.\dagger$

---

* Each line of the system is characterized by special values of a and b which we may call its coordinates, and the surface passes through those lines whose coordinates are connected by the perfectly arbitrary relation $f(a, b) = C$.

† These values of P, Q and R are known as the "Jacobians" of the pair of functions u, v with respect to the pairs of variables y, z; z, x; and x, y respectively. Owing to their analogy to the derivatives of a single function they are sometimes denoted thus :

$$P = \frac{\partial(u, v)}{\partial(y, z)}, \quad Q = \frac{\partial(u, v)}{\partial(z, x)}, \quad R = \frac{\partial(u, v)}{\partial(x, y)}.$$

The Jacobian vanishes if the functions u and v are not independent, that is to say, if u can be expressed identically as a function of v. In like manner,

As an illustration, let the given partial differential equation be

$$(mz - ny)p + (nx - lz)q = ly - mx, \tag{6}$$

for which Lagrange's Equations are

$$\frac{dx}{mz - ny} = \frac{dy}{nx - lz} = \frac{dz}{ly - mx}. \tag{7}$$

These equations were solved at the end of Art. 12, the two integrals there found being

$$lx + my + nz = a \quad \text{and} \quad x^2 + y^2 + z^2 = b^2. \tag{8}$$

Hence in this case the system of "Lagrangean lines" consists of the entire system of circles having the straight line

$$\frac{x}{l} = \frac{y}{m} = \frac{z}{n} \tag{9}$$

for axis.   The general integral of equation (6) is then

$$lx + my + nz = \phi(x^2 + y^2 + z^2), \tag{10}$$

which represents any surface passing through the circles just mentioned, that is, any surface of revolution of which (9) is the axis.*

Lagrange's solution extends to the linear equation containing n independent variables.   Thus the equation being

$$P_1\frac{\partial z}{\partial x_1} + P_2\frac{\partial z}{\partial x_2} + \ldots + P_n\frac{\partial z}{\partial x_n} = R,$$

the auxiliary equations are

$$\frac{dx_1}{P_1} = \frac{dx_2}{P_2} = \ldots = \frac{dx_n}{P_n} = \frac{dz}{R},$$

$\dfrac{\partial(\phi, u, v)}{\partial(x, y, z)} = 0$ is the condition that ϕ (a function of x, y and z) is expressible identically as a function of u and v, that is to say, that $\phi = 0$ shall be an integral of $Pp + Qq = R$.

* When the equation $Pdx + Qdy + Rdz = 0$ is integrable (as it is in the above example; see Prob. 76, Art. 20), its integral, which may be put in the form $V = C$, represents a singly infinite system of surfaces which the Lagrangean lines cut orthogonally ; therefore, in this case, the general integral may be defined as the general equation of the surfaces which cut orthogonally the system $V = C$.   Conversely, starting with a given system $V = C$, $u = f(v)$ is the general equation of the orthogonal surfaces, if $u = a$ and $v = b$ are integrals of

$$dx \left/ \frac{\partial V}{\partial x} \right. = dy \left/ \frac{\partial V}{\partial y} \right. = dz \left/ \frac{\partial V}{\partial z} \right. .$$

and if $u_1 = c_1, u_2 = c_2, \ldots u_n = c_n$ are independent integrals, the most general solution is

$$f(u_1, u_2, \ldots u_n) = 0,$$

where f is an arbitrary function.

Prob. 80. Solve $xz\dfrac{\partial z}{\partial x} + yz\dfrac{\partial z}{\partial y} = xy.$ $\left(\text{Ans. } xy - z^2 = f\left(\dfrac{x}{y}\right).\right)$

Prob. 81. Solve $(y + z)p + (z + x)q = x + y.$

Prob. 82. Solve $(x + y)(p - q) = z.$

$$\text{(Ans. } (x + y) \log z - x = f(x + y).)$$

Prob. 83. Solve $x(y - z)p + y(z - x)q = z(x - y).$

$$\text{(Ans. } x + y + z = f(xyz).)$$

ART. 22. COMPLETE AND GENERAL INTEGRALS.

We have seen in the preceding article that an equation between three variables containing an arbitrary function gives rise to a partial differential equation of the linear form. It follows that, when the equation is not linear in p and q, the general integral cannot be expressed by a single equation of the form $\varphi(u, v) = 0$; it will, however, still be found to depend upon a single arbitrary function.

It therefore becomes necessary to consider an integral having as much generality as can be given by the presence of arbitrary constants. Such an equation is called a " complete integral"; it contains two arbitrary constants (n arbitrary constants in the general case of n independent variables), because this is the number which can be eliminated from such an equation, considered as a primitive, and its two derived equations. For example, if

$$(x - a)^2 + (y - b)^2 + z^2 = k^2,$$

a and b being regarded as arbitrary, be taken as the primitive, the derived equations are

$$x - a + zp = 0, \qquad y - b + zq = 0,$$

and the elimination of a and b gives the differential equation

$$z^2(p^2 + q^2 + 1) = k^2,$$

of which therefore the given equation is a complete integral.

Geometrically, the complete integral represents a doubly infinite system of surfaces ; in this case they are spherical surfaces having a given radius and centers in the plane of xy.

In general, a partial differential equation of the first order with two independent variables is of the form

$$F(x, y, z, p, q) = 0, \tag{1}$$

and a complete integral is of the form

$$f(x, y, z, a, b) = 0. \tag{2}$$

In equation (1) suppose x, y and z to have special values, namely, the coordinates of a special point A ; the equation becomes a relation between p and q. Now consider any surface passing through A of which the equation is an integral of (1), or, as we may call it, a given "integral surface" passing through A. The tangent plane to this surface at A determines values of p and q which must satisfy the relation just mentioned. Consider also those of the complete integral surfaces [equation (2)] which pass through A. They form a singly infinite system whose tangent planes at A have values of p and q which also satisfy the relation. There is obviously among them one which has the same value of p, and therefore also the same value of q, as the given integral. Thus there is one of the complete integral surfaces which touches at A the given integral surface. It follows that every integral surface (not included in the complete integral) must at every one of its points touch a surface included in the complete integral.*

It is hence evident that every integral surface is the envelope of a singly infinite system selected from the complete integral system. Thus, in the example at the beginning of this article, a right cylinder whose radius is k and whose axis lies in the plane of xy is an integral, because it is the envelope

---

* Values of x, y, and z, determining a point, together with values of p and q, determining the direction of a surface at that point, are said to constitute an "element of surface." The theorem shows that the complete integral is "complete" in the sense of including all the surface elements which satisfy the differential equation. The method of grouping the "consecutive" elements to form an integral surface is to a certain extent arbitrary.

of those among the spheres represented by the complete integral whose centers are on the axis of the cylinder. If we make the center of the sphere describe an arbitrary curve in the plane of xy we shall have the general integral in this example.

In general, if in equation (2) an arbitrary relation between a and b, such as $b = \phi(a)$, be established, the envelope of the singly infinite system of surfaces thus defined will represent the general integral. By the usual process, the equation of the envelope is the result of eliminating a between the two equations

$$f(x, y, z, a, \phi(a)) = 0, \quad \frac{\partial}{\partial a} f(x, y, z, a, \phi(a)) = 0. \quad (3)$$

These two equations together determine a line, namely, the "ultimate intersection of two consecutive surfaces." Such lines are called the "characteristics" of the differential equation. They are independent of any particular form of the complete integral, being in fact lines along which all integral surfaces which pass through them touch one another. In the illustrative example above they are equal circles with centers in the plane of xy and planes perpendicular to it.*

The example also furnishes an instance of a "singular solution" analogous to those of ordinary differential equations.

---

* The characteristics are to be regarded not merely as lines, but as "linear elements of surface," since they determine at each of their points the direction of the surfaces passing through them. Thus, in the illustration, they are circles regarded as great-circle elements of a sphere, or as elements of a right cylinder, and may be likened to narrow hoops. They constitute in all cases a triply infinite system. The surfaces of a complete integral system contain them all, but they are differently grouped in different integral surfaces.

If we arbitrarily select a curve in space there will in general be at each of its points but one characteristic through which the selected curve passes; that is, whose tangent plane contains the tangent to the selected curve. These characteristics (for all points of the curve) form an integral surface passing through the selected curve ; and it is the only one which passes through it unless it be itself a characteristic. Integral surfaces of a special kind result when the selected curve is reduced to a point. In the illustration these are the results of rotating the circle about a line parallel to the axis of z.

For the planes $z = \pm k$ envelop the whole system of spheres represented by the complete integral, and indeed all the surfaces included in the general integral. When a singular solution exists it is included in the result of eliminating a and b from equation (2) and its derivatives with respect to a and b, that is, from

$$f = 0, \qquad \frac{\partial f}{\partial a} = 0, \qquad \frac{\partial f}{\partial b} = 0; \tag{4}$$

but, as in the case of ordinary equations, this result may include relations which are not solutions.

Prob. 84. Derive a differential equation from the primitive $lx + my + nz = a$, where l, m, n are connected by the relation $l^2 + m^2 + n^2 = 1$.

Prob. 85. Show that the singular solution of the equation found in Prob. 84 represents a sphere, that the characteristics consist of all the straight lines which touch this sphere, and that the general integral therefore represents all developable surfaces which touch the sphere.

Prob. 86. Find the integral which results from taking in the general integral above $l^2 + m^2 = \cos^2 \theta$ (a constant) for the arbitrary relation between the parameters.

ART. 23.   COMPLETE INTEGRAL FOR SPECIAL FORMS.

A complete integral of the partial differential equation

$$F(x, y, z, p, q) = 0 \tag{1}$$

contains two constants, a and b. If a be regarded as fixed and b as an arbitrary parameter, it is the equation of a singly infinite system of surfaces, of which one can be found passing through any given point. The ordinary differential equation of this system, which will be independent of b, may be put in the form

$$dz = p\,dx + q\,dy, \tag{2}$$

in which the coefficients p and q are functions of the variables and the constant a. Now the form of equation (2) shows that these quantities are the partial derivatives of z, in an integral of equation (1); therefore they are values of p and q which satisfy equation (1). Conversely, if values of p and q in terms of the variables and a constant a which satisfy equation (1) are such as to make equation (2) the differential equation of a system of surfaces, these surfaces will be integrals. In other words, if we can find values of p and q containing a constant a which satisfy equation (1) and make $dz = pdx + qdy$ integrable, we can obtain by direct integration a complete integral, the integration introducing a second constant.

There are certain forms of equations for which such values of p and q are easily found. In particular there are forms in which p and q admit of constant values, and these obviously make equation (2) integrable. Thus, if the equation contains p and q only, being of the form

$$F(p, q) = 0, \tag{3}$$

we may put $p = a$ and $q = b$, provided

$$F(a, b) = 0. \tag{4}$$

Equation (2) thus becomes

$$dz = adx + bdy,$$

whence we have the complete integral

$$z = ax + by + c, \tag{5}$$

in which a and b are connected by the relation (4) so that a, b and c are equivalent to two arbitrary constants.

In the next place, if the equation is of the form

$$z = px + qy + f(p, q), \tag{6}$$

which is analogous to Clairaut's form, Art. 10, constant values of p and q are again admissible if they satisfy

$$z = ax + by + f(a, b), \tag{7}$$

and this is itself the complete integral. For this equation is of the form $z = ax + by + c$, and expresses in itself the relations between the three constants. Problem 84 of the preceding article is an example of this form.

In the third place, suppose the equation to be of the form

$$F(z, p, q) = 0, \tag{8}$$

in which neither x nor y appears explicitly. If we assume $q = ap$, p will be a function of z determined from

$$F(z, p, ap) = 0, \quad \text{say} \quad p = \phi(z). \tag{9}$$

Then $dz = pdx + qdy = 0$ becomes $dz = \phi(z)(dx + ady)$, which is integrable, giving the complete integral

$$x + ay = \int \frac{dz}{\phi(z)} + b. \tag{10}$$

A fourth case is that in which, while z does not explicitly occur, it is possible to separate x and p from y and q, thus putting the equation in the form

$$f_1(x, p) = f_2(y, q). \tag{11}$$

If we assume each member of this equation equal to a constant a, we may determine p and q in the forms

$$p = \phi_1(x, a), \qquad q = \phi_2(y, a). \tag{12}$$

and $dz = pdx + qdy$ takes an integrable form giving

$$z = \int \phi_1(x, a)dx + \int \phi_2(y, a)dy + b. \tag{13}$$

It is frequently possible to reduce a given equation by transformation of the variables to one of the four forms considered in this article.* For example, the equation $x^2p^2 + y^2 q^2 = z^2$ may be written

$$\left(\frac{xdz}{zdx}\right)^2 + \left(\frac{ydz}{zdy}\right)^2 = 1 ;$$

---

* The general method, due to Charpit, of finding a proper value of p consists of establishing, by means of the condition of integrability, a linear partial differential equation for p, of which we need only a particular integral. This may be any value of p taken from the auxiliary equations employed in Lagrange's process. See Boole, Differential Equations (London 1865), p. 336 ; also Forsyth, Differential Equations (London 1885), p. 316, in which the auxiliary equations are deduced in a more general and symmetrical form, involving both p and q. These equations are in fact the equations of the characteristics regarded as in the concluding note to the preceding article. Denoting the partial derivatives of $F(x, y, z, p, q)$ by X, Y, Z, P, Q, they are

$$\frac{dx}{P} = \frac{dy}{Q} = \frac{dz}{Pp+Qq} = -\frac{dp}{X+Zp} = -\frac{dq}{Y+Zq}.$$

See Jordan's Course d'Analyse (Paris, 1887), vol. III, p. 318 ; Johnson's Differential Equations (New York, 1889), p. 300. Any relation involving one or both the quantities p and q, combined with $F = 0$, will furnish proper values of p

whence, putting $x' = \log x$, $y' = \log y$, $z' = \log z$, it becomes $p'^2 + q'^2 = 1$, which is of the form $F(p', q') = 0$, equation (3). Hence the integral is given by equation (5) when $a^2 + b^2 = 1$; it may therefore be written

$$z' = x' \cos \alpha + y' \sin \alpha + c,$$

and restoring x, y, and z, that of the given equation is

$$z = c x^{\cos \alpha} \, y^{\sin \alpha}.$$

Prob. 87. Find a complete integral for $p^2 - q^2 = 1$.
(Ans. $z = x \sec \alpha + y \tan \alpha + b$.)

Prob. 88. Find the singular solution of $z = px + qy + pq$.
(Ans. $z = - xy$.)

Prob. 89. Solve by transformation $q = 2yp^2$.
(Ans. $z = ax + a^2y^2 + b$.)

Prob. 90. Solve $z(p^2 - q^2) = x - y$.
(Ans. $z^{\frac{3}{2}} = (x + a)^{\frac{3}{2}} + (y + a)^{\frac{3}{2}} + b$.)

Prob. 91. Show that the solution given for the form $F(z, p, q) = 0$ represents cylindrical surfaces, and that $F(z, 0, 0) = 0$ is a singular solution.

Prob. 92. Deduce by the method quoted in the foot-note two complete integrals of $pq = px + qy$.

(Ans. $2z = \left(\dfrac{x}{\alpha} + \alpha y\right)^2 + \beta$, and $z = xy + y \sqrt{(x^2 + a)} + b$.)

ART. 24. Partial Equations of Second Order.

We have seen in the preceding articles that the general solution of a partial differential equation of the first order depends upon an arbitrary function; although it is only when the equation is linear in p and q that it is expressible by a single equation. But in the case of higher orders no general account can be given of the nature of a solution. Moreover, when we consider the equations derivable from a primitive containing arbitrary functions, there is no correspondence between their number and the order of the equation. For example, if

and q. Sometimes several such relations are readily found ; for example, for the equation $z = pq$ we thus obtain the two complete integrals

$$z = (y + a)(x + b) \quad \text{and} \quad 4z = \left(\frac{x}{\alpha} + \alpha y + \beta\right)^2.$$

the primitive with two independent variables contains two arbitrary functions, it is not generally possible to eliminate them and their derivatives from the primitive and its two derived equations of the first and three of the second order.

Instead of a primitive containing two arbitrary functions, let us take an equation of the first order containing a single arbitrary function. This may be put in the form $u = \phi(v)$, u and v now denoting known functions of x, y, z, p, and q. $\phi'(v)$ may now be eliminated from the two derived equations as in Art. 21. Denoting the second derivatives of z by

$$r = \frac{\partial^2 z}{\partial x^2}, \qquad s = \frac{\partial^2 z}{\partial x \partial y}, \qquad t = \frac{\partial^2 z}{\partial^2 y},$$

the result is found to be of the form

$$Rr + Ss + Tt + U(rt - s^2) = V, \qquad (1)$$

in which R, S, T, U, and V are functions of x, y, z, p, and q. With reference to the differential equation of the second order the equation $u = \phi(v)$ is called an "intermediate equation of the first order": it is analogous to the first integral of an ordinary equation of the second order. It follows that an intermediate equation cannot exist unless the equation is of the form (1); moreover, there are two other conditions which must exist between the functions R, S, T, and U.

In some simple cases an intermediate equation can be obtained by direct integration. Thus, if the equation contains derivatives with respect to one only of the variables, it may be treated as an ordinary differential equation of the second order, the constants being replaced by arbitrary functions of the other variable. Given, for example, the equation $xr - p = xy$, which may be written

$$x\,dp - p\,dx = xy\,dx.$$

This becomes exact with reference to x when divided by x^2, and gives the intermediate equation

$$p = yx \log x + x\phi(y).$$

A second integration (and change in the form of the arbitrary function) gives the general integral

$$z = \tfrac{1}{2}yx^2 \log x + x^2\phi(y) + \psi(y).$$

Again, the equation $p + r + s = 1$ is already exact, and gives the intermediate equation

$$z + p + q = x + \phi(y),$$

which is of Lagrange's form.   The auxiliary equations* are

$$dx = dy = \frac{dz}{x - z + \phi(y)},$$

of which the first gives $x - y = a$, and eliminating x from the second, its integral is of the form

$$z = a + \phi(y) + e^{-y}b.$$

Hence, putting $b = \psi(a)$, we have for the final integral

$$z = x + \phi(y) + e^{-y}\psi(x - y),$$

in which a further change is made in the form of the arbitrary function ϕ.

Prob. 93. Solve $t - q = e^x + e^y$.
$$\text{(Ans. } z = y(e^y - e^x) + \phi(x) + e^y\psi(x).)$$

Prob. 94. Solve $r + p^2 = y^2$.
$$\text{(Ans. } z = \log[e^{xy}\psi(y) - e^{-xy}] + \psi(y).)$$

Prob. 95. Solve $y^2(s - t) = x$.
$$\text{(Ans. } z = (x + y)\log y + \phi(x) + \psi(x + y).)$$

Prob. 96. Solve $ps - qr = 0$. $\text{(Ans. } x = \phi(y) + \psi(z).)$

Prob. 97. Show that Monge's equations (see foot-note) give for Prob. 96 the intermediate integral $p = \phi(z)$ and hence derive the solution.

---

* In Monge's method (for which the reader must be referred to the complete treatises) of finding an intermediate integral of

$$Rr + Ss + Tt = V$$

when one exists, the auxiliary equations

$$Rdy^2 - Sdy\,dx + Tdx^2 = 0, \qquad Rdp\,dy + Tdq\,dx = Vdx\,dy$$

are established.   These, in connection with

$$dz = pdx + qdy,$$

form an incomplete system of ordinary differential equations, between the five variables x, y, z, p, and q.   But if it is possible to obtain two integrals of the system in the form $u = a$, $v = b$, $u = \phi(v)$ will be the intermediate integral. The first of the auxiliary equations is a quadratic giving two values for the ratio $dy : dx$.   If these are distinct, and an intermediate integral can be found, for each, the values of p and q determined from them will make $dz = pdx + qdy$ integrable, and give the general integral at once.

Prob. 98. Derive by Monge's method for $q^2r - 2pqs + p^2t = 0$ the intermediate integral $p = q\,\phi(z)$, and thence the general integral.

(Ans. $y + x\phi(z) = \psi(z)$.)

ART. 25. LINEAR PARTIAL DIFFERENTIAL EQUATIONS.

Equations which are linear with respect to the dependent variable and its partial derivatives may be treated by a method analogous to that employed in the case of ordinary differential equations. We shall consider only the case of two independent variables x and y, and put

$$D = \frac{\partial}{\partial x}, \qquad D' = \frac{\partial}{\partial y},$$

so that the higher derivatives are denoted by the symbols D^2, DD', D'^2, D^3, etc. Supposing further that the coefficients are constants, the equation may be written in the form

$$f(D, D')z = F(x, y), \tag{1}$$

in which f denotes an algebraic function, or polynomial, of which the degree corresponds to the order of the differential equation. Understanding by an "integral" of this equation an explicit value of z in terms of x and y, it is obvious, as in Art. 15, that the sum of a particular integral and the general integral of

$$f(D, D')z = 0 \tag{2}$$

will constitute an equally general solution of equation (1). It is, however, only when $f(D, D')$ is a *homogeneous* function of D and D' that we can obtain a solution of equation (2) containing n arbitrary functions,* which solution is also the "complementary function" for equation (1).

Suppose then the equation to be of the form

$$A_0\frac{d^nz}{dx^n} + A_1\frac{d^nz}{dx^{n-1}dy} + \dots + A_n\frac{d^nz}{dy^n} = 0, \tag{3}$$

and let us assume $z = \phi(y + mx)$,   (4)

---

* It is assumed that such a solution constitutes the general integral of an equation of the nth order; for a primitive containing more than n independent arbitrary functions cannot give rise by their elimination to an equation of the nth order.

where m is a constant to be determined. From equation (4), $Dz = m\phi'(y + mx)$ and $D'z = \phi'(y + mx)$, so that $Dz = mD'z$, $D^2z = m^2D'^2z$, $DD'z = mD'^2y$, etc. Substituting in equation (3) and rejecting the factor D'^nz or $\phi^{(n)}(y + mx)$, we have

$$A_0m^n + A_1m^{n-1} + \ldots + A_n = 0 \qquad (5)$$

for the determination of m. If $m_1, m_2, \ldots m_n$ are distinct roots of this equation,

$$z = \phi_1(y + m_1x) + \phi_2(y + m_2x) + \ldots + \phi_n(y + m_nx) \quad (6)$$

is the general integral of equation (3).

For example, the general integral of $\dfrac{d^2z}{dx^2} - \dfrac{d^2z}{dy^2} = 0$ is thus found to be $z = \phi(y + x) + \psi(y - x)$. Any expression of the form $Axy + Bx + Cy + D$ is a particular integral; accordingly it is expressible as the sum of certain functions of $x + y$ and $x - y$ respectively.

The homogeneous equation (3) may now be written symbolically in the form

$$(D - m_1D')(D - m_2D') \ldots (D - m_nD')z = 0, \qquad (7)$$

in which the several factors correspond to the several terms of the general integral. If two of the roots of equation (5) are equal, say, to m_1, the corresponding terms in equation (6) are equivalent to a single arbitrary function. To form the general integral we need an integral of

$$(D - m_1D')^2z = 0 \qquad (8)$$

in addition to $\phi(y + m_1x)$. This will in fact be the solution of

$$(D - m_1D')z = \phi(y + m_1x); \qquad (9)$$

for, if we operate with $D - m_1D'$ upon both members of this equation, we obtain equation (8). Writing equation (9) in the form

$$p - m_1q = \phi(y + mx),$$

Lagrange's equations are

$$dx = -\frac{dy}{m_1} = \frac{dz}{\phi(y + m_1x)},$$

giving the integrals $y + m_1x = a$, $z = x\phi(a) + b$. Hence the integral of equation (9) is

$$z = x\phi(y + m_1x) + \psi(y + m_1x), \qquad (10)$$

and regarding ϕ also as arbitrary, these are the two independent terms corresponding to the pair of equal roots.

If equation (5) has a pair of imaginary roots $m = \mu \pm i\nu$, the corresponding terms of the integral take the form

$$\phi(y + \mu x + i\nu x) + \psi(y + \mu x - i\nu x),$$

which when ϕ and ψ are real functions contain imaginary terms. If we restrict ourselves to real integrals we cannot now say that there are two radically distinct classes of integrals; but if any real function of $y + \mu x + i\nu x$ be put in the form $X + iY$, either of the real functions X or Y will be an integral of the equation. Given, for example, the equation

$$\frac{d^2z}{dx^2} + \frac{d^2z}{dy^2} = 0,$$

of which the general integral is

$$z = \phi(y + ix) + \psi(y - ix);$$

to obtain a real integral take either the real or the coefficient of the imaginary part of any real form of $\phi(y + ix)$. Thus, if $\phi(t) = e^t$ we find $e^y \cos x$ and $e^y \sin x$, each of which is an integral.

As in the corresponding case of ordinary equations, the particular integral of equation (1) may be made to depend upon the solution of linear equations of the first order. The inverse symbol $\dfrac{1}{D - mD'} F(x, y)$ in the equation corresponding to equation (14), Art. 16, denotes the value of z in

$$(D - mD')z = F(x, y) \quad \text{or} \quad p - mq = F(x, y). \tag{11}$$

For this equation Lagrange's auxiliary equations give

$$y + mx = a, \quad z = \int F(x, a - mx)dx + b = F_1(x, a) + b,$$

and the general integral is

$$z = F_1(x, y + mx) + \phi(y + mx). \tag{12}$$

The first term, which is the particular integral, may therefore be found by subtracting mx from y in $F(x, y)$, integrating with respect to x, and then adding mx to y in the result.*

For certain forms of $F(x, y)$ there exist more expeditious methods, of which we shall here only notice that which applies to the form $F(ax + by)$. Since $DF(ax + by) = aF'(ax + by)$ and $D'F(ax + by) = bF'(ax + by)$, it is readily inferred that, when $f(D, D')$ is a homogeneous function of the nth degree,

$$f(D, D')F(ax + by) = f(a, b)F^{(n)}(ax + by). \qquad (13)$$

That is, if $t = ax + by$, the operation of $f(D, D')$ on $F(t)$ is equivalent to multiplication by $f(a, b)$ and taking the nth derivative, the final result being still a function of t. It follows that, conversely, the operation of the inverse symbol upon a function of t is equivalent to dividing by $f(a, b)$ and integrating n times. Thus,

$$\frac{1}{f(D, D')} F(ax + by) = \frac{1}{f(a, b)} \int \int \cdots \int F(t)dt^n. \qquad (14)$$

When $ax + by$ is a multiple of $y + m_1 x$, where m_1 is a root of equation (5), this method fails with respect to the corresponding symbolic factor, giving rise to an equation of the form (9), of which the solution is given in equation (10). Given, for example, the equation

$$\frac{d^2z}{dx^2} + \frac{d^2z}{dx\,dy} - 2\frac{d^2z}{dy^2} = \sin(x - y) + \sin(x + y)$$

or　　$(D - D')(D + 2D')z = \sin(x - y) + \sin(x + y).$

The complementary function is $\phi(y + x) + \psi(y - 2x)$. The part of the particular integral arising from $\sin(x - y)$, in which

$$a = 1, b = -1, \text{ is } -\frac{1}{2}\int \int \sin t\,dt^2 = \frac{1}{2}\sin(x - y). \text{ That aris-}$$

---

* The symbolic form of this theorem is

$$\frac{1}{D - mD'}F(x, y) = e^{mxD'}\int e^{-mxD'}F(x, y)dx$$

corresponding to equation (13), Art. 16. The symbol $e^{mxD'}$ here indicates the addition of mx to y in the operand. Accordingly, using the expanded form of the symbol,

$$e^{mxD'}F(y) = \left(1 + mx\frac{d}{dy} + \frac{m^2x^2}{2!}\frac{d^2}{dy^2} + \ldots\right)F(y) = F(y + mx),$$

the symbolic expression of Taylor's Theorem.

ing from $\sin (x + y)$ which is of the form of a term in the com-plementary function is $-\dfrac{1}{3} \dfrac{1}{D - D'} \cos (x + y)$, which by equa-tion (10) is $-\frac{1}{3} x \cos (x + y)$. Hence the general integral of the given equation is

$$z = \phi(y + x) + \psi(y - 2x) + \tfrac{1}{2} \sin (x - y) - \tfrac{1}{3} x \cos (x + y).$$

If in the equation $f(D, D')z = 0$ the symbol $f(D, D')$, though not homogeneous with respect to D and D', can be separated into factors, the integral is still the sum of those corresponding to the several symbolic factors. The integral of a factor of the first degree is found by Lagrange's process; thus that of

$$(D - mD' - a)z = 0 \tag{15}$$

is

$$z = e^{ax}\phi(y + mx). \tag{16}$$

But in the general case it is not possible to express the solution in a form involving arbitrary functions. Let us, how-ever, assume

$$z = ce^{hx + ky}, \tag{17}$$

where c, h, and k are constants. Since $De^{hx + ky} = he^{hx + ky}$ and $D'e^{hx + ky} = ke^{hx + ky}$, substitution in $f'(D, D')z = 0$ gives $cf(h, k)e^{hx + ky} = 0$. Hence we have a solution of the form (17) whenever h and k satisfy the relation

$$f(h, k) = 0, \tag{18}$$

c being altogether arbitrary. It is obvious that we may also write the more general solution

$$z = \Sigma c e^{hx + F(h)y}, \tag{19}$$

where $k = F(h)$ is derived from equation (18), and c and h admit of an infinite variety of arbitrary values.

Again, since the difference of any two terms of the form $e^{hx + F(h)y}$ with different values of h is included in expression (19), we infer that the derivative of this expression with respect to h is also an integral, and in like manner the second and higher derivatives are integrals.

For example, if the equation is

$$\frac{d^2z}{dx^2} - \frac{dz}{dy} = 0,$$

for which equation (18) is $h^2 - k = 0$, we have classes of integrals of the forms

$$e^{hx + h^2 y}, \quad e^{hx + h^2 y}(x + 2hy),$$

$$e^{hx + h^2 y}[(x + 2hy)^2 + 2y)], \quad e^{hx + h^2 y}[(x + 2hy)^3 + 6y(x + 2hy)].$$

. . . .

In particular, putting $h = 0$ we obtain the algebraic integrals $c_1 x$, $c_2(x^2 + 2y)$, $c_3(x^3 + 6xy)$, etc.

The solution of a linear partial differential equation with variable coefficients may sometimes be effected by a change of the independent variables as illustrated in some of the examples below.

Prob. 99. Show that if m_1 is a triple root the corresponding terms of the integral are $x^2\phi(y + m_1 x) + x\psi(y + m_1 x) + \chi(y + m_1 x)$.

Prob. 100. Solve $2\dfrac{\partial^2 z}{\partial x^2} - 3\dfrac{\partial^2 z}{\partial x \partial y} - 2\dfrac{\partial^2 z}{\partial y^2} = 0$.

Prob. 101. Solve $\dfrac{\partial^3 z}{\partial x^2 \partial y} + 2\dfrac{\partial^3 z}{\partial x \partial y^2} + \dfrac{\partial^3 z}{\partial y^3} = \dfrac{1}{x^2}$.

　　(Ans. $z = \phi(x) + \psi(x + y) + x\chi(x + y) - y \log x$.)

Prob. 102. Solve $(D^2 + 5DD' + 6D'^2)z = (y - 2x)^{-1}$.

　　(Ans. $z = \phi(y - 2x) + \psi(y - 3x) + x \log(y - 2x)$.)

Prob. 103. Solve $\dfrac{\partial^2 z}{\partial x^2} - \dfrac{\partial^2 z}{\partial x \partial y} + \dfrac{\partial z}{\partial y} - z = 0$.

Prob. 104. Show that for an equation of the form (15) the solution given by equation (19) is equivalent to equation (16).

Prob. 105. Solve $\dfrac{1}{x^2}\dfrac{\partial^2 z}{\partial x^2} - \dfrac{1}{x^3}\dfrac{\partial z}{\partial x} = \dfrac{1}{y^2}\dfrac{\partial^2 z}{\partial y^2} - \dfrac{1}{y^3}\dfrac{\partial z}{\partial y}$ by transposition to the independent variables x^2 and y^2.

Prob. 106. Solve $x^2\dfrac{\partial^2 z}{\partial x^2} + 2xy\dfrac{\partial^2 z}{\partial x \partial y} + y^2\dfrac{\partial^2 z}{\partial y^2} = 0$.

CHAPTER VIII.

GRASSMANN'S SPACE ANALYSIS.

By EDWARD W. HYDE,
Professor of Mathematics in the University of Cincinnati.

ART. 1. EXPLANATIONS AND DEFINITIONS.

The algebra with which the student is already familiar deals
directly with only one quality of the various geometric and
mechanical entities, such as lines, forces, etc., namely, with
their magnitude. Such questions as How much? How far?
How long? are answered by an algebraic operation or series of
operations. Questions of direction and position are dealt with
indirectly by means of systems of coordinates of various kinds.
In this chapter an algebra* will be developed which deals
directly with the three qualities of geometric and mechanical
quantities, viz., magnitude, position, and direction. A geometric quantity may possess one, two, or all three of these properties simultaneously; thus a straight line of given length has
all three, while a point has only one.

The geometric quantities with which we are to be concerned
are the point, the straight line, the plane, the vector, and the
plane-vector.

When the word "line" is used by itself, a "straight line"
will be always intended. A portion of a given straight line of
definite length will be called a "sect"; though when the length

---

* The algebra of this chapter is a particular case of the very general and
comprehensive theory developed by Hermann Grassmann, and published by
him in 1844 under the title "Die lineale Ausdehnungslehre, ein neuer Zweig
der Mathematik." He published also a second treatise on the subject in 1862.

of the sect is a matter of indifference, the word line will frequently be used instead. Similarly, a definite area of a given plane will be called a " plane-sect."

If a point recede to infinity, it has no longer any significance as regards position, but still indicates a direction, since all lines passing through finite points, and also through this point at infinity, are parallel. Similarly, a line wholly at infinity fixes a plane direction, that is, all planes passing through finite points, and also through this line at infinity, are parallel. Thus a point and line at infinity are respectively equivalent to a line direction and a plane direction.

A quantity. possessing magnitude only will be termed a " scalar " quantity. Such are the ordinary subjects of algebraic analysis, a, x, $\sin \theta$, $\log z$, etc., and they may evidently be intrinsically either positive or negative.

The letter T prefixed to a letter denoting some geometric quantity will be used to designate its absolute or numerical magnitude, always positive. Thus, if L be a sect, and P a plane-sect, then TL is the length of L, and TP is the area of P. That portion of a geometric quantity whose magnitude is unity will be called its " unit," and will be indicated by prefixing the letter U; thus UL = unit of L = sect one unit long on line L.* Hence we have $TL \cdot UL = L$.

ART. 2. SUM AND DIFFERENCE OF TWO POINTS.

In geometric addition and subtraction we shall use the ordinary symbols $+$, $-$, $=$, but with modified significance, as will appear in the development of the subject.

Every mathematical, or other, theory rests on certain fundamental assumptions, the justification for these assumptions

---

* The word "scalar" and the use of the letters T and U, as above, were introduced by Hamilton in his Quaternions. T stands for tensor, i.e., stretcher, and TL is the factor that stretches UL into L. The notation $| L |$ for absolute magnitude is not used, because the sign $|$ has been appropriated by Grassmann to another use.

lying in the harmony and reasonableness of the resulting theory, and its accordance with the ascertained facts of nature.

Our first assumption, then, will be that the associative and commutative laws hold for geometric addition and subtraction, that is, whatever A, B, C may represent, we have

$$A + B + C = (A + B) + C = A + (B + C)$$
$$= A + C + B = (A + C) + B.$$

We shall also assume that we always have $A - A = 0$, and that the same quantity may be added to or subtracted from both sides of an equation without affecting the equality.

Now let p_1, p_2 be two points, and consider the equation

$$p_2 + p_1 - p_1 = p_2 + (p_1 - p_1) = p_2. \tag{1}$$

In this form we have an identity. Write it, however, in the form

$$p_2 - p_1 + p_1 = (p_2 - p_1) + p_1 = p_2, \tag{2}$$

and it appears that $p_2 - p_1$ is an operator that changes p_1 into p_2 by being added to it. Conceive this change of p_1 into p_2 to take place along the straight line through p_1 and p_2; then the operation is that of moving a point through a definite length or distance in a definite direction, namely, from p_1 to p_2. This operator has been called by Hamilton "a vector," * that is, a carrier, because it carries p_1 rectilinearly to p_2. Grassmann gives to it the name Strecke, and some writers now use the word "stroke" in the same sense.

Again, $p_2 - p_1$ is the difference of two points, and the only difference that can exist between them is that of position, i.e. a certain distance in a certain direction.

Hence we may regard $p_2 - p_1$ as a directed length, and also as the operator which moves p_1 over this length in this direction. Writing $p_2 - p_1 = \epsilon$, equation (2) becomes

$$p_1 + \epsilon = p_2. \tag{3}$$

* See the first of Hamilton's Lectures on Quaternions, where a very full discussion of equation (2) will be found. Also Grassmann (1862), Art. 227.

Thus the sum of a point and a vector is a point distant from the first by the length of the vector and in its direction.

Since $p_2 - p_1 = -(p_1 - p_2)$, it appears that the negative of a vector is a vector of the same length in the opposite direction.

If $p_2 - p_1 = 0$, or $p_2 = p_1$, p_2 must coincide with p_1 because there is now no difference between the two points.

The question arises as to what, if any, effect the operator $p_2 - p_1$ should have on any other point p_3, that is, what is the value of the expression $p_2 - p_1 + p_3$?

We will assume that it is some point p_4, so that we have

$$p_2 - p_1 + p_3 = p_4,$$

or $$p_2 - p_1 = p_4 - p_3. \tag{4}$$

This implies that the transference from p_3 to p_4 is the same in amount and direction as that from p_1 to p_2, that is, that p_1, p_2, p_4, p_3 are the four corners of a parallelogram taken in order. Thus equal vectors have the same length and direction, and, conversely, vectors having the same length and direction are equal.

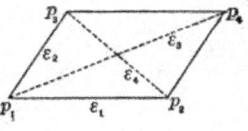

Note that parallel vectors of equal length are not necessarily equal, for their directions may be opposite.

Equation (4) may also be written

$$p_2 + p_3 = p_4 + p_1, \tag{5}$$

so that, whatever meaning may be assigned to the sum of two points, if we are to be consistent with assumptions already made, we must have the sum of either pair of opposite corner-points of a parallelogram equal to the sum of the other pair. The sum cannot therefore depend on the actual distances apart of the points forming the pairs, for the ratio of these two distances may be made as large or as small as we please.

If n be a scalar quantity, $n\epsilon$ will denote that the operation ϵ is to be performed n times on a point to which $n\epsilon$ is added, that is, the point will be moved n times the length of ϵ; hence

$n\epsilon$ is a vector n times as long as ϵ, and having the same or the opposite direction according to the sign of n.

In the figure above, let

$$p_2 - p_1 = \epsilon_1, \quad p_3 - p_1 = \epsilon_2, \quad p_4 - p_1 = \epsilon_3, \quad p_3 - p_2 = \epsilon_4.$$

Then

$$\epsilon_1 + \epsilon_2 = p_2 - p_1 + p_3 - p_1 = p_2 - p_1 + p_4 - p_2 = p_4 - p_1 = \epsilon_3, \quad (5)$$

since, by Art. 4, $p_3 - p_1 = p_4 - p_2$.

Also, $\qquad\qquad \epsilon_2 - \epsilon_1 = p_3 - p_2 = \epsilon_4.$ $\qquad\qquad\qquad$ (6)

Hence, if two vectors are drawn outwards from a point, and the parallelogram of which these are two adjacent sides is completed, then the two diagonals of this parallelogram will represent respectively the sum and difference of the two vectors, the sum being that diagonal which passes through the origin of the two vectors, and the difference that which passes through their extremities.[*]

Again, $p_2 - p_1 + p_3 - p_2 + p_1 - p_3 = 0 = \epsilon_1 + \epsilon_4 + (-\epsilon_2)$; hence the sum of three vectors represented by the sides of a triangle taken around in order is zero.

Similarly, if $p_1, p_2, \ldots p_n$ be any n points whatever taken as corners of a closed polygon, we shall have

$$(p_2 - p_1) + (p_3 - p_2) + (p_4 - p_3) + \cdots + (p_n - p_{n-1}) + (p_1 - p_n) = 0;$$

that is, the sum of vectors represented by the sides taken in order about the polygon is zero. By "taken in order" is not meant that any particular order of the points must be observed in forming the polygon, which is evidently unnecessary, but simply that, when the polygon is formed, the vectors will be the operators that will move a point from the starting position along the successive sides back to this position again, so that the final distance from the starting-point will be nothing.

ART. 3.   SUM OF TWO WEIGHTED POINTS.[†]

Consider the sum $m_1 p_1 + m_2 p_2$, in which m_1 and m_2 are scalars, that is, numbers, positive or negative, and p_1, p_2 are points.

---

[*] Grassmann (1844), § 15.

[†] Grassmann (1844), § 95, and (1862), Art. 227.

The scalars m_1 and m_2 will be regarded as values or weights assigned to the points p_1 and p_2. When any weight is of unit value the figure 1 will be omitted, so that p means $1p$, and is called a unit point. Occasionally, however, a letter may be used to denote a point whose weight is not unity.

To assist his thinking, the reader may consider the weights initially as like or unlike parallel forces acting at the points.

In order to arrive at a meaning for the above expression we shall make two reasonable assumptions, which will prove to be consistent with those already made, viz., first, that the sum is a point, and second, that its weight is the sum of the weights of the two given points. Denoting this sum-point by $\bar{p}$, we write

$$m_1 p_1 + m_2 p_2 = (m_1 + m_2)\bar{p}. \tag{7}$$

Transposing, we have $m_1(p_1 - \bar{p}) = m_2(\bar{p} - p_2)$, or

$$\frac{p_1 - \bar{p}}{m_2} = \frac{\bar{p} - p_2}{m_1}. \tag{8}$$

Both members of (8) are vectors, and, being equal, they must, by Art. 4, be parallel. This requires that $\bar{p}$ shall be collinear with p_1 and p_2. Also, since $p_1 - \bar{p}$ and $\bar{p} - p_2$ are vectors whose lengths are respectively the distances from p_1 to $\bar{p}$ and from $\bar{p}$ to p_2, it follows that these distances are in the ratio of m_2 to m_1. Hence, $\bar{p}$ is a point on the line $p_1 p_2$ whose distances from p_1 and p_2 are inversely proportional to the weights of these points. We shall call $\bar{p}$ the mean point of the two weighted points. If m_1 and m_2 are both positive, (8) shows that $\bar{p}$ must lie between p_1 and p_2; but if one, say m_2, is negative, let $m_2 = -m_2'$. Thus

$$m_1(p_1 - \bar{p}) = m_2'(p_2 - \bar{p}), \tag{9}$$

and $\bar{p}$ is on the same side of each point, that is, its direction from each point is the same. Also, since its distances from the two points are inversely as their weights, $\bar{p}$ must be nearest the point whose weight is greatest.

Case when $m_1 + m_2 = 0$, or $m_2 = -m$.*—With this condition equations (7) and (8) become

$$m_1 p_1 + m_2 p_2 = m_1 (p_1 - p_2) = 0 \cdot \bar{p}, \qquad (10)$$

and
$$\bar{p} - p_1 = \bar{p} - p_2. \qquad (11)$$

Thus $\bar{p}$ is in the same direction from each point, that is, not between them, and yet is equidistant from them. This requires either that the two points shall coincide, that is, $p_2 = p_1$, which evidently satisfies (10) and (11); or else, p_1 and p_2 being different points, that $\bar{p}$ shall be at an infinite distance. Thus the sum is in this case a point of zero weight at infinity.† Eq. (10) shows that a zero point at infinity is equivalent to a vector, or directed quantity, as stated in Art. 1. It has been shown in Art. 2 that $p_2 = p_1$ is the condition that p_1 and p_2 coincide; let us consider the equality of weighted points in general, say $m_1 p_1 = m_2 p_2$. Hence, by (7), there is found $m_1 p_1 - m_2 p_2 = (m_1 - m_2)\bar{p} = 0$; hence, since $\bar{p}$ cannot be zero, $m_1 - m_2 = 0$, or $m_1 = m_2$; and therefore $m_1(p_1 - p_2) = 0$, or, since $m_1 \gtrless 0$, $p_1 - p_2 = 0$, that is, $p_1 = p_2$. Therefore, if any two points are equal, their weights must be the same and their positions identical, that is, they are the same point.

Exercise 1.—To find the sum and difference of the two weighted points $3p_1$ and p_2:

$$3p_1 + p_2 = 4\bar{p}, \qquad 3p_1 - p_2 = 2\bar{p}',$$

and the mean points are as shown in the figure. The reciprocals of the distances of $\bar{p}$, p_1, and $\bar{p}'$ from p_2, viz., $\frac{1}{2}$, $\frac{1}{4}$, $\frac{1}{6}$, are in arithmetical progression, hence the points form a harmonic range.

Exercise 2.—Given a circular disk with a circular disk of

half its radius removed, as in the figure; to find the centroid of the remaining portion.

Take p_1 at center of large circle, p_2 at center of small circle, and p_3 at the point of contact; then $p_3 = \frac{1}{2}(p_1 + p_2)$. The areas of the two circles are as $1 : 4$; call them 1 and 4. Then it is as if there were a weight 4 at p_1, and a weight -1 at p_3; hence
$$\bar{p} = [4p_1 - \tfrac{1}{2}(p_1 + p_2)] \div 3 = (7p_1 - p_2) \div 6.$$

Prob. 1. Show that $p_1, p_2, m_1p_1 + m_2p_2$, and $m_1p_1 - m_2p_2$ are four points forming a harmonic range.

Prob. 2. An inscribed right-angled triangle is cut from a circular disk; show that the centroid of the remainder of the disk is at the point
$$\frac{(3\pi - 2\sin 2\alpha)\, p_1 - p_2 \sin 2\alpha}{3(\pi - \sin 2\alpha)},$$
if p_1 is the center of the circle, p_2 the opposite vertex of the triangle, and α one of its angles.

ART. 4.   SUM OF ANY NUMBER OF POINTS.

As in the last article we assume the sum to be a point whose weight is equal to the sum of the weights of the given points; thus,
$$\overset{n}{\underset{1}{\Sigma}}mp = \bar{p}\overset{n}{\underset{1}{\Sigma}}m. \tag{12}$$

Let e be some fixed point, and subtract $e\overset{n}{\underset{1}{\Sigma}}m$ from both sides of (12); thus we have
$$\overset{n}{\underset{1}{\Sigma}}m(p - e) = (\bar{p} - e)\overset{n}{\underset{1}{\Sigma}}m, \tag{13}$$
an equation which gives a simple construction for $\bar{p}$.

If $\overset{n}{\underset{1}{\Sigma}}m = 0$, then $m_1 = -\overset{n}{\underset{2}{\Sigma}}m$, and
$$\overset{n}{\underset{1}{\Sigma}}mp = m_1p_1 + \overset{n}{\underset{2}{\Sigma}}mp = m_1\left(p_1 - \frac{\overset{n}{\underset{2}{\Sigma}}mp}{\overset{n}{\underset{2}{\Sigma}}m}\right), \tag{14}$$

so that the sum becomes the difference of two unit points, or a vector whose direction is parallel to the line joining p_1 with the mean of all the other points of the system, and whose length is m_1 times the distance between these points. Since any point of the system may be designated as p_1, it follows that the line joining any point of the system to the mean of all the others is parallel to any other such line. If $\overset{n}{\underset{1}{\Sigma}} mp = 0$, equation (14) shows that p_1 is the mean of all the other points of the system, and, since any one of the points may be taken as p_1, any point of the system is the mean of all the others.

Let $n = 3$ in (12) and (13); then

$$m_1 p_1 + m_2 p_2 + m_3 p_3 = (m_1 + m_2 + m_3)\bar{p}, \qquad (15)$$

$$m_1(p_1 - e) + m_2(p_2 - e) + m_3(p_3 - e) = (m_1 + m_2 + m_3)(\bar{p} - e), \quad (16)$$

and $\bar{p}$ is on the line joining the point $m_1 p_1 + m_2 p_2$ with p_3, and therefore inside the triangle $p_1 p_2 p_3$, if the m's are all positive. If m_3 be negative and numerically less than $m_1 + m_2$, then $\bar{p}$ will have passed across the line $p_1 p_2$ to the outside of the triangle. If m_1 and m_2 are negative and their sum numerically less than m_3, then $\bar{p}$ will have passed outside the triangle through p_3, i.e., it will have crossed $p_2 p_3$ and $p_3 p_1$. The point e must evidently always be in the plane $p_1 p_2 p_3$.

As a numerical example let $m_1 = 3$, $m_2 = 4$, $m_3 = -5$, so that (16) becomes

$$\bar{p} - e = \tfrac{3}{2}(p_1 - e) + 2(p_2 - e) - \tfrac{5}{2}(p_3 - e).$$

Now, since e may be any point whatever, put $e = p_3$; then $\bar{p} - p_3 = \tfrac{3}{2}(p_1 - p_3) + 2(p_2 - p_3)$, and the construction is shown in the figure. $p_4 - p_3 = \tfrac{3}{2}(p_1 - p_3)$, and $\bar{p} - p_4 = 2(p_2 - p_3)$.

As another example take $\bar{p} = 4p_1 + 5p_2 - 2p_3 - 6p_4$, or, by (13), making $e = p_5$,

$$\bar{p} - p_5 = 4(p_1 - p_5) + 5(p_2 - p_5) - 2(p_3 - p_5)$$
$$= p_2 - p_4 + p_6 - p_5 + \bar{p} - p_6.$$

When any number of geometric quantities can be connected with each other by an equation of the form $\Sigma mp = 0$, in which the m's are finite and different from zero, then they are said to be mutually dependent, that is, any one can be expressed in terms of the others.   If no such relation can exist between the

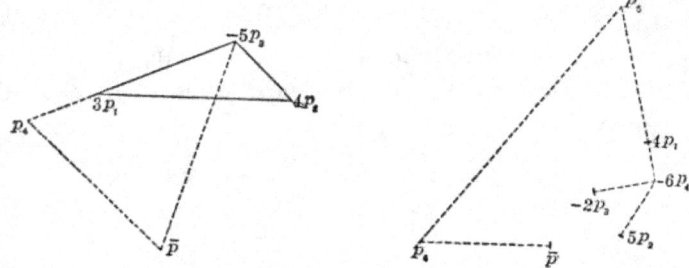

quantities, they are independent.   We obtain from what has preceded the following conditions:

That two points shall concide,

$$m_1 p_1 + m_2 p_2 = 0. \tag{17}$$

That three points shall be collinear,

$$m_1 p_1 + m_2 p_2 + m_3 p_3 = 0. \tag{18}$$

That four points shall be coplanar,

$$m_1 p_1 + m_2 p_2 + m_3 p_3 + m_4 p_4 = 0. \tag{19}$$

It follows that three non-collinear points cannot be connected by an equation like (18) unless each coefficient is separately zero.   Similarly four non-coplanar points cannot be connected by an equation like (19) unless each coefficient is separately zero.

The significance of these statements will be presently illustrated.

The following are corresponding equations of condition for vectors:

That two vectors shall be parallel,

$$n_1 \epsilon_1 + n_2 \epsilon_2 = 0. \tag{20}$$

That three vectors shall be parallel to one plane,

$$n_1\epsilon_1 + n_2\epsilon_2 + n_3\epsilon_3 = 0. \qquad (21)$$

These conditions follow from the results of Art. 2, or from equations (17) and (18) by regarding the ϵ's as points at infinity. If in addition to (21) we have

$$n_1 + n_2 + n_3 = 0, \qquad (22)$$

the extremities of the three vectors, if radiating from a point, will be collinear: for, let $e_0 \ldots e_3$ be four points so taken that $e_1 - e_0 = \epsilon_1, \; e_2 - e_0 = \epsilon_2, \; e_3 - e_0 = \epsilon_3$; then (21) becomes

$$n_1(e_1 - e_0) + n_2(e_2 - e_0) + n_3(e_3 - e_0) = 0,$$

or by (22)

$$n_1 e_1 + n_2 e_2 + n_3 e_3 = 0,$$

which by (18) requires e_1, e_2, e_3 to be collinear.

It may be shown similarly that

$$\overset{4}{\underset{1}{\Sigma}} n\epsilon = \overset{4}{\underset{1}{\Sigma}} n = 0 \qquad (23)$$

are the conditions that four vectors radiating from a point shall have their extremities coplanar.

Exercise 3.—Given a triangle $e_0 e_1 e_2$ and a point p in its plane; pe_0 cuts $e_1 e_2$ in q_0, pe_1 cuts $e_2 e_0$ in q_1, pe_2 cuts $e_0 e_1$ in q_2, $q_1 q_2$ cuts $e_1 e_2$ in p_0, $q_2 q_0$ cuts $e_2 e_0$ in p_1, and $q_0 q_1$ cuts $e_0 e_1$ in p_2: to show that p_0, p_1, and p_2 are collinear.

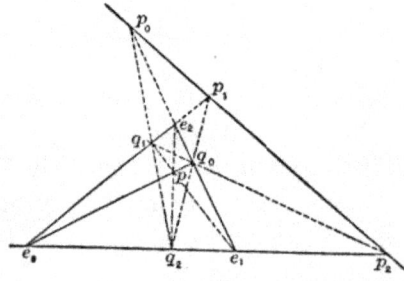

Let $p = n_0 e_0 + n_1 e_1 + n_2 e_2$; then q_0, q_1, q_2 coincide respectively with $n_1 e_1 + n_2 e_2$, $n_2 e_2 + n_0 e_0$, and $n_0 e_0 + n_1 e_1$ because p lies on the line joining e_0 with q_0, etc. Hence, if x_0, x_1, y_0, y_1 are scalars,

$$p_2 = x_0 e_0 + x_1 e_1 = y_0(n_1 e_1 + n_2 e_2) + y_1(n_2 e_2 + n_0 e_0);$$

hence $(x_0 - y_1 n_0)e_0 + (x_1 - y_0 n_1)e_1 - n_2(y_0 + y_1)e_2 = 0.$

Now the e's are not collinear, and yet are connected by a

relation of the form of equation (18); hence, as was there shown, each coefficient must be zero; accordingly

$$x_0 - y_1 n_0 = x_1 - y_0 n_1 = y_0 + y_1 = 0,$$

whence we find $\qquad x_0 : x_1 = n_0 : - n_1.$

hence $\qquad (n_0 - n_1)p_2 = n_0 e_0 - n_1 e_1,$ and similarly

$$(n_1 - n_2)p_0 = n_1 e_2 - n_2 e_2, \quad (n_2 - n_0)p_1 = n_2 e_2 - n_0 e_0.$$

Adding, we have

$$(n_1 - n_2)p_0 + (n_2 - n_0)p_1 + (n_0 - n_1)p_2 = 0;$$

therefore, by (18), p_0, p_1, p_2 are collinear.

Exercise 4.—Let $p = \overset{2}{\underset{0}{\Sigma}} ne \div \overset{2}{\underset{0}{\Sigma}} n$ be any point in the plane of the triangle $e_0 e_1 e_2$: show that lines through the middle points of the sides $e_1 e_2$, $e_2 e_0$, and $e_0 e_1$ of the triangle parallel to $e_0 p$, $e_1 p$, and $e_2 p$ meet in a point

$$p' = [(n_1 + n_2)e_0 + (n_2 + n_0)e_1 + (n_0 + n_1)e_2] \div 2 \overset{2}{\underset{0}{\Sigma}} n.$$

By the conditions the vector from the middle point of $e_1 e_2$ to p' is a multiple of the vector $e_0 - p$; hence

$$p' - \tfrac{1}{2}(e_1 + e_2) = x(e_0 - p) \quad \text{or}$$

$$p' = \tfrac{1}{2}(e_1 + e_2) + x(e_0 - p) = \tfrac{1}{2}(e_0 + e_1) + y(e_2 - p),$$

or, substituting value of p,

$$p' = \tfrac{1}{2}(e_1 + e_2) + x(e_0 - \Sigma ne \div \Sigma n) = \tfrac{1}{2}(e_0 + e_1) + y(e_2 - \Sigma ne \div \Sigma n).$$

hence $\qquad [(x - \tfrac{1}{2})\Sigma n + n_0(y - x)]e_0 + n_1(y - x)e_1$

$$+ [(\tfrac{1}{2} - y)\Sigma n + n_2(y - x)]e_2 = 0;$$

therefore, as in the previous exercise, each coefficient must be zero, whence $x = y = \tfrac{1}{2}$, and substituting we find p' as above. It follows also that the distances of p' from the middle points of the sides are the halves of the distances of p from the opposite vertices.

Prob. 3. Show that $\bar{e} = \tfrac{1}{3} \overset{2}{\underset{0}{\Sigma}} e$ is collinear with p and p' of Exercise 4. Also that, by properly choosing p, it follows that $\bar{e}$ is collinear with the common point of the perpendiculars from the vertices on the opposite sides, and the common point of the perpendiculars to the sides at their middle points.

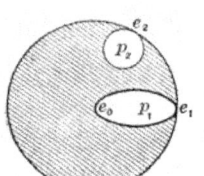

Prob. 4. Given two circles and an ellipse, as in the figure, with centers at e_0, p_2, and p_1. Radii of circles 4 and 1, axes of ellipse 2 and 4, small circle and ellipse touching large circle at e_2 and e_1 respectively, $e_0 e_1 e_2$ an equilateral triangle: show that the centroid of the remainder of the large circle, after the small areas are removed, will be at

$$\bar{p} = \tfrac{1}{13}(16e_0 - p_2 - 2p_1) = \tfrac{1}{52}(59e_0 - 4e_1 - 3e_2).$$

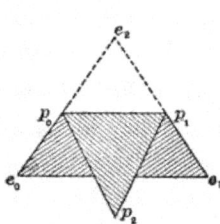

Prob. 5. If a sheet of tin in the shape of an isosceles triangle be folded over as in the figure, show that its centroid is given by

$$3\bar{p} = \tfrac{1}{27}[35(e_0 + e_1) + 11e_2].$$

Prob. 6. If a tetrahedron $e_0 e_1 e_2 e_3$ have a tetrahedron of $\tfrac{1}{8}$ of its volume cut off by a plane parallel to $e_0 e_1 e_2$, and one of $\tfrac{1}{64}$ of its volume cut off by a plane parallel to $e_1 e_2 e_3$, show that the centroid of the remaining solid is at

$$\bar{p} = \tfrac{1}{880}(227e_0 + 175e_3 + 239(e_1 + e_2)).$$

ART. 5. REFERENCE SYSTEMS.

Let p be any unit point, e_0, e_1, e_2 three fixed unit points, and w, x, y scalars; then, writing

$$p = we_0 + xe_1 + ye_2, \qquad (24)$$

we must have also, because p is a unit point,

$$w + x + y = 1, \qquad (25)$$

and p is the mean of the weighted points we_0, xe_1, ye_2. The point p may occupy any position whatever in the plane $e_0 e_1 e_2$; for it is on the line joining $we_0 + xe_1$ with e_2, and by varying y and $w + x$, $\dfrac{w}{x}$ remaining constant, p may be moved along this line from $-\infty$ to $+\infty$; while by varying the ratio $\dfrac{w}{x}$ the point $we_0 + xe_1$ may be moved from $-\infty$ to $+\infty$ along e_0e_1, and thus the first line will be rotated through 180 degrees, and p may thus be given any position whatever in the plane.

A system of unit points to which the positions of other points may be referred is called a reference system, and the triangle $e_0e_1e_2$ is a reference triangle. For reasons that will appear later, the double area of this triangle will be taken as the unit of measurement of area for a point system in two-dimensional space.

Similarly, in solid space, taking a fourth point e_3, we write

$$p = we_0 + xe_1 + ye_2 + ze_3, \tag{26}$$

which implies also $\quad w + x + y + z = 1;$ $\qquad$ (27)

and p may be shown as above to be capable of occupying any position whatever in space by properly assigning the values of w, x, y, z; so that $e_0, \ldots e_3$ form a reference system for points in three-dimensional space. The tetrahedron $e_0e_1e_2e_3$ is called the reference tetrahedron, and six times its volume will be taken as the unit of volume for a point system in three-dimensional space.

Eliminating w between (24) and (25), we have

$$p = e_0 + x(e_1 - e_0) + y(e_2 - e_0), \tag{28}$$

from which it may also be easily seen that p may be any point in the plane $e_0e_1e_2$. Writing $p - e_0 = \rho$, $e_1 - e_0 = \epsilon_1$, $e_2 - e_0 = \epsilon_2$, (28) becomes $\qquad\qquad \rho = x\epsilon_1 + y\epsilon_2, \qquad\qquad$ (29)

and ϵ_1, ϵ_2 form a plane reference system for vectors.

Similarly, from (26) and (27) we find

$$\rho = x\epsilon_1 + y\epsilon_2 + z\epsilon_3, \tag{30}$$

and ϵ_1, ϵ_2, ϵ_3 are a reference system for vectors in solid space, any vector whatever being expressible in terms of these three.

If, in equations (25) and (26), the reference vectors are of unit length and mutually perpendicular, we have unit, normal reference systems, and in this case ι_1, ι_2, ι_3 will generally be used instead of ϵ_1, ϵ_2, ϵ_3.

Exercise 5.—To change from one reference system to another, say from e_0, e_1, e_2 to e_0', e_1', e_2'.

The new reference points must be connected with the old ones by equations such as

$$e_0 = l_0 e_0' + l_1 e_1' + l_2 e_2', \quad e_1 = m_0 e_0' + m_1 e_1' + m_2 e_2',$$
$$e_2 = n_0 e_0' + n_1 e_1' + n_2 e_2'.$$

Then any point $p = x_0 e_0 + x_1 e_1 + x_2 e_2$ will be expressed in terms of the new reference points by substituting the values of e_0, etc., as given. If e_0', e_1', e_2' are given in terms of the old points, e_0, e_1, e_2 may be found by elimination. Thus, if $e_0' = \Sigma le$, $e_1' = \Sigma me$, $e_2' = \Sigma ne$, we have at once

$$\begin{vmatrix} l_0 & l_1 & l_2 \\ m_0 & m_1 & m_2 \\ n_0 & n_1 & n_2 \end{vmatrix} e_0 = \begin{vmatrix} e_0' & l_1 & l_2 \\ e_1' & m_1 & m_2 \\ e_2' & n_1 & n_2 \end{vmatrix},$$

with similar values for e_1 and e_2.

As a numerical example let the new reference triangle be formed by joining the middle points of the sides of the old one. Then $e_0' = \frac{1}{2}(e_1 + e_2)$, $e_1' = \frac{1}{2}(e_2 + e_0)$, $e_2' = \frac{1}{2}(e_0 + e_1)$; whence $e_0 = -e_0' + e_1' + e_2'$, $e_1 = e_0' - e_1' + e_2'$, $e_2 = e_0' + e_1' - e_2'$. Thus $p = x_0 e_0 + x_1 e_1 + x_2 e_2$

$$= (-x_0 + x_1 + x_2)e_0' + (x_0 - x_1 + x_2)e_1' + (x_0 + x_1 - x_2)e_2'.$$

Exercise 6.—Three points being given in terms of the reference points e_0, e_1, e_2, find the condition that must hold between their weights when they are collinear.

Let $p_0 = \overset{2}{\underset{0}{\Sigma}} le$, $p_1 = \overset{2}{\underset{0}{\Sigma}} me$, $p_2 = \overset{2}{\underset{0}{\Sigma}} ne$; then, k_0, k_1, k_2 being scalars, we must have for collinearity, by (18),

$$k_0 p_0 + k_1 p_1 + k_2 p_2 = 0,$$

that is, $k_0 \Sigma le + k \Sigma_1 me + k \Sigma ne = 0$,

whence $(k_0 l_0 + k_1 m_0 + k_2 n_0)e_0 + (k_0 l_1 + k_1 m_1 + k_2 n_1)e_1$

$$+ (k_0 l_2 + k_1 m_2 + k_2 n_2)e_2 = 0,$$

and, as e_0, e_1, e_2 are not collinear, the coefficients must be zero, by Art. 4; hence

$$k_0 l_0 + k_1 m_0 + k_2 n_0 = k_0 l_1 + k_1 m_1 + k_2 n_1 = k_0 l_2 + k_1 m_2 + k_2 n_2 = 0,$$

and, by elimination of the k's,

$$\begin{vmatrix} l_0 & m_0 & n_0 \\ l_1 & m_1 & n_1 \\ l_2 & m_2 & n_2 \end{vmatrix} = 0, \tag{31}$$

which is the required condition of collinearity.

Prob. 7. If $p = 3e_0 - e_1 - e_2$, $4e_0' = 3e_1 + e_2$, $4e_1' = 3e_2 + e_0$, $4e_2' = 3e_0 + e_1$, show that $7p = -19e_0' - 3e_1' + 29e_2'$.

Prob. 8. Find the condition that four points $\overset{3}{\underset{0}{\Sigma}}ke$, $\overset{3}{\underset{0}{\Sigma}}le$, $\overset{3}{\underset{0}{\Sigma}}me$, $\overset{3}{\underset{0}{\Sigma}}ne$ shall be coplanar. Ans. $[k_0, l_1, m_2, n_3] = 0$.

Prob. 9. If $p = we_0 + xe_1 + ye_2$, and there exist between the scalars w, x, y a linear relation such as $Aw + Bx + Cy = 0$, A, B, C being scalar constants, show that p will always lie on a straight line which cuts the reference lines in $Ae_1 - Be_0$, $Ae_2 - Ce_0$, and $Ce_1 - Be_2$. Consider the special cases when $A = B$, $B = C$, $C = A$, $A = B = C$, $A = 0$, $B = 0$, and $C = 0$.

Prob. 10. If $p = we_0 + xe_1 + ye_2 + ze_3$, and there exist also an equation $Aw + Bx + Cy + Dz = 0$, show that p will lie on a plane which cuts the edges of the reference tetrahedron in $\dfrac{e_1}{B} - \dfrac{e_0}{A}$, $\dfrac{e_2}{C} - \dfrac{e_0}{A}$, etc. Also, if a second relation between the variables, such as $A'w + B'x + C'y + D'z = 0$, be given, then p lies on a line which pierces the faces of the reference tetrahedron in

$$\begin{vmatrix} e_0 & e_1 & e_2 \\ A & B & C \\ A' & B' & C' \end{vmatrix}, \quad \begin{vmatrix} e_3 & e_0 & e_1 \\ D & A & B \\ D' & A' & B' \end{vmatrix}, \quad \text{etc.}$$

ART. 6.   NATURE OF GEOMETRIC MULTIPLICATION.[*]

The fundamental idea of geometric multiplication is, that a product of two or more factors is that which is determined by those factors.

Thus, two points determine a line passing through them, and also a length, viz., the shortest distance between them; hence $p_1 p_2 = L$ is the sect [†] drawn from p_1 to p_2, or generated by a point moving rectilinearly from p_1 to p_2.

The student should note carefully the difference between $p_1 p_2$ and $p_2 - p_1$; they have the same length and direction, but the sect $p_1 p_2$ is confined to the line through these two points, while the vector $p_2 - p_1$ is not.   The sect has position in addition to the direction and length possessed by the vector.

Again, in plane space, two sects determine a point, the intersection of the lines in which they lie, and also an area, as will appear later, so that $L_1 L_2 = p$, in which p is not in general a unit point.   In solid space, however, two lines do not, in general, meet, and hence cannot fix a point; but two sects, in this case, determine a tetrahedron of which they are opposite edges.

It appears, therefore, that a product may have different interpretations in spaces of different dimensions.   Hence we will consider separately products in plane space, or planimetric products, and those in solid space, or stereometric products.

Products of the kind here considered are termed "combinatory," because two or more factors combine to form a new quantity different from any one of them.   This is the fundamental difference between this algebra and the linear associative algebras of Peirce, of which quaternions are a special case.

Before discussing in detail the various products that may arise, we will give a table which will serve as a sort of bird's-eye view of the subject.

---

[*] Grassmann (1844), Chap. 2 ; (1862), Chap. 2.
[†] See Art. 1.

In this table and generally throughout the chapter we shall use p, p_1, p_2, etc., for points; ϵ, ϵ_1, ϵ_2, etc., for vectors; L, L_1, etc., for sects, or lines; η, η_1, etc., for plane-vectors; and P, P_1, etc., for plane-sects, or planes. Also p, p_1, etc., as used in this table will not generally be unit points.

The products are arranged in two columns, so as to bring out the geometric principle of duality.

PLANIMETRIC PRODUCTS.

$p_1 p_2 = L.$	$L_1 L_2 = p.$
$p_1 p_2 p_3 = $ area (scalar).	$L_1 L_2 L_3 = $ (area)^2(scalar).
$pL = $ area (scalar).	$Lp = $ area (scalar).
$p_1 . L_1 L_2 = L.$	$L_1 . p_1 p_2 = p.$
$p_1 p_2 . p_3 p_4 = p.$	$L_1 L_2 . L_3 L_4 = L.$
$p_1 p_2 . p_3 p_4 . p_5 p_6 = $(area)^2(scalar).	$L_1 L_2 . L_3 L_4 . L_5 L_6 = $(area)^4(scalar)

$$\epsilon_1 \epsilon_2 = \text{area (scalar)}.$$

STEREOMETRIC PRODUCTS.

$p_1 p_2 = L.$	$P_1 P_2 = L.$
$p_1 p_2 p_3 = P.$	$P_1 P_2 P_3 = p.$
$p_1 p_2 p_3 p_4 = $ volume (scalar).	$P_1 P_2 P_3 P_4 = $ (volume)^3 (scalar).
$pP = $ volume (scalar).	$Pp = $ volume (scalar).
$L_1 L_2 = $ volume (scalar).	$L_1 L_2 = $ volume (scalar).
$pL = Lp = P.$	$PL = LP = p.$
$p . P_1 P_2 = P.$	$P . p_1 p_2 = p.$
$p . P_1 P_2 P_3 = L.$	$P . p_1 p_2 p_3 = L.$
$L . p_1 p_2 p_3 = p.$	$L . P_1 P_2 P_3 = P.$
$\epsilon_1 \epsilon_2 = \eta.$	$\eta_1 \eta_2 = \epsilon.$
$\epsilon_1 \epsilon_2 \epsilon_3 = $ volume (scalar).	$\eta_1 \eta_2 \eta_3 = $ (volume)^2 (scalar).
$\epsilon_1 \epsilon_2 . \epsilon_3 \epsilon_4 = \epsilon.$	$\eta_1 \eta_2 . \eta_3 \eta_4 = \eta.$

Laws of Combinatory Multiplication. — All combinatory products are assumed to be subject to the distributive law expressed by the equation

$$A(B + C) = AB + AC.$$

The planimetric product of three points or of three lines, and the stereometric product of three points or planes, or of four points or planes, are subject to the associative law. That is,

In Plane Space :

$$p_1p_2p_3 = p_1p_2 \cdot p_3 = p_1 \cdot p_2p_3; \quad L_1L_2L_3 = L_1L_2 \cdot L_3 = L_1 \cdot L_2L_3.$$

In Solid Space :

$$p_1p_2p_3 = p_1 \cdot p_2p_3 = p_1p_2 \ p_3; \quad P_1P_2P_3 = P_1 \cdot P_2P_3 = P_1P_2 \cdot P_3.$$

$$p_1p_2p_3p_4 = p_1 \cdot p_2p_3p_4 = p_1p_2 \cdot p_3p_4;$$
$$P_1P_2P_3P_4 = P_1 \cdot P_2P_3P_4 = P_1P_2 \cdot P_3P_4.$$

The commutative law of scalar algebra does not, in general, hold. Instead of this, in the products just given as being associative, a law prevails which may be expressed by the equation

$$AB = - BA,$$

from which it follows that the interchange of any two single factors of those products changes the sign of the product.[*]

Since vectors are equivalent to points at ∞, the associative law holds for $\epsilon_1\epsilon_2\epsilon_3$ and $\eta_1\eta_2\eta_3$.

ART. 7.   PLANIMETRIC PRODUCTS.

Product of Two Points.[†]—This has been fully defined in Art. 6, and it is evident from its nature as there given that

$$p_1p_2 = -p_2p_1. \tag{32}$$

If $p_2 = p_1$, this becomes $p_1p_1 = 0$, which must evidently be true, since the sect is now of no length.

Also, $$p_1(p_2 - p_1) = p_1p_2 - p_1p_1 = p_1p_2. \tag{33}$$

---

[*] Grassmann (1862), Chap. 3.   [†] Grassmann (1862). Arts. 245, 246, 247.

But $p_2 - p_1$ is a vector, say, ϵ; hence

$$p_1\epsilon = p_1p_2;\tag{34}$$

or the product of a point and a vector is a sect having the direction and magnitude of the vector; or, again, multiplying a vector by a point fixes its position by making it pass through the point.

To find under what conditions pp' will be equal to p_1p_2. Take any other point p_3 in the plane space under consideration, and write $p = x_1p_1 + x_2p_2 + x_3p_3$, $p' = y_1p_1 + y_2p_2 + y_3p_3$, with the conditions for unit points $\Sigma x = \Sigma y = 0$.

Then $$pp' = \begin{vmatrix} x_1 & x_2 \\ y_1 & y_2 \end{vmatrix} p_1p_2 + \begin{vmatrix} x_2 & x_3 \\ y_2 & y_3 \end{vmatrix} p_2p_3 + \begin{vmatrix} x_3 & x_1 \\ y_3 & y_1 \end{vmatrix} p_3p_1.$$

If this is to reduce to p_1p_2, we must have the third condition $x_2y_3 - x_3y_2 = x_3y_1 - x_1y_3 = 0$, which requires that $x_3 = y_3 = 0$, unless the coefficient of p_1p_2 is to vanish also. Thus pp' must be in the same straight line with p_1p_2. If, moreover, in addition $x_1y_2 - x_2y_1 = 1$, we shall have $pp' = p_1p_2$. Hence pp' is equal to p_1p_2 when, and only when, the four points are collinear, and p' is distant from p by the same amount and in the same direction that p_2 is from p_1.

Product of Three Points.—By Art. 6 the product is what is determined by the three points. In solid space they would fix a plane, but, as we are now confined to plane space, this is not the case. The points evidently fix either a triangle or a parallelogram of twice its area, and the product $p_1p_2p_3$ will be taken as the area of this, or an equivalent, parallelogram.

This area is taken rather than that of the triangle, because it is what is generated by p_1p_2 as it is moved parallel to its initial position till it passes through p_3.

We have $p_1p_2p_3 = p_1 \cdot p_2p_3 = -p_1 \cdot p_3p_2 = -p_1p_3p_2$, so that if we go around the triangle in the opposite sense the sign is changed. As this product possesses only the properties of magnitude and sign it is scalar.

Write $p = \overset{3}{\underset{1}{\Sigma}}xp$, $p' = \overset{3}{\underset{1}{\Sigma}}yp$, $p'' = \overset{3}{\underset{1}{\Sigma}}zp$; then

$$pp'p'' = \begin{vmatrix} x_1 & x_2 & x_3 \\ y_1 & y_2 & y_3 \\ z_1 & z_2 & z_3 \end{vmatrix} p_1 p_2 p_3 ; \tag{35}$$

that is, any triple point product in plane space differs from any other only by a scalar factor.*

Finally, $\quad p_1 p_2 p_3 = p_1 (p_2 - p_1)(p_3 - p_1) = p_1 \epsilon \epsilon',$ $\qquad$ (36)

if $\epsilon = p_2 - p_1$ and $\epsilon' = p_3 - p_1$.

Product of Two Vectors.—Using the values of ϵ and ϵ' just given, we see that ϵ and ϵ' determine the same parallelogram that p_1, p_2, and p_3 do; hence the meaning of the product is the same in all respects in two-dimensional space.

We shall have $\epsilon \epsilon' = - \epsilon' \epsilon$, for

$$\epsilon \epsilon' = (p_2 - p_1)(p_3 - p_1) = - (p_3 - p_1)(p_2 - p_1) = - \epsilon' \epsilon ;$$

since we have shown that inverting the order changes the sign in a product of points. The result may be obtained also by regarding ϵ and ϵ' as points at infinity, or by consideration of a figure.

As we have seen that $\epsilon \epsilon'$ has, in plane space, precisely the same meaning as $p_1 p_2 p_3$ we may write

$$p_1 p_2 p_3 = p_1 \epsilon \epsilon' = \epsilon \epsilon'$$
$$= (p_2 - p_1)(p_3 - p_1) = p_2 p_3 + p_3 p_1 + p_1 p_2. \tag{37}$$

Thus the sum of three sects which form the sides of a triangle, all taken in the same sense as looked at from outside the triangle, is equal to the area of the triangle.

Product of Two Sects.—Any two sects in plane space,

L_1, L_2, determine a point, the intersection of the lines in which they lie, and an area, that of a parallelogram as in the figure. Let p_0 be the intersection, and take p_1 and p_2 so that $L_1 = p_0 p_1$ and $L_2 = p_0 p_2$. The area

* Grassmann (1862), Art. 255.

determined by L_1 and L_2 is then the same that we have given as the value of $p_0 p_1 p_2$. We write therefore

$$L_1 L_2 = p_0 p_1 \cdot p_0 p_2 = p_0 p_1 p_2 \cdot p_0. \tag{38}$$

The third member of (38) is not to be regarded as derived from the second by ordinary transposition and reassociation of the points, for the associative law does not hold for the four points taken together, since $p_0 p_1 p_0 \cdot p_2 = 0$. The third member simply results from the definition of $L_1 L_2$.* It may be taken as a model form which will be found to apply to several other cases, for instance to (38) when points and lines are interchanged throughout. Thus, if $p_1 = L_0 L_1$ and $p_2 = L_0 L_2$ we have

$$p_1 p_2 = L_0 L_1 \cdot L_0 L_2 = L_0 L_1 L_2 \cdot L_0. \tag{39}$$

For take p_1' and p_2' so that $p_1 p_1' = L_1$ and $p_2 p_2' = L_2$; $p_1 p_2$ is evidently some multiple of L_0, say $n L_0$; hence

$$p_1 p_2 = n L_0 = \frac{1}{n^2}(p_1 p_2 \cdot p_1 p_1') \cdot (p_1 p_2 \cdot p_2 p_2')$$

$$= \frac{1}{n^2}(p_1 p_2 p_1' \cdot p_1) \cdot (p_1 p_2 p_2' \cdot p_2), \text{ by (38)},$$

$$= \frac{1}{n^2} \cdot p_1 p_2 p_1' \cdot p_1 p_2 p_2' \cdot p_1 p_2, \text{ because } p_1 p_2 p_1' \text{ and}$$

$$\quad p_1 p_2 p_1' \text{ are scalar},$$

$$= \frac{1}{n} \cdot (p_1 p_2 \cdot p_1 p_1' \cdot p_2 p_2') \cdot L_0, \text{ by (38)},$$

$$= L_0 L_1 L_2 \cdot L_0, \text{ which was to be proved}.$$

Product of Three Sects.—The method has just been indicated, but we may also proceed thus: Let the lines be L_0, L_1, L_2, and let p_0, p_1, p_2 be their common points. Take scalars n_0, n_1, n_2 so that $L_0 = n_0 p_1 p_2$, etc., then

$$L_0 L_1 L_2 = n_0 n_1 n_2 \cdot p_1 p_2 \cdot p_2 p_0 \cdot p_0 p_1 = - n_0 n_1 n_2 \cdot p_2 p_1 p_2 p_0 \cdot p_0 p_1$$

$$= - n_0 n_1 n_2 \cdot p_1 p_1 p_0 \cdot p_2 p_0 p_1 = n_0 n_1 n_2 (p_0 p_1 p_2)^2. \tag{40}$$

---

* Grassmann applies the terms "eingewandt" and "regressiv" to a product of this kind, the first term being used in the Ausdehnungslehre of 1844, and the second in that of 1862. See Chapter 3 of the first, and Chapter 3, Art. 94, of the second.

Product of a Point and Two Sects.—Let p be any point and let L_1 and L_2 be as in (38); then

$$pL_1L_2 = p \cdot p_0p_1 \cdot p_0p_2 = p \cdot p_0p_1p_2 \cdot p_0 = p_0p_1p_2 \cdot pp_0. \quad (41)$$

It has been here assumed that $pL_1L_2 = p \cdot L_1L_2$. The product is not associative, for $pL_1 \cdot L_2$ is the line L_2 times the scalar pL_1, a different meaning from that assigned in (41). As a rule, to avoid ambiguity, the grouping of such products will be indicated by dots.

Product of Two Parallel Sects.—Let them be $p_1\epsilon$ and $np_2\epsilon$; then, as in (38),

$$p_1\epsilon \cdot np_2\epsilon = n \cdot p_1\epsilon \cdot p_2\epsilon = n \cdot \epsilon p_1 \cdot \epsilon p_2 = n \cdot \epsilon p_1 p_2 \cdot \epsilon, \quad (42)$$

that is, a scalar times the common point at ∞.

Addition and Subtraction of Sects.—Let L_1 and L_2 be two sects, p_0 their common point, and p_1 and p_2 so taken that $L_1 = p_0p_1$, $L_2 = p_0p_2$; then

$$L_1 + L_2 = p_0p_1 + p_0p_2 = p_0(p_1 + p_2) = 2p_0\bar{p}, \quad (43)$$

$\bar{p}$ being the mean of p_1 and p_2; hence the sum is that diagonal of the parallelogram which passes through p_0. Also

$$L_1 - L_2 = p_0(p_1 - p_2), \quad (44)$$

so that the difference of the two passes also through p_0 and is parallel to the other diagonal of the parallelogram determined by L_1 and L_2.

If the two sects are parallel let them be $n_1p_1\epsilon$ and $n_2p_2\epsilon$; then

$$n_1p_1\epsilon + n_2p_2\epsilon = (n_1p_1 + n_2p_2)\epsilon = (n_1 + n_2)\bar{p}\epsilon_1, \quad (45)$$

so that the sum is a sect parallel to each of them, having a length equal to the sum of their lengths, and at distances from them inversely proportional to their lengths.

If $n_2 = -n_1$ the two sects are oppositely directed and of equal length, and the sum is

$$n_1(p_1\epsilon - p_2\epsilon) = n_1(p_1 - p_2)\epsilon, \quad (46)$$

which, being the product of two vectors, is a scalar area.

Consider next n sects $p_1\epsilon_1, p_2\epsilon_2, \ldots p_n\epsilon_n$, and let ϵ_0 be some arbitrarily chosen point; then

$$\overset{n}{\underset{1}{\Sigma}} p\epsilon \equiv \epsilon_0 \overset{n}{\underset{1}{\Sigma}}\epsilon - \epsilon_0 \overset{n}{\underset{1}{\Sigma}}\epsilon + \overset{n}{\underset{1}{\Sigma}}p\epsilon \equiv \epsilon_0 \overset{n}{\underset{1}{\Sigma}}\epsilon + \overset{n}{\underset{1}{\Sigma}}(p - \epsilon_0)\epsilon. \qquad (47)$$

The second term of the third member of this equation, being a sum of double vector products, that is, a sum of areas, is itself an area, and is equal to the product of any two non-parallel vectors of suitable lengths. Therefore, α and β being such vectors, write $\Sigma\epsilon = \alpha$ and $\Sigma(p - \epsilon_0) = \alpha\beta$. Hence (47) become

$$\Sigma p\epsilon = \epsilon_0\alpha + \alpha\beta = (\epsilon_0 - \beta)\alpha. \qquad (48)$$

Let q be some point on the line $\Sigma p\epsilon$; then

$$q\Sigma p\epsilon = 0 = q\epsilon_0\alpha + q\alpha\beta = q\epsilon_0\alpha + \alpha\beta,$$

by (37), hence $q\epsilon_0\alpha = -\alpha\beta = \beta\alpha.$

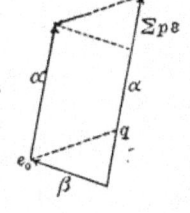

The figure presents the geometrical meaning of the equation, and hence it appears that $q\alpha (= \Sigma p\epsilon)$ is at a perpendicular distance from ϵ_0 of

$$\frac{\alpha\beta}{T\alpha} = \frac{\Sigma(p - \epsilon_0)\epsilon}{T\Sigma\epsilon}. \qquad (49)$$

It is easily seen that a sect possesses the exact geometrical properties of a force, namely, magnitude, direction, and position, and the discussion of the summation of sects which has just been given corresponds completely to the discussion of the resultant of a system of forces in a plane. In this algebra, then, the resultant of any system of forces is simply their sum, and this will be found hereafter to be equally true in three-dimensional space. The expression in (46) corresponds to a couple, as does also the $\Sigma(p - \epsilon_0)\epsilon$ of (47); and this equation proves the proposition that any system of forces in a plane is equivalent to a single force acting at an arbitrary point, ϵ_0, and a couple. Equation (49) gives the distance of the resultant from this arbitrary point.

Exercise 7.—To find x, y, z from the scalar equations

$$a_1x + b_1y + c_1z = d_1, \quad a_2x + b_2y + c_2z = d_2, \quad a_3x + b_3y + c_3z = d_3.$$

Multiply the equations by p_1, p_2, and p_3 respectively, and add; hence

$$x\sum_1^3 ap + y\sum_1^3 bp + z\sum_1^3 cp = \sum_1^3 dp.$$

Now Σap, Σbp, etc., are points: multiply the equation just written by $\Sigma ap . \Sigma bp$; thus

$$z\Sigma ap . \Sigma bp \Sigma cp = \Sigma ap . \Sigma bp . \Sigma dp,$$

because $\Sigma ap . \Sigma ap = 0$, etc.; therefore

$$z = \Sigma ap . \Sigma bp . \Sigma dp \div \Sigma ap . \Sigma bp \Sigma cp = [a_1, b_2, d_3] \div [a_1, b_2, c_3],$$

a very simple proof of the determinant solution. Of course x and y will be found by multiplying by the other pairs of points.

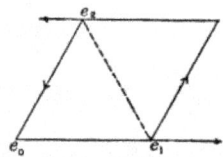

Exercise 8.—Forces are represented by given multiples of the sides of a parallelogram; determine their resultant.

Let the parallelogram be double the triangle $e_0 e_1 e_2$, and the forces

$$k_0 e_0 e_1 + k_1 e_1 (e_2 - e_0) + k_2 e_2 (e_0 - e_1) + k_3 e_2 e_0 = \Sigma pe$$
$$= (k_0 + k_1) e_0 e_1 + (k_1 + k_2) e_1 e_2 + (k_2 + k_3) e_2 e_0.$$

Multiply by $e_0 e_1$ to find where the resultant cuts this line; then

$$(k_1 + k_2) e_0 e_1 . e_1 e_2 + (k_2 + k_3) e_0 e_1 . e_2 e_0 = e_0 e_1 e_2 . [(k_1 + k_2) e_1 - (k_2 + k_3) e_0],$$

or $e_0 e_1$ cuts the resultant at the point

$$[(k_1 + k_2) e_1 - (k_2 + k_3) e_0] \div (k_1 - k_3).$$

Similarly the resultant cuts the other sides of the reference triangle at $[(k_2 + k_3) e_2 - (k_0 + k_1) e_1] \div (k_2 + k_3 - k_0 - k_1)$ and at $[(k_0 + k_1) e_0 - (k_1 + k_2) e_2] \div (k_0 - k_2)$.

Suppose $k_0 = k_1 = k_2 = k_3$; then each of the three points just found recedes to infinity; but in this case Σpe reduces to $2k_0 (e_0 e_1 + e_1 e_2 + e_2 e_0) = 2k_0 (e_1 - e_0)(e_2 - e_0)$, and the system is equivalent to a couple.

Prob. 11. Construct the resultant of Exercise 8 when $k_0 = 1$, $k_1 = 2$, $k_2 = 3$, $k_3 = 4$; when $k_0 = 1$, $k_1 = -2$, $k_2 = 3$, $k_3 = -4$; when $k_0 = 3$, $k_1 = k_2 = 2$, $k_2 = 1$; and when $k_1 = k_2 = 1$, $k_0 = k_3 = -2$.

Prob. 12. There are given n points $p_1 \ldots p_n$; to find a point e such that forces represented by the sects ep_1, ep_2, etc., shall be in equilibrium. (The equation of equilibrium is $\Sigma ep \equiv e\Sigma p \equiv \frac{1}{n} e\bar{p} = 0$. Hence e coincides with the mean point of the p's.)

Prob. 13. If a harmonic range e_1, p, e_2, p' be given, together with some point e_0 not collinear with these points, show that

$$e_0 e_1 p \cdot e_0 e_2 p' = - e_0 p e_2 \cdot e_0 p' e_1.$$

(Let $p = m_1 e_1 + m_2 e_2$ and $p' = m_1 e_1 - m_2 e_2$, as in Exercise 2 of Art. 3.)

Prob. 14. Show that the relation of Prob. 13 holds for any four points whatever taken respectively on the four lines $e_0 e_1$, $e_0 p$, $e_0 e_2$, $e_0 p'$. If the four points are all at the same distance from e_0, show that the areas $e_0 e_1 p$, etc., become proportional to the sines of the angles between $e_0 e_1$ and $e_0 p$, etc.

Art. 8. The Complement.*

Taking point reference systems, or unit normal vector reference systems, as in Art. 5, the product of the reference units taken in order being in any case unity, the complement of any reference unit is the product of all the others so taken that the unit times its complement is unity.

To find the complements of quantities other than reference units the following properties are assumed:

(a) The complement of a product is equal to the product of the complements of its factors.

(b) The complement of a sum is equal to the sum of the complements of the terms added together.

(c) The complement of a scalar quantity is the scalar itself.

Considering now the point system in plane space e_0, e_1, e_2 with the constant condition $e_0 e_1 e_2 = 1$, the sides of the reference triangle taken in order are the complements of the opposite vertices, and vice versâ.

The complement of a quantity is indicated by a vertical line, as $|p$, read, complement of p.

---

* See Ausdehnungslehre of 1862, Art. 89.

Thus $\qquad |e_0 = e_1 e_2, \qquad |e_1 e_2 = |(|e_0) = e_0,$

$\qquad\qquad\quad |e_1 = e_2 e_0, \qquad |e_2 e_0 = |(|e_1) = e_1,$

$\qquad\qquad\quad |e_2 = e_0 e_1, \qquad |e_0 e_1 = |(|e_2) = e_2.$

For $e_0 | e_0 = e_0 e_1 e_2 = 1$, which agrees with the definition ;

$|e_1 e_2 = |e_1 . |e_2 = e_2 e_0 . e_0 e_1 = - e_0 e_2 . e_0 e_1 = - e_0 e_2 e_1 . e_0 = e_0,$ by (a) and (38) ;

$|e_0 e_1 e_2 = |e_0 . |e_1 . |e_2 = e_1 e_2 . e_2 e_0 . e_0 e_1 = (e_0 e_1 e_2)^2 = 1 = e_0 e_1 e_2,$ which agrees with (c) ; $e_0 | e_1 = e_0 e_2 e_0 = 0 = e_0 | e_2 = e_1 | e_2.$

Next take any point $p_1 = \overset{2}{\underset{0}{\Sigma}} l e,$ and we have, by (b),

$$|p_1 = \overset{2}{\underset{0}{\Sigma}} l | e = l_0 e_1 e_2 + l_1 e_2 e_0 + l_2 e_0 e_1 = l_0 l_1 l_2 \left(\frac{e_1}{l_1} - \frac{e_0}{l_0}\right)\left(\frac{e_2}{l_2} - \frac{e_0}{l_0}\right) = L_1. \quad (50)$$

Thus the complement of a point is a line,* which may be easily constructed by the fourth member of (50), which expresses this line as the product of the points in which it cuts the sides $e_0 e_1$ and $e_0 e_2$ of the reference triangle. Comparing this equation with Ex. 3 in Art. 4, it appears that $|p_1$ above is related to the point $\overset{2}{\underset{0}{\Sigma}} \frac{e}{l}$ as the line $p_0 p_2$ of Ex. 3 is to the point $\Sigma n e.$ Hence $|p_1$ may be found by constructing this line corresponding to $\overset{2}{\underset{0}{\Sigma}} \frac{e}{l}$ as shown in the figure of Ex. 3, Art. 4.

Again, the line $|p_1$ may be shown to be the anti-polar of p with respect to an ellipse of such dimensions, and so placed upon $e_0 e_1 e_2$ that, with reference to it, each side of the reference triangle is the anti-polar of the opposite vertex.* From this it appears that complementary relations are polar reciprocal relations. Take any point $p_2 = \overset{2}{\underset{0}{\Sigma}} m e,$ and we have

$$p_1 | p_2 = (l_0 e_0 + l_1 e_1 + l_2 e_2)(m_0 e_1 e_2 + m_1 e_2 e_0 + m_2 e_0 e_1)$$

$$= \overset{2}{\underset{0}{\Sigma}} l m = \Sigma m e . \Sigma l | e = p_2 | p_1, \qquad (51)$$

* See Hyde's Directional Calculus, Arts. 41–43 and 121–123.

so that this product is commutative about the complement
sign, and scalar. This is true of all such products when the
quantities on each side of the complement sign are of the same
order in the reference units. Take for instance the product
$p_1 p_2 | p_3 p_4$. This is scalar because $| p_3 p_4$ is a point, so that the
whole quantity is equivalent to a triple-point product; and we
have $p_1 p_2 | p_3 p_4 = | p_3 p_4 \cdot p_1 p_2 = | (p_3 p_4 | p_1 p_2) = p_3 p_4 | p_1 p_2$, by (a) and
(c). If, however, such a quantity be taken as $p_1 p_2 \cdot | p_3$, it is neither
scalar nor commutative about the sign $|$; for, $| p_3$ being a line,
the product is that of two lines, that is, a point, and

$$p_1 p_2 \cdot | p_3 = - | p_3 \cdot p_1 p_2 = - | (p_3 \cdot | p_1 p_2). \qquad (52)$$

Such products as we have just been considering are called
by Grassmann " inner products," * and he regards the sign $|$
as a multiplication sign for this sort of product. Inasmuch,
however, as these products do not differ in nature from those
heretofore considered, it appears to the author to conduce to
simplicity not to introduce a nomenclature which implies a new
species of multiplication. For instance, $p | q$ will be treated as
the combinatory product of p into the complement of q, and
not as a different kind of product of p into q.

The term co-product may be applied to such expressions,
regarded as an abbreviation merely, after the analogy of cosine
for complement of the sine.

Consider next a unit normal vector system. By the definition we have

$$| \iota_1 = \iota_2, \ | \iota_2 = | (| \iota_1) = - \iota_1,$$

because $\quad \iota_1 | \iota_1 = \iota_1 \iota_2 = 1,$

$$\iota_2 | \iota_2 = \iota_2(- \iota_1) = - \iota_2 \iota_1 = \iota_1 \iota_2 = 1.$$

Also, $\quad \iota_1 | \iota_2 = - \iota_1 \iota_1 = 0 = \iota_2 | \iota_1.$

Next let

$$\epsilon_1 = m_1 \iota_1 + m_2 \iota_2 \quad \text{and} \quad \epsilon_2 = n_1 \iota_1 + n_2 \iota_2;$$

* Grassmann (1862), Chapter 4.

then, by (*b*) and (*c*),

$$|\epsilon_1 = m_1|\iota_1 + m_2|\iota_2 = m_1\iota_2 - m_2\iota_1. \tag{53}$$

By the figure it is evident that $|\epsilon_1$ is a vector of the same length as ϵ_1 and perpendicular to it, or, in other words, taking the complement of a vector in plane space rotates it positively through 90°.

The co-product $\epsilon_1|\epsilon_2$ is the area of the parallelogram, two of whose sides are ϵ_1 and $|\epsilon_2$ drawn outwards from a point; if ϵ_1 is parallel to $|\epsilon_2$, this area vanishes, or $\epsilon_1|\epsilon_2 = 0$; but, since $|\epsilon_2$ is perpendicular to ϵ_2, ϵ_1 must in this case be perpendicular to ϵ_2; hence the equation

$$\epsilon_1|\epsilon_2 = 0 \tag{54}$$

is the condition that two vectors ϵ_1 and ϵ_2 shall be perpendicular to each other.

The co-product $\epsilon_1|\epsilon_1$, which will usually be written ϵ_1^2, and called the co-square of ϵ_1, is the area of a square each of whose sides has the length $T\epsilon_1$; hence

$$T\epsilon_1 = \sqrt{\epsilon_1|\epsilon_1} = \sqrt{\epsilon_1^2}. \tag{55}$$

Let α_1 and α_2 be the angles between ι_1 and ϵ_1 and between ι_1 and ϵ_2 respectively, as in the figure. Then

$$\epsilon_1\epsilon_2 = m_1n_2 - m_2n_1 = T\epsilon_1 T\epsilon_2 \sin(\alpha_2 - \alpha_1), \tag{56}$$

the third member being the ordinary expression for the area of the parallelogram $\epsilon_1\epsilon_2$. Also

$$\epsilon_1|\epsilon_2 = (m_1\iota_1 + m_2\iota_2)(n_1\iota_2 - n_2\iota_1)$$
$$= m_1n_1 + m_2n_2 = T\epsilon_1 T\epsilon_2 \cos(\alpha_2 - \alpha_1), \tag{57}$$

the last member being found as before, remembering that $\sin(90° + \alpha_2 - \alpha_1) = \cos(\alpha_2 - \alpha_1)$.

If in (57) we let $\epsilon_2 = \epsilon_1$, whence $n_1 = m_1$ and $n_2 = m_2$, we have

$$T\epsilon_1 = \epsilon_1^2 = \sqrt{m_1^2 + m_2^2}. \tag{58}$$

If $T\epsilon_1 = T\epsilon_2 = 1$, then $m_1 = \cos\alpha_1$, $m_2 = \sin\alpha_1$, $n_1 = \cos\alpha_2$, $n_2 = \sin\alpha_2$, and equations (56) and (57) give the ordinary trigonometrical formulas $\sin(\alpha_2 - \alpha_1) = \sin\alpha_2 \cos\alpha_1 - \cos\alpha_2 \sin\alpha_1$, and $\cos(\alpha_2 - \alpha_1) = \cos\alpha_1\cos\alpha_2 + \sin\alpha_1\sin\alpha_2$. Squaring and adding (56) and (57), there results

$$T^2\epsilon_1 \cdot T^2\epsilon_2 = \epsilon_1^2\epsilon_2^2 = (\epsilon_1\epsilon_2)^2 + (\epsilon_1 | \epsilon_2)^2. \qquad (59)$$

Attention is called to the fact, which the student may have already noticed, that such an equation as $AB = AC$, in which AB and AC are combinatory products, does not, in general, imply that $B = C$, for the reason that the equation $A(B-C) = 0$ can usually be satisfied without either factor being itself zero. Thus $pL_1 = pL_2$ means simply that the two quantities which are equated have the same magnitude and sign, which permits L_2 to have an infinity of lengths and positions, when p and L_1 are given. The equation $p_1 p_2 = p_1 p_3$, or $p_1(p_2 - p_3) = 0$, p_2 and p_3 being unit points, implies, however, that $p_2 = p_3$, unless p_1 is at ∞, that is, a vector.

Exercise 9.—A triangle whose sides are of constant length moves so that two of its vertices remain on two fixed lines: find the locus of the other vertex.

Let $e_0\epsilon_1$ and $e_0\epsilon_2$ be the two fixed lines, and $pp'p''$ the triangle. Let pe be perpendicular to $p'p''$, $p' - e_0 = x\epsilon_1$ and $p'' - e_0 = y\epsilon_2$; then $p'' - p' = y\epsilon_2 - x\epsilon_1$, $T(y\epsilon_2 - x\epsilon_1) = c = $ constant, by the conditions. Also, $Tp'e = $ constant $= mc$, say, and $Tep = $ constant $= nc$, say. Hence

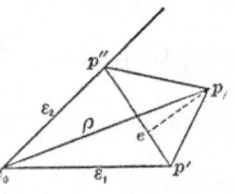

$$e - p' = Tp'e \cdot U(e - p') = mc \cdot \frac{y\epsilon_2 - x\epsilon_1}{T(y\epsilon_2 - x\epsilon_1)} = m(y\epsilon_2 - x\epsilon_1),$$

and similarly $p - e = n|(y\epsilon_2 - x\epsilon_1)$. Therefore

$$p - e_0 = \rho = x\epsilon_1 + m(y\epsilon_2 - x\epsilon_1) + n|(y\epsilon_2 - x\epsilon_1),$$

an equation which, with the condition $T(y\epsilon_2 - x\epsilon_1) = c$, or

$$y^2\epsilon_2^2 - 2xy\epsilon_1 | \epsilon_2 + x^2\epsilon_1^2 = c^2,$$

determines the locus to be a second-degree curve, which must in fact be an ellipse, since it can have no points at infinity. Let us rearrange the equation in ρ thus:

$$\rho = x[(1 - m)\epsilon_1 - n|\epsilon_1] + y[m\epsilon_2 + n|\epsilon_2] = x\epsilon + y\epsilon', \text{ say,}$$

so that $\epsilon = (1 - m)\epsilon_1 - n|\epsilon_1$, and $\epsilon' = m\epsilon_2 + n|\epsilon_2$; then multiply successively into ϵ and ϵ'; therefore $\rho\epsilon = y\epsilon'\epsilon$ and $\rho\epsilon' = x\epsilon\epsilon'$. Substituting these values of x and y in the equation of condition, we have

$$\epsilon_2^2 \cdot (\rho\epsilon)^2 + 2\epsilon_1|\epsilon_2 \cdot \rho\epsilon \cdot \rho\epsilon' + \epsilon_1^2(\rho\epsilon')^2 = c^2(\epsilon\epsilon')^2,$$

a scalar equation of the second degree in ρ.

Exercise 10.—There is given an irregular polygon of n sides: show that if forces act at the middle points of these sides, proportional to them in magnitude, and directed all outward or else all inward, these forces will be in equilibrium.

Let ϵ_0 be a vertex of the polygon, and let $2\epsilon_1$, $2\epsilon_2, \ldots 2\epsilon_n$ represent its sides in magnitude and direction. Then the middle points will be $\epsilon_0 + \epsilon_1$, $\epsilon_0 + 2\epsilon_1 + \epsilon_2$, etc., and, using the complement in a vector system, we have

$$\Sigma p\epsilon = (\epsilon_0+\epsilon_1)|\epsilon_1+(\epsilon_0+2\epsilon_1+\epsilon_2)|\epsilon_2+(\epsilon_0+2\epsilon_1+2\epsilon_2+\epsilon_3)|\epsilon_3+\ldots.$$

$$+ (\epsilon_0 + 2\epsilon_1 + \ldots + 2\epsilon_{n-1} + \epsilon_n)|\epsilon_n.$$

$$= \epsilon_0 \Big| \sum_1^n \epsilon + \sum_1^n \epsilon^2 + 2\epsilon_1 \Big| \sum_2^n \epsilon + 2\epsilon_2 \Big| \sum_3^n \epsilon + \ldots + 2\epsilon_{n-1}|\epsilon_n$$

$$= \epsilon_0 \Big| \sum_1^n \epsilon + \Big(\sum_1^n \epsilon \Big)^2 = 0, \text{ which was to be proved.}$$

Exercise 11.—A line passes through a fixed point and cuts two fixed lines; at the points of intersection perpendiculars to the fixed lines are erected; find the locus of the intersection of these perpendiculars.

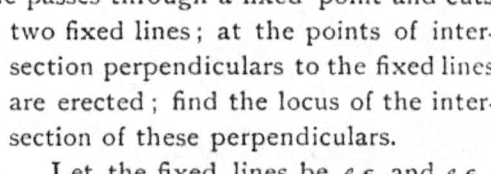

Let the fixed lines be $\epsilon_0\epsilon_1$ and $\epsilon_0\epsilon_2$, and the fixed point $\epsilon_0 + \epsilon_3$; the moving line cuts the fixed lines in p' and p'', at which points perpendiculars are erected meeting in p.

Let $p - \epsilon_0 = \rho$, $p' - \epsilon_0 = x\epsilon_1$, $p'' - \epsilon_0 = y\epsilon_2$, $T\epsilon_1 = T\epsilon_2 = 1$; then $\rho = x\epsilon_1 + x'|\epsilon_1 = y\epsilon_2 + y'|\epsilon_2$, whence $\rho|\epsilon_1 = x$ and $\rho|\epsilon_2 = y$.

Also, since $e_0 + \epsilon_2$, p', p'' are collinear points,
$$(x\epsilon_1 - \epsilon_3)(y\epsilon_2 - \epsilon_3) = 0 = xy\epsilon_1\epsilon_2 + y\epsilon_2\epsilon_3 + x\epsilon_3\epsilon_1;$$
or, substituting values of x and y,
$$\rho|\epsilon_1 . \rho|\epsilon_2 . \epsilon_1\epsilon_2 + \rho|\epsilon_2 . \epsilon_2\epsilon_3 + \rho|c_1 . \epsilon_3\epsilon_1 = 0,$$
an equation of the second degree in ρ, and hence representing a conic.

Prob. 15. If a, b, c are the lengths of the sides of a triangle, prove the formula $a^2 = b^2 + c^2 - 2bc \cos A$, by taking vectors ϵ_1, ϵ_2, and $\epsilon_2 - \epsilon_1$ equal to the respective sides.

Prob. 16. If $e_0\epsilon_1$ and $e_0\epsilon_2$ are two unit lines, show that the vector perpendicular from e_0 on the line $(e_0 + a\epsilon_1)(e_0 + be_2)$ is $\dfrac{ab\epsilon_1\epsilon_2}{(be_2 - a\epsilon_1)^2} . |(be_2 - a\epsilon_1)$, of which the length is $\dfrac{ab\epsilon_1\epsilon_2}{T(be_2 - a\epsilon_1)}$. From this derive the Cartesian expression for the perpendicular from the origin upon a straight line in oblique coordinates,
$ab \sin \omega \div (a^2 + b^2 - 2ab \cos \omega)^{\frac{1}{2}}$, ω being angle between the axes.

Prob. 17. If three points, $me_0 + ne_1$, $me_1 + ne_2$, $me_2 + ne_0$, be taken on the sides of the reference triangle, then the sides of the complementary triangle, $|(me_0 + ne_1)$, etc., will be respectively parallel to the corresponding sides of the triangle formed by the assumed points $(me_1 + ne_2)$, $(me_2 + ne_0)$, etc.

ART. 9. EQUATIONS OF CONDITION, AND FORMULAS.

Several equations of condition are placed here together for convenient reference: some have been already given; others follow from the results of Arts. 7 and 8.   When we have

$$\left.\begin{array}{l} p_1p_2 = 0, \\ \text{or} \quad n_1p_1 + n_2p_2 = 0, \end{array}\right\} \qquad \left.\begin{array}{l} L_1L_2 = 0, \\ \text{or} \quad n_1L_1 + n_2L_2 = 0, \end{array}\right\} \quad (60)$$

the two points coincide; | the two lines coincide;

$$\left.\begin{array}{l} p_1p_2p_3 = 0, \\ \text{or} \quad \overset{3}{\underset{1}{\Sigma}}np = 0, \end{array}\right\} \qquad \left.\begin{array}{l} L_1L_2L_3 = 0, \\ \text{or} \quad \overset{3}{\underset{1}{\Sigma}}nL = 0, \end{array}\right\} \quad (61)$$

the three points are collinear; | the three lines are confluent.

$$\epsilon_1\epsilon_2 = 0, \quad \text{or} \quad n_1\epsilon_1 + n_2\epsilon_2 = 0. \tag{62}$$

the two vectors are parallel (points at infinity coincide);

$$\epsilon_1|\epsilon_2 = 0, \tag{63}$$

the two vectors are perpendicular;

$$p_1 | p_2 = 0, \qquad\qquad L_1 | L_2 = 0, \qquad (64)$$

either point lies on the com- | either line passes through the
plementary line of the other. | complementary point of the
                                         | other.

If we write the equation

$$\rho = x_1\epsilon_1 + x_2\epsilon_2,$$

$x_1\epsilon_1$ is the projection of ρ on ϵ_1 parallel to ϵ_2, and $x_2\epsilon_2$ is the projection of ρ on ϵ_2 parallel to ϵ_1. Multiply both sides of the equation into ϵ_2; therefore $\rho\epsilon_2 = x_1\epsilon_1\epsilon_2$, or $x_1 = \rho\epsilon_2 \div \epsilon_1\epsilon_2$. Similarly, multiplying into ϵ_1, we have $\rho\epsilon_1 = x_2\epsilon_2\epsilon_1$, or $x_2 = \rho\epsilon_1 \div \epsilon_2\epsilon_1$, whence

$$\rho = \frac{\epsilon_1 \cdot \rho\epsilon_2}{\epsilon_1\epsilon_2} + \frac{\epsilon_2 \cdot \rho\epsilon_1}{\epsilon_2\epsilon_1}. \qquad (65)$$

The two terms of the second member of (65) are therefore the projections of ρ on ϵ_1 parallel to ϵ_2, and on ϵ_2 parallel to ϵ_1, respectively.*

Let ϵ_1 and ϵ_2 be unit normal vectors, say, ι and $|\iota$; then (65) becomes

$$\rho = \iota \cdot \rho|\iota - |\iota \cdot \rho\iota = \iota \cdot \rho|\iota + \iota\rho \cdot |\iota; \qquad (66)$$

or, if ι_1 and ι_2 be used instead of ι and $|\iota$,

$$\rho = \iota_1 \cdot \rho|\iota_1 + \iota_2 \cdot \rho|\iota_2. \qquad (67)$$

Again, in (65) let $\rho = \epsilon_3$, clear of fractions, and transpose; therefore

$$\epsilon_1\epsilon_2 \cdot \epsilon_3 + \epsilon_2\epsilon_3 \cdot \epsilon_1 + \epsilon_3\epsilon_1 \cdot \epsilon_2 = 0, \qquad (68)$$

a symmetrical relation between any three directions in plane space. Let $T\epsilon_1 = T\epsilon_2 = T\epsilon_3 = 1$, and multiply (68) into $|\epsilon_3$, thus $\qquad \epsilon_1\epsilon_2 + \epsilon_2\epsilon_3 \cdot \epsilon_1|\epsilon_3 + \epsilon_3\epsilon_1 \cdot \epsilon_2|\epsilon_3 = 0, \qquad (69)$

which is equivalent to

$$\sin(\alpha \pm \beta) = \sin\alpha\cos\beta \pm \cos\alpha\sin\beta,$$

the upper or lower sign corresponding to the case when ϵ_3 is

---

* Grassmann (1844), Chapter 5 (1862), Art. 129. Hyde's Directional Calculus, Arts. 46 and 47.

between ϵ_1 and ϵ_2, or outside, respectively. Writing in (69) $|\epsilon_2$ instead of ϵ_2, we have

$$\epsilon_1|\epsilon_2 - \epsilon_2|\epsilon_3 . \epsilon_3|\epsilon_1 + \epsilon_3\epsilon_1 . \epsilon_2\epsilon_3 = 0, \tag{70}$$

which gives the cos $(\alpha \pm \beta)$. These formulas being for any three directions in plane space, are independent of the magnitude of the angles involved.

There is given below a set of formulas for points and lines, arranged in complementary pairs, and all placed together for convenient reference, the derivation of them following after.

$$\left.\begin{array}{l} p=(p_0p_1p_2)^{-1}[p_0 . pp_1p_2 + p_1 . pp_2p_0 + p_2 . pp_0p_1], \\ L=(L_0L_1L_2)^{-1}[L_0 . LL_1L_2 + L_1 . LL_2L_0 + L_2 . LL_0L_1] \end{array}\right\} \tag{71}$$

$$\left.\begin{array}{l} p=(p_0p_1p_2)^{-1}[\,|\,p_1p_2 . p\,|\,p_0 + |\,p_2p_0 . p\,|\,p_1 + |\,p_0p_1 . p\,|\,p_2], \\ L=(L_0L_1L_2)^{-1}[\,|\,L_1L_2 . L\,|\,L_0 + |\,L_2L_0 . L\,|\,L_1 + |\,L_0L_1 . L\,|\,L_2] \end{array}\right\} \tag{72}$$

$$\left.\begin{array}{rl} p_1p_2 . p_3p_4 &= - p_1 . p_2p_3p_4 + p_2 . p_3p_4p_1 \\ &= p_3 . p_4p_1p_2 - p_4 . p_1p_2p_3, \\ L_1L_2 . L_3L_4 &= - L_1 . L_2L_3L_4 + L_2 . L_3L_4L_1 \\ &= L_3 . L_4L_1L_2 - L_4 . L_1L_2L_3 \end{array}\right\} \tag{73}$$

$$p_1p_2 . |\,q_1 = -\begin{vmatrix} p_1 & p_1|q_1 \\ p_2 & p_2|q_1 \end{vmatrix}, \quad L_1L_2|M_1 = -\begin{vmatrix} L_1 & L_1|M_1 \\ L_2 & L_2|M_1 \end{vmatrix}, \tag{74}$$

$$p_2|q_1q_2 = \begin{vmatrix} q_1 & p_2|q_1 \\ q_2 & p_2|q_2 \end{vmatrix}, \quad L_2|M_1M_2 = \begin{vmatrix} M_1 & L_2|M_1 \\ M_2 & L_2|M_2 \end{vmatrix}, \tag{75}$$

$$p_1p_2|q_1q_2 = \begin{vmatrix} p_1|q_1 & p_1|q_2 \\ p_2|q_1 & p_2|q_2 \end{vmatrix}, \quad L_1L_2|M_1M_2 = \begin{vmatrix} L_1|M_1 & L_1|M_2 \\ L_2|M_1 & L_2|M_2 \end{vmatrix}, \tag{76}$$

$$p_0p_1p_2 . q_0q_1q_2 = \begin{vmatrix} p_0|q_0 & p_0|q_1 & p_0|q_2 \\ p_1|q_0 & p_1|q_1 & p_1|q_2 \\ p_2|q_0 & p_2|q_1 & p_2|q_2 \end{vmatrix} \tag{77}$$

The complementary formula to (77) is not given, but may be obtained by putting L's and M's for p's and q's.

Derivation of Equations (71)–(77).—Equation (71). Write $p = x_0p_0 + x_1p_1 + x_2p_2$, and multiply this equation by p_1p_2; then $p_1p_2p = x_0p_1p_2p_0$, or $x_0 = pp_1p_2 \div p_0p_1p_2$.

Multiplying similarly by $p p_2$ and by $p_0 p_1$, we find $x_1 = pp_2p_0 \div p_0p_1p$ and $x_2 = pp_0p_1 \div p_0p_1p_2$. The substitution of these values gives the first of (71), and the second is similarly obtained or may be found by simply putting L's for p's in the first.

Equation (72). Write $p = x_0 |p_1 p_2 + x_1| p_2 p_0 + x_2 |p_0 p_1$, and multiply into $|p_0$; thus $p|p_0 = x_0 p_0 p_1 p_2$. Find in the same way values of x_1 and x_2, and substitute.

Equation (73). Write $p_1 p_2 \cdot p_3 p_4 = x p_1 + y p_2$, and multiply by $p p_2$; therefore $p p_2 \cdot p_1 p_2 \cdot p_3 p_4 = x p p_2 p_1$, or, by Art. 23, $p_2 p p_1 \cdot p_2 p_3 p_4 = x p p_2 p_1 = - x p_1 p p_2$; or, $x = - p_2 p_3 p_4$. Multiplying by $p p_1$, we find $y = p_3 p_4 p_1$, and on substituting obtain the first of (73). For the second put $p_1 p_2 \cdot p_3 p_4 = x p_3 + y p_4$, and proceed in a similar way.

Equation (74). In the first of (73) put $p_3 p_4 = |q_1$.

Equation (75). In the fourth of (73) put
$$L_1 L_2 = p_2, \ L_3 = |q_1, \ L_4 = |q_2.$$

Equation (76). Multiply (75) by p_1.

Equation (77). In the first of (72) put q_2 for p, and multiply by $p_0 p_1 p_2 \cdot q_0 q_1$; then

$$p_0 p_1 p_2 \cdot q_0 q_1 q_2 = q_0 q_1 |p_1 p_2 \cdot q_2| p_0 + q_0 q_1 |p_2 p_0 \cdot q_2| p_1 + q_0 q_1 |p_0 p_1 \cdot q_2| p_2$$
$$= p_0 |q_2 \cdot \begin{vmatrix} p_1|q_0 & p_1|q_1 \\ p_2|q_0 & p_2|q_1 \end{vmatrix} + p_1|q_2 \cdot \begin{vmatrix} p_2|q_0 & p_2|q_1 \\ p_0|q_0 & p_0|q_1 \end{vmatrix} + p_2|q_2 \cdot \begin{vmatrix} p_0|q_0 & p_0|q_1 \\ p_1|q_0 & p_1|q_1 \end{vmatrix},$$

by (76), which is equivalent to the third order determinant of equation (77).*

Exercise 12.—To show the product of two determinants as a determinant of the same order.

Let $p_0 = \overset{2}{\underset{0}{\Sigma}} le$, $p_1 = \Sigma me$, $p_2 = \Sigma ne$, $q_0 = \Sigma \lambda e$, $q_1 = \Sigma \mu e$, $q_2 = \Sigma \nu e$; then $p_0 p_1 p_2 = [l_0, m_1, n_2]$, $q_0 q_1 q_2 = [\lambda_0, \mu_1, \nu_2]$; also $p_0|q_0 = l_0 \lambda_0 + l_1 \lambda_1 + l_2 \lambda_2$, $p_1|q_0 = m_0 \lambda_0 + m_1 \lambda_1 + m_2 \lambda_2$, etc. Substituting these values in (77), we have the required result. A solution may also be obtained directly without the use of (77).

Let the q's be as above, but write $p_0 = \overset{2}{\underset{0}{\Sigma}} lq$, $p_1 = \Sigma mq$, $p_2 = \Sigma nq$. Then
$$p_0 p_1 p_2 = \Sigma lq \cdot \Sigma mq \cdot \Sigma nq = [l_0, m_1, n_2] q_0 q_1 q_2 = [l_0, m_1, n_2][\lambda_0, \mu_1, \nu_2].$$

* Grassmann (1862), Art. 173.

Also $p_0 = l_0 \Sigma \lambda e + l_1 \Sigma \mu e + l_2 \Sigma \nu e$

$= (l_0\lambda_0 + l_1\mu_0 + l_2\nu_0)e_0 + (l_0\lambda_1 + l_1\mu_1 + l_2\nu_1)e_1 + (l_0\lambda_2 + l_1\mu_2 + l_2\nu_2)e_2,$

with similar values for p_1 and p_2, which on being substituted in $p_0 p_1 p_2$ give the result. Equation (77), however, exhibits the product in a very compact, symmetrical, and easily remembered form.*

Exercise 13.—Show that the sides $p_1 p_2$, $p_2 p_3$, $p_3 p_1$ of the triangle $p_1 p_2 p_3$ cut the corresponding sides $|p_3, |p_1, |p_2$ of the complementary triangle in three collinear points.

The three points of intersection are, using (74),

$p_1 p_2 \cdot |p_3 = -p_1 \cdot p_2|p_3 + p_2 \cdot p_1|p_3,\ p_2 p_3 \cdot |p_1 = -p_2 \cdot p_3|p_1 + p_3 \cdot p_2|p_1,$
$p_3 p_1 \cdot |p_2 = -p_3 \cdot p_1|p_2 + p_1 \cdot p_3|p_2,$ of which the sum is zero, showing that the points are collinear. It may be shown in the same way that the lines joining corresponding vertices are confluent.

Exercise 14.—If the sides of a triangle pass through three fixed points, and two of the vertices slide on fixed lines, find the locus of the other vertex.

Let the fixed points and lines be p_1, p_2, p_3, L_1, L_2, and p, p', p'' the vertices of the triangle, as in the figure. Then $p'p_2p'' = 0$; p' coin-

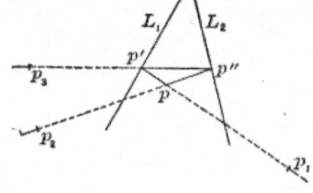

cides with $pp_1 \cdot L_1$ and p'' with $pp_2 \cdot L_2$; hence substituting $(pp_1 \cdot L_1)p_3(L_2 \cdot p_2 p) = 0$, the equation of the locus, which, being of the second degree in p, is that of a conic.

Prob. 18. Show that if the three fixed points of the last exercise are collinear, then the locus of p breaks up into two straight lines. Use equation (73).

Prob. 19. If the vertices of a triangle slide on three fixed lines, and two of the sides pass through fixed points, find the envelope of the other side. (This statement is reciprocally related to that of Exercise 14, that is, lines and points are replaced by points and

---

* These methods may be applied to determinants of any order by using a space of corresponding order.

lines respectively, and the resulting equation will be an equation of the second order in L, a variable line.)

Prob. 20. Show that if the three fixed lines of Exercise 5 are confluent, then the envelope of L reduces to two points and the line joining them.

ART. 10.   STEREOMETRIC PRODUCTS.

The product of two points in solid space is the same as in plane space. See Art. 7.

Product of Three Points.—Any three points determine a plane, and also, as in Art. 7, an area; hence $p_1p_2p_3$ is a plane-sect or a portion of the plane fixed by the three points whose area is double that of the triangle $p_1p_2p_3$. It may be shown, in the manner used in Art. 7 for the sect, that no plane-sect, not in this plane, can be equal to $p_1p_2p_3$, and that any plane-sect in this plane having the same area and sign will be equal to $p_1p_2p_3$.[*] Of course $p_1p_2p_3$ is not now scalar.

Product of Four Points.—Any four non-coplanar points

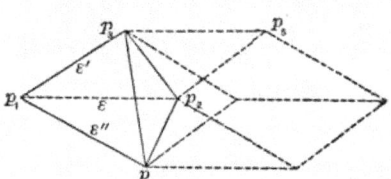

determine a tetrahedron, say $p_1p_2p_3p_4$, and six times the volume of this tetrahedron is taken for the value of the product, because this is the volume of the parallelepiped generated by the product $p_1p_2p_3$,—i.e. the parallelogram $p_1, p_2,$—when it moves parallel to its initial position from p_1 to p_4. Let $p_2 - p_1 = \epsilon,\ p_3 - p_1 = \epsilon',\ p_4 - p_1 = \epsilon''$, then

$$p_1p_2p_3p_4 = p_1p_2p_3\epsilon'' = p_1p_2\epsilon'\epsilon'' = p_1\epsilon\epsilon'\epsilon''. \qquad (78)$$

If $p_1 = \overset{3}{\underset{0}{\Sigma}}ke,\ p_2 = \overset{3}{\underset{0}{\Sigma}}le,\ p_3 = \overset{3}{\underset{0}{\Sigma}}me,\ p_4 = \overset{3}{\underset{0}{\Sigma}}ne$, then

$$p_1p_2p_3p_4 = \Sigma ke \Sigma le \Sigma me \Sigma ne = [k_0,\ l_1,\ m_2,\ n_3] \cdot e_0e_1e_2e_3; \qquad (79)$$

from which it appears that any two quadruple products of points differ from each other only by a scalar factor, that is, they differ only in magnitude, or sign, or both; hence such products are themselves scalar.[†] If $p_1p_2p_3p_4 = 0$, the volume of the tetrahedron vanishes, so that the four points are coplanar.

* Grassmann (1862), Art. 255.   † Grassmann (1862), Art. 263.

Product of Two Vectors.—The two vectors determine an area as in Art. 7, but they also determine now a plane direction, so that the product $\epsilon_1\epsilon_2$ is a plane-vector, and is not scalar as in plane space. Also, $\epsilon_1\epsilon_2$ differs from $p_1\epsilon_1\epsilon_2$ now just as ϵ differs from $p\epsilon$; namely, $\epsilon_1\epsilon_2$ has a definite area and plane direction, that is, toward a certain line at infinity, while $p_1\epsilon_1\epsilon_2$ is fixed in position by passing through p_1. Equation (37) therefore does not hold in solid space.

Product of Three Vectors.—Three vectors determine a parallelepiped as in the figure above, and $\epsilon\epsilon'\epsilon''$ is therefore the volume of this parallelepiped. Any other triple vector product can differ from this only in magnitude and sign. For let $\epsilon_1\epsilon_2\epsilon_3$ be such a product, and write

$$\epsilon = x_1\epsilon_1 + x_2\epsilon_2 + x_3\epsilon_3 = \overset{3}{\underset{1}{\Sigma}}x\epsilon, \ \epsilon' = \overset{3}{\underset{1}{\Sigma}}y\epsilon, \ \epsilon'' = \overset{3}{\underset{1}{\Sigma}}z\epsilon; \text{ then}$$

$$\epsilon\epsilon'\epsilon'' = \Sigma x\epsilon\Sigma y\epsilon\Sigma z\epsilon = \begin{vmatrix} x_1 & x_2 & x_3 \\ y_1 & y_2 & y_3 \\ z_1 & z_2 & z_3 \end{vmatrix}\epsilon_1\epsilon_2\epsilon_3, \tag{80}$$

so that the two products only differ by the scalar determinant factor. Hence the product of three vectors must be itself a scalar, by Art. 1. Since, then, the product of four points has precisely the same signification as that of three vectors, we may write

$$p_1p_2p_3p_4 = p_1\epsilon\epsilon'\epsilon'' = \epsilon\epsilon'\epsilon'' = (p_2 - p_1)(p_3 - p_1)(p_4 - p_1)$$
$$= p_2p_3p_4 - p_3p_4p_1 + p_4p_1p_2 - p_1p_2p_3. \tag{81}$$

Thus the sum of the plane-sects forming the doubles of the faces of a tetrahedron, all taken positively in the same sense as looked at from outside the tetrahedron, is equal to the volume of the tetrahedron. Compare equation (37).

If $\epsilon\epsilon'\epsilon'' = 0$, the volume of the parallelepiped vanishes, and the three vectors must be parallel to one plane.

Product of Two Sects.—In solid space two sects determine a tetrahedron of which they are opposite edges. Thus

$$p_1p_2p_3p_4 = p_1p_2 \cdot p_3p_4 = L_1L_2 = p_3p_4 \cdot p_1p_2 = L_2L_1, \tag{82}$$

so that the stereometric product of two sects is commutative, and has the same meaning as that of four points.

Product of a Sect and a Plane-Sect.—Let them be L and P, and let p_0 be their common point; take p_1, p_2, p_3 so that $L = p_0 p_1$, and $P = p_1 p_2 p_3$. L and P evidently determine the point p_0, and also the parallelepiped of which one edge is L and one face is P, so that the product should be made up of these two factors. Hence we write

$$\left.\begin{array}{l} LP = p_0 p_1 \cdot p_0 p_2 p_3 = p_0 p_1 p_2 p_3 \cdot p_0; \\ PL = p_0 p_2 p_3 \cdot p_0 p_1 = p_0 p_2 p_3 p_1 \cdot p_0 = LP. \end{array}\right\} \qquad (83)$$

If L is parallel to P, p_0 is at infinity, and, replacing it by ϵ, (83) becomes

$$PL = LP = \epsilon p_1 \cdot \epsilon p_2 p_3 = \epsilon p_1 p_2 p_3 \cdot \epsilon. \qquad (84)$$

Product of Two Plane-Sects.—Let them be P_1 and P_2, and let L be their intersection, while p_1 and p_2 are such points that $P_1 = L p_1$ and $P_2 = L p_2$; then P_1 and P_2 determine the line L and also a parallelepiped of which they are two adjacent faces, and

$$P_1 P_2 = L p_1 \cdot L p_2 = L p_1 p_2 \cdot L = - P_2 P_1. \qquad (85)$$

If P_1 and P_2 are parallel, L is at infinity, and is equivalent to a plane-vector, say to η; hence, substituting in (84),

$$P_1 P_2 = \eta p_1 \cdot \eta p_2 = \eta p_1 p_2 \cdot \eta = - P_2 P_1. \qquad (86)$$

Product of Three Plane-Sects.—By (85) and (83) this must be the square of a volume times the common point of the three planes; or, if p_0, p_1, p_2, p_3 be taken in such manner that $P_1 = p_0 p_2 p_3$, $P_2 = p_0 p_3 p_1$, $P_3 = p_0 p_1 p_2$, then

$$P_1 P_2 P_3 = 023 . 031 . 012 = 023 . 0123 . 01 = (p_0 p_1 p_2 p_3)^2 \cdot p_0; \quad (87)$$

the suffixes being used instead of the corresponding points. If p_0 be at infinity, the three planes are parallel to a single line, and may be written $P_1 = n_1 \epsilon p_2 p_3$, etc., and then treated as above.

Product of Four Plane-Sects.[*]—Let the planes be $P_0 \ldots P_3$, and let $p_0 \ldots p_3$ be the four common points of the planes taken three by three. $n_0 \ldots n_3$ may be so taken that $P_0 = n_0 p_1 p_2 p_3$, etc.; then

$$\begin{array}{l} P_0 P_1 P_2 P_3 = n_0 n_1 n_2 n_3 . 123 . 230 . 301 . 012 \\ \qquad = n_0 n_1 n_2 n_3 (p_0 p_1 p_2 p_3)^3. \end{array} \qquad (88)$$

[*] Grassmann (1862), Art. 300.

Product of Two Plane-Vectors.—Let η_1 and η_2 be two plane-vectors or lines at infinity; let ϵ be parallel to each of them, and ϵ_1 and ϵ_2 so taken that $\eta_1 = \epsilon\epsilon_1$, $\eta_2 = \epsilon\epsilon_2$, then

$$\eta_1\eta_2 = \epsilon\epsilon_1 \cdot \epsilon\epsilon_2 = \epsilon\epsilon_1\epsilon_2 \cdot \epsilon = -\eta_2\eta_1, \tag{89}$$

because η_1 and η_2 determine a common direction ϵ, and a parallelepiped of which three conterminous edges are equal to ϵ, ϵ_1, ϵ_2, respectively.

Product of Three Plane-Vectors.—Take ϵ_1, ϵ_2, ϵ_3 so that

$$\eta_1\eta_2\eta_3 = n \cdot \epsilon_2\epsilon_3 \cdot \epsilon_3\epsilon_1 \cdot \epsilon_1\epsilon_2 = n(\epsilon_1\epsilon_2\epsilon_3)^2. \tag{90}$$

The directions $\epsilon_1 \ldots \epsilon_3$ are common to the plane-vectors $\eta_1 \ldots \eta_3$ taken two by two.

Several conditions are given here together which follow from the results of this article.

$p_1p_2 = 0,$ | $P_1P_2 = 0,$ | $\quad(91)$
Two points coincide. | Two planes coincide. |

$p_1p_2p_3 = 0,$ | $P_1P_2P_3 = 0,$ | $\quad(92)$
Three points collinear. | Three planes collinear. |

$p_1p_2p_3p_4 = p_1p_2 \cdot p_3p_4$ | $P_1P_2P_3P_4 = P_1P_2 \cdot P_3P_4$
$\quad = L_1L_2 = 0,$ | $\quad = L_1L_2 = 0, \quad(93)$
Four points coplanar; two | Four planes confluent; two
lines intersect. | lines intersect.

$\epsilon_1\epsilon_2 = 0,$ | $\eta_1\eta_2 = 0,$ | $\quad(94)$
Vectors parallel. | Plane-vectors parallel. |

$\epsilon_1\epsilon_2\epsilon_3 = 0,$ | $\eta_1\eta_2\eta_3 = 0,$ | $\quad(95)$
Three vectors parallel to | Three plane-vectors parallel to
one plane. | one iine.

Sum of Two Planes.—Let them be P_1 and P_2, let L be a sect in their common line, and take p_1 and p_2 so that $P_1 = Lp_1$, $P_2 = Lp_2$; then

$$P_1 + P_2 = L(p_1 + p_2) = 2L\overline{p}, \tag{96}$$

$\overline{p}$ being the mean of p_1 and p_2. Also

$$P_1 - P_2 = L(p_1 - p_2); \tag{97}$$

whence the sum and difference are the diagonal plane through L, and a plane through L parallel to the diagonal plane which is itself parallel to L, of the parallelepiped determined by P_1

and P_2. If $TP_1 = TP_2$, $P_1 \pm P_2$ will evidently be the two bisecting planes of the angle between them. The bisecting planes may also be written

$$\frac{P_1}{TP_1} \pm \frac{P_2}{TP_2} \quad \text{or} \quad P_1 TP_2 \pm P_2 TP_1. \tag{98}$$

If the two planes are parallel, let η be a plane-vector parallel to each of them, that is, their common line at infinity, and let p_1 and p_2 be points in the respective planes; then we may write $P_1 = n_1 p_1 \eta$, $P_2 = n_2 p_2 \eta$, whence

$$P_1 + P_2 = (n_1 p_1 + n_2 p_2)\eta = (n_1 + n_2)\bar{p}\eta. \tag{99}$$

If $n_1 + n_2 = 0$, this becomes

$$P_1 + P_2 = n_2(p_2 - p_1)\eta, \tag{100}$$

the product of a vector into a plane-vector and therefore a scalar, by (80).

Two plane-vectors may be added similarly, since they will have a common direction, namely, that of the vector parallel to both of them.

Exercise 15.—If two tetrahedra $e_0 e_1 e_2 e_3$ and $e_0' e_1' e_2' e_3'$ are so situated that the right lines through the pairs of corresponding vertices all meet in one point, then will the corresponding faces cut each other in four coplanar lines.

The given conditions are equivalent to $e_0 e_0' \cdot e_1 e_1' = 0$
$= e_0 e_0' \cdot e_2 e_2' = e_0 e_0' \cdot e_3 e_3' = e_1 e_1' \cdot e_2 e_2' = e_2 e_2' \cdot e_3 e_3' = e_3 e_3' \cdot e_1 e_1'$.
Two of the intersecting lines of faces are $e_0 e_1 e_2 \cdot e_0' e_1' e_2'$ and
$e_1 e_2 e_3 \cdot e_1' e_2' e_3'$, and, if these intersect, we must accordingly have, by (92), $012 \cdot 0'1'2' \cdot 123 \cdot 1'2'3' = 0 = 012 \cdot 123 \cdot 0'1'2' \cdot 1'2'3$
$= 0123 \cdot 0'1'2'3' \cdot 121'2'$, the last factor of which is equivalent to the fourth condition above, since quadruple-point products in solid space are associative. Similarly all the other pairs of intersections may be treated.

Exercise 16.—The twelve bisecting planes of the diedral angles of a tetrahedron fix eight points, the centers of the inscribed and escribed spheres, through which they pass six by six.

The sum and difference of two unit planes are their two

bisecting planes, by (97). Let the tetrahedron be $e_0e_1e_2e_3$, and
let the double areas of its faces be $A_0 = Te_1e_2e_3$, etc.; then a
pair of bisecting planes will be $\dfrac{e_0e_1e_2}{A_3} \pm \dfrac{e_0e_1e_3}{A_2}$ or $e_0e_1(A_2e_2 \pm A_3e_3)$.
The pair through the opposite edge will be $e_2e_3(A_0e_0 \pm A_1e_1)$.
If there be a point through which the six internal bisecting
planes pass, it must be on the intersection of these two planes
taken with the upper signs, and we infer by symmetry that it
must be the point $\overset{3}{\underset{0}{\Sigma}}Ae$. Another internal bisecting plane is
$e_1e_0(A_1e_1 + A_2e_2)$, which gives zero when multiplied into ΣAe,
as do also the other three.

To obtain all the points we have only to use the double
signs, so that they are $\pm A_0e_0 \pm A_1e_1 \pm A_2e_2 \pm A_3e_3$. This
gives eight cases, namely,

$$
\begin{array}{cccc}
+ & + & + & + \\
+ & + & + & - \\
+ & + & - & + \\
+ & - & + & +
\end{array}
\qquad
\begin{array}{cccc}
- & + & + & + \\
+ & + & - & - \\
+ & - & - & + \\
+ & - & + & -
\end{array}
$$

The eight apparent cases that would arise by changing all the
signs are included in these because the points must be essentially positive. Moreover, no positive point could have three
negative signs, because the sum of any three faces of the tetrahedron must be greater than the fourth face. It will be found
on trial that six of the bisecting planes will pass through
$\Sigma(\pm Ae)$ with any one of the above arrangements of sign.

Prob. 21. The twelve points in which the edges of a tetrahedron
are cut by the bisecting planes of the opposite diedral angles fix
eight planes, each of which passes through six of them.

Prob. 22. The centroid of the faces of a tetrahedron coincides
with the center of the sphere inscribed within the tetrahedron
whose vertices are the centroids of the respective faces of the first
tetrahedron.

Prob. 23. If any plane be passed through the middle points of
two opposite edges of a tetrahedron, it will divide the volume of the
tetrahedron into two equal parts.

ART. 11.   THE COMPLEMENT IN SOLID SPACE.

According to the definitions of Art. 8 the complementary relations in a unit normal vector system are as follows:

$$\left. \begin{array}{ll} |\iota_1 = \iota_2\iota_3, & |\iota_2\iota_3 = |(|\iota_1) = \iota_1, \\ |\iota_2 = \iota_3\iota_1, & |\iota_3\iota_1 = |(|\iota_2) = \iota_2, \\ |\iota_3 = \iota_1\iota_2, & |\iota_1\iota_2 = |(|\iota_3) = \iota_3 \end{array} \right\}. \tag{101}$$

Let $\epsilon = \overset{3}{\underset{1}{\Sigma}} l\iota$; then

$$|\epsilon = l_1\iota_2\iota_3 + l_2\iota_3\iota_1 + l_3\iota_1\iota_2 = \frac{1}{l_1}(l_1\iota_2 - l_2\iota_1)(l_1\iota_3 - l_3\iota_1), \tag{102}$$

so that $|\epsilon$ is a plane-vector.   The figure, which is drawn in

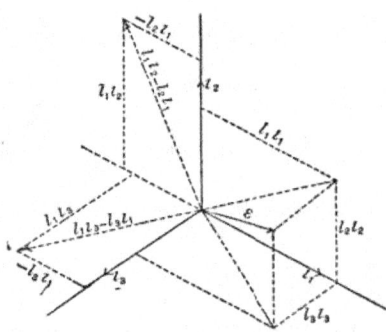

isometric projection, shows that the two vectors $l_1\iota_2 - l_2\iota_1$ and $l_1\iota_3 - l_3\iota_1$, whose product is $l_1 \cdot |\epsilon$, are both perpendicular to ϵ; for the first is perpendicular to $l_1\iota_1 + l_2\iota_2$, which is the orthogonal projection of ϵ upon $\iota_1\iota_2$, and to ι_3, and therefore is also perpendicular to ϵ, while the second is perpendicular to $l_1\iota_1 + l_3\iota_3$ and to ι_2, and therefore to ϵ.   Hence $|\epsilon$ is a plane-vector perpendicular to ϵ; and, since $|(|\epsilon) = \epsilon$, the converse is also true, i.e. the complement of a plane-vector is a line-vector normal to it.

The figure shows that ϵ is equal to the vector diagonal of the rectangular parallelepiped whose edges have the lengths l_1, l_2, l_3, hence

$$T\epsilon = \sqrt{l_1^2 + l_2^2 + l_3^2}. \tag{103}$$

Multiply equation (102) by ϵ; therefore

$$\epsilon|\epsilon = (l_1\iota_1 + l_2\iota_2 + l_3\iota_3)(l_1\iota_2\iota_3 + l_2\iota_3\iota_1 + l_3\iota_1\iota_2)$$
$$= l_1^2 + l_2^2 + l_3^2 = T^2\epsilon = \epsilon^2, \tag{104}$$

so that the co-square of a vector is equal to the square of its tensor.   The product $\epsilon|\epsilon$ is that of a vector ϵ into a plane-vector perpendicular to it, as has just been shown; it is therefore a volume which is equivalent to $T\epsilon \cdot T|\epsilon$; hence, by (104), $\epsilon|\epsilon = T\epsilon \cdot T|\epsilon = T^2\epsilon$, or $T\epsilon = T|\epsilon$. Hence, the complement of a vector in solid space is a plane-vector perpendicular to it and having the same tensor, or numerical measure of magnitude.*

Let a second vector be $\epsilon' = \overset{3}{\underset{1}{\Sigma}}m\iota$; then

$$\epsilon|\epsilon' = l_1m_1 + l_2m_2 + l_3m_3 = \epsilon'|\epsilon. \tag{105}$$

Now $\epsilon|\epsilon'$, being the product of ϵ into the plane-vector $|\epsilon'$, is the volume of the parallelepiped in the figure, that is, $T\epsilon T\epsilon'$ sin (angle between ϵ and $|\epsilon'$) $= T\epsilon T\epsilon' \cos \frac{\epsilon'}{\epsilon}$. Hence

$$\epsilon|\epsilon' = \epsilon'|\epsilon = l_1m_1 + l_2m_2 + l_3m_3 = T\epsilon T\epsilon' \cos \frac{\epsilon'}{\epsilon}. \tag{106}$$

If $T\epsilon = T\epsilon' = 1, l_1 \ldots l_3, m_1 \ldots m_3$ are direction cosines, and (105) gives a proof of the formula for the cosine of the angle between two lines in terms of the direction cosines of the lines. We have also in this case

$\epsilon\epsilon' = (l_1m_2 - l_2m_1)|\iota_3 + (l_2m_3 - l_3m_2)|\iota_1 + (l_3m_1 - l_1m_3)|\iota_2$, and, taking the co-square,

$$(\epsilon\epsilon')^2 = (\sin \tfrac{\epsilon'}{\epsilon})^2 = (l_1m_2 - l_2m_1)^2 + (l_2m_3 - l_3m_2)^2 + (l_3m_1 - l_1m_3)^2. \tag{107}$$

If

$$\epsilon|\epsilon' = 0, \tag{108}$$

ϵ is parallel to the plane-vector perpendicular to ϵ', that is, ϵ is perpendicular to ϵ', as is also shown by (106).

Let $\eta = |\epsilon, \eta' = |\epsilon'$; then

$$\eta|\eta' = |\epsilon \cdot \epsilon' = \epsilon'|\epsilon = \epsilon|\epsilon' = T\epsilon T\epsilon' \cos \tfrac{\epsilon'}{\epsilon} = T\eta T\eta' \cos \tfrac{\eta'}{\eta}, \tag{109}$$

and

$$\eta|\eta' = 0 \tag{110}$$

is the condition of perpendicularity of two plane-vectors. Also either

$$\epsilon|\eta' = 0, \quad \text{or} \quad \eta'|\epsilon = 0, \tag{111}$$

is the condition that a vector shall be perpendicular to a plane-vector, for the first means that ϵ is parallel to a vector which is

* Grassmann (1862), Art. 335.

perpendicular to η', and the second that η' is parallel to a plane-vector which is perpendicular to ϵ.

Equations (71)–(77) of Art. 9 become stereometric vector formulæ if ϵ_1, ϵ_2, etc., be substituted for p_1, p_2, etc., and η_1, η_2, etc., for L_1, L_2, etc. For instance, (76) gives the vector formulas

$$\epsilon_1\epsilon_2|\epsilon_1'\epsilon_2' = \begin{vmatrix} \epsilon_1|\epsilon_1' & \epsilon_1|\epsilon_2' \\ \epsilon_2|\epsilon_1' & \epsilon_2|\epsilon_2' \end{vmatrix}, \quad \eta_1\eta_2|\eta_1'\eta_2' = \begin{vmatrix} \eta_1|\eta_1' & \eta_1|\eta_2' \\ \eta_2|\eta_1' & \eta_2|\eta_2' \end{vmatrix}. \quad (112)$$

For lack of space no treatment of the complement in a point system in solid space is given.

Exercise 17.—To prove the formulas of spherical trigonometry $\cos a = \cos b \cos c + \sin b \sin c \cos A$, and

$$\frac{\sin a}{\sin A} = \frac{\sin b}{\sin B} = \frac{\sin c}{\sin C}.$$

Take three unit vectors ϵ_1, ϵ_2, ϵ_3 parallel to the radii to the vertices of the spherical triangle, then $a=$(angle bet. ϵ_2 and ϵ_3), $A=$(angle bet. $\epsilon_1\epsilon_2$ and $\epsilon_1\epsilon_3$), etc. In eq. (112) put $\epsilon_1\epsilon_3$ for $\epsilon_1'\epsilon_2'$;

hence $\epsilon_1\epsilon_2|\epsilon_1\epsilon_3 = \sin b \sin c \cos A = \epsilon_1^2 . \epsilon_2|\epsilon_3 - \epsilon_1|\epsilon_2 . \epsilon_1|\epsilon_3$
$$= \cos a - \cos b \cos c.$$

Again,

$$T(\epsilon_1\epsilon_2 . \epsilon_1\epsilon_3) = T(\epsilon_1\epsilon_2\epsilon_3 . \epsilon_1) = T\epsilon_1\epsilon_2\epsilon_3 = T(\epsilon_2\epsilon_3 . \epsilon_2\epsilon_1) = T(\epsilon_3\epsilon_1 . \epsilon_3\epsilon_2);$$

or $\quad \sin b \sin c \sin A = \sin a \sin c \sin B = \sin a \sin b \sin C,$

whence we have the second result by dividing by $\sin a \sin b \sin c$.

Exercise 18.—Show that in a spherical triangle taken as in Exercise 17, $\cos \dfrac{A}{2} = \dfrac{U\epsilon_1\epsilon_2|(U\epsilon_1\epsilon_2 + U\epsilon_1\epsilon_3)}{T(U\epsilon_1\epsilon_2 + U\epsilon_1\epsilon_3)}$, whence derive

the ordinary value $\sqrt{\dfrac{\sin s \sin (s - a)}{\sin b \sin c}}$.

Expanding, the numerator becomes $1 + U\epsilon_1\epsilon_2| U\epsilon_1\epsilon_3$, and the denominator $\sqrt{2(1 + U\epsilon_1\epsilon_2| U\epsilon_1\epsilon_3)}$. Also there is obtained

$U\epsilon_1\epsilon_2| U\epsilon_1\epsilon_3 = \dfrac{\epsilon_1\epsilon_2|\epsilon_1\epsilon_3}{T\epsilon_1\epsilon_2 T\epsilon_1\epsilon_3}$. The remainder is left to the student.

Prob. 24. If ϵ_1, ϵ_2, ϵ_3, drawn outward from a point, are taken as three edges of a tetrahedron, show that the six planes perpendicular to the edges at their middle points all pass through the end

of the vector $\rho = \dfrac{1}{2\epsilon_1\epsilon_2\epsilon_3}(\,|\,\epsilon_2\epsilon_3\,.\,\epsilon_1^2 + |\,\epsilon_3\epsilon_1\,.\,\epsilon_2^2 + |\,\epsilon_1\epsilon_2\,.\,\epsilon_3^2)$. (Suggestion. We must have $(\rho - \tfrac{1}{2}\epsilon_1)\,|\,\epsilon_1 = 0$, with two other similar expressions.)

Prob. 25. Show that ϵ, $|\,\epsilon\epsilon'$ and $\epsilon\epsilon'.\,|\,\epsilon$ are three mutually perpendicular vectors, no matter what the directions of ϵ and ϵ' may be.

Prob. 26. Let ϵ_1, ϵ_2, ϵ_3 be taken as in Prob. 24 ; let A_0 be the area of the face of the tetrahedron formed by joining the ends of these vectors, and $2A_1 = T\epsilon_2\epsilon_3$, etc.; also $\theta_1 =$ Angle between $\epsilon_1\epsilon_2$ and $\epsilon_1\epsilon_3$, etc.: then show that we have the relation, analogous to that of Prob. 15, Art. 8,

$$A_0^2 = A_1^2 + A_2^2 + A_3^2 - 2A_2A_3\cos\theta_1 - 2A_3A_1\cos\theta_2 - 2A_1A_2\cos\theta_3.$$

If $\theta_1 \ldots \theta_3$ are right angles, this becomes the space-analog of the proposition regarding the hypotenuse and sides of a right-angled triangle. (Suggestion. $2A_0 = T(\epsilon_2 - \epsilon_1)(\epsilon_3 - \epsilon_1)$.)

Prob. 27. There are given three non-coplanar lines $e_0\epsilon_1$, $e_0\epsilon_2$, $e_0\epsilon_3$; planes cut these lines at right angles, the sum of the squares of their distances from e_0 being constant. Show that the locus of the common point of these three planes is $(\rho\,|\,\epsilon_1)^2 + (\rho\,|\,\epsilon_2)^2 + (\rho\,|\,\epsilon_3)^2 = c^2$, if $T\epsilon_1 = T\epsilon_2 = T\epsilon_3 = 1$.

Art. 12.   Addition of Sects in Solid Space.

Two lines in solid space will not in general intersect, so that their sum will not be, as in eq. (43), a definite line. For let $p_1\epsilon_1$ and $p_2\epsilon_2$ be any two sects: then

$$p_1\epsilon_1 + p_2\epsilon_2 = p_1\epsilon_1 + p_2\epsilon_2 + e_0(\epsilon_1 + \epsilon_2) - e_0(\epsilon_1 + \epsilon_2)$$
$$= e_0(\epsilon_1 + \epsilon_2) + (p_1 - e_0)\epsilon_1 + (p_2 - e_0)\epsilon_2;$$

that is, the sum is a sect passing through an arbitrary point e_0, and a plane-vector, the sum of the two in the equation. The sum cannot be a single sect unless the two are coplanar; for let $p_2 = p_1 + x\epsilon_1 + y\epsilon_2 + z\epsilon_3$, ϵ_3 being a vector not parallel to $\epsilon_1\epsilon_2$;

hence $p_1\epsilon_1 + p_2\epsilon_2 = p_1\epsilon_1 + (p_1 + x\epsilon_1 + y\epsilon_2 + z\epsilon_3)\epsilon_2$
$$= p_1(\epsilon_1 + \epsilon_2) + x\epsilon_1(\epsilon_1 + \epsilon_2) + z\epsilon_3\epsilon_2$$
$$= (p_1 + x\epsilon_1)(\epsilon_1 + \epsilon_2) + z\epsilon_3\epsilon_2;$$

and this cannot reduce to a single sect unless $z = 0$, that is, unless $p_1\epsilon_1$ and $p_2\epsilon_2$ are coplanar. Since a plane-vector is a line at

∞, the sum of two lines may always be presented as the sum of a finite line and a line at ∞.

If the sum of any two sects is equal to the sum of any other two, their products will also be equal, that is, the two pairs will determine tetrahedra of equal volumes. For let $L_1 + L_2 = L_3 + L_4$; then squaring we have $L_1 L_2 = L_3 L_4$, since $L_1 L_1 = 0$, etc.

An infinite number of pairs of sects can be found such that the sum of each pair is equal to the sum of any given pair; for let a given pair be $p_1 \epsilon_1 + p_2 \epsilon_2$, and take a new pair

$$(x_1 p_1 + x_2 p_2)(u_1 \epsilon_1 + u_2 \epsilon_2) + (y_1 r_1 + y_2 p_2)(v_1 \epsilon_1 + v_2 \epsilon_2)$$
$$= (x_1 u_1 + y_1 v_1) p_1 \epsilon_1 + (x_2 u_2 + y_2 v_2) p_2 \epsilon_2 +$$
$$(x_1 u_2 + y_1 v_2) p_1 \epsilon_2 + (x_2 u_1 + y_2 v_1) p_2 \epsilon_1.$$

This will be equal to the given pair if we have

$$x_1 u_1 + y_1 v_1 = x_2 u_2 + y_2 v_2 = 1, \text{ and } x_1 u_2 + y_1 v_2 = x_2 u_1 + y_2 v_1 = 0).$$

Since there are eight arbitrary quantities with only four equations of condition, the desired result can evidently be accomplished in an infinite number of ways.

Let $p_1 \epsilon_1, p_2 \epsilon_2 \ldots p_n \epsilon_n$ be n sects, and let S be their sum, and e_0 any point, then

$$S = \sum_1^n p\epsilon \equiv e_0 \Sigma\epsilon - e_0 \Sigma\epsilon + \Sigma p\epsilon = e_0 \Sigma\epsilon + \Sigma(p - e_0)\epsilon, \ldots \ldots (113)$$

the sum of a sect and a plane-vector as before.

If $\Sigma(p - e_0)\epsilon$ is parallel to $\Sigma\epsilon$ it may be written as the product of some vector ϵ' into $\Sigma\epsilon$, that is, $\epsilon' \Sigma\epsilon$, when the sum becomes $S = e_0 \Sigma\epsilon + \epsilon' \Sigma\epsilon = (e_0 + \epsilon')\Sigma\epsilon$, a sect, because $e_0 + \epsilon'$ is a point. In no other case does S reduce to a single sect. If $\Sigma\epsilon = 0$, S becomes a plane-vector. Of the two parts composing S, the sect will be unchanged in magnitude and direction if e_0 be moved to a new position, while the plane-vector will in general be altered. It is proposed to show that a point q may be substituted for e_0 such that the plane-vector will be perpendicular to $\Sigma\epsilon$. Writing

$$S \equiv q\Sigma\epsilon - (q - e_0)\Sigma\epsilon + \Sigma(p - e_0)\epsilon,$$

and, for brevity, putting $q - e_0 = \rho$, $\Sigma\epsilon = \alpha$, $\Sigma(p - e_0)\epsilon = |\beta$, so that

$$S \equiv q\alpha - \rho\alpha + |\beta, \tag{114}$$

we must have for perpendicularity, by (111),

$$(|\beta - \rho\alpha)|\alpha = 0 = |\beta\alpha - \rho\alpha.|\alpha,$$

or $\qquad \rho\alpha.|\alpha \equiv \alpha.\rho|\alpha - \rho.\alpha^2 = |\beta\alpha. \qquad$ (115)

The second member is obtained from the first by substituting in eq. (74) ρ for p_1 and α for p_2 and q_1, in accordance with the statement at the end of Art. 11. If in (115) we make $\rho|\alpha = 0$, ρ will be the vector from e_0 to q taken perpendicularly to α, say

$$\rho_1 = |\alpha\beta \div \alpha^2 = q_1 - e_0. \qquad (116)$$

Since α and β are known, the required point has been found. Multiply (115) by α; then, using (75),

$$- \alpha\rho.\alpha^2 \equiv \rho\alpha.\alpha^2 = \alpha|\beta\alpha = |\beta.\alpha^2 - |\alpha.\alpha|\beta,$$

whence, substituting in (114),

$$S = q\alpha + \frac{\alpha|\beta}{\alpha^2}.|\alpha = q\Sigma\epsilon + \frac{\Sigma\epsilon\Sigma(p - e_0)\epsilon}{(\Sigma\epsilon)^2}.\Sigma\epsilon. \qquad (117)$$

This may be called the normal form of S.*

The sects of this article represent completely the geometric properties of forces, hence all that has been shown applies immediately to a system of forces in solid space. We have only to substitute the words force and couple for sect and plane-vector. The resultant action of any system of forces is S, called by Ball in his Theory of Screws "a wrench." The condition for equilibrium is $S = 0$, which gives at once

$$\Sigma\epsilon = 0 \quad \text{and} \quad \Sigma(p - e_0)\epsilon = 0; \qquad (118)$$

since otherwise we must have $e_0\Sigma\epsilon = - \Sigma(p - e_0)\epsilon$, which is an impossibility. The line $q\Sigma\epsilon$ is the central axis of the system of forces S.

Lack of space forbids a further development of the subject, but what has been given in this article will indicate the perfect adaptability of this method to the requirements of mechanics.

Exercise 19.—Reduce $p_1\epsilon_1 + p_2\epsilon_2 = S$ to its normal form. $S \equiv e_0(\epsilon_1 + \epsilon_2) + (p_1 - e_0)\epsilon_1 + (p_2 - e_0)\epsilon_2.$ For convenience suppose p_1 and p_2 to be taken at the ends of the common per-

* Grassmann (1862), Art. 346.

pendicular on $p_1\epsilon_1$ and $p_2\epsilon_2$, and moreover let $e_0 = \frac{1}{2}(p_1 + p_2)$, $p_1 - e_0 = \iota = -(p_2 - e_0)$; then $\iota | \epsilon_1 = \iota | \epsilon_2 = 0$. Accordingly

$$S \equiv e_0(\epsilon_1 + \epsilon_2) + \iota(\epsilon_1 - \epsilon_2) = q(\epsilon_1 + \epsilon_2) + \frac{(\epsilon_1 + \epsilon_2)\iota(\epsilon_1 - \epsilon_2)}{(\epsilon_1 + \epsilon_2)^2} \cdot |(\epsilon_1 + \epsilon_2)$$

$$= q(\epsilon_1 + \epsilon_2) + \frac{\iota\epsilon_1\epsilon_2}{(\epsilon_1 + \epsilon_2)^2} \cdot |(\epsilon_1 + \epsilon_2).$$

By (116), $q - e_0 = -\dfrac{|\beta \cdot |\alpha}{\alpha^2} = -\dfrac{\iota(\epsilon_1 - \epsilon_2 \cdot |(\epsilon_1 + \epsilon_2)}{(\epsilon_1 + \epsilon_2)^2}$

$$= \frac{\iota \cdot (\epsilon_1 - \epsilon_2)|(\epsilon_1 + \epsilon_2)}{(\epsilon_1 + \epsilon_2)^2}, \text{ by (74), } = \frac{\epsilon_1^2 - \epsilon_2^2}{(\epsilon_1 + \epsilon_2)^2} \cdot \iota.$$

Hence the normal form of S is

$$S = \left(e_0 + \frac{\epsilon_1^2 - \epsilon_2^2}{(\epsilon_1 + \epsilon_2)^2} \cdot \iota\right)(\epsilon_1 + \epsilon_2) + \frac{\iota\epsilon_1\epsilon_2}{(\epsilon_1 + \epsilon_2)^2} \cdot |(\epsilon_1 + \epsilon_2).$$

Exercise 20.—Forces are represented by the six edges of a tetrahedron e_0e_1, e_0e_2, e_0e_3, e_2e_3, e_3e_1, e_1e_2; find the S, reduce to normal form, and consider the special case when three diedral angles are right angles. $S \equiv e_0(e_1 + e_2 + e_3) + e_2e_3 + e_3e_1 + e_1e_2$
$\equiv e_0(\epsilon_1 + \epsilon_2 + \epsilon_3) + (e_2 - e_1)(e_3 - e_1) \equiv e_0(\epsilon_1 + \epsilon_2 + \epsilon_3) + (\epsilon_2 - \epsilon_1)(\epsilon_3 - \epsilon_1)$
$\equiv e_0(\epsilon_1 + \epsilon_2 + \epsilon_3) + \epsilon_2\epsilon_3 + \epsilon_3\epsilon_1 + \epsilon_1\epsilon_2$, in which $\epsilon_1 = e_1 - e_0$, etc. Hence

$$S \equiv \left(e_0 + \frac{(\epsilon_2\epsilon_3 + \epsilon_3\epsilon_1 + \epsilon_1\epsilon_2)|(\epsilon_1 + \epsilon_2 + \epsilon_3)}{(\epsilon_1 + \epsilon_2 + \epsilon_3)^2}\right)(\epsilon_1 + \epsilon_2 + \epsilon_3)$$

$$+ \frac{3\epsilon_1\epsilon_2\epsilon_3}{(\epsilon_1 + \epsilon_2 + \epsilon_3)^2} \cdot |(\epsilon_1 + \epsilon_2 + \epsilon_3).$$

For the rectangular tetrahedron let $\epsilon_1 = a\iota_1$, $\epsilon_2 = b\iota_2$, $\epsilon_3 = c\iota_3$, ι_1, ι_2, ι_3 being unit normal vectors. Then we find

$$S \equiv \left(e_0 + \frac{a(c^2 - b^2)\iota_1 + b(a^2 - c^2)\iota_2 + c(b^2 - a^2)\iota_3}{a^2 + b^2 + c^2}\right)(a\iota_1 + b\iota_2 + c\iota_3)$$

$$+ \frac{3abc}{a^2 + b^2 + c^2} \cdot |(a\iota_1 + b\iota_2 + c\iota_3).$$

Exercise 21.—A pole 50 feet high stands on the ground and is held erect by three guy-ropes symmetrically arranged about it, attached to its top and to pegs in the ground 50 feet from the pole. The wind blows against the pole with a pressure of 50 pounds in the direction $e_0 - p$, when e_0 is at the bottom of the pole, and p divides the distance between two of the pegs

in the ratio $\dfrac{m}{n}$: find the tension on the guys and the pressure

on the ground.

Evidently only two of the guys will be in tension; let their pegs be at e_1 and e_3, and let e_2 be at the top of the pole, and w

the weight of the pole. Then $p = \dfrac{me_1 + ne_3}{m + n}$, and the equation

of equilibrium is

$$50. \ \frac{(e_0 + e_2)(e_0 - p)}{2T(e_0 - p)} + \frac{25e_0(p - e_0)}{T(e_0 - p)} + \frac{(x + w)e_0 e_2}{Te_0 e_2} + \frac{ye_0 e_1}{Te_2 e_1} + \frac{ze_0 e_3}{Te_2 e_3} = 0.$$

$$Te_0 e_2 = 50, \ Te_2 e_1 = Te_2 e_3 = 50 \sqrt{2}, \ T(p - e_0) = T\left(\frac{me_1 + ne_3}{m + n} - e_0\right)$$

$$= T\left(\frac{m(e_1 - e_0) + n(e_3 - e_0)}{m + n}\right) = \frac{50}{m + n} T(m\epsilon_1 + n\epsilon_3), \ \text{if} \ \epsilon_1 = U(e_1 - e_0)$$

and $\epsilon_3 = U(e_3 - e_0)$; then $T(p - e_0) = \dfrac{50}{m + n} \sqrt{m^2 + n^2 - mn}$,

because $\epsilon_1{}^2 = \epsilon_3{}^2 = 1$, and $\epsilon_1 | \epsilon_3 = \cos 120° = -\frac{1}{2}$. Hence the equation of equilibrium becomes

$$\frac{25e_0((m + n)e_0 - me_1 - ne_3)}{\sqrt{m^2 + n^2 - mn}} + (x + w)e_0 e_2 + \frac{y}{\sqrt{2}}e_2 e_1 + \frac{z}{\sqrt{2}}e_2 e_3 = 0.$$

Multiply successively by $e_3 e_1$, $e_0 e_3$, and $e_0 e_1$, and we obtain

$$\frac{x + w}{m + n} = \frac{y}{m \sqrt{2}} = \frac{z}{n \sqrt{2}} = \frac{25}{\sqrt{m^2 + n^2 - mn}},$$

y and z being the tensions, and $x + w$ the upward pressure.

Prob. 28. Three equal poles are set up so as to form a tripod, and are mutually perpendicular; a weight w hangs upon a rope which passes over a pulley at the top of the tripod, and thence

down under a pulley at the ground at a point $p = \overset{3}{\underset{1}{\Sigma}} le$, in which

$e_1 \ldots e_3$ are at the feet of the poles, and $\overset{3}{\underset{1}{\Sigma}} l = 1$; if the rope is pulled so as to raise w, show that the pressures on the poles, supposing the pulleys frictionless, are

$$w\left(\frac{l_1}{\sqrt{\Sigma l^2}} + \frac{1}{\sqrt{3}}\right), \quad w\left(\frac{l_2}{\sqrt{\Sigma l^2}} + \frac{1}{\sqrt{3}}\right), \quad w\left(\frac{l_3}{\sqrt{\Sigma l^2}} + \frac{1}{\sqrt{3}}\right).$$

Prob. 29. Six equal forces act along six successive edges of a cube which do not meet a given diagonal; show that if the edges of the cube be parallel to ι_1, ι_2, ι_3, and F be the magnitude of each force, then $S = - 2F|\,(\iota_1 + \iota_2 + \iota_3)$, if the diagonal taken be parallel to $\iota_1 + \iota_2 + \iota_3$.

Prob. 30. Three forces whose magnitudes are 1, 2, and 3 act along three successive non-coplanar edges of a cube; show that the normal form of S is

$$S = (e_0 + \tfrac{13}{14}\iota_1 + \tfrac{1}{2}\iota_2 - \tfrac{9}{14}\iota_3)(\iota_1 + 2\iota_2 + 3\iota_3) + \tfrac{3}{14}|\,(\iota_1 + 2\iota_2 + 3\iota_3).$$

Prob. 31. Forces act at the centroids of the faces of a tetrahedron, perpendicular and proportional to the faces on which they act, and all directed inwards, or else all outwards; show that they are in equilibrium.

CHAPTER IX.

VECTOR ANALYSIS AND QUATERNIONS.

By ALEXANDER MACFARLANE,
Lecturer in Electrical Engineering in Lehigh University.

ART. 1.   INTRODUCTION.

By " Vector Analysis " is meant a space analysis in which the vector is the fundamental idea ; by " Quaternions " is meant a space-analysis in which the quaternion is the fundamental idea.   They are in truth complementary parts of one whole; and in this chapter they will be treated as such, and developed so as to harmonize with one another and with the Cartesian Analysis.*   The subject to be treated is the analysis of quantities in space, whether they are vector in nature, or quaternion in nature, or of a still different nature, or are of such a kind that they can be adequately represented by space quantities.

Every proposition about quantities in space ought to remain true when restricted to a plane ; just as propositions about quantities in a plane remain true when restricted to a straight line.   Hence in the following articles the ascent to the algebra of space is made through the intermediate algebra of the plane.   Arts. 2–4 treat of the more restricted analysis, while Arts. 5–10 treat of the general analysis.

This space analysis is a universal Cartesian analysis, in the same manner as algebra is a universal arithmetic.   By providing an explicit notation for directed quantities, it enables their general properties to be investigated independently of any particular system of coordinates, whether rectangular, cylindrical, or polar.   It also has this advantage that it can express

---

* For a discussion of the relation of Vector Analysis to Quaternions, see Nature, 1891–1893.

the directed quantity by a linear function of the coordinates, instead of in a roundabout way by means of a quadratic function.

The different views of this extension of analysis which have been held by independent writers are briefly indicated by the titles of their works:

Argand, Essai sur une maniére de représenter les quantités imaginaires dans les constructions géométriques, 1806.

Warren, Treatise on the geometrical representation of the square roots of negative quantities, 1828.

Moebius, Der barycentrische Calcul, 1827.

Bellavitis, Calcolo delle Equipollenze, 1835.

Grassmann, Die lineale Ausdehnungslehre, 1844.

De Morgan, Trigonometry and Double Algebra, 1849.

O'Brien, Symbolic Forms derived from the conception of the translation of a directed magnitude. Philosophical Transactions, 1851.

Hamilton, Lectures on Quaternions, 1853, and Elements of Quaternions, 1866.

Tait, Elementary Treatise on Quaternions, 1867.

Hankel, Vorlesungen über die complexen Zahlen und ihre Functionen, 1867.

Schlegel, System der Raumlehre, 1872.

Hoüel, Théorie des quantités complexes, 1874.

Gibbs, Elements of Vector Analysis, 1881–4.

Peano, Calcolo geometrico, 1888.

Hyde, The Directional Calculus, 1890.

Heaviside, Vector Analysis, in " Reprint of Electrical Papers," 1885–92.

Macfarlane, Principles of the Algebra of Physics, 1891. Papers on Space Analysis, 1891–3.

An excellent synopsis is given by Hagen in the second volume of his " Synopsis der höheren Mathematik."

ART. 2. ADDITION OF COPLANAR VECTORS.

By a " vector " is meant a quantity which has magnitude and direction. It is graphically represented by a line whose length represents the magnitude on some convenient scale, and whose direction coincides with or represents the direction of the vector. Though a vector is represented by a line, its physical dimensions may be different from that of a line. Examples are a linear velocity which is of one dimension in length, a directed area which is of two dimensions in length, an axis which is of no dimensions in length.

A vector will be denoted by a capital italic letter, as B,* its magnitude by a small italic letter, as b, and its direction by a small Greek letter, as β. For example, $B = b\beta$, $R = r\rho$. Sometimes it is necessary to introduce a dot or a mark $\angle$ to separate the specification of the direction from the expression for the magnitude; † but in such simple expressions as the above, the difference is sufficiently indicated by the difference of type. A system of three mutually rectangular axes will be indicated, as usual, by the letters i, j, k.

The analysis of a vector here supposed is that into magnitude and direction. According to Hamilton and Tait and other writers on Quaternions, the vector is analyzed into tensor and unit-vector, which means that the tensor is a mere ratio destitute of dimensions, while the unit-vector is the physical magnitude. But it will be found that the analysis into magnitude and direction is much more in accord with physical ideas, and explains readily many things which are difficult to explain by the other analysis.

A vector quantity may be such that its components have a common point of application and are applied simultaneously; or it may be such that its components are applied in succession, each component starting from the end of its predecessor. An example of the former is found in two forces applied simultaneously at the same point, and an example of the latter in

---

* This notation is found convenient by electrical writers in order to harmonize with the Hospitalier system of symbols and abbreviations.

† The dot was used for this purpose in the author's Note on Plane Algebra, 1883; Kennelly has since used $\angle$ for the same purpose in his electrical papers

two rectilinear displacements made in succession to one another.

Composition of Components having a common Point of Application.—Let OA and OB represent two vectors of the same kind simultaneously applied at the point O. Draw BC parallel to OA, and AC parallel to OB, and 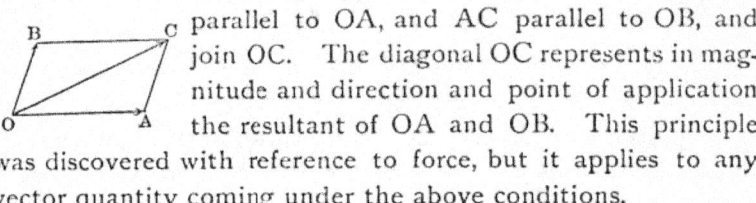 join OC. The diagonal OC represents in magnitude and direction and point of application the resultant of OA and OB. This principle was discovered with reference to force, but it applies to any vector quantity coming under the above conditions.

Take the direction of OA for the initial direction; the direction of any other vector will be sufficiently denoted by the angle round which the initial direction has to be turned in order to coincide with it. Thus OA may be denoted by $f_1 /0$, OB by f_2/θ_2. OC by f/θ. From the geometry of the figure it follows that

$$f^2 = f_1^2 + f_2^2 + 2f_1 f_2 \cos \theta_2$$

and

$$\tan \theta = \frac{f_2 \sin \theta_2}{f_1 + f_2 \cos \theta_2};$$

hence $OC = \sqrt{f_1^2 + f_2^2 + 2f_1 f_2 \cos \theta_2} \ /\tan^{-1} \dfrac{f_2 \sin \theta_2}{f_1 + f_2 \cos \theta_2}.$

Example.—Let the forces applied at a point be $2/0°$ and $3/60°$. Then the resultant is $\sqrt{4 + 9 + 12 \times \frac{1}{2}} \ /\tan^{-1} \dfrac{3\sqrt{3}}{7}$
$= 4.36/36° \ 30'.$

If the first component is given as f_1 /θ_1, then we have the more symmetrical formula

$OC = \sqrt{f_1^2 + f_2^2 + 2f_1 f_2 \cos(\theta_2 - \theta_1)} \ /\tan^{-1} \dfrac{f_1 \sin \theta_1 + f_2 \sin \theta_2}{f_1 \cos \theta_1 + f_2 \cos \theta_2}.$

When the components are equal, the direction of the resultant bisects the angle formed by the vectors; and the magnitude of the resultant is twice the projection of either component on the bisecting line. The above formula reduces to

$$OC = 2f_1 \cos \frac{\theta_2}{2} \ / \frac{\theta_2}{2}.$$

Example.—The resultant of two equal alternating electromotive forces which differ 120° in phase is equal in magnitude to either and has a phase of 60°.

Given a vector and one component, to find the other component.—Let OC represent the resultant, and OA the component. Join AC and draw OB equal and parallel to AC. The line OB represents the component required, for it is the only line which combined with OA gives OC as resultant. The line OB is identical with the diagonal of the parallelogram formed by OC and OA reversed; hence the rule is, "Reverse the direction of the component, then compound it with the given resultant to find the required component." Let $f\underline{/\theta}$ be the vector and $f_1\underline{/o}$ one component; then the other component is

$$f_2\underline{/\theta_2} = \sqrt{f^2 + f_1^2 - 2ff_1 \cos\theta}\underline{/\tan^{-1}\frac{f\sin\theta}{-f_1 + f\cos\theta}}.$$

Given the resultant and the directions of the two components, to find the magnitude of the components.—The resultant is represented by OC, and the directions by OX and OY.

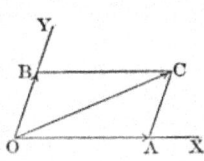

From C draw CA parallel to OY, and CB parallel to OX; the lines OA and OB cut off represent the required components. It is evident that OA and OB when compounded produce the given resultant OC, and there is only one set of two components which produces a given resultant; hence they are the only pair of components having the given directions.

Let $f\underline{/\theta}$ be the vector and $\underline{/\theta_1}$ and $\underline{/\theta_2}$ the given directions. Then

$$f_1 + f_2 \cos(\theta_2 - \theta_1) = f\cos(\theta - \theta_1),$$
$$f_1 \cos(\theta_2 - \theta_1) + f_2 = f\cos(\theta_2 - \theta),$$

from which it follows that

$$f_1 = f\frac{\{\cos(\theta - \theta_1) - \cos(\theta_2 - \theta)\cos(\theta_2 - \theta_1)\}}{1 - \cos^2(\theta_2 - \theta_1)}.$$

For example, let $100/60°$, $/30°$, and $/90°$ be given; then

$$f_1 = 100 \frac{\cos 30°}{1 + \cos 60°}.$$

Composition of any Number of Vectors applied at a common Point.—The resultant may be found by the following graphic construction : Take the vectors in any order, as A, B, C.

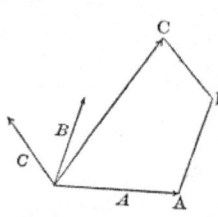

From the end of A draw B' equal and parallel to B, and from the end of B' draw C' equal and parallel to C; the vector from the beginning of A to the end of C' is the resultant of the given vectors. This follows by continued application of the parallelogram construction. The resultant obtained is the same, whatever the order; and as the order is arbitrary, the area enclosed has no physical meaning.

The result may be obtained analytically as follows :

Given $f_1/\theta_1 + f_2/\theta_2 + f_3/\theta_3 + \cdots + f_n/\theta_n.$

Now $f_1/\theta_1 = f_1 \cos \theta_1/0 + f_1 \sin \theta_1 /\dfrac{\pi}{2}.$

Similarly $f_2/\theta_2 = f_2 \cos \theta_2/0 + f_2 \sin \theta_2 /\dfrac{\pi}{2},$

and $f_n/\theta_n = f_n \cos \theta_n/0 + f_n \sin \theta_n /\dfrac{\pi}{2}.$

Hence $\Sigma\{f/\theta\} = \{\Sigma f \cos \theta\}/0 + \{\Sigma f \sin \theta\} /\dfrac{\pi}{2}$

$$= \sqrt{(\Sigma f \cos \theta)_2 + (\Sigma f \sin \theta)^2} \cdot \tan^{-1}\frac{\Sigma f \sin \theta}{\Sigma f \cos \theta}.$$

In the case of a sum of simultaneous vectors applied at a common point, the ordinary rule about the transposition of a term in an equation holds good. For example, if $A + B + C = 0$, then $A + B = -C$, and $A + C = -B$, and $B + C = -A$, etc. This is permissible because there is no real order of succession among the given components.*

* This does not hold true of a sum of vectors having a real order of succession. It is a mistake to attempt to found space-analysis upon arbitrary formal

Composition of Successive Vectors.—The composition of successive vectors partakes more of the nature of multiplication than of addition.   Let A be a vector start-
ing from the point O, and B a vector starting from the end of A.   Draw the third side OP, and from O draw a vector equal to B, and from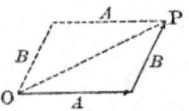
its extremity a vector equal to A.   The line OP is not the complete equivalent of $A + B$; if it were so, it would also be the complete equivalent of $B + A$.   But $A + B$ and $B + A$ determine different paths; and as they go oppositely around, the areas they determine with OP have different signs.   The diagonal OP represents $A + B$ only so far as it is considered independent of path.   For any number of successive vectors, the sum so far as it is independent of path is the vector from the initial point of the first to the final point of the last.   This is also true when the successive vectors become so small as to form a continuous curve.   The area between the curve OPQ and the vector OQ depends on the path, and has a physical meaning.

Prob. 1. The resultant vector is $123/45°$, and one component is $100/0°$; find the other component.

Prob. 2. The velocity of a body in agiven plane is $200 /75°$, and one component is $100/25°$; find the other component.

Prob. 3. Three alternating magnetomotive forces are of equal virtual value, but each pair differs in phase by $120°$; find the resultant.   (Ans. Zero.)

Prob. 4. Find the components of the vector $100/70°$ in the directions $20°$ and $100°$.

Prob. 5. Calculate the resultant vector of $1/10°$, $2/20°$, $3/30°$, $4/40°$.

Prob. 6. Compound the following magnetic fluxes: $h \sin nt + h \sin (nt - 120°)/120° + h \sin (nt - 240°)/240°$.   (Ans. $\frac{3}{2}h/nt$.)

laws; the fundamental rules must be made to express universal properties of the thing denoted.   In this chapter no attempt is made to apply formal laws to directed quantities.   What is attempted is an analysis of these quantities.

Prob. 7. Compound two alternating magnetic fluxes at a point, $a \cos nt \underline{/\text{o}}$ and $a \sin nt \underline{\Big/ \frac{\pi}{2}}$.   (Ans. $a \underline{/nt}$.)

Prob 8. Find the resultant of two simple alternating electromotive forces $100\underline{/20°}$ and $50\underline{/75°}$.

Prob. 9. Prove that a uniform circular motion is obtained by compounding two equal simple harmonic motions which have the space-phase of their angular positions equal to the supplement of the time-phase of their motions.

ART. 3.   PRODUCTS OF COPLANAR VECTORS.

When all the vectors considered are confined to a common plane, each may be expressed as the sum of two rectangular components. Let i and j denote two directions in the plane at right angles to one another; then $A = a_1 i + a_2 j$, $B = b_1 i + b_2 j$, $R = xi + yj$. Here i and j are not unit-vectors, but rather signs of direction.

Product of two Vectors.—Let $A = a_1 i + a_2 j$ and $B = b_1 + b_2 j$ be any two vectors, not necessarily of the same kind physically. We assume that their product is obtained by applying the distributive law, but we do not assume that the order of the factors is indifferent. Hence

$$AB = (a_1 i + a_2 j)(b_1 i + b_2 j) = a_1 b_1 ii + a_2 b_2 jj + a_1 b_2 ij + a_2 b_1 ji.$$

If we assume, as suggested by ordinary algebra, that the square of a sign of direction is $+$, and further that the product of two directions at right angles to one another is the direction normal to both, then the above reduces to

$$AB = a_1 b_1 + a_2 b_2 + (a_1 b_2 - a_2 b_1)k.$$

Thus the complete product breaks up into two partial products, namely, $a_1 b_1 + a_2 b_2$ which is independent of direction, and $(a_1 b_2 - a_2 b_1)k$ which has the axis of the plane for direction.*

---

* A common explanation which is given of $ij = k$ is that i is an operator, j an operand, and k the result. The kind of operator which i is supposed to denote is a quadrant of turning round the axis i; it is supposed not to be an axis, but a quadrant of rotation round an axis. This explains the result $ij = k$, but unfortunately it does not explain $ii = +$; for it would give $ii = i$.

Scalar Product of two Vectors.—By a scalar quantity is meant a quantity which has magnitude and may be positive or negative but is destitute of direction. The former partial product is so called because it is of such a nature. It is denoted by SAB where the symbol S, being in Roman type, denotes, not a vector, but a function of the vectors A and B. The geometrical mean- ing of SAB is the product of A and the orthogonal projection of B upon A. Let OP and OQ represent the vectors A and B; draw QM and NL perpendicular to OP. Then

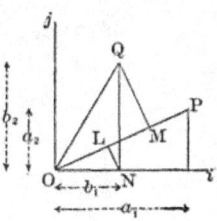

$$(OP)(OM) = (OP)(OL) + (OP)(LM),$$

$$= a \left\{ b_1 \frac{a_1}{a} + b_2 \frac{a_2}{a} \right\},$$

$$= a_1 b_1 + a_2 b_2.$$

Corollary 1.—$SBA = SAB$. For instance, let A denote a force and B the velocity of its point of application; then SAB denotes the rate of working of the force. The result is the same whether the force is projected on the velocity or the velocity on the force.

Example 1.—A force of 2 pounds East + 3 pounds North is moved with a velocity of 4 feet East per second + 5 feet North per second; find the rate at which work is done.

$$2 \times 4 + 3 \times 5 = 23 \text{ foot-pounds per second.}$$

Corollary 2.—$A^2 = a_1^2 + a_2^2 = a^2$. The square of any vector is independent of direction; it is an essentially positive or signless quantity; for whatever the direction of A, the direction of the other A must be the same; hence the scalar product cannot be negative.

Example 2.—A stone of 10 pounds mass is moving with a velocity 64 feet down per second + 100 feet horizontal per second. Its kinetic energy then is

$$\frac{10}{2} (64^2 + 100^2) \text{ foot-poundals,}$$

a quantity which has no direction. The kinetic energy due to the downward velocity is $10 \times \dfrac{64^2}{2}$ and that due to the horizontal velocity is $\dfrac{10}{2} \times 100^2$; the whole kinetic energy is obtained, not by vector, but by simple addition, when the components are rectangular.

Vector Product of two Vectors.—The other partial product from its nature is called the vector product, and is denoted by

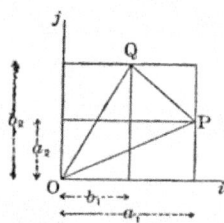

VAB. Its geometrical meaning is the product of A and the projection of B which is perpendicular to A, that is, the area of the parallelogram formed upon A and B. Let OP and OQ represent the vectors A and B, and draw the lines indicated by the figure. It is then evident that the area of the triangle OPQ $= a_1 b_2 - \frac{1}{2} a_1 a_2 - \frac{1}{2} b_1 b_2 - \frac{1}{2}(a_1 - b_1)(b_2 - a_2),$
$= \frac{1}{2}(a_1 b_2 - a_2 b_1).$

Thus $(a_1 b_2 - a_2 b_1)k$ denotes the magnitude of the parallelogram formed by A and B and also the axis of the plane in which it lies.

It follows that $VBA = - VAB$. It is to be observed that the coordinates of A and B are mere component vectors, whereas A and B themselves are taken in a real order.

Example.—Let $A = (10i + 11j)$ inches and $B = (5i + 12j)$ inches, then $VAB = (120 - 55)k$ square inches; that is, 65 square inches in the plane which has the direction k for axis.

If A is expressed as $a\alpha$ and B as $b\beta$, then $SAB = ab \cos \alpha\beta$, where $\alpha\beta$ denotes the angle between the directions α and β.

Example.—The effective electromotive force of 100 volts per inch $/90°$ along a conductor 8 inch $/45°$ is $SAB = 8 \times 100 \cos /45° /90°$ volts, that is, 800 cos 45° volts. Here $/45°$ indicates the direction α and $/90°$ the direction β, and $/45° /90°$ means the angle between the direction of 45° and the direction of 90°.

Also $VAB = ab \sin \alpha\beta . \overline{\alpha\beta}$, where $\overline{\alpha\beta}$ denotes the direction which is normal to both α and β, that is, their pole.

Example.—At a distance of 10 feet $/30°$ there is a force of 100 pounds $/60°$. The moment is VAB

$$= 10 \times 100 \sin /30° \ /60° \text{ pound-feet } \overline{90°/} \ /90°$$

$$= 1000 \sin 30° \text{ pound-feet } \overline{90°/} \ /90°.$$

Here $\overline{90°/}$ specifies the plane of the angle and $/90°$ the angle. The two together written as above specify the normal k.

Reciprocal of a Vector.—By the reciprocal of a vector is meant the vector which combined with the original vector produces the product $+ 1$. The reciprocal of A is denoted by A^{-1}. Since $AB = ab (\cos \alpha\beta + \sin \alpha\beta \cdot \overline{\alpha\beta})$, b must equal a^{-1} and β must be identical with α in order that the product may be 1. It follows that

$$A^{-1} = \frac{1}{a}\alpha = \frac{a\alpha}{a^2} = \frac{a_1 i + a_2 j}{a_1^2 + a_2^2}.$$

The reciprocal and opposite vector is $- A^{-1}$. In the figure let $OP = 2\beta$ be the given vector; then $OQ = \frac{1}{2}\beta$ is its reciprocal, and $OR = \frac{1}{2}(-\beta)$ is its reciprocal and opposite.*

$$\overset{\leftarrow\ \ O}{\underset{R\ \ \ \ Q}{\rule{2.5cm}{0.4pt}}} \overset{}{\underset{P}{}}$$

Example.—If $A = 10$ feet East $+ 5$ feet North, $A^{-1} = \frac{10}{125}$ feet East $+ \frac{5}{125}$ feet North and $- A^{-1} = \frac{-10}{125}$ feet East $- \frac{5}{125}$ feet North.

Product of the reciprocal of a vector and another vector.—

$$A^{-1}B = \frac{1}{a^2}AB,$$

$$= \frac{1}{a^2}\{a_1 b_1 + a_2 b_2 + (a_1 b_2 - a_2 b_1)\overline{\alpha\beta}\},$$

$$= \frac{b}{a}(\cos \alpha\beta + \sin \alpha\beta \cdot \overline{\alpha\beta}).$$

---

* Writers who identify a vector with a quadrantal versor are logically led to define the reciprocal of a vector as being opposite in direction as well as reciprocal in magnitude.

Hence $SA^{-1}B = \dfrac{b}{a}\cos \alpha\beta$ and $VA^{-1}B = \dfrac{b}{a}\sin \alpha\beta \cdot \overline{\alpha\beta}$.

Product of three Coplanar Vectors.—Let $A = a_1 i + a_2 j$, $B = b_1 i + b_2 j$, $C = c_1 i + c_2 j$ denote any three vectors in a common plane. Then

$$(AB)C = \{(a_1 b_1 + a_2 b_2) + (a_1 b_2 - a_2 b_1)k\}(c_1 i + c_2 j)$$
$$= (a_1 b_1 + a_2 b_2)(c_1 i + c_2 j) + (a_1 b_2 - a_2 b_1)(-c_2 i + c_1 j).$$

The former partial product means the vector C multiplied by the scalar product of A and B; while the latter partial product means the complementary vector of C multiplied by the magnitude of the vector product of A and B.

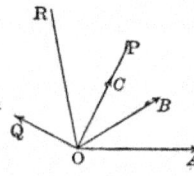

If these partial products (represented by OP and OQ) unite to form a total product, the total product will be represented by OR, the resultant of OP and OQ.

The former product is also expressed by $SAB \cdot C$, where the point separates the vectors to which the S refers; and more analytically by $abc \cos \alpha\beta \cdot \gamma$.

The latter product is also expressed by $(VAB)C$, which is equivalent to $V(VAB)C$, because VAB is at right angles to C. It is also expressed by $abc \sin \alpha\beta \cdot \overline{\alpha\beta\gamma}$, where $\overline{\alpha\beta\gamma}$ denotes the direction which is perpendicular to the perpendicular to α and β and γ.

If the product is formed after the other mode of association we have

$$A(BC) = (a_1 i + a_2 j)(b_1 c_1 + b_2 c_2) + (a_1 i + a_2 j)(b_1 c_2 - b_2 c_1)k$$
$$= (b_1 c_1 + b_2 c_2)(a_1 i + a_2 j) + (b_1 c_2 - b_2 c_1)(a_2 i - a_1 j)$$
$$= SBC \cdot A + VA(VBC).$$

The vector $a_2 i - a_1 j$ is the opposite of the complementary vector of $a_1 i + a_2 j$. Hence the latter partial product differs with the mode of association.

Example.—Let $A = 1/\underline{0} + 2/\underline{90°}$, $B = 3/\underline{0°} + 4/\underline{90°}$, $C = 5/\underline{0°} + 6/\underline{90°}$. The fourth proportional to A, B, C is

$$(A^{-1}B)C = \frac{1 \times 3 + 2 \times 4}{1^2 + 2^2} \{ 5\underline{/0^\circ} + 6\underline{/90^\circ} \}$$

$$+ \frac{1 \times 4 - 2 \times 3}{1^2 + 2^2} \{ -6 \underline{/0^\circ} + 5\underline{/90^\circ} \}$$

$$= 13.4\underline{/0^\circ} + 11.2\underline{/90^\circ}.$$

Square of a Binomial of Vectors.—If $A + B$ denotes a sum of non-successive vectors, it is entirely equivalent to the resultant vector C. But the square of any vector is a positive scalar, hence the square of $A + B$ must be a positive scalar. Since A and B are in reality components of one vector, the square must be formed after the rules for the products of rectangular components (p. 432). Hence

$$(A + B)^2 = (A + B)(A + B),$$
$$= A^2 + AB + BA + B^2,$$
$$= A^2 + B^2 + SAB + SBA + VAB + VBA,$$
$$= A^2 + B^2 + 2SAB.$$

This may also be written in the form

$$a^2 + b^2 + 2ab \cos \alpha\beta.$$

But when $A + B$ denotes a sum of successive vectors, there is no third vector C which is the complete equivalent; and consequently we need not expect the square to be a scalar quantity. We observe that there is a real order, not of the factors, but of the terms in the binomial; this causes both product terms to be AB, giving

$$(A + B)^2 = A^2 + 2AB + B^2$$
$$= A^2 + B^2 + 2SAB + 2VAB.$$

The scalar part gives the square of the length of the third side, while the vector part gives four times the area included between the path and the third side.

Square of a Trinomial of Coplanar Vectors.—Let $A + B + C$ denote a sum of successive vectors. The product terms must be formed so as to preserve the order of the vectors in the trinomial; that is, A is prior to B and C, and B is prior to C.

Hence

$$(A + B + C)^2 = A^2 + B^2 + C^2 + 2AB + 2AC + 2BC,$$
$$= A^2 + B^2 + C^2 + 2(SAB + SAC + SBC), \quad (1)$$
$$+ 2(VAB + VAC + VBC). \quad (2)$$

Hence $$S(A + B + C)^2 = (1)$$

$$= a^2 + b^2 + c^2 + 2ab \cos \alpha\beta + 2ac \cos \alpha\gamma + 2bc \cos \beta\gamma$$

and $$V(A + B + C)^2 = (2)$$

$$= \{2ab \sin \alpha\beta + 2ac \sin \alpha\gamma + 2bc \sin \beta\gamma\} . \overline{\alpha\beta}$$

The scalar part gives the square of the vector from the beginning of A to the end of C and is all that exists when the vectors are non-successive. The vector part is four times the area included between the successive sides and the resultant side of the polygon.

Note that it is here assumed that $V(A + B)C = VAC + VBC$, which is the theorem of moments. Also that the product terms are not formed in cyclical order, but in accordance with the order of the vectors in the trinomial.

Example.—Let $A = 3/0$, $B = 5/30°$, $C = 7/45°$; find the area of the polygon.

$$\tfrac{1}{2} V(AB + AC + BC),$$
$$= \tfrac{1}{2}\{15 \sin /0 /30° + 21 \sin /0 /45° + 35 \sin /30° /45°\},$$
$$= 3.75 + 7.42 + 4.53 = 15.7.$$

Prob. 10. At a distance of 25 centimeters $/20°$ there is a force of 1000 dynes $/80°$; find the moment.

Prob. 11. A conductor in an armature has a velocity of 240 inches per second $/300°$ and the magnetic flux is 50,000 lines per square inch $/0$; find the vector product.

(Ans. 1.04×10^7 lines per inch per second.)

Prob. 12. Find the sine and cosine of the angle between the directions 0.8141 E. $+ 0.5807$ N., and 0.5060 E. $+ 0.8625$ N.

Prob. 13. When a force of 200 pounds $/270°$ is displaced by 10 feet $/30°$, what is the work done (scalar product)? What is the meaning of the negative sign in the scalar product?

Prob. 14. A mass of 100 pounds is moving with a velocity of 30 feet E. per second $+$ 50 feet SE. per second; find its kinetic energy.

Prob. 15. A force of 10 pounds $/45°$ is acting at the end of 8 feet $/200°$; find the torque, or vector product.

Prob. 16. The radius of curvature of a curve is $2/0° + 5/90°$; find the curvature.   (Ans. $.03/0° + .17/90°$.)

Prob. 17. Find the fourth proportional to $10/0° + 2/90°$ $8/0° - 3/90°$, and $6/0° + 5/90°$.

Prob. 18. Find the area of the polygon whose successive sides are $10/30°$, $9/100°$, $8/180°$, $7/225°$.

ART. 4. COAXIAL QUATERNIONS.

By a "quaternion" is meant the operator which changes one vector into another. It is composed of a magnitude and a turning factor. The magnitude may or may not be a mere ratio, that is, a quantity destitute of physical dimensions; for the two vectors may or may not be of the same physical kind. The turning is in a plane, that is to say, it is not conical. For the present all the vectors considered lie in a common plane; hence all the quaternions considered have a common axis.*

Let A and R be two coinitial vectors; the direction normal to the plane may be denoted by β. The operator which changes A into R consists of a scalar multiplier and a turning round the axis β. Let the former be denoted by r and the latter by β^θ, where θ denotes the angle in radians. Thus $R = r\beta^\theta A$ and reciprocally $A = \frac{1}{r}\beta^{-\theta}R$. Also $\frac{1}{A}R = r\beta^\theta$ and $\frac{1}{R}A = \frac{1}{r}\beta^{-\theta}$.

The turning factor β^θ may be expressed as the sum of two component operators, one of which has a zero angle and the other an angle of a quadrant. Thus

$$\beta^\theta = \cos\theta . \beta^0 + \sin\theta . \beta^{\pi/2}.$$

---

* The idea of the "quaternion" is due to Hamilton. Its importance may be judged from the fact that it has made solid trigonometrical analysis possible. It is the most important key to the extension of analysis to space. Etymologically "quaternion" means defined by four elements; which is true in space · in plane analysis it is defined by two.

When the angle is naught, the turning-factor may be omitted; but the above form shows that the equation is homogeneous, and expresses nothing but the equivalence of a given quaternion to two component quaternions.*

Hence
$$r\beta^\theta = r\cos\theta + r\sin\theta \cdot \beta^{\pi/2}$$
$$= p + q \cdot \beta^{\pi/2}$$

and
$$\boldsymbol{r}\beta^\theta A = pA + q\beta^{\pi/2}A$$
$$= pa \cdot \alpha + qa \cdot \beta^{\pi/2}\alpha.$$

The relations between r and θ, and p and q, are given by

$$r = \sqrt{p^2 + q^2}, \quad \theta = \tan^{-1}\frac{p}{q}.$$

Example.—Let E denote a sine alternating electromotive force in magnitude and phase, and I the alternating current in magitude and phase, then

$$E = (r + 2\pi nl \cdot \beta^{\pi/2})I,$$

where r is the resistance, l the self-induction, n the alternations per unit of time, and β denotes the axis of the plane of representation. It follows that $E = rI + 2\pi nl \cdot \beta^{\pi/2}I$; also that

$$I^{-1}E = r + 2\pi nl \cdot \beta^{\pi/2},$$

that is, the operator which changes the current into the electromotive force is a quaternion. The resistance is the scalar part of the quaternion, and the inductance is the vector part.

Components of the Reciprocal of a Quaternion.—Given

$$R = (p + q \cdot \beta^{\pi/2})A,$$

then
$$A = \frac{1}{p + q \cdot \beta^{\pi/2}} R$$
$$= \frac{p - q \cdot \beta^{\pi/2}}{(p + q \cdot \beta^{\pi/2})(p - q \cdot \beta^{\pi/2})} R$$
$$= \frac{p - q \cdot \beta^{\pi/2}}{p^2 + q^2} R$$
$$= \left\{ \frac{p}{p^2 + q^2} - \frac{q}{p^2 + q^2} \cdot \beta^{\pi/2} \right\} R.$$

* In the method of complex numbers $\beta^{\pi/2}$ is expressed by i, which stands for $\sqrt{-1}$. The advantages of using the above notation are that it is capable of being applied to space, and that it also serves to specify the general turning factor β^θ as well as the quadrantal turning factor $\beta^{\pi/2}$.

Example.—Take the same application as above. It is important to obtain I in terms of E. By the above we deduce that from $E = (r + 2\pi nl . \beta^{\pi/2})I$

$$I = \left\{ \frac{r}{r^2 + (2\pi nl)^2} - \frac{2\pi nl}{r^2 + (2\pi nl)^2} . \beta^{\pi/2} \right\} E.$$

Addition of Coaxial Quaternions.—If the ratio of each of several vectors to a constant vector A is given, the ratio of their resultant to the same constant vector is obtained by taking the sum of the ratios. Thus, if

$$R_1 = (p_1 + q_1 . \beta^{\pi/2})A,$$
$$R_2 = (p_2 + q_2 . \beta^{\pi/2})A,$$
$$\vdots \quad \vdots \quad \vdots \quad \vdots \quad \vdots$$
$$R_n = (p_n + q_n . \beta^{\pi/2})A,$$

then

$$\Sigma R = \{\Sigma p + (\Sigma q) . \beta^{\pi/2}\}A,$$

and reciprocally

$$A = \frac{\Sigma p - (\Sigma q) . \beta^{\pi/2}}{(\Sigma p)^2 + (\Sigma q)^2} \Sigma R.$$

Example.—In the case of a compound circuit composed of a number of simple circuits in parallel

$$I_1 = \frac{r_1 - 2\pi nl_1 . \beta^{\pi/2}}{r_1^2 + (2\pi n)^2 l_1^2} E, \qquad I_2 = \frac{r_2 - 2\pi nl_2 . \beta^{\pi/2}}{r_2^2 + (2\pi n)^2 l_2^2} E, \text{ etc.,}$$

therefore, $\Sigma I = \Sigma \left\{ \dfrac{r - 2\pi nl . \beta^{\pi/2}}{r^2 + (2\pi n)^2 l^2} \right\} E$

$$= \left\{ \Sigma\left(\frac{r}{r^2 + (2\pi n)^2 l^2}\right) - 2\pi n \Sigma \frac{l}{r^2 + (2\pi n)^2 l^2} . \beta^{\pi/2} \right\} E,$$

and reciprocally

$$E = \frac{\Sigma\left(\dfrac{r}{r^2 + (2\pi n)^2 l^2}\right) + 2\pi n \Sigma\left(\dfrac{l}{r^2 + (2\pi n)^2 l^2}\right) . \beta^{\pi/2}}{\left(\Sigma\dfrac{r}{r^2 + (2\pi n)^2 l^2}\right)^2 + (2\pi n)^2 \left(\Sigma\dfrac{l}{r^2 + (2\pi n)^2 l^2}\right)^2} \Sigma I.^*$$

Product of Coaxial Quaternions.—If the quaternions which change A to R, and R to R', are given, the quaternion which changes A to R' is obtained by taking the product of the given quaternions.

---

* This theorem was discovered by Lord Rayleigh; Philosophical Magazine, May, 1886. See also Bedell & Crehore's Alternating Currents, p. 238.

Given $\quad R = r\beta^\theta A = (p + q \cdot \beta^{\pi/2})A$

and $\qquad R' = r'\beta^{\theta'}R = (p' + q' \cdot \beta^{\pi/2})R,$

then $R' = rr'\beta^{\theta+\theta'}\, A = \{(pp' - qq') + (pq' + p'q) \cdot \beta^{\pi/2}\}A.$

Note that the product is formed by taking the product of the magnitudes, and likewise the product of the turning factors. The angles are summed because they are indices of the common base β.*

Quotient of two Coaxial Quaternions.—If the given quaternions are those which change A to R, and A to R', then that which changes R to R' is obtained by taking the quotient of the latter by the former.

Given $\quad R = r\beta^\theta A = (p + q \cdot \beta^{\pi/2})A$

and $\qquad R' = r'\beta^{\theta'}A = (p' + q' \cdot \beta^{\pi/2})A,$

then $\qquad R' = \dfrac{r'}{r}\beta^{\theta'-\theta}R,$

$$= (p' + q' \cdot \beta^{\pi/2})\frac{1}{p + q \cdot \beta^{\pi/2}}R,$$

$$= (p' + q' \cdot \beta^{\pi/2})\frac{(p - q \cdot \beta^{\pi/2})}{p^2 + q^2}R,$$

$$= \frac{(pp' + qq') + (pq' - p'q) \cdot \beta^{\pi/2}}{p^2 + q^2}R.$$

Prob. 19. The impressed alternating electromotive force is 200 volts, the resistance of the circuit is 10 ohms, the self-induction is $\frac{1}{100}$ henry, and there are 60 alternations per second ; required the current.   (Ans. 18.7 amperes $\underline{/-20°\ 42'}$.)

Prob. 20. If in the above circuit the current is 10 amperes, find the impressed voltage.

Prob. 21. If the electromotive force is 110 volts $\underline{/\theta}$ and the current is 10 amperes $\underline{/\theta - \frac{1}{4}\pi}$, find the resistance and the self-induction, there being 120 alternations per second.

Prob. 22.. A number of coils having resistances r_1, r_2, etc., and self-inductions l_1, l_2, etc., are placed in series : find the impressed electromotive force in terms of the current, and reciprocally.

---

* Many writers, such as Hayward in '' Vector Algebra and Trigonometry,'' and Stringham in '' Uniplanar Algebra,'' treat this product of coaxial quaternions as if it were the product of vectors. This is the fundamental error in the Argand method.

ART. 5.   ADDITION OF VECTORS IN SPACE.

A vector in space can be expressed in terms of three inde-pendent components, and when these form a rectangular set the directions of resolution are expressed by i, j, k. Any vari-able vector R may be expressed as $R = r\rho = xi + yj + zk$, and any constant vector B may be expressed as

$$B = b\beta = b_1 i + b_2 j + b_3 k.$$

In space the symbol ρ for the direction involves two ele-ments. It may be specified as

$$\rho = \frac{xi + yj + zk}{x^2 + y^2 + z^2},$$

where the three values are subject to the condition that their sum is unity. Or it may be specified by this notation, $\overline{\phi/}/\theta$, a generalization of the notation for a plane. The additional angle $\overline{\phi/}$ is introduced to specify the plane in which the angle from the initial line lies.

If we are given R in the form $r\,\overline{\phi/}/\theta$, then we deduce the other form thus:

$$R = r\cos\theta \cdot i + r\sin\theta\cos\phi \cdot j + r\sin\theta\sin\phi \cdot k.$$

If R is given in the form $xi + yj + zk$, we deduce

$$R = \sqrt{x^2 + y^2 + z^2}\ \overline{\tan^{-1}\frac{z}{y}}\ /\!/\ \tan^{-1}\frac{\sqrt{y^2 + z^2}}{x}.$$

For example, $B = 10\ \overline{30°}/\!/45°$
$= 10\cos 45° \cdot i + 10\sin 45°\cos 30° \cdot j + 10\sin 45°\sin 30° \cdot k.$

Again, from $C = 3i - 4j + 5k$ we deduce

$$C = \sqrt{9 + 16 + 25}\ \overline{\tan^{-1}\frac{5}{4}}\ /\!/\ \tan^{-1}\frac{\sqrt{41}}{3}.$$

$$= 7.07\ \overline{51^{c}.4}/\!/64°.9.$$

To find the resultant of any number of component vectors applied at a common point, let $R_1, R_2, \ldots R_n$ represent the n vectors or,

$$R_1 = x_1 i + y_1 j + z_1 k,$$
$$R_2 = x_2 i + y_2 j + z_2 k,$$
$$\cdot \quad \cdot \quad \cdot \quad \cdot$$
$$R_n = x_n i + y_n j + z_n k;$$

then
$$\Sigma R = (\Sigma x)i + (\Sigma y)j + (\Sigma z)k$$

and
$$r = \sqrt{(\Sigma x)^2 + (\Sigma y)^2 + (\Sigma z)^2},$$

$$\tan \phi = \frac{\Sigma z}{\Sigma y} \quad \text{and} \quad \tan \theta = \frac{\sqrt{(\Sigma y)^2 + (\Sigma z)^2}}{\Sigma x}.$$

Successive Addition.—When the successive vectors do not lie in one plane, the several elements of the area enclosed will lie in different planes, but these add by vector addition into a resultant directed area.

Prob. 23. Express $A = 4i - 5j + 6k$ and $B = 5i + 6j - 7k$ in the form $r\overline{\phi}//\underline{\theta}$. (Ans. 8.8 $\overline{130°}//\underline{63°}$ and 10.5 $\overline{311°}//\underline{61°.5}$.)

Prob. 24. Express $C = 123\ \overline{57°}//\underline{142°}$ and $D = 456\ \overline{65°}//\underline{200°}$ in the form $xi + yj + zk$.

Prob. 25. Express $E = 100\ \overline{\dfrac{\pi}{4}}//\underline{\dfrac{\pi}{3}}$ and $F = 1000\ \overline{\dfrac{\pi}{6}}//\underline{3\dfrac{\pi}{4}}$ in the form $xi + yj + zk$.

Prob. 26. Find the resultant of 10 $\overline{20°}//\underline{30°}$, 20 $\overline{30°}//\underline{40°}$, and 30 $\overline{40°}//\underline{50°}$.

Prob. 27. Express in the form $r\overline{\phi}//\underline{\theta}$ the resultant vector of $1i + 2j - 3k$, $4i - 5j + 6k$, and $-7i + 8j + 9k$.

ART. 6.   PRODUCT OF TWO VECTORS.

Rules of Signs for Vectors in Space.—By the rules $i^2 = +$, $j^2 = +$, $ij = k$, and $ji = -k$ we obtained (p. 432) a product of two vectors containing two partial products, each of which has the highest importance in mathematical and physical analysis. Accordingly, from the symmetry of space we assume that the following rules are true for the product of two vectors in space:

$$i^2 = +, \qquad j^2 = +, \qquad k^2 = +,$$
$$ij = k, \qquad jk = i, \qquad ki = j,$$
$$ji = -k, \qquad kj = -i, \qquad ik = -j.$$

The square combinations give results which are indepen-

dent of direction, and consequently are summed by simple
addition. The area vector determined by
i and j can be represented in direction by k,
because k is in tri-dimensional space the axis
which is complementary to i and j. We also
observe that the three rules $ij = k$, $jk = i$,
$ki = j$ are derived from one another by cyclical permutation; likewise the three rules

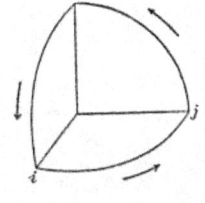

$ji = -k$, $kj = -i$, $ik = -j$. The figure shows that these
rules are made to represent the relation of the advance to the
rotation in the right-handed screw. The physical meaning of
these rules is made clearer by an application to the dynamo and
the electric motor. In the dynamo three principal vectors have
to be considered: the velocity of the conductor at any instant,
the intensity of magnetic flux, and the vector of electromotive
force. Frequently all that is demanded is, given two of these
directions to determine the third. Suppose that the direction
of the velocity is i, and that of the flux j, then the direction of
the electromotive force is k. The formula $ij = k$ becomes

$$\text{velocity flux} = \text{electromotive-force,}$$

from which we deduce

$$\text{flux electromotive-force} = \text{velocity,}$$

and

$$\text{electromotive-force velocity} = \text{flux.}$$

The corresponding formula for the electric motor is

$$\text{current flux} = \text{mechanical-force,}$$

from which we derive by cyclical permutation

$$\text{flux force} = \text{current, and force current} = \text{flux.}$$

The formula velocity flux = electromotive-force is much
handier than any thumb-and-finger rule; for it compares the
three directions directly with the right-handed screw.

Example.—Suppose that the conductor is normal to the
plane of the paper, that its velocity is towards the bottom, and
that the magnetic flux is towards the left; corresponding to
the rotation from the velocity to the flux in the right-handed
screw we have advance into the paper: that then is the direction of the electromotive force.

Again, suppose that in a motor the direction of the current

along the conductor is up from the paper, and that the magnetic flux is to the left; corresponding to current flux we have
advance towards the bottom of the page, which therefore must
be the direction of the mechanical force which is applied to
the conductor.

Complete Product of two Vectors.—Let $A = a_1i + a_2j + a_3k$
and $B = b_1i + b_2j + b_3k$ be any two vectors, not necessarily
of the same kind physically, Their product, according to the
rules (p. 444), is

$$AB = (a_1i + a_2j + a_3k)(b_1i + b_2j + b_3k),$$
$$= a_1b_1ii + a_2b_2jj + a_3b_3kk,$$
$$+ a_2b_3jk + a_3b_2kj + a_3b_1ki + a_1b_3ik + a_1b_2ij + a_2b_1ji$$
$$= a_1b_1 + a_2b_2 + a_3b_3,$$
$$+ (a_2b_3 - a_3b_2)i + (a_3b_1 - a_1b_3)j + (a_1b_2 - a_2b_1)k$$
$$= a_1b_1 + a_2b_2 + a_3b_3 + \begin{vmatrix} a_1 & a_2 & a_3 \\ b_1 & b_2 & b_3 \\ i & j & k \end{vmatrix}.$$

Thus the product breaks up into two partial products,
namely, $a_1b_1 + a_2b_2 + a_3b_3$, which is independent of direction, and
$\begin{vmatrix} a_1 & a_2 & a_3 \\ b_1 & b_2 & b_3 \\ i & j & k \end{vmatrix}$, which has the direction normal to the plane of
A and B. The former is called the scalar product, and the
latter the vector product.

In a sum of vectors, the vectors are necessarily homogeneous, but in a product the vectors may be heterogeneous. By
making $a_3 = b_3 = 0$, we deduce the results already obtained
for a plane.

Scalar Product of two Vectors.—The scalar product is denoted as before by SAB. Its geometrical
meaning is the product of A and the orthogonal projection of B upon A. Let OP represent A, and OQ represent B, and let OL,
LM, and MN be the orthogonal projections
upon OP of the coordinates b_1i, b_2j, b_3k respectively. Then ON is the orthogonal projection of OQ, and

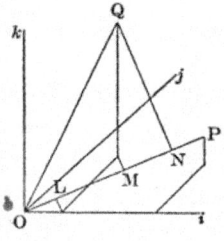

$$OP \times ON = OP \times (OL + LM + MN),$$
$$= a\left(b_1\frac{a_1}{a} + b_2\frac{a_2}{a} + b_3\frac{a_3}{a}\right),$$
$$= a_1b_1 + a_2b_2 + a_3b_3 = SAB.$$

Example. — Let the intensity of a magnetic flux be $B = b_1i + b_2j + b_3k$, and let the area be $S = s_1i + s_2j + s_3k$; then the flux through the area is $SSB = b_1s_1 + b_2s_2 + b_3s_3$.

Corollary 1.—Hence $SBA = SAB$. For
$$b_1a_1 + b_2a_2 + b_3a_3 = a_1b_1 + a_2b_2 + a_3b_3.$$

The product of B and the orthogonal projection on it of A is equal to the product of A and the orthogonal projection on it of B. The product is positive when the vector and the projection have the same direction, and negative when they have opposite directions.

Corollary 2.—Hence $A^2 = a_1^2 + a_2^2 + a_3^2 = a^2$. The square of A must be positive; for the two factors have the same direction.

Vector Product of two Vectors.—The vector product as before is denoted by VAB. It means the product of A and the component of B which is perpendicular to A, and is represented by the area of the parallelogram formed by A and B. The orthogonal projections of this area upon the planes of jk, ki, and ij represent the respective components of the product. For, let OP and OQ (see second figure of Art. 3) be the orthogonal projections of A and B on the plane of i and j; then the triangle OPQ is the projection of half of the parallelogram formed by A and B. But it is there shown that the area of the triangle OPQ is $\frac{1}{2}(a_1b_2 - a_2b_1)$. Thus $(a_1b_2 - a_2b_1)k$ denotes the magnitude and direction of the parallelogram formed by the projections of A and B on the plane of i and j. Similarly $(a_2b_3 - a_3b_2)i$ denotes in magnitude and direction the projection on the plane of j and k, and $(a_3b_1 - a_1b_3)j$ that on the plane of k and i.

Corollary 1.—Hence $VBA = - VAB$.

Example.—Given two lines $A = 7i - 10j + 3k$ and $B = -9i + 4j - 6k$; to find the rectangular projections of the parallelogram which they define:

$$VAB = (60 - 12)i + (- 27 + 42)j + (28 - 90)k$$
$$= 48i + 15j - 62k.$$

Corollary 2.—If A is expressed as $a\alpha$ and B as $b\beta$, then $SAB = ab \cos \alpha\beta$ and $VAB = ab \sin \alpha\beta . \overline{\alpha\beta}$, where $\overline{\alpha\beta}$ denotes the direction which is normal to both α and β, and drawn in the sense given by the right-handed screw.

Example.—Given $A = r\overline{\phi}//\underline{\theta}$ and $B = r'\overline{\phi'}//\underline{\theta'}$. Then

$$SAB = rr' \cos \overline{\phi}//\underline{\theta}\ \overline{\phi'}//\underline{\theta'}$$
$$= rr'\{\cos \theta \cos \theta' + \sin \theta \sin \theta' \cos (\phi' - \phi)\}.$$

Product of two Sums of non-successive Vectors.—Let A and B be two component vectors, giving the resultant $A + B$, and let C denote any other vector having the same point of application.

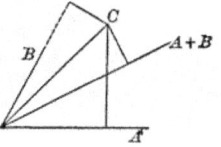

Let
$$A = a_1 i + a_2 j + a_3 k,$$
$$B = b_1 i + b_2 j + b_3 k,$$
$$C = c_1 i + c_2 j + c_3 k.$$

Since A and B are independent of order,

$$A + B = (a_1 + b_1)i + (a_2 + b_2)j + (a_3 + b_3)k,$$

consequently by the principle already established

$$S(A + B)C = (a_1 + b_1)c_1 + (a_2 + b_2)c_2 + (a_3 + b_3)c_3$$
$$= a_1 c_1 + a_2 c_2 + a_3 c_3 + b_1 c_1 + b_2 c_2 + b_3 c_3$$
$$= SAC + SBC.$$

Similarly $V(A + B)C = \{(a_2 + b_2)c_3 - (a_3 + b_3)c_2\}i + \text{etc.}$
$$= (a_2 c_3 - a_3 c_2)i + (b_2 c_3 - b_3 c_2)i + \ldots$$
$$= VAC + VBC.$$

Hence $(A + B)C = AC + BC.$

In the same way it may be shown that if the second factor consists of two components, C and D, which are non-successive in their nature, then

$$(A + B)(C + D) = AC + AD + BC + BD.$$

When $A + B$ is a sum of component vectors
$$(A + B)^2 = A^2 + B^2 + AB + BA$$
$$= A^2 + B^2 + 2SAB.$$

Prob. 28. The relative velocity of a conductor is S.W., and the magnetic flux is N.W.; what is the direction of the electromotive force in the conductor?

Prob. 29. The direction of the current is vertically downward, that of the magnetic flux is West; find the direction of the mechanical force on the conductor.

Prob. 30. A body to which a force of $2i + 3j + 4k$ pounds is applied moves with a velocity of $5i + 6j + 7k$ feet per second; find the rate at which work is done.

Prob. 31. A conductor $8i + 9j + 10k$ inches long is subject to an electromotive force of $11i + 12j + 13k$ volts per inch; find the difference of potential at the ends.   (Ans. 326 volts.)

Prob. 32. Find the rectangular projections of the area of the parallelogram defined by the vectors $A = 12i - 23j - 34k$ and $B = -45i - 56j + 67k$.

*Prob. 33. Show that the moment of the velocity of a body with respect to a point is equal to the sum of the moments of its component velocities with respect to the same point.

Prob. 34. The arm is $9i + 11j + 13k$ feet, and the force applied at either end is $17i + 19j + 23k$ pounds weight; find the torque.

Prob. 35. A body of 1000 pounds mass has linear velocities of 50 feet per second $\overline{30°}//\underline{45°}$, and 60 feet per second $\overline{60°}//\underline{22°.5}$; find its kinetic energy.

Prob. 36. Show that if a system of area-vectors can be represented by the faces of a polyhedron, their resultant vanishes.

Prob. 37. Show that work done by the resultant velocity is equal to the sum of the works done by its components.

ART. 7. PRODUCT OF THREE VECTORS.

Complete Product.—Let us take $A = a_1 i + a_2 j + a_3 k$, $B = b_1 i + b_2 j + b_3 k$, and $C = c_1 i + c_2 j + c_3 k$. By the product of A, B, and C is meant the product of the product of A and B with C, according to the rules p. 444). Hence

$$ABC = (a_1 b_1 + a_2 b_2 + a_3 b_3)(c_1 i + c_2 j + c_3 k)$$
$$+ \{(a_2 b_3 - a_3 b_2)i + (a_3 b_1 - a_1 b_3)j + (a_1 b_2 - a_2 b_1)k\}(c_1 i + c_2 j + c_3 k)$$
$$= (a_1 b_1 + a_2 b_2 + a_3 b_3)(c_1 i + c_2 j + c_3 k) \qquad (1)$$

$$+ \left\| \begin{array}{cc} a_2 & a_3 \\ b_2 & b_3 \\ c_1 & \\ i & \end{array} \quad \begin{array}{cc} a_3 & a_1 \\ b_3 & b_1 \\ c_2 & \\ j & \end{array} \quad \begin{array}{cc} a_1 & a_2 \\ b_1 & b_2 \\ c_3 & \\ k & \end{array} \right\| \quad (2) \qquad + \left| \begin{array}{ccc} a_1 & a_2 & a_3 \\ b_1 & b_2 & b_3 \\ c_1 & c_2 & c_3 \end{array} \right| \quad (3)$$

Example.—Let $A = 1i + 2j + 3k$, $B = 4i + 5j + 6k$, and $C = 7i + 8j + 9k$.   Then

$$(1) = (4 + 10 + 18)(7i + 8j + 9k) = 32(7i + 8j + 9k).$$

$$(2) = \left| \begin{array}{ccc} -3 & 6 & -3 \\ 7 & 8 & 9 \\ i & j & k \end{array} \right| = 78i + 6j - 66k.$$

$$(3) = \left| \begin{array}{ccc} 1 & 2 & 3 \\ 4 & 5 & 6 \\ 7 & 8 & 9 \end{array} \right| = 0.$$

If we write $A = a\alpha$, $B = b\beta$, $C = c\gamma$, then

$$ABC = abc \cos \alpha\beta . \gamma \tag{1}$$

$$+ \, abc \sin \alpha\beta \sin \overline{\alpha\beta\gamma} . \overline{\overline{\alpha\beta\gamma}} \tag{2}$$

$$+ \, abc \sin \alpha\beta \cos \overline{\alpha\beta\gamma}, \tag{3}$$

where $\cos \overline{\alpha\beta\gamma}$ denotes the cosine of the angle between the directions $\overline{\alpha\beta}$ and γ, and $\overline{\overline{\alpha\beta\gamma}}$ denotes the direction which is normal to both $\overline{\alpha\beta}$ and γ.

We may also write

$$ABC = SAB.C + \mathrm{V}(\mathrm{V}AB)C + \mathrm{S}(\mathrm{V}AB)C.$$
$$\qquad\quad (1) \qquad\qquad (2) \qquad\qquad (3)$$

First Partial Product.—It is merely the third vector multiplied by the scalar product of the other two, or weighted by that product as an ordinary algebraic quantity.   If the directions are kept constant, each of the three partial products is proportional to each of the three magnitudes.

Second Partial Product.—The second partial product may be expressed as the difference of two products similar to the first.   For

$$\mathrm{V}(\mathrm{V}AB)C = \{ - (b_2c_2 + b_3c_3)a_1 + (c_2a_2 + c_3a_3)b_1 \} i$$
$$+ \{ - (b_3c_3 + b_1c_1)a_2 + (c_3a_3 + c_1a_1)b_2 \} j$$
$$+ \{ - (b_1c_1 + b_2c_2)a_3 + (c_1a_1 + c_2a_2)b_3 \} k.$$

By adding to the first of these components the null term $(b_1c_1a_1 - c_1a_1b_1)i$ we get $- SBC \cdot a_1i + SCA \cdot b_1i,$ and by treating the other two components similarly and adding the results we obtain

$$V(VAB)C = - SBC \cdot A + SCA \cdot B.$$

The principle here proved is of great use in solving equations (see p. 455).

Example.—Take the same three vectors as in the preceding example. Then

$$V(VAB)C = - (28 + 40 + 54)(1i + 2j + 3k)$$
$$+ (7 + 16 + 27)(4i + 5j + 6k)$$
$$= 78i + 6j - 66k.$$

The determinant expression for this partial product may also be written in the form

$$\begin{vmatrix} a_1 & a_2 \\ b_1 & b_2 \end{vmatrix}\begin{vmatrix} c_1 & c_2 \\ i & j \end{vmatrix} + \begin{vmatrix} a_2 & a_3 \\ b_2 & b_3 \end{vmatrix}\begin{vmatrix} c_2 & c_3 \\ j & k \end{vmatrix} + \begin{vmatrix} a_3 & a_1 \\ b_3 & b_1 \end{vmatrix}\begin{vmatrix} c_3 & c_1 \\ k & i \end{vmatrix}$$

It follows that the frequently occurring determinant expression

$$\begin{vmatrix} a_1 & a_2 \\ b_1 & b_2 \end{vmatrix}\begin{vmatrix} c_1 & c_2 \\ d_1 & d_2 \end{vmatrix} + \begin{vmatrix} a_2 & a_3 \\ b_2 & b_3 \end{vmatrix}\begin{vmatrix} c_2 & c_3 \\ d_2 & d_3 \end{vmatrix} + \begin{vmatrix} a_3 & a_1 \\ b_3 & b_1 \end{vmatrix}\begin{vmatrix} c_3 & c_1 \\ d_3 & d_1 \end{vmatrix}$$

means $S(VAB)(VCD).$

Third Partial Product.—From the determinant expression for the third product, we know that

$$S(VAB)C = S(VBC)A = S(VCA)B$$
$$= - S(VBA)C = - S(VCB)A = - S(VAC)B.$$

Hence any of the three former may be expressed by $SABC,$ and any of the three latter by $- SABC.$

The third product $S(VAB)C$ is represented by the volume of the parallelepiped formed by the vectors A, B, C taken in that order. The line VAB represents in magnitude and direction the area formed by A and B, and the product of VAB with the projection of C upon it is the measure of the 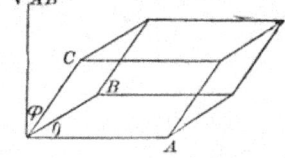 volume in magnitude and sign. Hence the volume formed by the three vectors has no direction in space, but it is positive or negative according to the cyclical order of the vectors.

In the expression $abc \sin \alpha\beta \cos \alpha\beta\gamma$ it is evident that $\sin \alpha\beta$ corresponds to $\sin \theta$, and $\cos \alpha\beta\gamma$ to $\cos \phi$, in the usual formula for the volume of a parallelepiped.

Example.—Let the velocity of a straight wire parallel to itself be $V = 1000 \,\underline{/30^\circ}$ centimeters per second, let the intensity of the magnetic flux be $B = 6000 \,\underline{/90^\circ}$ lines per square centimeter, and let the straight wire $L = 15$ centimeters $\overline{60^\circ}/\,\underline{/45^\circ}$. Then $V\,VB = 6000000 \sin 60^\circ \,\overline{90^\circ}/\,\underline{/90^\circ}$ lines per centimeter per second. Hence $S(VVB)L = 15 \times 6000000 \sin 60^\circ \cos\phi$ lines per second where $\cos\phi = \sin 45^\circ \sin 60^\circ$.

Sum of the Partial Vector Products.—By adding the first and second partial products we obtain the total vector product of ABC, which is denoted by $V(ABC)$. By decomposing the second product we obtain
$$V(ABC) = SAB.\,C - SBC.\,A + SCA.\,B.$$
By removing the common multiplier abc, we get
$$V(\alpha\beta\gamma) = \cos \alpha\beta.\,\gamma - \cos \beta\gamma.\,\alpha + \cos \gamma\alpha.\,\beta.$$
Similarly $V(\beta\gamma\alpha) = \cos \beta\gamma.\,\alpha - \cos \gamma\alpha.\,\beta + \cos \alpha\beta.\,\gamma$
and $\qquad V(\gamma\alpha\beta) = \cos \gamma\alpha.\,\beta - \cos \alpha\beta.\,\gamma + \cos \beta\gamma.\,\alpha.$

These three vectors have the same magnitude, for the square of each is
$$\cos^2 \alpha\beta + \cos^2 \beta\gamma + \cos^2 \gamma\alpha - 2 \cos \alpha\beta \cos \beta\gamma \cos \gamma\alpha,$$
that is, $1 - \{S(\alpha\beta\gamma)\}^2$.

They have the directions respectively of α', β', γ', which are the corners of the triangle whose sides are bisected by the corners α, β, γ of the given triangle.

Prob. 38. Find the second partial product of $9\,\overline{20^\circ}/\,\underline{/30^\circ}$, $10\,\overline{30^\circ}/\,\underline{/40^\circ}$, $11\,\overline{45^\circ}/\,\underline{/45^\circ}$. Also the third partial product.

Prob. 39. Find the cosine of the angle between the plane of $l_1 i + m_1 j + n_1 k$ and $l_2 i + m_2 j + n_2 k$ and the plane of $l_3 i + m_3 j + n_3 k$ and $l_4 i + m_4 j + n_4 k$.

Prob. 40. Find the volume of the parallelepiped determined by the vectors $100i + 50j + 25k$, $50i + 10j + 80k$, and $-75i + 40j - 80k$.

Prob. 41. Find the volume of the tetrahedron determined by the extremities of the following vectors : $3i - 2j + 1k$, $- 4i + 5j - 7k$, $3i - 7j - 2k$, $8i + 4j - 3k$.

Prob. 42. Find the voltage at the terminals of a conductor when its velocity is 1500 centimeters per second, the intensity of the magnetic flux is 7000 lines per square centimeter, and the length of the conductor is 20 centimeters, the angle between the first and second being 30°, and that between the plane of the first two and the direction of the third 60°.   (Ans. .91 volts.)

Prob. 43. Let $\alpha = \overline{20°}//\underline{10}°$, $\beta = \overline{30°}//\underline{25}°$, $\gamma = \overline{40°}//\underline{35}°$. Find $V\alpha\beta\gamma$, and deduce $V\beta\gamma\alpha$ and $V\gamma\alpha\beta$.

ART. 8.   COMPOSITION OF QUANTITIES.

A number of homogeneous quantities are simultaneously located at different points; it is required to find how to add or compound them.

Addition of a Located Scalar Quantity.—Let m_A denote a mass m situated at the extremity of the radius-vector A. A mass $m - m$ may be introduced at the extremity of any radius-vector R, so that

$$m_A = (m - m)_R + m_A$$
$$= m_R + m_A - m_R$$
$$= m_R + m(A - R).$$

Here $A - R$ is a simultaneous sum, and denotes the radius-vector from the extremity of R to the extremity of A. The product $m(A - R)$ is what Clerk Maxwell called a mass-vector, and means the directed moment of m with respect to the extremity of R. The equation states that the mass m at the extremity of the vector A is equivalent to the equal mass at the extremity of R, together with the said mass-vector applied at the extremity of R. The equation expresses a physical or mechanical principle.

Hence for any number of masses, m_1 at the extremity of A_1, m_2 at the extremity of A_2, etc.,

$$\Sigma m_A = \Sigma m_R + \Sigma\{m(A - R)\},$$

where the latter term denotes the sum of the mass-vectors treated as simultaneous vectors applied at a common point.

Since
$$\Sigma\{m(A - R)\} = \Sigma mA - \Sigma mR$$
$$= \Sigma mA - R\Sigma m,$$

the resultant moment will vanish if

$$R = \frac{\Sigma mA}{\Sigma m}, \quad \text{or} \quad R\Sigma m = \Sigma mA$$

Corollary.—Let $\quad R = xi + yj + zk,$
and $\quad A = a_i i + b_i j + c_i k;$

then the above condition may be written as

$$xi + yj + zk = \frac{\Sigma\{m(ai + bj + ck)\}}{\Sigma m}$$

$$= \frac{\Sigma(ma) \cdot i}{\Sigma m} + \frac{(\Sigma mb) \cdot j}{\Sigma m} + \frac{\Sigma(mc) \cdot k}{\Sigma m};$$

therefore $\quad x = \dfrac{\Sigma(ma)}{\Sigma m}, \quad y = \dfrac{\Sigma(mb)}{\Sigma m}, \quad z = \dfrac{\Sigma mc}{\Sigma m}.$

Example.—Given 5 pounds at 10 feet $\overline{45°}//\underline{30°}$ and 8 pounds at 7 feet $\overline{60°}//\underline{45°}$; find the moment when both masses are transferred to 12 feet $\overline{75°}//\underline{60°}$.

$$m_1 A_1 = 50(\cos 30°i + \sin 30° \cos 45°j + \sin 30° \sin 45°k),$$
$$m_2 A_2 = 56(\cos 45°i + \sin 45° \cos 60°j + \sin 45° \sin 60°k),$$
$$(m_1 + m_2)R = 126(\cos 60°i + \sin 60° \cos 75°j + \sin 60° \sin 75°k),$$
$$\text{moment} = m_1 A_1 + m_2 A_2 - (m_1 + m_2)R.$$

Composition of a Located Vector Quantity.—Let F_A denote a force applied at the extremity of the radius-vector A. As a force $F - F$ may introduced at the extremity of any radius-vector R, we have

$$F_A = (F - F)_R + F_A$$
$$= F_R + V(A - R)F.$$

This equation asserts that a force F applied at the extremity of A is equivalent to an equal force applied at the extremity of R together with a couple whose magnitude and direction are given by the vector product of the radius-vector from the extremity of R to the extremity of A and the force.

Hence for a system of forces applied at different points, such as F_1 at A_1, F_2 at A_2, etc., we obtain

$$\Sigma(F_A) = \Sigma(F_R) + \Sigma V(A - R)F$$
$$= (\Sigma F)_R + \Sigma V(A - R)F.$$

Since
$$\Sigma V(A - R)F = \Sigma VAF - \Sigma VRF$$
$$= \Sigma VAF - VR\Sigma F$$

the condition for no resultant couple is

$$VR\Sigma F = \Sigma VAF,$$

which requires ΣF to be normal to ΣVAF.

Example.—Given a force $1i + 2j + 3k$ pounds weight at $4i + 5j + 6k$ feet, and a force of $7i + 9j + 11k$ pounds weight at $10i + 12j + 14k$ feet; find the torque which must be supplied when both are transferred to $2i + 5j + 3k$, so that the effect may be the same as before.

$$VA_1F_1 = 3i - 6j + 3k,$$
$$VA_2F_2 = 6i - 12j + 6k,$$
$$\Sigma VAF = 9i - 18j + 9k,$$
$$\Sigma F = 8i + 11j + 14k,$$
$$VR\Sigma F = 37i - 4j - 18k,$$
$$\text{Torque} = -28i - 14j + 27k.$$

By taking the vector product of the above equal vectors with the reciprocal of ΣF we obtain

$$V\left\{ (VR\Sigma F)\frac{1}{\Sigma F} \right\} = V\left\{ (\Sigma VAF)\frac{1}{\Sigma F} \right\}.$$

By the principle previously established the left member resolves into $-R + SR\frac{1}{\Sigma F} \cdot \Sigma F$; and the right member is equivalent to the complete product on account of the two factors being normal to one another; hence

$$-R + SR\frac{1}{\Sigma F} \cdot \Sigma F = \Sigma(VAF)\frac{1}{\Sigma F};$$

that is,
$$R = \frac{1}{\Sigma F}\Sigma(\mathrm{V}AF) + SR\frac{1}{\Sigma F}\cdot \Sigma F.$$

$$(1) \qquad\qquad (2)$$

The extremity of R lies on a straight line whose perpendicular is the vector (1) and whose direction is that of the resultant force. The term (2) means the projection of R upon that line.

The condition for the central axis is that the resultant force and the resultant couple should have the same direction; hence it is given by

$$\mathrm{V}\{\Sigma\mathrm{V}AF - \mathrm{V}R\Sigma F\}\Sigma F = 0;$$

that is,
$$\mathrm{V}(\mathrm{V}R\Sigma F)\Sigma F = \mathrm{V}(\Sigma AF)\Sigma F.$$

By expanding the left member according to the same principle as above, we obtain

$$-(\Sigma F)^2 R + SR\Sigma F \cdot \Sigma F = \mathrm{V}(\Sigma AF)\Sigma F;$$

therefore
$$R = \frac{1}{(\Sigma F)^2}\mathrm{V}\Sigma F(\mathrm{V}\Sigma AF) + \frac{SR\Sigma F}{(\Sigma F)^2}\cdot \Sigma F$$

$$= \mathrm{V}\left(\frac{1}{\Sigma F}\right)(\mathrm{V}\Sigma AF) + SR\frac{1}{\Sigma F}\cdot \Sigma F.$$

This is the same straight line as before, only no relation is now imposed on the directions of ΣF and $\Sigma\mathrm{V}AF$; hence there always is a central axis.

Example.—Find the central axis for the system of forces in the previous example. Since $\Sigma F = 8i + 11j + 14k$, the direction of the line is

$$\frac{8i + 11j + 14k}{\sqrt{64 + 121 + 196}}.$$

Since $\dfrac{1}{\Sigma F} = \dfrac{8i + 11j + 14k}{381}$ and $\Sigma\mathrm{V}AF = 9i - 18j + 9k$, the perpendicular to the line is

$$\mathrm{V}\frac{8i + 11j + 14k}{381}\,9i - 18j + 9k = \frac{1}{381}\{351i + 54j - 243k\}.$$

Prob. 44. Find the moment at $\overline{90°}//270°$ of 10 pounds at 4 feet $\overline{10°}//20°$ and 20 pounds at 5 feet $30°//120°$.

Prob. 45. Find the torque for $4i + 3j + 2k$ pounds weight at $2i - 3j + 1k$ feet, and $2i - 1j - 1k$ pounds weight at $-3i + 4j + 5k$ feet when transferred to $-3i + 2j - 4k$ feet.

Prob. 46. Find the central axis in the above case.

Prob. 47. Prove that the mass-vector drawn from any origin to a mass equal to that of the whole system placed at the center of mass of the system is equal to the sum of the mass-vectors drawn from the same origin to all the particles of the system.

ART. 9. SPHERICAL TRIGONOMETRY.

Let i, j, k denote three mutually perpendicular axes. In order to distinguish clearly between an axis and a quadrantal version round it, let $i^{\pi/2}, j^{\pi/2}, k^{\pi/2}$ denote quadrantal versions in the positive sense about the axes i, j, k respectively. The directions of positive version are indicated by the arrows.

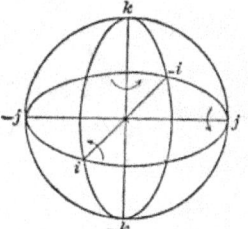

By $i^{\pi/2}i^{\pi/2}$ is meant the product of two quadrantal versions round i; it is equivalent to a semicircular version round i; hence $i^{\pi/2}i^{\pi/2} = i^\pi = -$. Similarly $j^{\pi/2}j^{\pi/2}$ means the product of two quadrantal versions round j, and $j^{\pi/2}j^{\pi/2} = j^\pi = -$. Similarly $k^{\pi/2}k^{\pi/2} = k^\pi = -$.

By $i^{\pi/2}j^{\pi/2}$ is meant a quadrant round i followed by a quadrant round j; it is equivalent to the quadrant from j to i, that is, to $-k^{\pi/2}$. But $j^{\pi/2}i^{\pi/2}$ is equivalent to the quadrant from $-i$ to $-j$, that is, to $k^{\pi/2}$. Similarly for the other two pairs of products. Hence we obtain the following

Rules for Versors.

$$i^{\pi/2}i^{\pi/2} = -, \qquad j^{\pi/2}j^{\pi/2} = -, \qquad k^{\pi/2}k^{\pi/2} = -,$$
$$i^{\pi/2}j^{\pi/2} = -k^{\pi/2}, \qquad j^{\pi/2}i^{\pi/2} = k^{\pi/2},$$
$$j^{\pi/2}k^{\pi/2} = -i^{\pi/2}, \qquad k^{\pi/2}j^{\pi/2} = i^{\pi/2},$$
$$k^{\pi/2}i^{\pi/2} = -j^{\pi/2}, \qquad i^{\pi/2}k^{\pi/2} = j^{\pi/2}.$$

The meaning of these rules will be seen from the following application. Let $li + mj + nk$ denote any axis, then

$(li + mj + nk)^{\pi/2}$ denotes·a quadrant of angle round that axis. This quadrantal version can be decomposed into the three rectangular components $li^{\pi/2}, mj^{\pi/2}, nk^{\pi/2}$; and these components are not successive versions, but the parts of one version. Similarly any other quadrantal version $(l'i + m'j + n'k)^{\pi/2}$ can be resolved into $l'i^{\pi/2}, m'j^{\pi/2}, n'k^{\pi/2}$. By applying the above rules, we obtain

$$(li + mj + nk)^{\pi/2}(l'i + m'j + n'k)^{\pi/2}$$
$$= (li^{\pi/2} + mj^{\pi/2} + nk^{\pi/2})(l'i^{\pi/2} + m'j^{\pi/2} + n'k^{\pi/2})$$
$$= -(ll' + mm' + nn')$$
$$\quad - (mn' - m'n)i^{\pi/2} - (nl' - n'l)j^{\pi/2} - (lm' - l'm)k^{\pi/2}$$
$$= -(ll' + mm' + nn')$$
$$\quad - \{(mn' - m'n)i + (nl' - n'l)j + (lm' - l'm)k\}^{\pi/2}.$$

Product of Two Spherical Versors.—Let β denote the axis and b the ratio of the spherical versor PA, then the versor itself is expressed by β^b. Similarly let γ denote the axis and c the ratio of the spherical versor AQ, then the versor itself is expressed by γ^c.

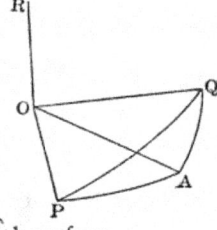

Now $\beta^b = \cos b + \sin b \cdot \beta^{\pi/2}$,

and $\gamma^c = \cos c + \sin c \cdot \gamma^{\pi/2}$;

therefore

$$\beta^b \gamma^c = (\cos b + \sin b \cdot \beta^{\pi/2})(\cos c + \sin c \cdot \gamma^{\pi/2})$$
$$= \cos b \cos c + \cos b \sin c \cdot \gamma^{\pi/2} + \cos c \sin b \cdot \beta^{\pi/2}$$
$$\quad + \sin b \sin c \cdot \beta^{\pi/2} \gamma^{\pi/2}.$$

But from the preceding paragraph

$$\beta^{\pi/2} \gamma^{\pi/2} = -\cos \beta\gamma - \sin \beta\gamma \cdot \overline{\beta\gamma}^{\pi/2};$$

therefore $\beta^b \gamma^c = \cos b \cos c - \sin b \sin c \cos \beta\gamma$ (1)

$$+ \{\cos b \sin c \cdot \gamma + \cos c \sin b \cdot \beta - \sin b \sin c \sin \beta\gamma \cdot \overline{\beta\gamma}\}^{\pi/2}. \quad (2)$$

The first term gives the cosine of the product versor; it is equivalent to the fundamental theorem of spherical trigonometry, namely,

$$\cos a = \cos b \cos c + \sin b \sin c \cos A,$$

where A denotes the external angle instead of the angle included by the sides.

The second term is the directed sine of the angle; for the square of (2) is equal to 1 minus the square of (1), and its direction is normal to the plane of the product angle.*

Example.—Let $\beta = \overline{30°//45°}$ and $\gamma = \overline{60°//30°}$. Then

$$\cos \beta\gamma = \cos 45° \cos 30° + \sin 45° \sin 30° \cos 30°,$$

and
$$\sin \beta\gamma \cdot \overline{\beta\gamma} = V\beta\gamma;$$

but
$$\beta = \cos 45° i + \sin 45° \cos 30° j + \sin 45° \sin 30° k,$$

and
$$\gamma = \cos 30° i + \sin 30° \cos 60° j + \sin 30° \sin 60° k;$$

therefore

$$V\beta\gamma = \{\sin 45° \cos 30° \sin 30° \sin 60°$$
$$- \sin 45° \sin 30° \sin 30° \cos 60°\}i$$
$$+ \{\sin 45° \sin 30° \cos 30° - \cos 45° \sin 30° \sin 60°\}j$$
$$+ \{\cos 45° \sin 30° \cos 60° - \sin 45° \cos 30° \cos 30°\}k.$$

Quotient of Two Spherical Versors.—The reciprocal of a given versor is derived by changing the sign of the index; γ^{-c} is the reciprocal of γ^{c}. As $\beta^{b} = \cos b + \sin b \cdot \beta^{\pi/2}$, and
$$\gamma^{-c} = \cos c - \sin c \cdot \gamma^{\pi/2},$$
$$\beta^{b}\gamma^{-c} = \cos b \cos c + \sin b \sin c \cos \beta\gamma$$
$$+ \{\cos c \sin b \cdot \beta - \cos b \sin c \cdot \gamma + \sin b \sin c \sin \beta\gamma \cdot \overline{\beta\gamma}\}^{\pi/2}.$$

Product of Three Spherical Versors.—Let α^{a} denote the versor PQ, β^{b} the versor QR, and γ^{c} the versor RS; then $\alpha^{a}\beta^{b}\gamma^{c}$ denotes PS. Now $\alpha^{a}\beta^{b}\gamma^{c}$

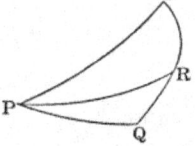

$$= (\cos a + \sin a \cdot \alpha^{\pi/2})(\cos b + \sin b \cdot \beta^{\pi/2})(\cos c + \sin c \cdot \gamma^{\pi/2})$$
$$= \cos a \cos b \cos c \qquad\qquad (1)$$
$$+ \cos a \cos b \sin c \cdot \gamma^{\pi/2} + \cos a \cos c \sin b \cdot \beta^{\pi/2}$$
$$+ \cos b \cos c \sin a \cdot \alpha^{\pi/2} \qquad (2)$$
$$+ \cos a \sin b \sin c \cdot \beta^{\pi/2}\gamma^{\pi/2} + \cos b \sin a \sin c \cdot \alpha^{\pi/2}\gamma^{\pi/2}$$
$$+ \cos c \sin a \sin b \cdot \alpha^{\pi/2}\beta^{\pi/2} \quad (3)$$

---

* Principles of Elliptic and Hyperbolic Analysis, p. 2.

$$+ \sin a \sin b \sin c \cdot \alpha^{\pi/2} \beta^{\pi/2} \gamma^{\pi/2}. \tag{4}$$

The versors in (3) are expanded by the rule already obtained, namely,

$$\beta^{\pi/2} \gamma^{\pi/2} = - \cos \beta\gamma - \sin \beta\gamma \cdot \overline{\beta\gamma}^{\pi/2}.$$

The versor of the fourth term is

$$\alpha^{\pi/2} \beta^{\pi/2} \gamma^{\pi/2} = - (\cos \alpha\beta + \sin \alpha\beta \cdot \overline{\alpha\beta}^{\pi/2}) \gamma^{\pi/2}$$

$$= - \cos \alpha\beta \cdot \gamma^{\pi/2} + \sin \alpha\beta \cos \overline{\alpha\beta}\gamma + \sin \alpha\beta \sin \overline{\alpha\beta}\gamma \cdot \overline{\alpha\beta}\gamma^{\pi/2}.$$

Now $\sin \alpha\beta \sin \overline{\alpha\beta}\gamma \cdot \overline{\alpha\beta}\gamma = \cos \alpha\gamma \cdot \beta - \cos \beta\gamma \cdot \alpha$ (p. 451), hence the last term of the product, when expanded, is

$$\sin a \sin b \sin c \{ - \cos \alpha\beta \cdot \gamma^{\pi/2} + \cos \alpha\gamma \cdot \beta^{\pi/2}$$

$$- \cos \beta\gamma \cdot \alpha^{\pi/2} + \cos \overline{\alpha\beta}\gamma \}.$$

Hence

$$\cos \alpha^a \beta^b \gamma^c = \cos a \cos b \cos c - \cos a \sin b \sin c \cos \beta\gamma$$

$$- \cos b \sin a \sin c \cos \alpha\gamma - \cos c \sin a \sin b \cos \alpha\beta$$

$$+ \sin a \sin b \sin c \sin \alpha\beta \cos \overline{\alpha\beta}\gamma,$$

and, letting Sin denote the directed sine,

$$\text{Sin } \alpha^a \beta^b \gamma^c = \cos a \cos b \sin c \cdot \gamma + \cos a \cos c \sin b \cdot \beta$$

$$+ \cos b \cos c \sin a \cdot \alpha - \cos a \sin b \sin c \sin \beta\gamma \cdot \overline{\beta\gamma}$$

$$- \cos b \sin a \sin c \sin \alpha\gamma \cdot \overline{\alpha\gamma}$$

$$- \cos c \sin a \sin b \sin \alpha\beta \cdot \overline{\alpha\beta}$$

$$- \sin a \sin b \sin c \{ \cos \alpha\beta \cdot \gamma - \cos \alpha\gamma \cdot \beta + \cos \beta\gamma \cdot \alpha \}.*$$

Extension of the Exponential Theorem to Spherical Trigonometry.—It has been shown (p. 458) that

$$\cos \beta^b \gamma^c = \cos b \cos c - \sin b \sin c \cos \beta\gamma$$

and

$$(\sin \beta^b \gamma^c)^{\pi/2} = \cos c \sin b \cdot \beta^{\pi/2} + \cos b \sin c \cdot \gamma^{\pi/2}$$

$$- \sin b \sin c \sin \beta\gamma \cdot \overline{\beta\gamma}^{\pi/2}.$$

Now $\qquad \cos b = 1 - \dfrac{b^2}{2!} + \dfrac{b^4}{4!} - \dfrac{b^6}{6!} + \text{etc.}$

* In the above case the three axes of the successive angles are not perfectly independent, for the third angle must begin where the second leaves off. But the theorem remains true when the axes are independent ; the factors are then quaternions in the most general sense.

and $$\sin b = b - \frac{b^3}{3!} + \frac{b^5}{5!} - \text{etc.}$$

Substitute these series for $\cos b$, $\sin b$, $\cos c$, and $\sin c$ in the above equations, multiply out, and group the homogeneous terms together. It will be found that

$$\cos \beta^b \gamma^c = 1 - \frac{1}{2!}\{b^2 + 2bc \cos \beta\gamma + c^2\}$$

$$+ \frac{1}{4!}\{b^4 + 4b^3c \cos \beta\gamma + 6b^2c^2 + 4bc^3 \cos \beta\gamma + c^4\}$$

$$- \frac{1}{6!}\{b^6 + 6b^5c \cos \beta\gamma + 15b^4c^2 + 20b^3c^3 \cos \beta\gamma$$

$$+ 15b^2c^4 + 6bc^5 \cos \beta\gamma + c^6\} + \ldots,$$

where the coefficients are those of the binomial theorem, the only difference being that $\cos \beta\gamma$ occurs in all the odd terms as a factor. Similarly, by expanding the terms of the sine, we obtain

$$(\mathrm{Sin}\, \beta^b \gamma^c)^{\pi/2} = b \cdot \beta^{\pi/2} + c \cdot \gamma^{\pi/} - bc \sin \beta\gamma \cdot \overline{\beta\gamma}^{\pi/2}$$

$$- \frac{1}{3!}\{b^3 \cdot \beta^{\pi/2} + 3b^2c \cdot \gamma^{\pi/2} + 3bc^2 \cdot \beta^{\pi/2} + c^3 \cdot \gamma^{\pi/2}\}$$

$$+ \frac{1}{3!}\{bc^2 + b^2c\} \sin \beta\gamma \cdot \overline{\beta\gamma}^{\pi/2}$$

$$+ \frac{1}{5!}\{b^5 \cdot \beta^{\pi/2} + 5b^4c \cdot \gamma^{\pi/2} + 10b^3c^2 \cdot \beta^{\pi/2}$$

$$+ 10b^2c^3 \cdot \gamma^{\pi/2} + 5bc^4 \cdot \beta^{\pi/2} + c^5 \cdot \gamma^{\pi/2}\}$$

$$- \frac{1}{5!}\left\{ b^4c + \frac{5 \cdot 4}{2 \cdot 3}b^3c^3 + bc^5 \right\} \sin \beta\gamma \cdot \overline{\beta\gamma}^{\pi/2} - \ldots$$

By adding these two expansions together we get the expansion for $\beta^b \gamma^c$, namely,

$$\beta^b \gamma^c = 1 + b \cdot \beta^{\pi/2} + c \cdot \gamma^{\pi/2}$$

$$- \frac{1}{2!}\{b^2 + 2bc(\cos \beta\gamma + \sin \beta\gamma \cdot \overline{\beta\gamma}^{\pi/2}) + c^2\}$$

$$- \frac{1}{3!}\{b^3 \cdot \beta^{\pi/2} + 3b^2c \cdot \gamma^{\pi/2} + 3bc^2 \cdot \beta^{\pi/2} + c^3 \cdot \gamma^{\pi/2}\}$$

$$+ \frac{1}{4!}\{b^4 + 4b^3c(\cos \beta\gamma + \sin \beta\gamma \cdot \overline{\beta\gamma}^{\pi/2}) + 6b^2c^2$$

$$+ 4bc^3(\cos \beta\gamma + \sin \beta\gamma \cdot \overline{\beta\gamma}^{\pi/2}) + c^4\} + \ldots$$

By restoring the minus, we find that the terms on the second line can be thrown into the form

$$\frac{1}{2!}\{b^2 \cdot \beta^\pi + 2bc \cdot \beta^{\pi/2}\gamma^{\pi/2} + c^2 \cdot \gamma^\pi\},$$

and this is equal to

$$\frac{1}{2!}\{b \cdot \beta^{\pi/2} + c \cdot \gamma^{\pi/2}\}^2,$$

where we have the square of a sum of successive terms. In a similar manner the terms on the third line can be restored to

$$b^3 \cdot \beta^{3\pi/2} + 3b^2c \cdot \beta^\pi \gamma^{\pi/2} + 3bc^2 \cdot \beta^{\pi/2}\gamma^\pi + c^3 \cdot \gamma^{3(\pi/2)},$$

that is,

$$\frac{1}{3!}\{b \cdot \beta^{\pi/2} + c \cdot \gamma^{\pi/2}\}^3.$$

Hence

$$\beta^b\gamma^c = 1 + b \cdot \beta^{\pi/2} + c \cdot \gamma^{\pi/2} + \frac{1}{2!}\{b \cdot \beta^{\pi/2} + c \cdot \gamma^{\pi/2}\}^2$$

$$+ \frac{1}{3!}\{b \cdot \beta^{\pi/2} + c \cdot \gamma^{\pi/2}\}^3 + \frac{1}{4!}\{b \cdot \beta^{\pi/2} + c \cdot \gamma^{\pi/2}\}^4 +$$

$$= e^{b \cdot \beta^{\pi/} + c \cdot \gamma^{\pi/2}}.*$$

Extension of the Binomial Theorem.—We have proved above that $e^{b\beta^{\pi/2}}e^{c\gamma^{\pi/2}} = e^{b\beta^{\pi/2} + c\gamma\pi/2}$ provided that the powers of the binomial are expanded as due to a successive sum, that is, the order of the terms in the binomial must be preserved. Hence the expansion for a power of a successive binomial is given by

$$\{b \cdot \beta^{\pi/2} + c \cdot \gamma^{\pi/2}\}^n = b^n \cdot \beta^{n\pi/2} + nb^{n-1}c \cdot \beta^{(n-1)(\pi/2)}\gamma^{\pi/2}$$

$$+ \frac{n(n-1)}{1 \cdot 2}b^{n-2}c^2 \cdot \beta^{(n-2)(\pi/2)}\gamma^\pi + \text{etc.}$$

---

* At page 386 of his Elements of Quaternions, Hamilton says : "In the present theory of diplanar quaternions we cannot expect to find that the sum of the logarithms of any two proposed factors shall be generally equal to the logarithm of the product ; but for the simpler and earlier case of coplanar quaternions, that algebraic property may be considered to exist, with due modification for multiplicity of value." He was led to this view by not distinguishing between vectors and quadrantal quaternions and between simultaneous and successive addition. The above demonstration was first given in my paper on "The Fundamantal Theorems of Analysis generalized for Space." It forms the key to the higher development of space analysis.

Example.—Let $b = \frac{1}{10}$ and $c = \frac{1}{5}$, $\beta = \overline{30°}//\underline{45°}$, $\gamma = \overline{60°}//\underline{30°}$.

$$(b . \beta^{\pi/2} + c . \gamma^{\pi/2})^2 = -\{b^2 + c^2 + 2bc \cos \beta\gamma + 2bc(\sin \beta\gamma)^{\pi/2}\}$$

$$= -(\tfrac{1}{100} + \tfrac{1}{25} + \tfrac{2}{50} \cos \beta\gamma) - \tfrac{2}{50}(\sin \beta\gamma)^{\pi/2}.$$

Substitute the calculated values of $\cos \beta\gamma$ and $\sin \beta\gamma$.

Prob. 48. Find the equivalent of a quadrantal version round $\frac{\sqrt{3}}{2}i + \frac{1}{2\sqrt{2}}j + \frac{1}{2\sqrt{2}}k$ followed by a quadrantal version round $\frac{1}{2}i + \frac{\sqrt{3}}{4}j + \frac{3}{4}k.$

Prob. 49. In the example on p. 459 let $b = 25°$ and $c = 50°$; calculate out the cosine and the directed sine of the product angle.

Prob. 50. In the above example calculate the cosine and the directed sine up to and inclusive of the fourth power of the binomial. (Ans. $\cos = .9735$.)

Prob. 51. Calculate the first four terms of the series when $b = \frac{1}{80}, c = \frac{1}{100}, \beta = \overline{0}//\underline{0}, \gamma = \overline{90°}//\underline{90°}$.

Prob. 52. From the fundamental theorem of spherical trigonometry deduce the polar theorem with respect to both the cosine and the directed sine.

Prob. 53. Prove that if α^a, β^b, γ^c denote the three versors of a spherical triangle, then

$$\frac{\sin \beta\gamma}{\sin a} = \frac{\sin \gamma\alpha}{\sin b} = \frac{\sin \alpha\beta}{\sin c}.$$

ART. 10. COMPOSITION OF ROTATIONS.

A version refers to the change of direction of a line, but a rotation refers to a rigid body. The composition of rotations is a different matter from the composition of versions.

Effect of a Finite Rotation on a Line.—Suppose that a rigid body rotates θ radians round the axis β passing through the point O, and that R is the radius-vector from O to some particle. In the diagram OB represents the axis β, and OP the vector R. Draw OK and OL, the rectangular components of R.

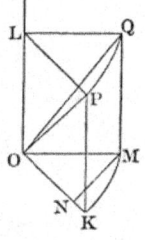

$$\beta^\theta R = (\cos \theta + \sin \theta . \beta^{\pi/2})r\rho$$

$$= r(\cos\theta + \sin\theta \cdot \beta^{\pi/2})(\cos\beta\rho \cdot \beta + \sin\beta\rho \cdot \overline{\overline{\beta\rho\beta}})$$
$$= r\{\cos\beta\rho \cdot \beta + \cos\theta\sin\beta\rho \cdot \overline{\overline{\beta\rho\beta}} + \sin\theta\sin\beta\rho \cdot \overline{\beta\rho}\}.$$

When $\cos\beta\rho = 0$, this reduces to

$$\beta^\theta R = \cos\theta R + \sin\theta V(\beta R).$$

The general result may be written

$$\beta^\theta R = S\beta R \cdot \beta + \cos\theta(V\beta R)\beta + \sin\theta V\beta R.$$

Note that $(V\beta R)\beta$ is equal to $V(V\beta R)\beta$ because $S\beta R\beta$ is 0, for it involves two coincident directions.

Example.—Let $\beta = li + mj + nk$, where $l^2 + m^2 + n^2 = 1$ and $R = xi + yj + zk$; then $S\beta R = lx + my + nz$

$$V(\beta R)\beta = \begin{vmatrix} mz - ny & nx - lz & ly - mx \\ l & m & n \\ i & j & k \end{vmatrix}$$

and

$$V\beta R = \begin{vmatrix} l & m & n \\ x & y & z \\ i & j & k \end{vmatrix}.$$

Hence

$$\beta^\theta R = (lx + my + nz)(li + mj + nk)$$
$$+ \cos\theta \begin{vmatrix} mz - ny & nx - lz & ly - mx \\ l & m & n \\ i & j & k \end{vmatrix}$$
$$+ \sin\theta \begin{vmatrix} l & m & n \\ x & y & z \\ i & j & k \end{vmatrix}.$$

To prove that $\beta^b\rho$ coincides with the axis of $\beta^{-b/2}\rho^{\pi/2}\beta^{b/2}$. Take the more general versor ρ^θ. Let OP represent the axis

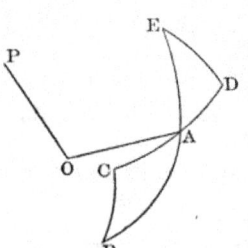

β, AB the versor $\beta^{-b/2}$, BC the versor ρ^θ. Then $(AB)(BC) = AC = DA$, therefore $(AB)(BC)(AE) = (DA)(AE) = DE$. Now DE has the same angle as BC, but its axis has been rotated round P by the angle b. Hence if $\theta = \pi/2$, the axis of $\beta^{-b/2}\rho^{\pi/2}\beta^{b/2}$ will coincide with $\beta^b\rho$.*

The exponential expression for

* This theorem was discovered by Cayley. It indicates that quaternion multiplication in the most general sense has its physical meaning in the composition of rotations.

$\beta^{-b/2}\rho^{\pi/2}\beta^{b/2}$ is $e^{-\frac{1}{2}\delta\beta^{\pi/2}+\frac{1}{2}\pi\rho^{\pi/2}+\frac{1}{2}\delta\beta^{\pi/2}}$, which may be expanded according to the exponential theorem, the successive powers of the trinomial being formed according to the multinomial theorem.

Composition of Finite Rotations round Axes which Intersect.—Let β and γ denote the two axes in space round which the successive rotations take place, and let β^b denote the first and γ^c the second. Let $\beta^b \times \gamma^c$ denote the single rotation which is equivalent to the two given rotations applied in succession; the sign $\times$ is introduced to distinguish from the product of versors. It has been shown in the preceding paragraph that

$$\beta^b\rho = \beta^{-b/2}\rho^{\pi/2}\beta^{b/2};$$

and as the result is a line, the same principle applies to the subsequent rotation. Hence

$$\gamma^c(\beta^b\rho) = \gamma^{-c/2}(\beta^{-b/2}\rho^{\pi/2}\beta^{b/2})\gamma^{c/2}$$
$$= (\gamma^{-c/2}\beta^{-b/2})\rho^{\pi/2}(\beta^{b/2}\gamma^{c/2}),$$

because the factors in a product of versors can be associated in any manner. Hence, reasoning backwards,

$$\beta^b \times \gamma^c = (\beta^{b/2}\gamma^{c/2})^2.$$

Let m denote the cosine of $\beta^{b/2}\gamma^{c/2}$, namely,

$$\cos b/2 \cos c/2 - \sin b/2 \sin c/2,$$

and $n \cdot \nu$ their directed sine, namely,

$$\cos b/2 \sin c/2 . \gamma + \cos c/2 \sin b/2 . \beta - \sin b/2 \sin c/2 \sin \beta\gamma . \overline{\beta\gamma};$$

then $\beta^b \times \gamma^c = m^2 - n^2 + 2mn . \nu.$

Observation.—The expression $(\beta^{b/2}\gamma^{c/2})^2$ is not, as might be supposed, identical with $\beta^b\gamma^c$. The former reduces to the latter only when β and γ are the same or opposite. In the figure β^b is represented by PQ, γ^c by QR, $\beta^b\gamma^c$ by PR, $\beta^{b/2}\gamma^{c/2}$ by ST, and $(\beta^{b/2}\gamma^{c/2})^2$ by SU, which is twice ST. The cosine of SU differs from the cosine of PR by the term $-(\sin b/2 \sin c/2 \sin \beta\gamma)^2$. It is evident from the figure that their axes are also different.

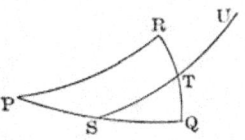

Corollary.—When b and c are infinitesimals, $\cos \beta^b \times \gamma^c = 1$, and $\mathrm{Sin}\ \beta^b \times \gamma^c = b \cdot \beta + c \cdot \gamma$, which is the parallelogram rule for the composition of infinitesimal rotations.

Prob. 54. Let $\beta = \overline{30°}/\underline{/45°}$, $\theta = \pi/3$, and $R = 2i - 3j + 4k$; calculate $\beta^\theta R$.

Prob. 55. Let $\beta = \overline{90°}/\underline{/90°}$, $\theta = \pi/4$, $R = -i + 2j - 3k$; calculate $\beta^\theta R$.

Prob. 56. Prove by multiplying out that $\beta^{-b/2}\rho^{\pi/2}\beta^{b/2} = \{\beta^b\rho\}^{\pi/2}$;

Prob. 57. Prove by means of the exponential theorem that $\gamma^{-c}\beta^b\gamma^c$ has an angle b, and that its axis is $\gamma^{2c}\beta$.

Prob. 58. Prove that the cosine of $(\beta^{b/2}\gamma^{c/2})^2$ differs from the cosine of $\beta^b\gamma^c$ by $-\left(\sin\dfrac{b}{2}\ \sin\dfrac{c}{2}\ \sin\beta\gamma\right)^2$.

Prob. 59. Compare the axes of $(\beta^{b/2}\gamma^{c/2})^2$ and $\beta^b\gamma^c$.

Prob. 60. Find the value of $\beta^b \times \gamma^c$ when $\beta = \overline{0°}/\underline{/90°}$ and $\gamma = \overline{90°}/\underline{/90°}$.

Prob. 61. Find the single rotation equivalent to $i^{\pi/2} \times j^{\pi/2} \times k^{\pi/2}$.

Prob. 62. Prove that successive rotations about radii to two corners of a spherical triangle and through angles double of those of the triangle are equivalent to a single rotation about the radius to the third corner, and through an angle double of the external angle of the triangle.

Chapter X.

PROBABILITY AND THEORY OF ERRORS.

By Robert S. Woodward,
Professor of Mechanics in Columbia University.

Art. 1.   Introduction.

It is a curious circumstance that a science so profoundly mathematical as the theory of probability should have originated in the games of chance which occupy the thoughtless and profligate.* That such is the case is sufficiently attested by the fact that much of the terminology of the science and many of its familiar illustrations are drawn directly from the vocabulary and the paraphernalia of the gambler and the trickster.   It is somewhat surprising, also, considering the antiquity of games of chance, that formal reasoning on the simpler questions in probability did not begin before the time of Pascal and Fermat.   Pascal was led to consider the subject during the year 1654 through a problem proposed to him by the Chevalier de Méré, a reputed gambler.† The problem in question is known as the problem of points and may be stated as follows: two players need each a given number of points to win at a certain stage of their game; if they stop at this stage, how should the stakes be divided?   Pascal corresponded with his friend Fermat on this question; and it appears that the letters which passed between them contained the earliest distinct formulation of principles falling within the theory of probability.   These

---

* The historical facts referred to in this article are drawn mostly from Todhunter's History of the Mathematical Theory of Probability from the time of Pascal to that of Laplace (Cambridge and London, 1865).

† "Un problème relatif aux jeux de hasard, proposé à un austère janséniste par un homme du monde, a été l'origine du calcul des probabilités." Poisson, Recherches sur la Probabilité des Jugements (Paris, 1837).

acute thinkers, however, accomplished little more than a correct start in the science.   Each seemed to rest content at the time with the approbation of the other.   Pascal soon renounced such mundane studies altogether ; Fermat had only the scant leisure of a life busy with affairs to devote to mathematics; and both died soon after the epoch in question,—Pascal in 1662, and Fermat in 1665.

A subject which had attracted the attention of such distinguished mathematicians could not fail to excite the interest of their contemporaries and successors.   Amongst the former Huygens is the most noted.   He has the honor of publishing the first treatise* on the subject.   It contains only fourteen propositions and is devoted entirely to games of chance, but it gave the best account of the theory down to the beginning of the eighteenth century, when it was superseded by the more elaborate works of James Bernoulli,† Montmort,‡ and De Moivre.§ Through the labors of the latter authors the mathematical theory of probability was greatly extended.   They attacked, quite successfully in the main, the most difficult problems ; and great credit is due them for the energy and ability displayed in developing a science which seemed at the time to have no higher aim than intellectual diversion.‖   Their names, undoubtedly, with one exception, that of Laplace, are the most important in the history of probability.

Since the beginning of the eighteenth century almost every mathematician of note has been a contributor to or an expositor of the theory of probability.   Nicolas, Daniel, and John Bernoulli, Simpson, Euler, d'Alembert, Bayes, Lagrange, Lambert, Condorcet, and Laplace are the principal names which figure in the history of the subject during the hundred years

* De Ratiociniis in Ludo Aleæ, 1657.

† Ars Conjectandi, 1713.

‡ Essai d'Analyse sur les Jeux de Hazards, 1708.

§ The Doctrine of Chances, 1718.

‖ Todhunter says of Montmort, for example, " In 1708 he published his work on Chances, where with the courage of Columbus he revealed a new world to Mathematicians."

ending with the first quarter of the present century. Of the contributions from this brilliant array of mathematical talent, the Théorie Analytique des Probabilités of Laplace is by far the most profound and comprehensive. It is, like his Mécanique Céleste in dynamical astronomy, still the most elaborate treatise on the subject. An idea of the grand scale of the work in its present form* may be gained by the facts that the non-mathematical introduction† covers about one hundred and fifty quarto pages; and that, in spite of the extraordinary brevity of mathematical language, the pure theory and its accessories and applications require about six hundred and fifty pages.

From the epoch of Laplace down to the present time the extensions of the science have been most noteworthy in the fields of practical applications, as in the adjustment of observations, and in problems of insurance, statistics, etc. Amongst the most important of the pioneers in these fields should be mentioned Poisson, Gauss, Bessel, and De Morgan. Numerous authors, also, have done much to simplify one or another branch of the subject and thus bring it within the range of elementary presentation. The fundamental principles of the theory are, indeed, now accessible in the best text-books on algebra : and there are many excellent treatises on the pure theory and its various applications.

Of all the applications of the doctrine of probability none is of greater utility than the theory of errors. In astronomy, geodesy, physics, and chemistry, as in every science which attains precision in measuring, weighing, and computing, a knowledge of the theory of errors is indispensable. By the aid of this theory the exact sciences have made great progress dur-

---

*The form of the third edition published in 1820, and of Vol. VII of the complete works of Laplace recently republished under the auspices of the Académie des Sciences by Gauthier-Villars. This Vol. VII bears the date 1886.

† "Cette Introduction," writes Laplace, "est le developpement d'une Leçon sur les Probabilités, que je donnai en 1795, aux Écoles Normales, où je fus appelé comme professeur de Mathématiques avec Lagrange, par un décret de la Convention nationale."

ing the present century, not only in the actual determination
of the constants of nature, but also in the fixation of clear
ideas as to the possibilities of future conquests in the same direction.   Nothing, for example, is more satisfactory and instructive in the history of science than the success with which
the unique method of least squares has been applied to the
problems presented by the earth and the other members of the
solar system.   So great, in fact, are the practical value and
theoretical importance of the method of least squares, that it is
frequently mistaken for the whole theory of errors, and is
sometimes regarded as embodying the major part of the doctrine of probability itself.

As may be inferred from this brief sketch, the theory of
probability and its more important applications now constitute
an extensive body of mathematical principles and precepts.
Obviously, therefore, it will be impossible within the limits of
a single chapter of this volume to do more than give an outline of the salient features of the subject.   It is hoped, however, in accordance with the general plan of the volume, that
such outline will prove suggestive and helpful to those who
may come to the science for the first time, and also to those
who, while somewhat familiar with the difficulties to be overcome, have not acquired a working knowledge of the subject.
Effort has been made especially to clear up the difficulties of
the theory of errors by presenting a somewhat broader view of
the elements of the subject than is found in the standard
treatises, which confine attention almost exclusively to the
method of least squares.   This chapter stops short of that
method, and seeks to supply those phases of the theory which
are either notably lacking or notably erroneous in works
hitherto published.   It is believed, also, that the elements here
presented are essential to an adequate understanding of the
well-worked domain of least squares.*

* The author has given a brief but comprehensive statement of the method
of least squares in the volume of Geographical Tables published by the Smith-
sonion Institution, 1894.

ART. 2. PERMUTATIONS.

The formulas and results of the theory of permutations and combinations are often needed for the statement and solution of problems in probabilities. This theory is now to be found in most works on algebra, and it will therefore suffice here to state the principal formulas and illustrate their meaning by a few numerical examples.

The number of permutations of n things taken r in a group is expressed by the formula

$$(n)_r = n(n-1)(n-2)\ldots(n-r+1). \tag{1}$$

Thus, to illustrate, the number of ways the four letters a, b, c, d can be arranged in groups of two is $4 \cdot 3 = 12$, and the groups are

$$ab, \quad ba, \quad ac, \quad ca, \quad ad, \quad da, \quad bc, \quad cb, \quad bd, \quad db, \quad cd, \quad dc.$$

Similarly, the formula gives for

$$n = 3 \text{ and } r = 2, \quad (3)_2 = 3 \cdot 2 \qquad\qquad = 6,$$
$$n = 7 \text{ “ } r = 3, \quad (7)_3 = 7 \cdot 6 \cdot 5 \qquad\qquad = 210,$$
$$n = 10 \text{ “ } r = 6, \quad (10)_6 = 10 \cdot 9 \cdot 8 \cdot 7 \cdot 6 \cdot 5 = 151200.$$

The results which follow from equation (1) when n and r do not exceed 10 each are embodied in the following table:

VALUES OF PERMUTATIONS.

	10	9	8	7	6	5	4	3	2	1
1	10	9	8	7	6	5	4	3	2	1
2	90	72	56	42	30	20	12	6	2	
3	720	504	336	210	120	60	24	6		
4	5040	3024	1680	840	360	120	24			
5	30240	15120	6720	2520	720	120				
6	151200	60480	20160	5040	720					
7	604800	181440	40320	5040						
8	18:4400	362880	40320							
9	3628800	362880								
10	3628800									
S_p	9864100	986409	109600	13699	1956	325	64	15	4	1

The use of this table is obvious. Thus, the number of permutations of eight things in groups of five each is found in the fifth line of the column headed with the number 8. It will be noticed that the last two numbers in each column (excepting that headed with 1) are the same. This accords with the formula, which gives for the number of permutations of n things in groups of n the same value as for n things in groups of $(n - 1)$. It will also be remarked that the last number in each column of the table is the factorial, $n!$, of the number n at the head of the column. For example, in the column under 7, the last number is $5040 = 1.2.3.4.5.6.7 = 7!$.

The total number of permutations of n things taken singly, in groups of two, three, etc., is found by summing the numbers given by equation (1) for all values of r from 1 to n. Calling this total or sum S_p, it will be given by

$$S_p = \Sigma(n)_r. \tag{2}$$

To illustrate, suppose $n = 3$, and, to fix the ideas, let the three things be the three digits 1, 2, 3. Then from the above table it is seen that $S_p = 3 + 6 + 6 = 15$; or, that the number of numbers (all different) which can be formed from those digits is fifteen. These numbers are 1, 2, 3; 12, 13, 21, 23, 31, 32; 123, 132, 213, 231, 312, 321.

The values of S_p for $n = 1, 2, \ldots$ 10 are given under the corresponding columns of the above table. But when n is large the direct summation indicated by (2) is tedious, if not impracticable. Hence a more convenient formula is desirable. To get this, observe that (1) may be written

$$(n)_r = \frac{n!}{(n - r)!}, \tag{1'}$$

if r is restricted to integer values between 1 and $(n - 1)$, both inclusive. This suffices to give all terms which appear in the right-hand member of (2), since the number of permutations for $r = (n - 1)$ is the same as for $r = n$. Hence it appears that

$$S_p = n! + \frac{n!}{1} + \frac{n!}{1.2} + \cdots \frac{n!}{(n - 1)!}$$

$$= n!\left(1 + \frac{1}{1} + \frac{1}{1.2} + \cdots \frac{1}{(n - 1)!}\right).$$

But as n increases, the series by which $n!$ is here multiplied approximates rapidly towards the base of natural logarithms; that is, towards

$$e = 2.7182818 +, \qquad \log e = 0.4342945.$$

Hence for large values of n

$$S_p = n!\, e, \text{ approximately.}* \qquad (3)$$

To get an idea of the degree of approximation of (3), suppose $n = 9$. Then the computation runs thus (see values in the above table):

$$
\begin{array}{lll}
& & \text{log} \\
9! = 362880 & & 5.5597630 \\
e & & 0.4342945 \\
\hline
9!\,e = 986410 & & 5.9940575 \\
S_p = 986409 & \text{by equation (2).}
\end{array}
$$

The error in this case is thus seen to be only one unit, or about one-millionth of S_p.†

Prob. 1. Tabulate a list of the numbers of three figures each which can be formed from the first five digits $1, \ldots 5$. How many numbers can be formed from the nine digits?

Prob. 2. Is S_p always an odd number for n odd? Observe values of S_p in the table above.

ART. 3.   COMBINATIONS.

In permutations attention is given to the order of arrangement of the things considered. In combinations no regard is paid to the order of arrangement. Thus, the permutations of the letters a, b, c, d in groups of three are

$$
\begin{array}{llllllll}
(abc) & (abd) & bac & bad & acb & (acd) & cab & cad \\
adb & adc & dab & dac & bca & (bcd) & cba & cbd \\
bda & bdc & dba & dbc & cda & cdb & dca & dcb
\end{array}
$$

* See Art. 6 for a formula for computing $n!$ when n is a large number.

† When large numbers are to be dealt with, equations (1)′ and (3) are easily managed by logarithms, especially if a table of values of $\log (n!)$ is available. Such tables are given to six places in De Morgan's treatise on Probability in the Encyclopædia Metropolitana, and to five places in Shortrede's Tables (Vol. I, 1849).

But if the order of arrangement is ignored all of these are seen to be repetitions of the groups enclosed in parentheses, namely, (abc), (abd), (acd), (bcd). Hence in this case out of twenty-four permutations there are only four combinations.

A general formula for computing the number of combinations of n things taken in groups of r things is easily derived. For the number of permutations of n things in groups of r is by (1) of Art. 2

$$(n)_r = n(n-1)(n-2)\ldots(n-r+1);$$

and since each group of r things gives $1.2.3\ldots r = r!$ permutations, the number of combinations must be the quotient of $(n)_r$ by $r!$. Denote this number by $C(n)_r$. Then the general formula is

$$C(n)_r = \frac{n(n-1)(n-2)\ldots(n-r+1)}{r!} \qquad (1)$$

This formula gives, for example, in the case of the four letters a, b, c, d taken in groups of three, as considered above,

$$C(4)_3 = \frac{4\cdot3\cdot2}{1\cdot2\cdot3} = 4.$$

Multiply both numerator and denominator of the right-hand member of (1) by $(n-r)!$ The result is

$$C(n)_r = \frac{n!}{r!(n-r)!}, \qquad (1)'$$

which shows that the number of combinations of n things in groups of r is the same as the number of combinations of n things in groups of $(n-r)$. Thus, the number of combinations of the first ten letters a, b, $c \ldots j$ in groups of three or seven is

$$\frac{10!}{3!7!} = 120.$$

The following table gives the values $C(n)_r$ for all values of n and r from 1 to 10.

The mode of using this table is evident. For example, the number of combinations of eight things in sets of five each is found on the fifth line of the column headed 8 to be 56.

VALUES OF COMBINATIONS.

	10	9	8	7	6	5	4	3	2	1
1	10	9	8	7	6	5	4	3	2	1
2	45	36	28	21	15	10	6	3	1	
3	120	84	56	35	20	10	4	1		
4	210	126	70	35	15	5	1			
5	252	126	56	21	6	1				
6	210	84	28	7	1					
7	120	36	8	1						
8	45	9	1							
9	10	1								
10	1									
S_c	1023	511	255	127	63	31	15	7	3	

It will be observed that the numbers in any column show a maximum value when n is even and two equal maximum values when n is odd. That this should be so is easily seen from (1)′, which shows that $C(n)_r$ will be a maximum for any value of n when $r!(n-r)!$ is a minimum. For n even this is a minimum for $r = \frac{1}{2}n$; while for n odd it has equal minimum values for $r = \frac{1}{2}(n-1)$ and $r = \frac{1}{2}(n+1)$. Thus,

$$\text{maximum of } C(n)_r = \frac{n!}{\left(\frac{n}{2}!\right)^2} \text{ for } n \text{ even,}$$

$$= \frac{n!}{\frac{n+1}{2}! \frac{n-1}{2}!} \text{ for } n \text{ odd.}$$

(2)

The total number of combinations of n things taken singly, in groups of two, three, etc., is found by summing the numbers given by (1) for all values of r from 1 to n both inclusive. Calling this total or sum S_c,

$$S_c = \Sigma C(n)_r.$$

The same sum will also come from (1)′ by giving to r all values from 1 to $(n-1)$, both inclusive, summing the results, and increasing their aggregate by unity. Thus by either process

$$S_c = n + \frac{n(n-1)}{1 \cdot 2} + \frac{n(n-1)(n-3)}{1 \cdot 2 \cdot 3} + \ldots + n + 1.$$

The second member of this equation is evidently equal to $(1 + 1)^n - 1$. Hence

$$S_c = \Sigma C(n)_r = 2^n - 1. \tag{3}$$

The values of S_c for values of n and r from 1 to 10 are given under the corresponding columns of the above table.

Prob. 3. How many different squads of ten men each can be formed from a company of 100 men?

Prob. 4 How many triangles are formed by six straight lines each of which intersects the other five?

Prob. 5. Examine this statement: "In dealing a pack of cards the number of hands, of thirteen cards each, which can be produced is 635 013 559 600. But in whist four hands are simultaneously held, and the number of distinct deals . . . would require twenty-eight figures to express it."*

Prob. 6. Assuming combination always possible, and disregarding the question of proportions, find how many different substances could be produced by combining the seventy-three chemical elements.

ART. 4. DIRECT PROBABILITIES.

If it is known that one of two events must occur in any trial or instance, and that the first can occur in a ways and the second in b ways, all of which ways are equally likely to happen, then the probability that the first will happen is expressed mathematically by the fraction $a/(a + b)$, while the probability that the second will happen is $b/(a + b)$. Such events are said to be mutually exclusive. Denote their probabilities by p and q respectively. Then there result

$$p = \frac{a}{a + b}, \qquad q = \frac{b}{a + b}, \qquad p + q = 1, \tag{1}$$

the last equation following from the first two and being the mathematical expression for the certainty that one of the two events must happen.

Thus, to illustrate, in tossing a coin it must give "head" or "tail"; $a = b = 1$, and $p = q = 1/2$. Again, if an urn contain $a = 5$ white and $b = 8$ black balls, the probability of drawing

* Jevons, Principles of Science, p. 217.

a white ball in one trial is $p = 5/13$ and that of drawing a black one $q = 8/13$.

Similarly, if there are several mutually exclusive events which can occur in $a, b, c \ldots$ ways respectively, their probabilities $p, q, r \ldots$ are given by

$$p = \frac{a}{a+b+c+\ldots}, \quad q = \frac{b}{a+b+c+\ldots}, \quad r = \frac{c}{a+b+c+\ldots},$$

$$p+q+r+\ldots = 1. \tag{2}$$

For example, if an urn contain $a = 4$ white, $b = 5$ black, and $c = 6$ red balls, the probabilities of drawing a white, black, and red ball at a single trial are $p = 4/15$, $q = 5/15$, and $r = 6/15$, respectively.

Formulas (1) and (2) may be applied to a wide variety of cases, but it must suffice here to give only a few such. As a first illustration, consider the probability of drawing at random a number of three figures from the entire list of numbers which can be formed from the first seven digits. A glance at the table of Art. 1 shows that the symbols of formula (1) have in this case the values $a = 210$, and $a + b = 13699$. Hence $b = 13489$, and $p = 210/13699$; that is, the probability in question is about $1/65$.

Secondly, what is the probability of holding in a hand of whist all the cards of one suit? Formula (1) of Art. 3 shows that the number of different hands of thirteen cards each which may be formed from a pack of fifty-two cards is

$$\frac{52 \cdot 51 \cdot 50 \ldots 40}{1 \cdot 2 \cdot 3 \ldots 13} = 635\,013\,559\,600,$$

and the probability required is the reciprocal of this number. The probability against this event is, therefore, very nearly unity.

Thirdly, consider the probabilities presented by the case of an urn containing 4 white, 5 black, and 6 red balls, from which at a single trial three balls are to be drawn. Evidently the triad of balls drawn may be all white, all black, all red, partly white and black, partly white and red, partly black and red, or one each of the white, black, and red. There are thus seven different probabilities to be taken into account. The theory of combinations shows (see equation (1), Art. 3) that the total number of

White triads $\qquad = \dfrac{4 \cdot 3 \cdot 2}{1 \cdot 2 \cdot 3} \qquad = \ 4 = a$

Black triads $\qquad = \dfrac{5 \cdot 4 \cdot 3}{6} \qquad = 10 = b$

Red triads $\qquad = \dfrac{6 \cdot 5 \cdot 4}{6} \qquad = 20 = c$

White and black triads $\quad = \dfrac{9 \cdot 8 \cdot 7}{6} -(4+10)= 70 = d$

White and red triads $\quad = \dfrac{10 \cdot 9 \cdot 8}{6} -(4+20)= 96 = e$

Black and red triads $\quad = \dfrac{11 \cdot 10 \cdot 9}{6}-(10+20)=135 = f$

White, black, and red triads $= 4 \cdot 5 \cdot 6 \qquad =120 = g$

$$\text{Sum} = \overline{455}$$

The total number of these triads is 455, and is, as it should be, the number of combinations in groups of three each of the whole number of balls. Hence formulas (2) give the seven different probabilities which follow, using the initial letters w, b, r to indicate the colors represented in a triad :

For a triad		www	$p =$	$4/455$,
"	"	" bbb	$q =$	$10/455$,
"	"	" rrr	$r =$	$20/455$,
"	"	" wwb or wbb	$s =$	$70/455$,
"	"	" wwr or wrr	$t =$	$96/455$,
"	"	" bbr or brr	$u =$	$135/455$,
"	"	" wbr	$v =$	$120/455$.

Prob. 7. When three dice are thrown together, what is the probability that the throw will be greater than 9 ?

Prob. 8. Write down a literal formula for the probabilities of the several possible triads considered in the above question of the balls, supposing the numbers of white, black, and red balls to be l, m, n, respectively.

ART. 5. PROBABILITY OF CONCURRENT EVENTS.

If the probabilities of two independent events are p_1 and p_2, respectively, the probability of their concurrence in any single instance is p_1p_2. Thus, suppose there are two urns U_1 and U_2, the first of which contains a_1 white and b_1 black balls, and the second a_2 white and b_2 black balls. Then the probability of drawing a white ball from U_1 is $p_1 = a_1/(a_1 + b_1)$, while that of drawing a white ball from U_2 is $p_2 = a_2/(a_2 + b_2)$. The total number of different pairs of balls which can be formed from the entire number of balls is $(a_1 + b_1)(a_2 + b_2)$. Of these pairs a_1a_2 are favorable to the concurrence of white in simultaneous or successive drawings from the two urns. Hence the probability of a concurrence of

white with white $= a_1a_2/(a_1 + b_1)(a_2 + b_2)$,
white with black $= (a_1b_2 + a_2b_1)/(a_1 + b_1)(a_2 + b_2)$,
black with black $= b_1b_2/(a_1 + b_1)(a_2 + b_2)$,

and the sum of these is unity, as required by equations (2) of Article 4.

In general, if $p_1, p_2, p_3 \ldots$ denote the probabilities of several independent events, and P denote the probability of their concurrence,

$$P = p_1p_2p_3. \tag{1}$$

To illustrate this formula, suppose there is required the probability of getting three aces with three dice thrown simultaneously. In this case $p_1 = p_2 = p_3 = 1/6$ and $P = (1/6)^3 = 1/216$.

Similarly, if two dice are thrown simultaneously the probability that the sum of the numbers shown will be 11 is $2/36$; and the probability that this sum 11 will appear in two successive throws of the same pair of dice is $4/36.36$.

The probability that the alternatives of a series of events will concur is evidently given by

$$Q = q_1q_2q_3 \ldots = (1 - p_1)(1 - p_2)(1 - p_3) \tag{2}$$

Thus, in the case of the three dice mentioned above, the probability that each will show something other than an ace is

$q_1 = q_2 = q_3 = 5/6$, and the probability that they will concur in this is $Q = 125/216$.

Many cases of interest occur for the application of (1) and (2). One of the most important of these is furnished by successive trials of the same event. Consider, for example, what may happen in n trials of an event for which the probability is p and against which the probability is q. The probability that the event will occur every time is p^n. The probability that the event will occur $(n - 1)$ times in succession and then fail is $p^{n-1}q$. But if the order of occurrence is disregarded this last combination may arrive in n different ways; so that the probability that the event will occur $(n - 1)$ times and fail once is $np^{n-1}q$. Similarly, the probability that the event will happen $(n - 2)$ times and fail twice is $\frac{1}{2}n(n - 1)p^{n-2}q^2$; etc. That is, the probabilities of the several possible occurrences are given by the corresponding terms in the development of $(p + q)^n$.

By the same reasoning used to get equations (2) of Art. 3 it may be shown that the maximum term in the expansion of $(p + q)^n$ is that in which the exponent m, say, of q is the whole number lying between $(n + 1)q - 1$ and $(n + 1)q$. In other words, the most probable result in n trials is the occurrence of the event $(n - m)$ times and its failure m times. When n is large this means that the most probable of all possible results is that in which the event occurs $n - nq = n(1 - q) = np$ times and fails nq times. Thus, if the event be that of throwing an ace with a single die the most probable of the possible results in 600 throws is that of 100 aces and 500 failures.

Since q^n is the probability that the event will fail every time in n trials, the probability that it will occur at least once in n trials is $1 - q^n$. Calling this probability r,[*]

$$r = 1 - q^n = 1 - (1 - p)^n. \tag{3}$$

If r in this equation be replaced by $1/2$, the corresponding value of n is the number of trials essential to render the

[*] See Poisson's Probabilité des Jugements, pp. 40, 41.

chances even that the event whose probability is p will occur at least once. Thus, in this case, the value of n is given by

$$n = - \frac{\log 2}{\log (1-p)}.$$

This shows, for example, if the event be the throwing of double sixes with two dice, for which $p = 1/36$, that the chances are even ($r = 1/2$) that in 25 throws ($n = 24.614$ by the formula) double sixes will appear at least once.

Equation (3) shows that however small p may be, so long as it is finite, n may be taken so large as to make r approach indefinitely near to unity; that is, n may be so large as to render it practically certain that the event will occur at least once.

When n is large

$$(1-p)^n = 1 - np + \frac{n(n-1)}{1 \cdot 2}p^2 - \frac{n(n-1)(n-2)}{1 \cdot 2 \cdot 3}p^3 + \ldots$$

$$= 1 - np + \frac{(np)^2}{1 \cdot 2} - \frac{(np)^3}{1 \cdot 2 \cdot 3} + \ldots$$

$$= e^{-np} \text{ approximately.}$$

Thus an approximate value of r is

$$r = 1 - e^{-np}, \qquad \log e = 0.4342495. \tag{4}$$

This formula gives, for example, for the probability of drawing the ace of spades from a pack of fifty-two cards at least once in 104 trials $r = 1 - e^{-2} = 0.865$, while the exact formula (3) gives 0.867.

Similarly, the probability of the occurrence of the event at least t times in n trials will be given by the sum of the terms of $(p + q)^n$ from p^n up to that in $p^t q^{n-t}$ inclusive. This probability must be carefully distinguished from the probability that the event will occur t times only in the n trials, the latter being expressed by the single term in $p^t q^{n-t}$.

Prob. 9. Compare the probability of holding exactly four aces in five hands of whist with the probability of holding at least four aces in the same number of hands.

Prob. 10. What is the probability of an event if the chances are even that it occurs at least once in a million trials? See equation (4).

ART. 6. BERNOULLI'S THEOREM.

Denote the exponents of p and q in the maximum term of $(p + q)^n$ by μ and m respectively, and denote this term by T. Then

$$T = \frac{n(n - 1)(n - 2) \ldots (\mu + 1)}{m!} p^\mu q^m = \frac{n!}{\mu!\, m!} p^\mu q^m. \quad (1)$$

As shown in Art. 5, μ in this formula is the greatest whole number in $(n + 1)p$, and m the greatest whole number in $(n + 1)q$; so that when n is large, μ and m are sensibly equal to np and nq respectively.

The direct calculation of T by (1) is impracticable when n is large. To overcome this difficulty the following expression is used:*

$$n! = n^n e^{-n} \sqrt{2\pi n}\left(1 + \frac{1}{12n} + \frac{1}{288n^2} + \ldots\right). \quad (2)$$

$$\log e = 0.4342495, \quad \log 2\pi = 0.7981799.$$

This expression approaches $n^n e^{-n} \sqrt{2\pi n}$ as a limit with the increase of n, and in this approximate form is known as Stirling's theorem. Although a rude approximation to $n!$ for small values of n this theorem suffices in nearly all cases wherein such probabilities as T are desired. Making use of the theorem in (1) it becomes

$$T = \frac{1}{\sqrt{2\pi npq}}. \quad (3)$$

That this formula affords a fair approximation even when n is small is seen from the case of a die thrown 12 times. The probability that any particular face will appear in one throw is $p = 1/6$, whence $q = 5/6$; and the most probable result in 12 throws is that in which the particular face appears twice and fails to appear ten times. The probability of this result computed from (3) is 0.309, while the exact formula (1) gives 0.296.

The probability that the event will occur a number of times

---

* This expression is due to Laplace, Théorie Analytique des Probabilités. See also De Morgan's Calculus, pp. 600–604.

comprised between $(\mu - \alpha)$ and $(\mu + \alpha)$ in n trials is evidently expressed by the sum of the terms in $(p + q)^n$ for which the exponent of p has the specified range of values. Calling this probability R, putting

$$\mu = np + u, \quad \text{and} \quad m = nq - u,$$

and using Stirling's theorem (which implies that n is a large number),*

$$R = \Sigma \frac{1}{\sqrt{2\pi npq}} \left(1 + \frac{u}{np}\right)^{-(np+u)} \left(1 - \frac{u}{nq}\right)^{-(nq-u)},$$

very nearly; and the summation is with respect to u from $u = -\alpha$ to $u = +\alpha$. But expansion shows that the natural logarithm of the product of the two binomial factors in this equation is approximately $-u^2/2npq$. Hence

$$R = \Sigma \frac{1}{\sqrt{2\pi npq}} e^{-u^2/2npq};$$

and, since n is supposed large, this may be replaced by a definite integral, putting

$$dz = 1/\sqrt{2npq}, \quad \text{and} \quad z^2 = u^2/2npq.$$

Thus

$$R = \frac{1}{\sqrt{\pi}} \int_{-\alpha/\sqrt{2npq}}^{+\alpha/\sqrt{2npq}} e^{-z^2} dz = \frac{2}{\sqrt{\pi}} \int_{0}^{\alpha/\sqrt{2npq}} e^{-z^2} dz. \qquad (4)$$

This equation expresses the theorem of James Bernoulli, given in his Ars Conjectandi, published in 1713.

The value of the right-hand member of (4) varies, as it should, between 0 and 1, and approaches the latter limit rapidly as z increases. Thus, writing for brevity

$$I = \frac{2}{\sqrt{\pi}} \int_{0}^{z} e^{-z^2} dz,$$

* See Bertrand, Calcul des Probabilités, Paris, 1889, for an extended discussion of the questions considered in this Article.

the following table shows the march of the integral:

z	I	z	I	z	I
0.00	0.000	0.75	0.711	1.50	0.966
.25	.276	1.00	.843	1.75	.987
.50	.520	1.25	.923	2.00	.995

To illustrate the use of (4), suppose there is required the probability that in 6000 throws of a die the ace will appear a number of times which shall be greater than $1/6 \times 6000 - 10$ and less than $1/6 \times 6000 + 10$, or a number of times lying between 990 and 1010. In this case $\alpha = 10$, $n = 6000$, $p = 1/6$, $q = 5/6$. Thus, $\alpha/\sqrt{2npq} = 10/\sqrt{2 \cdot 6000 \cdot 1/6 \cdot 5/6} = 0.245$. Hence, by (4) and the table, $R = 0.27$.

Prob. 11. If the ratio of males to females at birth is 105 to 100, what is the probability that in the next 10,000 births the number of males will fall within two per cent of the most probable number?

Prob. 12. If the chance is even for head and tail in tossing a coin, what is the probability that in a million throws the difference between heads and tails will exceed 1500?

ART. 7.   INVERSE PROBABILITIES.*

If an observed event can be attributed to any one of several causes, what is the probability that any particular one of these causes produced the event? To put the question in a concrete form, suppose a white ball has been drawn from one of two urns, U_1 containing 3 white and 5 black balls, and U_2 containing 2 white and 4 black balls; and that the probability in favor of each urn is required. If U_1 is as likely to have been chosen as U_2, the probability that U_1 was chosen is $1/2$. After such choice the probability of drawing a white ball from U_1 is $3/8$. Before drawing, therefore, the probability of getting a white ball from U_1 was $1/2 \times 3/8 = 3/16$, by Art. 5. Similarly, before drawing the probability of getting a white ball from U_2 was $1/2 \times 2/6 = 1/6$. These probabilities will remain unchanged if the number of balls in either urn be increased or

---

* See Poisson, Probabilité des Jugements, pp. 81–83.

diminished so long as the ratio of white to black balls is kept
constant. Make these numbers the same for the two urns.
Thus let the first contain 9 white and 15 black, and the second
8 white and 16 black; whence the above probabilities may be
written $1/2 \times 9/24$ and $1/2 \times 8/24$. It is now seen that there
are $(9 + 8)$ cases favorable to the production of a white ball,
each of which has the same antecedent probability, namely, $1/2$.
Since the fact that a white ball was drawn excludes consideration of the black balls, the probability that the white ball came
from U_1 is $9/17$ and that it came from U_2 is $8/17$; and the sum
of these is unity, as it should be.

To generalize this result, let there be m causes, $C_1, C_2, \ldots C_m$.
Denote their direct probabilities by $q_1, q_2, \ldots q_m$; their antecedent
probabilities by $r_1, r_2, \ldots r_m$; and their resultant probabilities
on the supposition of separate existence by $p_1, p_2, \ldots p_m$.
That is,

$$p_1 = q_1 r_1, \quad p_2 = q_2 r_2, \ldots p_m = q_m r_m. \tag{1}$$

Let D be the common denominator of the right-hand members in (1), and denote the corresponding numerators of the
several fractions by $s_1, s_2, \ldots s_m$. Then

$$p_1 = s_1/D, \quad p_2 = s_2/D, \ldots p_m = s_m/D;$$

and it is seen that there are in all $(s_1 + s_2 + \ldots s_m)$ equally
possible cases, and that of these s_1 are favorable to C_1, s_2 to
$C_2, \ldots$ Hence, if $P_1, P_2, \ldots P_m$ denote the probabilities of
the several causes on the supposition of their coexistence,

$$P_1 = s_1/(s_1 + s_2 + \ldots s_m) = p_1/(p_1 + p_2 + \ldots p_m).$$

Thus in general

$$P_1 = p_1/\Sigma p, \quad P_2 = p_2/\Sigma p, \ldots P_m = p_m/\Sigma p. \tag{2}$$

To illustrate the meaning of these formulas by the above
concrete case of the urns it suffices to observe that

for U_1, $q_1 = 3/8$ and $r_1 = 1/2$,
for U_2, $q_2 = 1/3$ and $r_2 = 1/2$;
whence $p_1 = 3/16, \quad p_2 = 1/6, \quad p_1 + p_2 = 17/48$;
and $P_1 = 9/17, \quad P_2 = 8/17$.

As a second illustration, suppose it is known that a white

ball has been drawn from an urn which originally contained m balls, some of them being black, if all are not white. What is the probability that the urn contained exactly n white balls? The facts are consistent with m different and equally probable hypotheses (or causes), namely, that there were 1 white and $(m-1)$ black balls, 2 white and $(m-2)$ black balls, etc. Hence in (1), $q_1 = q_2 = \ldots = 1$, and

$$p_1 = 1/m, \quad p_2 = 2/m, \ldots p_n = n/m, \ldots p_m = m/m.$$

Thus
$$\Sigma p = (1/2)(m+1),$$

and
$$P_n = p_n/\Sigma p = \frac{2n}{m(m+1)}.$$

This shows, as it evidently should, that $n = m$ is the most probable number of white balls in the urn. The probability for this number is $P_m = 2/(m+1)$, which reduces, as it ought, to 1 for $m = 1$.

Formulas (1) and (2) may also be applied to the problem of estimating the probability of the occurrence of an event from the concurrent testimony of several witnesses, $X_1, X_2, \ldots$ Denote the probabilities that the witnesses tell the truth by $x_1, x_2, \ldots$ Then, supposing them to testify independently, the probability that they will concur in the truth concerning the event is $x_1 x_2 \ldots$; while the probability that they will concur in the only other alternative, falsehood, is $(1-x_1)(1-x_2)\ldots$ The two alternatives are equally possible. Hence by equations (1) and (2)

$$p_1 = x_1 x_2 \ldots, \quad p_2 = (1 - x_1)(1 - x_2) \ldots,$$
$$P_1 = \frac{x_1 x_2 \ldots}{x_1 x_2 \ldots + (1 - x_1)(1 - x_2) \ldots},$$
$$P_2 = \frac{(1 - x_1)(1 - x_2) \ldots}{x_1 x_2 \ldots + (1 - x_1)(1 - x_2) \ldots}, \tag{3}$$

P_1 being the probability for and P_2 that against the event.

To illustrate (3), if the chances are 3 to 1 that X_1 tells the truth and 5 to 1 that X_2 tells the truth, $x_1 = 3/4$, $x_2 = 5/6$, and $P_1 = 15/16$; or, the chances are 15 to 1 that an event occurred if they agree in asserting that it did.*

* For some interesting applications of equations (3) see note E of Appendix to the Ninth Bridgewater Treatise by Charles Babbage (London, 1838).

It is of theoretical interest to observe that if $x_1, x_2, \ldots$ in (3) are each greater than $1/2$, P_1 approaches unity as the number of witnesses is indefinitely increased.

Prob. 13. The groups of numbers of one figure each, two figures each, three figures each, etc., which it is possible to form from the nine digits $1, 2, \ldots 9$ are printed on cards and placed severally in nine similar urns. What is the probability that the number 777 will be drawn in a single trial by a person unaware of the contents of the urns?

Prob. 14. How many witnesses whose credibilities are each $3/4$ are essential to make $P_1 = 0.999$ in equation (3)?

ART. 8.   PROBABILITIES OF FUTURE EVENTS.

Equations (2) of Art. 7 may be written in the following manner:

$$\frac{P_1}{p_1} = \frac{P_2}{p_2} = \ldots \frac{P_m}{p_m} = \frac{1}{\Sigma p}. \tag{1}$$

If $p_1, p_2, \ldots p_m$ are found by observation, $P_1, P_2, \ldots P_m$ will express the probabilities of the corresponding causes or their effects. When, as in the case of most physical facts, the number of causes and events is indefinitely great, the value of any p or P in (1) becomes indefinitely small, and the value of Σp must be expressed by means of a definite integral. Let x denote the probability of any particular cause, or of the event to which it gives rise. Then, supposing this and all the other causes mutually exclusive, $(1 - x)$ will be the probability against the event. Now suppose it has been observed that in $(m + n)$ cases the event in question has occurred m times and failed n times. The probability of such a concurrence is, by Art. 5, $cx^m(1 - x)^n$, where c is a constant. Since x is unknown, it may be assumed to have any value within the limits 0 and 1; and all such values are à priori equally possible. Put

$$y = cx^m(1 - x)^n.$$

Then evidently the probability that x will fall within any assigned possible limits a and b is expressed by the fraction

$$\int_a^b y\,dx \bigg/ \int_0^1 y\,dx \,;$$

so that the probability of any particular x is given by

$$P = \frac{x^m(1 - x)^n dx}{\int_0^1 x^m(1 - x)^n dx}. \qquad (2)$$

This may be regarded as the antecedent probability of the cause or event in question.

What then is the probability that in the next $(r + s)$ trials the event will occur r times and fail s times, if no regard is had of the order of occurrence? If x were known, the answer would be by Arts. 2 and 5

$$\frac{(r + s)!}{r!\,s!} x^r(1 - x)^s. \qquad (3)$$

But since x is restricted only by the condition (2), the required probability will be found by taking the product of (2) and (3) and integrating throughout the range of x. Thus, calling the required probability Q,

$$Q = \frac{(r+s)!}{r!\,s!} \cdot \frac{\int_0^1 x^{m+r}(1 - x)^{n+s} dx}{\int_0^1 x^m(1 - x)^n dx}. \qquad (4)$$

The definite integrals which appear here are known as Gamma functions. They are discussed in all of the higher treatises on the Integral Calculus. Applying the rules derived in such treatises there results *

$$Q = \frac{(r + s)!\,(m + r)!\,(n + s)!\,(m + n + 1)!}{r!\,s!\,m!\,n!\,(m + n + r + s + 1)!}. \qquad (5)$$

If regard is had to the order of occurrence of the event; that is, if the probability required is that of the event happening r times in succession and then failing s times in succession,

---

* It is a remarkable fact that formula (5) is true without restriction as to values of m, n, r, s. The formula may be established by elementary considerations, as was done by Prevost and Lhuilier, 1795. See Todhunter's History of the Theory of Probability, pp. 453–457.

the factor $(r + s)!/r!s!$ in (3), (4), (5) must be replaced by unity.

To illustrate these formulas, suppose first that the event has happened m times and failed no times. What is the probability that it will occur at the next trial? In this case (4) gives

$$Q = \int_0^1 x^{m+1} dx \Big/ \int_0^1 x^m dx = (m + 1)/(m + 2).$$

When m is large this probability is nearly unity. Thus, the sun has risen without failure a great number of times m; the probability that it will rise to-morrow is

$$\left(1 + \frac{1}{m}\right)\left(1 + \frac{2}{m}\right)^{-1} = 1 + \frac{1}{m} - \frac{2}{m} + \ldots$$

which is practically 1.

Secondly, suppose an urn contains white and black balls in an unknown ratio. If in ten trials 7 white and 3 black balls are drawn, what is the probability that in the next five trials 2 white and 3 black balls will be drawn? The application of (5) supposes the ratio of the white and black balls in the urn to remain constant. This will follow if the balls are replaced after each drawing, or if the number of balls in the urn is supposed infinite. The data give

$$m = 7, \qquad n = 3, \qquad r = 2, \qquad\qquad s = 3,$$
$$m + r = 9, \quad n + s = 6, \quad r + s = 5, \quad m + n + 1 = 11,$$
$$m + n + r + s + 1 = 16.$$

Thus by (5)

$$Q = \frac{5!\,9!\,6!\,11!}{2!\,3!\,7!\,3!\,16!} = 30/91.$$

Suppose there are two mutually exclusive events, the first of which has happened m times and the second n times in $m + n$ trials. What is the probability that the chance of the occurrence of the first exceeds $1/2$? The answer to this question is given directly by equation (2) by integrating the numerator between the specified limits of x. That is,

$$P = \frac{\int_{0.5}^{1} x^m (1 - x)^n dx}{\int_{0}^{1} x^m (1 - x)^n dx}. \tag{6}$$

Thus, if $m = 1$ and $n = 0$, $P = 3/4$; or the odds are three to one that the event is more likely to happen than not. Similarly, if the event has occurred m times in succession,

$$P = 1 - (1/2)^{m+1},$$

which approaches unity rapidly with increase of n.

ART. 9.   THEORY OF ERRORS.

The theory of errors may be defined as that branch of mathematics which is concerned, first, with the expression of the resultant effect of one or more sources of error to which computed and observed quantities are subject; and, secondly, with the determination of the relation between the magnitude of an error and the probability of its occurrence. In the case of computed quantities which depend on numerical data, such as tables of logarithms, trigonometric functions, etc., it is usually possible to ascertain the actual values of the resultant errors. In the case of observed quantities, on the other hand, it is not generally possible to evaluate the resultant actual error, since the actual errors of observation are usually unknown. In either case, however, it is always possible to write down a symbolical expression which will show how different sources of error enter and affect the aggregate error; and the statement of such an expression is of fundamental importance in the theory of errors.

To fix the ideas, suppose a quantity Q to be a function of several independent quantities $x, y, z \ldots$; that is,

$$Q = f(x, y, z \ldots),$$

and let it be required to determine the error in Q due to errors in $x, y, z \ldots$ Denote such errors by $\Delta Q, \Delta x, \Delta y, \Delta z \ldots$ Then, supposing the errors so small that their squares, products, and higher powers may be neglected, Taylor's series gives

$$\Delta Q = \frac{\partial Q}{\partial x}\Delta x + \frac{\partial Q}{\partial y}\Delta y + \frac{\partial Q}{\partial z}\Delta z + \ldots \qquad (1)$$

This equation may be said to express the resultant actual error of the function in terms of the component actual errors, since the actual value of ΔQ is known when the actual errors of $x, y, z \ldots$ are known. It should be carefully noted that the quantities $x, y, z \ldots$ are supposed subject to errors which are independent of one another. The discovery of the independent sources of error is sometimes a matter of difficulty, and in general requires close attention on the part of the student if he would avoid blunders and misconceptions. Every investigator in work of precision should have a clear notion of the error-equation of the type (1) appertaining to his work; for it is thus only that he can distinguish between the important and unimportant sources of error.

Prob. 15. Write out the error-equation in accordance with (1) for the function $Q = xyz + x^2 \log(y/z)$.

Prob. 16. In a plane triangle $a/b = \sin A/\sin B$. Find the error in a due to errors in b, A, and B.

Prob. 17. Suppose in place of the data of problem 16 that the angles used in computation are given by the following equations: $A = A_1 + \frac{1}{3}(180° - A_1 - B_1 - C_1)$, $B = B_1 + \frac{1}{3}(180° - A_1 - B_1 - C_1)$, where A_1, B_1, C_1 are observed values. What then is Δa?

Prob. 18. If w denote the weight of a body and r the radius of the earth, show that for small changes in altitude, $\Delta w/w = -\Delta r/r$; whence, if a precision of one part in 500 000 000 is attainable in comparing two nearly equal masses, the effect of a difference in altitude of one centimeter in the scale-pans of a balance will be noticeable.*

ART. 10. LAWS OF ERROR.

A law of error is a function which expresses the relative frequency of occurrence of errors in terms of their magnitudes. Thus, using the customary notation, let ϵ denote the magni-

---

* This problem arose with the International Bureau of Weights and Measures, whose work of intercomparison of the Prototype Kilogrammes attained a precision indicated by a probable error of 1/500 000 000th part of a kilogramme.

tude o. any error in a system of possible errors. Then the law of such system may be expressed by an equation of the form

$$y = \phi(\epsilon). \tag{1}$$

Representing ϵ as abscissa and y as ordinate, this equation gives a curve called the curve of frequency, the nature of which, as is evident, depends on the form of the function ϕ. This equation gives the relative frequency of occurrence of errors in the system ; so that if ϵ is continuous the probability of the occurrence of any particular error is expressed by $yd\epsilon = \phi(\epsilon)d\epsilon$; which is infinitesimal, as it plainly should be, since in any continuous system the number of different values of ϵ is infinite.

Consider the simplest form of $\phi(\epsilon)$, namely, that in which $\phi(\epsilon) = c$, a constant. This form of $\phi(\epsilon)$ obtains in the case of the errors of tabular logarithms, natural trigonometric functions, etc. In this case all errors between minus a half-unit and plus a half-unit of the last tabular place are equally likely to occur. Suppose, to cover the class of cases to which that just cited belongs, all errors between the limits $-a$ and $+a$ are equally likely to occur. The probability of any individual error will then be $\phi(\epsilon)d\epsilon = cd\epsilon$, and the sum of all such probabilities, by equation (2), Art. 4, must be unity. That is,

$$\int_{-a}^{+a} \phi(\epsilon)d\epsilon = c\int_{-a}^{+a} d\epsilon = 1.$$

This gives $c = 1/2a$, or by (1) $y = 1/2a$. The curve of frequency in this case is shown in the figure, AB being the axis of ϵ and OQ that of y. It is evident from this diagram that if the errors of the system be considered with respect to magnitude only, half of them should be greater and half less than $a/2$. This is easily found to be so in the case of tabular logarithms, etc.

As a second illustration of (1), suppose y and ϵ connected by the relation $y = c\sqrt{a^2 - \epsilon^2}$, where a is the radius of a circle,

c a constant, and ϵ may have any value between $- a$ and $+ a$.
Then the condition

$$c \int_{-a}^{+a} d\epsilon \sqrt{a^2 - \epsilon^2} = 1$$

gives $c = 2/(a^2 \pi)$. In this, as in the preceding case, $\phi(+ \epsilon) = \phi(- \epsilon)$, the meaning of which is that positive and negative errors of the same magnitude are equally likely to occur. It will be noticed, however, that in the latter case small errors have a much higher probability than those near the limit a, while in the former case all errors have the same probability.

In general, when ϵ is continuous $\phi(\epsilon)$ must satisfy the condition $\int \phi(\epsilon) d\epsilon = 1$, the limits being such as to cover the entire range of values of ϵ. The cases most commonly met with are those in which $\phi(\epsilon)$ is an even function, or those in which $\phi(+ \epsilon) = \phi(- \epsilon)$. In such cases, if $\pm a$ denote the limiting value of ϵ,

$$\int_{-a}^{+a} \phi(\epsilon) d\epsilon = 2 \int_{0}^{a} \phi(\epsilon) d\epsilon = 1. \qquad (3)$$

ART. 11. TYPICAL ERRORS OF A SYSTEM.

Certain typical errors of a system have received special designations and are of constant use in the literature of the theory of errors. These special errors are the probable error, the mean error, and the average error. The first is that error of the system of errors which is as likely to be exceeded as not ; the second is the square root of the mean of the squares of all the errors ; and the third is the mean of all the errors regardless of their signs. Confining attention to systems in which positive and negative errors of the same magnitude are equally probable, these typical errors are defined mathematically as follows. Let

$\epsilon_p = $ the probable error,
$\epsilon_m = $ the mean error,
$\epsilon_a = $ the average error.

Then, observing (2), of Art. 10,

$$\left.\begin{array}{c}
\displaystyle\int_{-a}^{-\epsilon_p}\phi(\epsilon)d\epsilon = \int_{-\epsilon_p}^{0}\phi(\epsilon)d\epsilon = \int_{0}^{+\epsilon_p}\phi(\epsilon)d\epsilon = \int_{+\epsilon_p}^{+a}\phi(\epsilon)d\epsilon = \tfrac{1}{4}. \\[2mm]
\displaystyle\epsilon_m^{2} = \int_{-a}^{+a}\phi(\epsilon)\epsilon^{2}d\epsilon, \quad \epsilon_a = 2\int_{0}^{+a}\phi(\epsilon)\epsilon d\epsilon.
\end{array}\right\} \quad (1)$$

The student should seek to avoid the very common misapprehension of the meaning of the probable error. It is not "the most probable error," nor "the most probable value of the actual error"; but it is that error which, disregarding signs, would occupy the middle place if all the errors of the system were arranged in order of magnitude. A few illustrations will suffice to fix the ideas as to the typical errors. Thus, take the simple case wherein $\phi(\epsilon) = c = 1/2a$, which applies to tabular logarithms, etc. Equations (1) give at once

$$\epsilon_p = \pm\frac{1}{2}a, \quad \epsilon_m = \pm\frac{a}{3}\sqrt{3}, \quad \epsilon_a = \pm\frac{1}{2}a.$$

For the case of tabular values, $a = 0.5$ in units of the last tabular place. Hence for such values

$$\epsilon_p = \pm 0.25, \quad \epsilon_m = \pm 0.29, \quad \epsilon_a = \pm 0.25.$$

Prob. 19. Find the typical errors for the cases in which the law of error is $\phi(\epsilon) = c\sqrt{a^2 - \epsilon^2}$, $\phi(\epsilon) = c(\pm a \mp \epsilon)$, $\phi(\epsilon) = c\cos^2(\pi\epsilon/2a)$; c being a constant to be determined in each case and ϵ having any value between $-a$ and $+a$.

ART. 12. LAWS OF RESULTANT ERROR.

When several independent sources of error conspire to produce a resultant error, as specified by equation (1) of Art. 9, there is presented the problem of determining the law of the resultant error by means of the laws of the component errors. The algebraic statement of this problem is obtained as follows for the case of continuous errors:

In the equation (1), Art. 9, write for brevity

$$\epsilon = \Delta Q, \quad \epsilon_1 = \frac{\partial Q}{\partial x}\Delta x, \quad \epsilon_2 = \frac{\partial Q}{\partial y}\Delta y, \ldots;$$

and let the laws of error of ϵ, ϵ_1, ϵ_2, ... be denoted by $\phi(\epsilon)$, $\phi_1(\epsilon_1)$, $\phi_2(\epsilon_2)$ ... Then the value of ϵ is given by

$$\epsilon = \epsilon_1 + \epsilon_2 + \ldots \tag{1}$$

The probabilities of the occurrence of any particular values of ϵ_1, ϵ_2, ... are given by $\phi_1(\epsilon_1)d\epsilon_1$, $\phi_2(\epsilon_2)d\epsilon_2$, ...; and the probability of their concurrence is the probability of the corresponding value of ϵ. But since this value may arise in an infinite number of ways through the variations of ϵ_1, ϵ_2, ... over their ranges, the probability of ϵ, or $\phi(\epsilon)d\epsilon$, will be expressed by the integral of $\phi_1(\epsilon_1)d\epsilon_1$, $\phi_2(\epsilon_2)d\epsilon_2$ ... subject to the restriction (1). This latter gives $\epsilon_1 = \epsilon - \epsilon_2 - \epsilon_3 \ldots$, and $d\epsilon_1 = d\epsilon$ for the multiple integration with respect to ϵ_2, ϵ_3, ... Hence there results

$$\phi(\epsilon)d\epsilon = d\epsilon \int \phi_1(\epsilon - \epsilon_2 - \epsilon_3 - \ldots)\phi_2(\epsilon_2)d\epsilon_2 \ldots,$$

or

$$\phi(\epsilon) = \int \phi_1(\epsilon - \epsilon_1 - \epsilon_2 - \ldots)\phi_2(\epsilon_2)d\epsilon_2 \int \phi_3(\epsilon_3)d\epsilon_3, \ldots \tag{2}$$

It is readily seen that this formula will increase rapidly in complexity with the number of independent sources of error.* For some of the most important practical applications, however, it suffices to limit equation (2) to the case of two independent sources of error, each of constant probability within assigned limits. Thus, to consider this case, let ϵ_1 vary over the range $-a$ to $+a$, and ϵ_2 vary over the range $-b$ to $+b$. Then by equation (2), Art. 10,

$$\phi_1(\epsilon_1) = 1/(2a), \quad \phi_2(\epsilon_2) = 1/(2b).$$

Hence equation (2) becomes

$$\phi(\epsilon) = \frac{1}{4ab} \int d\epsilon_2.$$

In evaluating this integral ϵ_2 must not surpass $\pm b$ and $\epsilon_1 = \epsilon - \epsilon_2$ must not surpass $\pm a$. Assuming $a > b$, the limits of the integral for any value of $\epsilon = \epsilon_1 + \epsilon_2$ lying between $-(a + b)$ and $-(a - b)$ are $-b$ and $+(\epsilon + a)$. This fact is

---

* The reader desirous of pursuing this phase of the subject should consult Bessel's Untersuchungen ueber die Wahrscheinlichkeit der Beobachtungsfehler; Abhandlungen von Bessel (Leipzig, 1876), Vol. II.

made plain by a numerical example. For instance, suppose $a = 5$ and $b = 3$. Then $-(a+b) = -8$ and $-(a-b) = -3$. Take $\epsilon = -6$, a number intermediate to -8 and -3. Then the following are the possible integer values of ϵ_1 and ϵ_2 which will produce $\epsilon = -6$:

$$
\begin{array}{cccc}
\epsilon & \epsilon_1 \quad \epsilon_2 & & \text{limits of } \epsilon_2 \\
-6 = & -5 - 1, & -1 = +(\epsilon+a), \\
 = & -4 - 2, & \\
 = & -3 - 3, & -3 = -b.
\end{array}
$$

Similarly, the limits of ϵ_2 for values of ϵ lying between $-(a-b)$ and $+(a-b)$ are $-b$ and $+b$; and the limits of ϵ_2 for values of ϵ between $+(a-b)$ and $+(a+b)$ are $+(\epsilon-a)$ and $+b$. Hence

$$
\left.
\begin{aligned}
\phi(\epsilon) &= \frac{1}{4ab}\int_{-b}^{\epsilon+a} d\epsilon_2 = \frac{\epsilon+a+b}{4ab} \text{ for } -(a+b) < \epsilon < -(a-b), \\[2ex]
\phi(\epsilon) &= \frac{1}{4ab}\int_{-b}^{+b} d\epsilon_2 = \frac{2b}{4ab} \text{ for } -(a-b) < \epsilon < +(a-b), \\[2ex]
\phi(\epsilon) &= \frac{1}{4ab}\int_{\epsilon-a}^{+b} d\epsilon_2 = \frac{-\epsilon+a+b}{4ab} \text{ for } +(a-b) < \epsilon < +(a+b).
\end{aligned}
\right\} \quad (3)
$$

Thus it appears that in this case the graph of the resultant law of error is represented by the upper base and the two sides of a trapezoid, the lower base being the axis of ϵ and the line joining the middle points of the bases being the axis of $\phi(\epsilon)$. (See the first figure in Art. 13.) The properties of (3), including the determination of the limits, are also illustrated by the adjacent trapezoid of numerals arranged to represent the case wherein $a = 0.5$ and

```
        11011
       1110111
      111101111
     11111011111
    1111110111111
   111111101111111
  11111111011111111
```

$b = 0.3$. The vertical scale, or that for $\phi(\epsilon)$, does not, however, conform exactly to that for ϵ.

Prob. 20. Prove that the values of $\phi(\epsilon)$ as given by equation (3) satisfy the condition specified in equation (3), Art. 10.

Prob. 21. Examine equations (3) for the case wherein $a = b$ and $b = 0$; and interpret for the latter case the first and last of (3).

Prob. 22. Find from (3), and (1) of Art. 11, the probable error of the sum of two tabular logarithms.

ART. 13.   ERRORS OF INTERPOLATED VALUES.

Case I.—One of the most instructive cases to which formulas (3) of Art. 12 are applicable is that of interpolated logarithms, trigonometric functions, etc., dependent on first differences. Thus, suppose that v_1 and v_2 are two tabular logarithms, and that it is required to get a value v lying t tenths of the interval from v_1 towards v_2. Evidently

$$v = v_1 + (v_2 - v_1) t = (1 - t)v_1 + tv_2 ;$$

and hence if e, e_1, e_2 denote the actual errors of v, v_1, v_2, respectively,

$$e = (1 - t)e_1 + te_2. \tag{1}$$

It is to be carefully noted here that e as given by (1) requires the retention in v of at least one decimal place beyond the last tabular place. For example, let $v = \log (24373)$ from a 5-place table. Then $v_1 = 4.38686$, $v_2 = 4.38703$, $v_2 - v_1 = +0.00017$, $t = 0.3$, and $v = 4.38691.1$. Likewise, as found from a 7-place table, $e_1 = -0.45$, $e_2 = +0.37$ in units of the fifth place; and hence by (1) $e = -0.20$. That is, the actual error of $v = 4.38691.1$ is $= 0.20$, and this is verified by reference to a 7-place table.

The reader is also cautioned against mistaking the species of interpolated values here considered for the species commonly used by computers, namely, that in which the interpolated value is rounded to the nearest unit of the last tabular place. The latter species is discussed under Case II below.

Confining attention now to the class of errors specified by equation (1), there result in the notation of the preceding article

$$\epsilon_1 = (1 - t)e_1, \quad \epsilon_2 = te_2, \quad \text{and} \quad \epsilon = e = \epsilon_1 + \epsilon_2 ;$$

and since e_1 and e_2 each vary continuously between the limits

± 0.5 of a unit of the last tabular place, a and b in equations (3) of that article have the values

$$a = 0.5(1 - t), \quad b = 0.5t.$$

Hence the law of error of the interpolated values is expressed as follows:

$$
\begin{aligned}
\phi(\epsilon) &= \frac{0.5 + \epsilon}{(1 - t)t} \text{ for values of } \epsilon \text{ betw. } -0.5 \text{ and } -(0.5 - t), \\
&= \frac{1}{1 - t} \text{ for values of } \epsilon \text{ betw. } -(0.5 - t) \text{ and } +(0.5 - t), \\
&= \frac{0.5 - \epsilon}{(1 - t)t} \text{ for values of } \epsilon \text{ betw. } +(0.5 - t) \text{ and } +0.5.
\end{aligned}
\quad (2)
$$

The graph of $\phi(\epsilon)$ for $t = 1/3$ is shown by the trapezoid AB, BC, CD in the figure on page 500. Evidently the equations (2) are in general represented by a trapezoid, which degenerates to an isosceles triangle when $t = 1/2$.

The probable, mean, and average errors of an interpolated value of the kind in question are readily found from (2), and from equations (1) of Art. 11, to be

$$
\begin{aligned}
\epsilon_p &= (1/4)(1 - t) && \text{for} \quad 0 < t < 1/3, \\
&= 1/2 - (1/2)\sqrt{2t(1 - t)} && \text{for} \quad 1/3 < t < 2/3, \\
&= 1/4t && \text{for} \quad 2/3 < t < 1. \\
\epsilon_m &= \left\{ \frac{1 - (1 - 2t)^4}{96(1 - t)t} \right\}^{1/2}. \\
\epsilon_a &= \frac{1 - (1 - 2t)^3}{24(1 - t)t} && \text{for} \quad 0 < t < 1/2, \\
&= \frac{1 - (2t - 1)^3}{24(1 - t)t} && \text{for} \quad 1/2 < t < 1.
\end{aligned}
\quad (3)
$$

It is thus seen that the probable error of the interpolated value here considered decreases from 0.25 to 0.15 of a unit of the last tabular place as t increases from 0 to 0.5. Hence such values are more precise than tabular values; and the computer who desires to secure the highest attainable precision with a given table of logarithms should retain one additional figure beyond the last tabular place in interpolated values.

Case II.—Recurring to the equation $v = v_1 + t(v_2 - v_1)$ for an interpolated value v in terms of two consecutive tabular values v_1 and v_2, it will be observed that if the quantity $t(v_2 - v_1)$ is rounded to the nearest unit of the last tabular place, a new error is introduced. For example, if $v_1 = \log 1633 = 3.21299$, and $v_2 = \log 1634 = 3.21325$ from a 5-place table, $v_2 - v_1 = +26$ units of the last tabular place; and if $t = 1/3$, $t(v_2 - v_1) = 8\frac{2}{3}$; so that by the method of interpolation in question there results $v = 3.21299 + 9 = 3.21308$. Now the actual errors of v_1 and v_2 are, as found from a 7-place table, -0.38 and $+0.21$ in units of the fifth place. Hence the actual error of v is by equation (1), $\frac{2}{3} \times -0.38 + \frac{1}{3} \times +0.21 - \frac{1}{3} = -0.52$, as is shown directly by a 7-place table.

It appears, then, that in this case the error-equation corresponding to (1) is

$$e = (1 - t)e_1 + te_2 + e_3, \qquad (4)$$

wherein e_1 and e_2 are the same as in (1) and e_3 is the actual error that comes from rounding $t(v_2 - v_1)$ to the nearest unit of the last tabular place.

The error e_3, however, differs radically in kind from e_1 and e_2. The two latter are continuous, that is, they may each have any value, between the limits -0.5 and $+0.5$; while e_3 is discontinuous, being limited to a finite number of values dependent on the interpolating factor t. Thus, for $t = 1/2$ the only possible values of e_3 are $0 + 1/2$, and $-1/2$; likewise for $t = 1/3$, the only possible values of e_3 are 0, $+1/3$, and $-1/3$. It is also clear that the maximum value of e, which is constant and equal to $1/2$ for (1), is variable for (4) in a manner dependent on t. For example, in (4),

The maximum of $e = 1/2 + 1/2 = 1$,   for $t = 1/2$,
"   "   " $e = 1/2 + 1/3 = 5/6$, " $t = 1/3$,
"   "   " $e = 1/2 + 1/2 = 1$,   " $t = 1/4$,
"   "   " $e = 1/2 + 2/5 = 9/10$ " $t = 1/5$.

The determination of the law of error for this case presents some novelty, since it is essential to combine the continuous errors $(1 - t)e_1$ and te_2 with the discontinuous error e_3. The simplest mode of attacking the problem seems to be the following quasi-geometrical one. In the notation of Arts. 12 and 13, put in (4) $e = \epsilon$, $(1 - t)e_1 = \epsilon_1$, $te_2 = \epsilon_2$, and $e_3 = \epsilon_3$. Then

$$\epsilon = (\epsilon_1 + \epsilon_2) + \epsilon_3. \tag{5}$$

The law of error for $(\epsilon_1 + \epsilon_2)$ is given by equation (2) for any value of t. Hence for a given value of t there will be as many expressions of $\phi(\epsilon)$ as there are different values of ϵ_3. The graphs of $\phi(\epsilon)$ will all be of the same form but will be differently placed with reference to the axis of $\phi(\epsilon)$. Thus, if $t = 1/3$ the

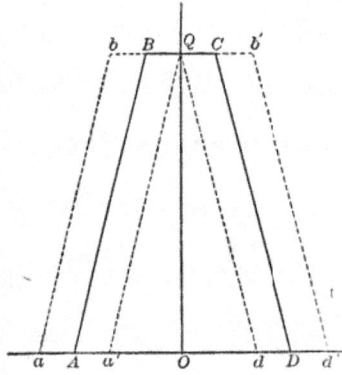

values of ϵ_3 are $- 1/3$, o, and $+ 1/3$, and these are equally likely to occur. For $\epsilon_3 = $ o the graph is given directly by (2), and is the trapezoid $ABCD$ symmetrical with respect to OQ. For $\epsilon_3 = - 1/3$ the graph is $abQd$, of the same form as $ABCD$ but shifted to the left by the amount of $\epsilon_3 = - 1/3$. Similarly, the graph for the case of $\epsilon_3 = + 1/3$ is $a'Qb'd'$, and is produced by shifting $ABCD$ to the right by an amount equal to $+ 1/3$.

Now, since the three systems of errors for this case are equally likely to occur, they may be combined into one system by simple addition of the corresponding element areas of the several graphs. Inspection of the diagram shows* that the resultant law of error is expressed by

$$\begin{aligned}
\phi(\epsilon) &= (1/4)(5 + 6\epsilon) & \text{for } &- 5/6 < \epsilon < - 1/6, \\
&= 1 & \text{for } &- 1/6 < \epsilon < + 1/6, \\
&= (1/4)(5 - 6\epsilon) & \text{for } &+ 1/6 < \epsilon < + 5/6.
\end{aligned} \right\} \tag{6}$$

This is represented by a trapezoid whose lower base is 10/6, upper base 2/6, and altitude 1.

---

* Sum the three areas and divide by 3 to make resultant area = 1, as required by equation (3), Art. 10.

As a second illustration, consider equation (5) for the case $t = 1/2$. In this case ϵ_s must be either o or $1/2$, the sign of which latter is arbitrary. For $\epsilon_s = 0$, equations (2) give

$$\phi(\epsilon) = 2 + 4\epsilon \quad \text{for} \quad -1/2 < \epsilon < 0, \\ = 2 - 4\epsilon \quad \text{for} \quad 0 < \epsilon < +1/2. \tag{7}$$

This function is represented by the isosceles triangle AQE whose altitude OQ is twice the base AE.

Similarly $\phi(\epsilon)$ for $\epsilon_s = +1/2$ would be represented by the triangle AQE displaced to the right a distance $1/2$; and if the two systems for $\epsilon_s = 0$ and $\epsilon_s = +1/2$ be combined into one system, their resultant law of error is evidently

$$\phi(\epsilon) = 1 + 2\epsilon \quad \text{for} \quad -1/2 < \epsilon < 0, \\ = 1 \qquad\quad \text{for} \quad 0 < \epsilon < +1/2, \\ = 2 - 2\epsilon \quad \text{for} \quad +1/2 < \epsilon < 1; \tag{8}$$

the graph of which is $ABCD$. On the other hand, if the errors in this combined system be considered with respect to magnitude only, the law of error is

$$\phi(\epsilon) = 2(1 - \epsilon) \quad \text{for} \quad 0 < \epsilon < 1, \tag{9}$$

the graph of which is OQD.

The student should observe that (6), (7), (8), and (9) satisfy the condition $\int \phi(\epsilon)d\epsilon = 1$ if the integration embraces the whole range of ϵ.

The determination of the general form of $\phi(\epsilon)$ in terms of the interpolating factor t for the present case presents some difficulties, and there does not appear to be any published solution of this problem.* The results arising from one phase of the problem have been given, however, by the author in the Annals of Mathematics,† and may be here stated without proof. The phase in question is that wherein t is of the form $1/n$, n being any positive integer less than twice the greatest

---

* The author explained a general method of solution in a paper read at the summer meeting of the American Mathematical Society, August, 1895.

† Vol. II, pp. 54–59.

tabular difference of the table to which the formulas are applied. For this restricted form of t the possible maximum value of ϵ as given by equation (5) is, in units of the last tabular place, $(2n - 1)/n$ for n odd and 1 for n even.

The possible values of ϵ, of equation (5) are

$$0, \quad \pm \frac{1}{n}, \quad \pm \frac{2}{n}, \cdots \pm \frac{n-1}{2n} \qquad \text{for } n \text{ odd,}$$

$$0, \quad \pm \frac{1}{n}, \quad \pm \frac{2}{n}, \cdots \pm \frac{n-2}{2n}, \quad \pm \frac{1}{2} \quad \text{for } n \text{ even.}$$

An important fact with regard to the error $1/2$ for n even is that its sign is arbitrary, or is not fixed by the computation as is the case with all the other errors. However, the computer's rule, which makes the rounded last figure of an interpolated value even when half a unit is to be disposed of, will, in the long-run, make this error as often plus as minus.

The laws of error which result are then as follows:

For n odd.

$$\phi(\epsilon) = 1 \qquad \text{for } \epsilon \text{ between } -1/2n \text{ and } +1/2n,$$

$$\phi(\epsilon) = \frac{n}{n-1}\left(\frac{2n-1}{2n} \pm \epsilon\right) \text{ for } \epsilon \text{ betw. } \mp 1/2n \text{ and } \mp(2n-1)/2n.$$

For n even.

$$\phi(\epsilon) = \frac{n}{2(n-1)}\left(\frac{2n-2}{n} \pm \epsilon\right) \text{ for } \epsilon \text{ between } 0 \text{ and } \mp 1/n,$$

$$= \frac{n}{n-1}\left(\frac{2n-1}{2n} \pm \epsilon\right) \text{ for } \epsilon \text{ betw. } \mp 1/n \text{ and } \mp(n-1)/n,$$

$$= \frac{n}{2(n-1)}(1 \pm \epsilon) \qquad \text{for } \epsilon \text{ between } \mp(n-1/n) \text{ and } \mp 1.$$

By means of these formulas and (1) of Art. 11 the probable, mean, and average errors for any value of n can be readily found. The following table contains the results of such a computation for values of n ranging from 1 to 10. The maximum actual error for each value of n is also added. The verification of the tabular quantities will afford a useful exercise to the student.

Typical Errors of Interpolated Logarithms, etc.

Interpotating Factor. $t = 1/n$	Probable Error. ϵ_p	Mean Error. ϵ_m	Average Error. ϵ_a	Maximum Actual Error.
1	0.250	0.289	0.250	1/2
1/2	.292	.408	.333	1
1/3	.256	.347	.287	5/6
1/4	.276	.382	.313	1
1/5	.268	.370	.303	9/10
1/6	.277	.385	.315	1
1/7	.274	.380	.311	13/14
1/8	.279	.389	.318	1
1/9	.278	.386	.316	17/18
1/10	.281	.392	.320	1

When the interpolating factor t has the more general form m/n, wherein m and n are integers with no common factor, the possible values of ϵ, are the same as for $t = 1/n$. But equations (3) of Art. 12 are not the same for $t = m/n$ as for $t = 1/n$, and hence for the more general form of t, $\phi(\epsilon)$ assumes a new type which is somewhat more complex than that discussed above. The limits of this work render it impossible to extend the investigation to these more complex forms of $\phi(\epsilon)$. It may suffice, therefore, to give a single instance of such a function, namely, that for which $t = 2/5$. For this case

$\phi(\epsilon) = 1$ for ϵ between 0 and $\mp 1/10$,
    $= (5/6)(13/10 \pm \epsilon)$ for ϵ between $\mp 1/10$ and $\mp 3/10$,
    $= (5/3)(4/5 \pm \epsilon)$ for ϵ between $\mp 3/10$ and $\mp 7/10$,
    $= (5/6)(9/10 \pm \epsilon)$ for ϵ between $\mp 7/10$ and $\mp 9/10$.

The graph of the right-hand half of this function is shown in the accompanying diagram, the whole graph being symmetrical with respect to OA, or the axis of $\phi(\epsilon)$.

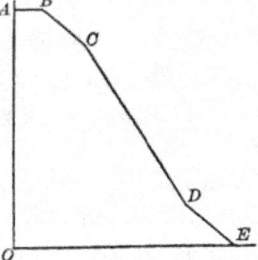

Attention may be called to the striking resemblance of this graph to that of the law of error of least squares.

Prob. 23. Show from equations (3) that ϵ_m varies from $1/\sqrt{12} = 0.29 -$, for $t = 0$, to $1/\sqrt{24} = 0.20 +$, for $t = 0.5$; and that ϵ_a varies from 0.25 to 1/6 for the same limits.

Prob. 24. Show that the probable, mean, and average errors for the case of $t = 2/5$ cited above (p. 503) are ± 0.261, ± 0.251, and ± 0.290, respectively.

ART. 14. STATISTICAL TEST OF THEORY.

A statistical test of the theory developed in Art. 13 may be readily drawn from any considerable number of actual errors of interpolated values dependent on the same interpolating factor. The application of such a test, if carried out fully by the student, will go far also towards fixing clear notions as to the meaning of the critical errors.

Consider first the case in which an interpolated value falls midway between two consecutive values, and suppose this interpolated value retains two additional figures beyond the last tabular place. Then by equations (2), Art. 13, the law of error of this interpolated value is

$$\phi(\epsilon) = 2 + 4\epsilon \text{ for } \epsilon \text{ between } -0.5 \text{ and } 0$$
$$= 2 - 4\epsilon \text{ for } \epsilon \text{ between } 0 \text{ and } +0.5.$$

Hence by equation (1) of Art. 11, or equation (3) of Art. 12, the probable error in this system of errors is $\frac{1}{2} - (\frac{1}{4}) \sqrt{2} = 0.15$. It follows, therefore, that in any large number of actual errors of this system, half should be less and half greater than 0.15. Similarly, of the whole number of such errors the percentage falling between the values 0.0 and 0.2 should be

$$\int_{-0.2}^{+0.2} \phi(\epsilon)d\epsilon = 2\int_{0}^{+0.2} (2 - 4\epsilon)d\epsilon = 0.64;$$

that is, sixty-four per cent of the errors in question should be less numerically than 0.2.

To afford a more detailed comparison in this case, the actual errors of five hundred interpolated values from a 5-place table have been computed by means of a 7-place table. The arguments used were the following numbers: 20005, 20035, 20065, 20105, 20135, etc., in the same order to 36635. The actual and theoretical percentages of the whole number of errors falling between the limits 0.0 and 0.1, 0.1 and 0.2, etc., are shown in the tabular form following:

Limits of Errors.	Actual Percentage.	Theoretical Percentage.
0.0 and 0.1	33.2	36
0.1 and 0.2	30.2	28
0.2 and 0.3	19.0	20
0.3 and 0.4	13.2	12
0.4 and 0.5	4.4	4
0.0 and 0.15	51.4	50

The agreement shown here between the actual and theoretical percentages is quite close, the maximum discrepancy being 2.8 and the average 1.5 per cent.

Secondly, consider the case of interpolated mid-values of the species treated under Case II of Art. 13. The law of error for this case is given by the single equation (9) of Art. 13, namely, $\phi(\epsilon) = 2(1 - \epsilon)$, no regard being paid to the signs of the errors. The probable error is then found from

$$2\int_0^{\epsilon_p} (1 - \epsilon)d\epsilon = \tfrac{1}{2},$$

whence $\epsilon_p = 1 - \tfrac{1}{2}\sqrt{2} = 0.29$. Similarly, the percentage of the whole number of errors which may be expected to lie, for example, between 0.0 and 0.2 in this system is

$$2\int_0^{0.2} (1 - \epsilon)d\epsilon = 0.36.$$

Using the same five hundred interpolated values cited above, but rounding them to the nearest unit of the last tabular place and computing their actual errors by means of a 7-place table, the following comparison is afforded:

Limits of Errors.	Actual Percentage.	Theoretical Percentage.
0.0 and 0.2	35.8	36
0.2 and 0.4	27.8	28
0.4 and 0.6	18.6	20
0.6 and 0.8	12.2	12
0.8 and 1.0	5.6	4
0.0 and 0.29	49.8	50

The agreement shown here between the actual and theoretical percentages is somewhat closer than in the preceding case, the maximum discrepancy being only 1.6 and the average only 0.6 per cent.

Finally, the following data derived from one thousand actual errors may be cited. The errors of one hundred interpolated values rounded to the nearest unit of the last tabular place were computed * for each of the interpolating factors 0.1, 0.2, . . . 0.9. The averages of these several groups of actual errors are given along with the corresponding theoretical errors in the parallel columns below:

Interpolating Factor.	Actual Average Error.	Theoretical Average Error.
0.1	0.338	0.320
0.2	0.288	0.303
0.3	0.321	0.304
0.4	0.268	0.290
0.5	0.324	0.333
0.6	0.276	0.290
0.7	0.321	0.304
0.8	0.289	0.303
0.9	0.347	0.320

The average discrepancy between the actual and theoretical values shown here is 0.017. It is, perhaps, somewhat smaller than should be expected, since the computation of the actual errors to three places of decimals is hardly warranted by the assumption of dependence on first differences only.

The average of the whole number of actual errors in this case is 0.308, which agrees to the same number of decimals with the average of the theoretical errors. †

* By Prof. H. A. Howe. See Annals of Mathematics, Vol. III, p. 74. The theoretical averages were furnished to Prof. Howe by the author.

† The reader who is acquainted with the elements of the method of least squares will find it instructive to apply that method to equation (1), Art. 13, and derive the probable error of e. This is frequently done without reserve by

Prob. 25.   Apply formulas (3) of Art. 12 to the case of the sum or difference of two tabular logarithms and derive the corresponding values of the probable, mean, and average errors.   The graph of $\varphi(\epsilon)$ is in this case an isosceles triangle whose base, or axis of ϵ, is 2, and whose altitude, or axis of $\varphi(\epsilon)$, is 1.

those familiar with least squares.   Thus, the probable error of e_1 or e_2 being 0.25, the probable error of e is found to be

$$0.25 \ \sqrt{1 - 2t + 2t^2}.$$

This varies between 0.25 for $t = 0$ and 0.18 for $t = \frac{1}{2}$; while the true value of the probable error, as shown by equations (3), Art. 13, varies from 0.25 to 0.15 for the same values of t.   It is, indeed, remarkable that the method of least squares, which admits infinite values for the actual errors e_1 and e_2, should give so close an approximate formula as the above for the probable error of e.

Similarly, one accustomed to the method of least squares would be inclined to apply it to equation (4), Art. 13, to determine the probable error of e.   The natural blunder in this case is to consider e_1, e_2, and e_3 independent, and e_3 like e_1 and e_2 continuous between the limits 0.0 and 0.5; and to assign a probable error of 0.25 to each.   In this manner the value

$$0.25 \ \sqrt{2(1 - t + t^2)}$$

is derived.   But this is absurd, since it gives 0.25 $\sqrt{2}$ instead of 0.25 for $t = 0$. The formula fails then to give even approximate results except for values of t near 0.5.

CHAPTER XI.

HISTORY OF MODERN MATHEMATICS.

By DAVID EUGENE SMITH,
Professor of Mathematics in the Michigan State Normal School.

ART. 1. INTRODUCTION.

Modern Mathematics is a term by no means well defined. Algebra cannot be called modern, and yet the theory of equations has received some of its most important additions during the nineteenth century, while the theory of forms is a recent creation. Similarly with elementary geometry; the labors of Lobachevsky and Bolyai during the second quarter of the century threw a new light upon the whole subject, and more recently the study of the triangle has added another chapter to the theory. Thus the history of modern mathematics must also be the modern history of ancient branches, while subjects which seem the product of late generations have root in other centuries than the present.

How unsatisfactory must be so brief a sketch may be inferred from a glance at the Index du Répertoire Bibliographique des Sciences Mathématiques (Paris, 1893), whose seventy-one pages contain the mere enumeration of subjects in large part modern, or from a consideration of the twenty-six volumes of the Jahrbuch über die Fortschritte der Mathematik, which now devotes over a thousand pages a year to a record of the progress of the science.*

The seventeenth and eighteenth centuries laid the founda-

* The foot-notes give only a few of the authorities which might easily be cited. They are thought to include those from which considerable extracts have been made, the necessary condensation of these extracts making any other form of acknowledgment impossible.

tions of much of the subject as known to-day. The discovery of the analytic geometry by Descartes, the contributions to the theory of numbers by Fermat, to algebra by Harriot, to geometry and to mathematical physics by Pascal, and the discovery of the differential calculus by Newton and Leibniz, all contributed to make the seventeenth century memorable. The eighteenth century was naturally one of great activity. Euler and the Bernoulli family in Switzerland, d'Alembert, Lagrange, and Laplace in Paris, and Lambert in Germany, popularized Newton's great discovery, and extended both its theory and its applications. Accompanying this activity, however, was a too implicit faith in the calculus and in the inherited principles of mathematics, which left the foundations insecure and necessitated their strengthening by the succeeding generation.

The nineteenth century has been a period of intense study of first principles, of the recognition of necessary limitations of various branches, of a great spread of mathematical knowledge, and of the opening of extensive fields for applied mathematics. Especially influential has been the establishment of scientific schools and journals and university chairs. The great renaissance of geometry is not a little due to the foundation of the École Polytechnique in Paris (1794–5), and the similar schools in Prague (1806), Vienna (1815), Berlin (1820), Karlsruhe (1825), and numerous other cities. About the middle of the century these schools began to exert a still a greater influence through the custom of calling to them mathematicians of high repute, thus making Zürich, Karlsruhe, Munich, Dresden, and other cities well known as mathematical centers.

In 1796 appeared the first number of the Journal de l'École Polytechnique. Crelle's Journal für die reine und angewandte Mathematik appeared in 1826, and ten years later Liouville began the publication of the Journal de Mathématiques pures et appliquées, which has been continued by Resal and Jordan. The Cambridge Mathematical Journal was established in 1839, and merged into the Cambridge and Dublin Mathematical

Journal in 1846.   Of the other periodicals which have contributed to the spread of mathematical knowledge, only a few can be mentioned: the Nouvelles Annales de Mathématiques (1842), Grunert's Archiv der Mathematik (1843), Tortolini's Annali di Scienze Matematiche e Fisiche (1850), Schlömilch's Zeitschrift für Mathematik und Physik (1856), the Quarterly Journal of Mathematics (1857), Battaglini's Giornale di Matematiche (1863), the Mathematische Annalen (1869), the Bulletin des Sciences Mathématiques (1870), the American Journal of Mathematics (1878), the Acta Mathematica (1882), and the Annals of Mathematics (1884).*   To this list should be added a recent venture, unique in its aims, namely, L'Intermédiaire des Mathématiciens (1894), and two annual publications of great value, the Jahrbuch already mentioned (1868), and the Jahresbericht der deutschen Mathematiker-Vereinigung (1892).

To the influence of the schools and the journals must be added that of the various learned societies † whose published proceedings are widely known, together with the increasing liberality of such societies in the preparation of complete works of a monumental character.

The study of first principles, already mentioned, was a natural consequence of the reckless application of the new calculus and the Cartesian geometry during the eighteenth century.   This development is seen in theorems relating to infinite series, in the fundamental principles of number, rational,

* For a list of current mathematical journals see the Jahrbuch über die Fortschritte der Mathematik.   A small but convenient list of standard periodicals is given in Carr's Synopsis of Pure Mathematics, p. 843 ; Mackay, J. S., Notice sur le journalisme mathématique en Angleterre, Association française pour l'Avancement des Sciences, 1893, II, 303 ; Cajori, F., Teaching and History of Mathematics in the United States, pp. 94, 277 ; Hart, D. S., History of American Mathematical Periodicals, The Analyst, Vol. II, p. 131.

† For a list of such societies consult any recent number of the Philosophical Transactions of Royal Society of London.   Dyck, W., Einleitung zu dem für den mathematischen Teil der deutschen Universitätsausstellung ausgegebenen Specialkatalog, Mathematical Papers Chicago Congress (New York, 1896), p. 44.

irrational, and complex, and in the concepts of limit, continuity, function, the infinite, and the infinitesimal. But the nineteenth century has done more than this. It has created new and extensive branches of an importance which promises much for pure and applied mathematics. Foremost among these branches stands the theory of functions founded by Cauchy, Riemann, and Weierstrass, followed by the descriptive and projective geometries, and the theories of groups, of forms, and of determinants.

The nineteenth century has naturally been one of specialization. At its opening one might have hoped to fairly compass the mathematical, physical, and astronomical sciences, as did Lagrange, Laplace, and Gauss. But the advent of the new generation, with Monge and Carnot, Poncelet and Steiner, Galois, Abel, and Jacobi, tended to split mathematics into branches between which the relations were long to remain obscure. In this respect recent years have seen a reaction, the unifying tendency again becoming prominent through the theories of functions and groups.*

ART. 2. THEORY OF NUMBERS.

The Theory of Numbers,† a favorite study among the Greeks, had its renaissance in the sixteenth and seventeenth centuries in the labors of Viète, Bachet de Méziriac, and especially Fermat. In the eighteenth century Euler and Lagrange contributed to the theory, and at its close the subject began to take scientific form through the great labors of Legendre (1798), and Gauss (1801). With the latter's Disquisitiones Arithmeticæ (1801) may be said to begin the modern theory of numbers. This theory separates into two branches, the one dealing with integers, and concerning itself especially

* Klein, F., The Present State of Mathematics, Mathematical Papers of Chicago Congress (New York, 1896), p. 133.

† Cantor, M., Geschichte der Mathematik, Vol. III, p. 94; Smith, H. J. S., Report on the theory of numbers; Collected Papers, Vol. I; Stolz, O., Grössen und Zahlen, Leipzig, 1891.

with (1) the study of primes, of congruences, and of residues, and in particular with the law of reciprocity, and (2) the theory of forms, and the other dealing with complex numbers.

The Theory of Primes* has attracted many investigators during the nineteenth century, but the results have been detailed rather than general. Tchébichef (1850) was the first to reach any valuable conclusions in the way of ascertaining the number of primes between two given limits. Riemann (1859) also gave a well-known formula for the limit of the number of primes not exceeding a given number.

The Theory of Congruences may be said to start with Gauss's Disquisitiones. He introduced the symbolism $a \equiv b$ (mod c), and explored most of the field. Tchébichef published in 1847 a work in Russian upon the subject, and in France Serret has done much to make the theory known.

Besides summarizing the labors of his predecessors in the theory of numbers, and adding many original and noteworthy contributions, to Legendre may be assigned the fundamental theorem which bears his name, the Law of Reciprocity of Quadratic Residues. This law, discovered by induction by Euler, was enunciated by Legendre and first proved in his Théorie des Nombres (1798) for special cases. Independently of Euler and Legendre, Gauss discovered the law about 1795, and was the first to give a general proof. To the subject have also contributed Cauchy, perhaps the most versatile of French mathematicians of the century; Dirichlet, whose Vorlesungen uber Zahlentheorie, edited by Dedekind, is a classic; Jacobi, who introduced the generalized symbol which bears his name; Liouville, Zeller, Eisenstein, Kummer, and Kronecker. The theory has been extended to include cubic and biquadratic reciprocity, notably by Gauss, by Jacobi, who first proved the law of cubic reciprocity, and by Kummer.

---

* Brocard, H., Sur la fréquence et la totalité des nombres premiers; Nouvelle Correspondence de Mathématiques, Vols. V and VI; gives ecent history to 1879.

To Gauss is also due the representation of numbers by binary quadratic forms. Cauchy, Poinsot (1845), Lebesgues (1859, 1868), and notably Hermite have added to the subject. In the theory of ternary forms Eisenstein has been a leader, and to him and H. J. S. Smith is also due a noteworthy advance in the theory of forms in general. Smith gave a complete classification of ternary quadratic forms, and extended Gauss's researches concerning real quadratic forms to complex forms. The investigations concerning the representation of numbers by the sum of 4, 5, 6, 7, 8 squares were advanced by Eisenstein and the theory was completed by Smith.

In Germany, Dirichlet was one of the most zealous workers in the theory of numbers, and was the first to lecture upon the subject in a German university. Among his contributions is the extension of Fermat's theorem on $x^n + y^n = z^n$, which Euler and Legendre had proved for $n = 3$, 4, Dirichlet showing that $x^5 + y^5 \neq az^5$. Among the later French writers are Borel; Poincaré, whose memoirs are numerous and valuable; Tannery, and Stieltjes. Among the leading contributors in Germany are Kronecker, Kummer, Schering. Bachmann, and Dedekind. In Austria Stolz's Vorlesungen über allgemeine Arithmetik (1885–86), and in England Mathews' Theory of Numbers (Part I, 1892) are among the most scholarly of general works. Genocchi, Sylvester, and J. W. L. Glaisher have also added to the theory.

ART. 3.   IRRATIONAL AND TRANSCENDENT NUMBERS.

The sixteenth century saw the final acceptance of negative numbers, integral and fractional. The seventeenth century saw decimal fractions with the modern notation quite generally used by mathematicians. The next hundred years saw the imaginary become a powerful tool in the hands of De Moivre, and especially of Euler. For the nineteenth century it remained to complete the theory of complex numbers, to separate irrationals into algebraic and transcendent, to prove the existence of transcendent numbers, and to make a scientific study of a subject which had remained almost dormant since Euclid, the theory of irrationals. The year 1872 saw the publication of the theories of Weierstrass (by his pupil Kossak), Heine (Crelle, 74), G. Cantor (Annalen, 5), and Dedekind. Méray had taken in 1869 the same point of departure as Heine, but the theory is generally referred to the year 1872. Weierstrass's method has been completely set forth by Pincherle (1880), and Dedekind's has received additional prominence through the author's later work (1888) and the recent indorsement by Tannery (1894). Weierstrass, Cantor, and Heine base their theories on infinite series, while Dedekind founds his on the idea of a cut (Schnitt) in the system of real numbers, separating all rational numbers into two groups having certain characteristic properties. The subject has received later contributions at the hands of Weierstrass, Kronecker (Crelle, 101), and Méray.

Continued Fractions, closely related to irrational numbers (and due to Cataldi, 1613),* received attention at the hands of Euler, and at the opening of the nineteenth century were brought into prominence through the writings of Lagrange. Other noteworthy contributions have been made by Druckenmüller (1837), Kunze (1857), Lemke (1870), and Günther (1872). Ramus (1855) first connected the subject with determinants, resulting, with the subsequent contributions of Heine, Möbius, and Günther, in the theory of Kettenbruchdeterminanten. Dirichlet also added to the general theory, as have numerous contributors to the applications of the subject.

Transcendent Numbers† were first distinguished from algebraic irrationals by Kronecker. Lambert proved (1761) that π cannot be rational, and that e^n (n being zero or rational) is irrational, a proof, however, which left much to be desired.

---

* But see Favaro, A., Notizie storiche sulle frazioni continue dal secolo decimoterzo al decimosettimo, Boncompagni's Bulletino, Vol. VII, 1874, pp. 451, 533.

† Klein, F., Vorträge über ausgewählte Fragen der Elementargeometrie, 1895, p. 38 ; Bachman, P., Vorlesungen über die Natur der Irrationalzahlen, 1892.

Legendre (1794) completed Lambert's proof, and showed that π is not the square root of a rational number. Liouville (1840) showed that neither e nor e^2 can be a root of an integral quadratic equation. But the existence of transcendent numbers was first established by Liouville (1844, 1851), the proof being subsequently displaced by G. Cantor's (1873). Hermite (1873) first proved e transcendent, and Lindemann (1882), starting from Hermite's conclusions, showed the same for π. Lindemann's proof was much simplified by Weierstrass (1885), still further by Hilbert (1893), and has finally been made elementary by Hurwitz and Gordan.

Art. 4. Complex Numbers.

The Theory of Complex Numbers * may be said to have attracted attention as early as the sixteenth century in the recognition, by the Italian algebraists, of imaginary or impossible roots. In the seventeenth century Descartes distinguished between real and imaginary roots, and the eighteenth saw the labors of De Moivre and Euler. To De Moivre is due (1730) the well-known formula which bears his name, $(\cos \phi + i \sin \phi)^n = \cos n\phi + i \sin n\phi$, and to Euler (1748) the formula $\cos \phi + i \sin \phi = e^{\phi i}$.

The geometric notion of complex quantity now arose, and as a result the theory of complex numbers received a notable expansion. The idea of the graphic representation of complex numbers had appeared, however, as early as 1685, in Wallis's De Algebra tractatus. In the eighteenth century Kühn (1750) and Wessel (about 1795) made decided advances towards the present theory. Wessel's memoir appeared in the Proceedings of the Copenhagen Academy for 1799, and is exceedingly

---

* Riecke, F., Die Rechnung mit Richtungszahlen, 1856, p. 161 ; Hankel, H., Theorie der komplexen Zahlensysteme, Leipzig, 1867 ; Holzmüller, G., Theorie der isogonalen Verwandtschaften, 1882, p. 21; Macfarlane, A., The Imaginary of Algebra, Proceedings of American Association 1892, p. 33 ; Baltzer, R., Einführung der komplexen Zahlen, Crelle, 1882 ; Stolz, O., Vorlesungen über allgemeine Arithmetik, 2. Theil, Leipzig, 1886.

clear and complete, even in comparison with modern works. He also considers the sphere, and gives a quaternion theory from which he develops a complete spherical trigonometry. In 1804 the Abbé Buée independently came upon the same idea which Wallis had suggested, that $\pm \sqrt{-1}$ should represent a unit line, and its negative, perpendicular to the real axis. Buée's paper was not published until 1806, in which year Argand also issued a pamphlet on the same subject. It is to Argand's essay that the scientific foundation for the graphic representation of complex numbers is now generally referred. Nevertheless, in 1831 Gauss found the theory quite unknown, and in 1832 published his chief memoir on the subject, thus bringing it prominently before the mathematical world. Mention should also be made of an excellent little treatise by Mourey (1828), in which the foundations for the theory of directional numbers are scientifically laid. The general acceptance of the theory is not a little due to the labors of Cauchy and Abel, and especially the latter, who was the first to boldly use complex numbers with a success that is well known.

The common terms used in the theory are chiefly due to the founders. Argand called $\cos\phi + i\sin\phi$ the "direction factor", and $r = \sqrt{a^2 + b^2}$ the "modulus"; Cauchy (1828) called $\cos\phi + i\sin\phi$ the "reduced form" (l'expression réduite); Gauss used i for $\sqrt{-1}$, introduced the term "complex number" for $a + bi$, and called $a^2 + b^2$ the "norm." The expression "direction coefficient", often used for $\cos\phi + i\sin\phi$, is due to Hankel (1867), and "absolute value," for "modulus," is due to Weierstrass.

Following Cauchy and Gauss have come a number of contributors of high rank, of whom the following may be especially mentioned: Kummer (1844), Kronecker (1845), Scheffler (1845, 1851, 1880), Bellavitis (1835, 1852), Peacock (1845), and De Morgan (1849). Möbius must also be mentioned for his numerous memoirs on the geometric applications of complex numbers, and Dirichlet for the expansion of the theory to include primes, congruences, reciprocity, etc., as in the case of real numbers.

Other types* have been studied, besides the familiar $a + bi$, in which i is the root of $x^2 + 1 = 0$. Thus Eisenstein has studied the type $a + bj$, j being a complex root of $x^3 - 1 = 0$. Similarly, complex types have been derived from $x^k - 1 = 0$ (k prime). This generalization is largely due to Kummer, to whom is also due the theory of Ideal Numbers,† which has recently been simplified by Klein (1893) from the point of view of geometry. A further complex theory is due to Galois, the basis being the imaginary roots of an irreducible congruence, $F(x) \equiv 0$ (mod p, a prime). The late writers (from 1884) on the general theory include Weierstrass, Schwarz, Dedekind, Hölder, Berloty, Poincaré, Study, and Macfarlane.

ART. 5. QUATERNIONS AND AUSDEHNUNGSLEHRE.

Quaternions and Ausdehnungslehre‡ are so closely related to complex quantity, and the latter to complex number, that the brief sketch of their development is introduced at this point. Caspar Wessel's contributions to the theory of complex quantity and quaternions remained unnoticed in the proceedings of the Copenhagen Academy. Argand's attempts to extend his method of complex numbers beyond the space of two dimensions failed. Servois (1813), however, almost trespassed on the quaternion field. Nevertheless there were fewer traces of the theory anterior to the labors of Hamilton than is usual in the case of great discoveries. Hamilton discovered the principle of quaternions in 1843, and the next year his first contribution to the theory appeared, thus extending the Argand idea to three-dimensional space. This step neces-

* Chapman, C. H., Weierstrass and Dedekind on General Complex Numbers, in Bulletin New York Mathematical Society, Vol. I, p. 150; Study, E., Aeltere und neuere Untersuchungen über Systeme complexer Zahlen, Mathematical Papers Chicago Congress, p. 367; bibliography, p. 381.

† Klein, F., Evanston Lectures, Lect. VIII.

‡ Tait, P. G., on Quaternions, Encyclopædia Britannica; Schlegel, V., **Die** Grassmann'sche Ausdehnungslehre, Schlömilch's Zeitschrift, Vol. XLI.

sitated an expansion of the idea of $r(\cos \phi + j \sin \phi)$ such that while r should be a real number and ϕ a real angle, $i, j,$ or k should be any directed unit line such that $i^2 = j^2 = k^2 = -1$. It also necessitated a withdrawal of the commutative law of multiplication, the adherence to which obstructed earlier discovery. It was not until 1853 that Hamilton's Lectures on Quarternions appeared, followed (1866) by his Elements of Quaternions.

In the same year in which Hamilton published his discovery (1844), Grassmann gave to the world his famous work, Die lineale Ausdehnungslehre, although he seems to have been in possession of the theory as early as 1840. Differing from Hamilton's Quaternions in many features, there are several essential principles held in common which each writer discovered independently of the other.*

Following Hamilton, there have appeared in Great Britain numerous papers and works by Tait (1867), Kelland and Tait (1873), Sylvester, and McAulay (1893). On the Continent Hankel (1867), Hoüel (1874), and Laisant (1877, 1881) have written on the theory, but it has attracted relatively little attention. In America, Benjamin Peirce (1870) has been especially prominent in developing the quaternion theory, and Hardy (1881) Macfarlane, and Hathaway (1896) have contributed to the subject. The difficulties have been largely in the notation. In attempting to improve this symbolism Macfarlane has aimed at showing how a space analysis can be developed embracing algebra, trigonometry, complex numbers, Grassmann's method, and quaternions, and has considered the general principles, of vector and versor analysis, the versor being circular, elliptic logarithmic, or hyperbolic. Other recent contributors to the algebra of vectors are Gibbs (from 1881) and Heaviside (from 1885).

The followers of Grassmann † have not been much more

---

* These are set forth in a paper by J. W. Gibbs, Nature, Vol. XLIV, p. 79.

† For bibliography see Schlegel, V., Die Grassmann'sche Ausdehnungs-lehre, Schlömilch's Zeitschrift, Vol. XLI.

numerous than those of Hamilton.   Schlegel has been one of the chief contributors in Germany, and Peano in Italy.   In America, Hyde (Directional Calculus, 1890) has made a plea for the Grassmann theory.*

Along lines analogous to those of Hamilton and Grassmann have been the contributions of Scheffler.   While the two former sacrificed the commutative law, Scheffler (1846, 1851, 1880) sacrificed the distributive.   ·This sacrifice of fundamental laws has led to an investigation of the field in which these laws are valid, an investigation to which Grassmann (1872), Cayley, Ellis, Boole, Schröder (1890–91), and Kraft (1893) have contributed.   Another great contribution of Cayley's along similar lines is the theory of matrices (1858).

ART. 6. THEORY OF EQUATIONS.

The Theory of Numerical Equations † concerns itself first with the location of the roots, and then with their approximation.   Neither problem is new, but the first noteworthy contribution to the former in the nineteenth century was Budan's (1807).   Fourier's work was undertaken at about the same time, but appeared posthumously in 1831.   All processes were, however, exceedingly cumbersome until Sturm (1829) communicated to the French Academy the famous theorem which bears his name and which constitutes one of the most brilliant discoveries of algebraic analysis.

The Approximation of the Roots, once they are located, can be made by several processes.   Newton (1711), for example, gave a method which Fourier perfected ; and Lagrange (1767) discovered an ingenious way of expressing the root as a continued fraction, a process which Vincent (1836) elaborated.   It

* For Macfarlane's Digest of views of English and American writers, see Proceedings American Association for Advancement of Science, 1891.

† Cayley, A., Equations, and Kelland. P., Algebra, in Encyclopædia Britannica; Favaro, A., Notizie storico-critiche sulla costruzione delle equazioni. Modena, 1878; Cantor, M., Geschichte der Mathematik, Vol. III, p. 375.

was, however, reserved for Horner (1819) to suggest the most
practical method yet known, the one now commonly used.
With Horner and Sturm this branch practically closes. The
calculation of the imaginary roots by approximation is still an
open field.

The Fundamental Theorem* that every numerical equation
has a root was generally assumed until the latter part of the
eighteenth century. D'Alembert (1746) gave a demonstration,
as did Lagrange (1772), Laplace (1795), Gauss (1799) and Argand
(1806). The general theorem that every algebraic equation of
the nth degree has exactly n roots and no more follows as a
special case of Cauchy's proposition (1831) as to the number of
roots within a given contour. Proofs are also due to Gauss,
Serret, Clifford (1876), Malet (1878), and many others.

The Impossibility of Expressing the Roots of an equation
as algebraic functions of the coefficients when the degree exceeds 4 was anticipated by Gauss and announced by Ruffini,
and the belief in the fact became strengthened by the failure
of Lagrange's methods for these cases. But the first strict
proof is due to Abel, whose early death cut short his labors in
this and other fields.

The Quintic Equation has naturally been an object of
special study. Lagrange showed that its solution depends on
that of a sextic, "Lagrange's resolvent sextic," and Malfatti
and Vandermonde investigated the construction of resolvents.
The resolvent sextic was somewhat simplified by Cockle and
Harley (1858–59) and by Cayley (1861), but Kronecker (1858)
was the first to establish a resolvent by which a real simplifi-
cation was effected. The transformation of the general quintic
into the trinomial form $x^5 + ax + b = 0$ by the extraction of
square and cube roots only, was first shown to be possible by

---

* Loria, Gino, Esame di alcune ricerche concernenti l'esistenza di radici
nelle equazioni algebriche; Bibliotheca Mathematica, 1891, p. 99; bibliography
on p. 107. Pierpont, J., On the Ruffini-Abelian theorem, Bulletin of American
Mathematical Society, Vol. II, p. 200.

Bring (1786) and independently by Jerrard * (1834). Hermite (1858) actually effected this reduction, by means of Tschirnhausen's theorem, in connection with his solution by elliptic functions.

The Modern Theory of Equations may be said to date from Abel and Galois. The latter's special memoir on the subject, not published until 1846, fifteen years after his death, placed the theory on a definite base. To him is due the discovery that to each equation corresponds a group of substitutions (the "group of the equation") in which are reflected its essential characteristics.† Galois's untimely death left without sufficient demonstration several important propositions, a gap which Betti (1852) has filled. Jordan, Hermite, and Kronecker were also among the earlier ones to add to the theory. Just prior to Galois's researches Abel (1824), proceeding from the fact that a rational function of five letters having less than five values cannot have more than two, showed that the roots of a general quintic equation cannot be expressed in terms of its coefficients by means of radicals. He then investigated special forms of quintic equations which admit of solution by the extraction of a finite number of roots. Hermite, Sylvester, and Brioschi have applied the invariant theory of binary forms to the same subject.

From the point of view of the group the solution by radicals, formerly the goal of the algebraist, now appears as a single link in a long chain of questions relative to the transformation of irrationals and to their classification. Klein (1884) has handled the whole subject of the quintic equation in a simple manner by introducing the icosahedron equation as the normal form, and has shown that the method can be generalized so as to embrace the whole theory of higher equations.‡ He and Gordan (from 1879) have attacked those equations of

* Harley, R., A contribution of the history . . . of the general equation of the fifth degree, Quarterly Journal of Mathematics, Vol. VI, p. 38.

† See Art. 7.

‡ Klein, F., Vorlesungen über das Ikosaeder, 1884.

the sixth and seventh degrees which have a Galois group of 168 substitutions, Gordan performing the reduction of the equation of the seventh degree to the ternary problem. Klein (1888) has shown that the equation of the twenty-seventh degree occurring in the theory of cubic surfaces can be reduced to a normal problem in four variables, and Burkhardt (1893) has performed the reduction, the quaternary groups involved having been discussed by Maschke (from 1887).

Thus the attempts to solve the quintic equation by means of radicals has given place to their treatment by transcendents. Hermite (1858) has shown the possibility of the solution, by the use of elliptic functions, of any Bring quintic, and hence of any equation of the fifth degree. Kronecker (1858), working from a different standpoint, has reached the same results, and his method has since been simplified by Brioschi. More recently Kronecker, Gordan, Kiepert, and Klein, have contributed to the same subject, and the sextic equation has been attacked by Maschke and Brioschi through the medium of hyperelliptic functions.

Binomial Equations, reducible to the form $x^n - 1 = 0$, admit of ready solution by the familiar trigonometric formula

$$x = \cos \frac{2k\pi}{n} + i \sin \frac{2k\pi}{n};$$ but it was reserved for Gauss (1801)

to show that an algebraic solution is possible. Lagrange (1808) extended the theory, and its application to geometry is one of the leading additions of the century. Abel, generalizing Gauss's results, contributed the important theorem that if two roots of an irreducible equation are so connected that the one can be expressed rationally in terms of the other. the equation yields to radicals if the degree is prime and otherwise depends on the solution of lower equations. The binomial equation, or rather the equation $\sum_{0}^{n-1} x^m = 0$, is one of this class

considered by Abel, and hence called (by Kronecker) Abelian Equations. The binomial equation has been treated notably by Richelot (1832), Jacobi (1837), Eisenstein (1844, 1850), Cayley (1851), and Kronecker (1854), and is the subject of a treatise by Bachmann (1872). Among the most recent writers on Abelian equations is Pellet (1891).

Certain special equations of importance in geometry have been the subject of study by Hesse, Steiner, Cayley, Clebsch, Salmon, and Kummer. Such are equations of the ninth degree determining the points of inflection of a curve of the third degree, and of the twenty-seventh degree determining the points in which a curve of the third degree can have contact of the fifth order with a conic.

Symmetric Functions of the coefficients, and those which remain unchanged through some or all of the permutations of the roots, are subjects of great importance in the present theory. The first formulas for the computation of the symmetric functions of the roots of an equation seem to have been worked out by Newton, although Girard (1629) had given, without proof, a formula for the power sum. In the eighteenth century Lagrange (1768) and Waring (1770, 1782) contributed to the theory, but the first tables, reaching to the tenth degree, appeared in 1809 in the Meyer-Hirsch Aufgabensammlung. In Cauchy's celebrated memoir on determinants (1812) the subject began to assume new prominence, and both he and Gauss (1816) made numerous and valuable contributions to the theory. It is, however, since the discoveries by Galois that the subject has become one of great importance. Cayley (1857) has given simple rules for the degree and weight of symmetric functions, and he and Brioschi have simplified the computation of tables.

Methods of Elimination and of finding the resultant (Bezout) or eliminant (De Morgan) occupied a number of eighteenth-century algebraists, prominent among them being Euler (1748), whose method, based on symmetric functions, was improved by Cramer (1750) and Bezout (1764). The leading steps in the development are represented by Lagrange (1770–71), Jacobi, Sylvester (1840), Cayley (1848, 1857), Hesse (1843, 1859), Bruno (1859), and Katter (1876). Sylvester's dialytic method appeared in 1841, and to him is also due (1851) the name and a portion of the theory of the discriminant. Among recent writers on the general theory may be mentioned Burnside and Pellet (from 1887).

ART. 7.   SUBSTITUTIONS AND GROUPS.

The Theories of Substitutions and Groups* are among the most important in the whole mathematical field, the study of groups and the search for invariants now occupying the attention of all mathematicians. The first recognition of the importance of the combinatory analysis occurs in the problem of forming an mth-degree equation having for roots m of the roots of a given nth-degree equation $(m < n)$. For simple cases the problem goes back to Hudde (1659). Saunderson (1740) noted that the determination of the quadratic factors of a biquadratic expression necessarily leads to a sextic equation, and Le Sœur (1748) and Waring (1762 to 1782) still further elaborated the idea.

Lagrange† first undertook a scientific treatment of the theory of substitutions. Prior to his time the various methods of solving lower equations had existed rather as isolated artifices than as a unified theory.‡ Through the great power of analysis possessed by Lagrange (1770, 1771) a common foundation was discovered, and on this was built the theory of substitutions. He undertook to examine the methods then known, and to show a priori why these succeeded below the quintic, but otherwise failed. In his investigation he discovered the important fact that the roots of all resolvents (résolvantes, réduites) which he examined are rational functions of the roots of the respective equations. To study the properties of these functions he invented a "Calcul des Combinaisons," the first

* Netto, E., Theory of Substitutions, translated by Cole; Cayley, A., Equations, Encyclopædia Britannica, 9th edition.

† Pierpont, James, Lagrange's Place in the Theory of Substitutions, Bulletin of American Mathematical Society, Vol. I, p. 196.

‡ Matthiessen, L., Grundzüge der antiken und modernen Algebra der litteralen Gleichungen, Leipzig, 1878.

important step towards a theory of substitutions. Mention should also be made of the contemporary labors of Vandermonde (1770) as foreshadowing the coming theory.

The next great step was taken by Ruffini* (1799). Beginning like Lagrange with a discussion of the methods of solving lower equations, he attempted the proof of the impossibility of solving the quintic and higher equations. While the attempt failed, it is noteworthy in that it opens with the classification of the various "permutations" of the coefficients, using the word to mean what Cauchy calls a "système des substitutions conjuguées," or simply a " système conjugué," and Galois calls a " group of substitutions." Ruffini distinguishes what are now called intransitive, transitive and imprimitive, and transitive and primitive groups, and (1801) freely uses the group of an equation under the name "l'assieme della permutazioni." He also publishes a letter from Abbati to himself, in which the group idea is prominent.

To Galois, however, the honor of establishing the theory of groups is generally awarded. He found that if $r_1, r_2, \ldots r_n$ are the n roots of an equation, there is always a group of permutations of the r's such that (1) every function of the roots invariable by the substitutions of the group is rationally known, and (2), reciprocally, every rationally determinable function of the roots is invariable by the substitutions of the group. Galois also contributed to the theory of modular equations and to that of elliptic functions. His first publication on the group theory was made at the age of eighteen (1829), but his contributions attracted little attention until the publication of his collected papers in 1846 (Liouville, Vol. XI).

Cayley and Cauchy were among the first to appreciate the importance of the theory, and to the latter especially are due a number of important theorems. The popularizing of the subject is largely due to Serret, who has devoted section IV of his

* Burkhardt, H., Die Anfänge der Gruppentheorie und Paolo Ruffini, Abhandlungen zur Geschichte der Mathematik, VI, 1892, p. 119. Italian by E. Pascal, Brioschi's Annali di Matematici, 1894.

algebra to the theory; to Camille Jordan, whose Traité des Substitutions is a classic; and to Netto (1882), whose work has been translated into English by Cole (1892). Bertrand, Hermite, Frobenius, Kronecker, and Mathieu have added to the theory. The general problem to determine the number of groups of n given letters still awaits solution.

But overshadowing all others in recent years in carrying on the labors of Galois and his followers in the study of discontinuous groups stand Klein, Lie, Poincaré, and Picard. Besides these discontinuous groups there are other classes, one of which, that of finite continuous groups, is especially important in the theory of differential equations. It is this class which Lie (from 1884) has studied, creating the most important of the recent departments of mathematics, the theory of transformation groups. Of value, too, have been the labors of Killing on the structure of groups, Study's application of the group theory to complex numbers, and the work of Schur and Maurer.

ART. 8.   DETERMINANTS.

The Theory of Determinants * may be said to take its origin with Leibniz (1693), following whom Cramer (1750) added slightly to the theory, treating the subject, as did his predecessor, wholly in relation to sets of equations. The recurrent law was first announced by Bezout (1764). But it was Vandermonde (1771) who first recognized determinants as independent functions. To him is due the first connected exposition of the theory, and he may be called its formal founder. Laplace (1772) gave the general method of expanding a determinant in terms of its complementary minors, although Vandermonde had already given a special case. Immediately following, Lagrange (1773) treated determinants of the second

---

* Muir, T., Theory of Determinants in the Historical Order of its Development, Part I, 1890; Baltzer, R., Theorie und Anwendung der Determinanten, 1881. The writer is under obligations to Professor Weld, who contributes Chap. II, for valuable assistance in compiling this article.

and third order, possibly stopping here because the idea of hyperspace was not then in vogue. Although contributing nothing to the general theory, Lagrange was the first to apply determinants to questions foreign to eliminations, and to him are due many special identities which have since been brought under well-known theorems. During the next quarter of a century little of importance was done. Hindenburg (1784) and Rothe (1800) kept the subject open, but Gauss (1801) made the next advance. Like Lagrange, he made much use of determinants in the theory of numbers. He introduced the word "determinants" (Laplace had used "resultant"), though not in the present signification,* but rather as applied to the discriminant of a quantic. Gauss also arrived at the notion of reciprocal determinants, and came very near the multiplication theorem. The next contributor of importance is Binet (1811, 1812), who formally stated the theorem relating to the product of two matrices of m columns and n rows, which for the special case of $m = n$ reduces to the multiplication theorem. On the same day (Nov. 30, 1812) that Binet presented his paper to the Academy, Cauchy also presented one on the subject. In this he used the word "determinant" in its present sense, summarized and simplified what was then known on the subject, improved the notation, and gave the multiplication theorem with a proof more satisfactory than Binet's. He was the first to grasp the subject as a whole; before him there were determinants, with him begins their theory in its generality.

The next great contributor, and the greatest save Cauchy, was Jacobi (from 1827). With him the word "determinant" received its final acceptance. He early used the functional determinant which Sylvester has called the "Jacobian," and in his famous memoirs in Crelle for 1841 he specially treats this subject, as well as that class of alternating functions which Sylvester has called "Alternants." But about the time of Jacobi's closing memoirs, Sylvester (1839) and Cayley began

* "Numerum $bb-ac$, cuius indole proprietates formæ (a, b, c) imprimis pendere in sequentibus docebimus, determinantem huius uocabimus."

their great work, a work which it is impossible to briefly sum-
marize, but which represents the development of the theory to
the present time.

The study of special forms of determinants has been the
natural result of the completion of the general theory. Axi-
symmetric determinants have been studied by Lebesgue, Hesse,
and Sylvester; per-symmetric determinants by Sylvester and
Hankel; circulants by Catalan, Spottiswoode, Glaisher, and
Scott; skew determinants and Pfaffians, in connection with the
theory of orthogonal transformation, by Cayley; continuants
by Sylvester; Wronskians (so called by Muir) by Christoffel
and Frobenius; compound determinants by Sylvester, Reiss,
and Picquet; Jacobians and Hessians by Sylvester; and symmetric gauche determinants by Trudi. Of the text-books on
the subject Spottiswoode's was the first. In America, Hanus
(1886) and Weld (1893) have published treatises.

ART. 9. QUANTICS.

The Theory of Quantics or Forms * appeared in embryo in
the Berlin memoirs of Lagrange (1773, 1775), who considered
binary quadratic forms of the type $ax^2 + bxy + cy^2$, and established the invariance of the discriminant of that type when
$x + \lambda y$ is put for x. He classified forms of that type according to the sign of $b^2 - 4ac$, and introduced the ideas of transformation and equivalence. Gauss † (1801) next took up the
subject, proved the invariance of the discriminants of binary
and ternary quadratic forms, and systematized the theory of
binary quadratic forms, a subject elaborated by H. J. S.
Smith, Eisenstein, Dirichlet, Lipschitz, Poincaré, and Cayley.
Galois also entered the field, in his theory of groups (1829), and

---

* Meyer, W. F., Bericht über den gegenwärtigen Stand der Invarianten-
theorie. Jahresbericht der deutschen Mathematiker-Vereinigung, Vol. I,
1890–91: Berlin 1892, p. 97. See also the review by Franklin in Bulletin New
York Mathematical Society, Vol. III, p. 187; Biography of Cayley, Collected
Papers, VIII, p. ix, and Proceedings of Royal Society, 1895.

† See Art. 2.

the first step towards the establishment of the distinct theory is sometimes attributed to Hesse in his investigations of the plane curve of the third order.

It is, however, to Boole (1841) that the real foundation of the theory of invariants is generally ascribed. He first showed the generality of the invariant property of the discriminant, which Lagrange and Gauss had found for special forms. Inspired by Boole's discovery Cayley took up the study in a memoir "On the Theory of Linear Transformations" (1845), which was followed (1846) by investigations concerning co-variants and by the discovery of the symbolic method of find-ing invariants. By reason of these discoveries concerning invariants and covariants (which at first he called "hyperdeter-minants") he is regarded as the founder of what is variously called Modern Algebra, Theory of Forms, Theory of Quantics, and the Theory of Invariants and Covariants. His ten memoirs on the subject began in 1854, and rank among the greatest which have ever been produced upon a single theory. Syl-vester soon joined Cayley in this work, and his originality and vigor in discovery soon made both himself and the subject prominent. To him are due (1851–54) the foundations of the general theory, upon which later writers have largely built, as well as most of the terminology of the subject.

Meanwhile in Germany Eisenstein (1843) had become aware of the simplest invariants and covariants of a cubic and bi-quadratic form, and Hesse and Grassmann had both (1844) touched upon the subject. But it was Aronhold (1849) who first made the new theory known. He devised the symbolic method now common in Germany, discovered the invariants of a ternary cubic and their relations to the discriminant, and, with Cayley and Sylvester, studied those differential equations which are satisfied by invariants and covariants of binary quantics. His symbolic method has been carried on by Clebsch, Gordan, and more recently by Study (1889) and Stroh (1890), in lines quite different from those of the English school.

In France Hermite early took up the work (1851). He

discovered (1854) the law of reciprocity that to every covariant or invariant of degree ρ and order r of a form of the mth order corresponds also a covariant or invariant of degree m and of order r of a form of the ρth order. At the same time (1854) Brioschi joined the movement, and his contributions have been among the most valuable. Salmon's Higher Plane Curves (1852) and Higher Algebra (1859) should also be mentioned as marking an epoch in the theory.

Gordan entered the field, as a critic of Cayley, in 1868. He added greatly to the theory, especially by his theorem on the Endlichkeit des Formensystems, the proof for which has since been simplified. This theory of the finiteness of the number of invariants and covariants of a binary form has since been extended by Peano (1882), Hilbert (1884), and Mertens (1886). Hilbert (1890) succeeded in showing the finiteness of the complete systems for forms in n variables, a proof which Story has simplified.

Clebsch * did more than any other to introduce into Germany the work of Cayley and Sylvester, interpreting the projective geometry by their theory of invariants, and correlating it with Riemann's theory of functions. Especially since the publication of his work on forms (1871) the subject has attracted such scholars as Weierstrass, Kronecker, Mansion, Noether, Hilbert, Klein, Lie, Beltrami, Burkhardt, and many others. On binary forms Faà di Bruno's work is well known, as is Study's (1889) on ternary forms. De Toledo (1889) and Elliott (1895) have published treatises on the subject.

Dublin University has also furnished a considerable corps of contributors, among whom MacCullagh, Hamilton, Salmon, Michael and Ralph Roberts, and Burnside may be especially mentioned. Burnside, who wrote the latter part of Burnside and Panton's Theory of Equations, has set forth a method of transformation which is fertile in geometric interpretation and binds together binary and certain ternary forms.

---

* Klein's Evanston Lectures, Lect. I.

The equivalence **problem of quadratic** and bilinear forms has attracted the attention of Weierstrass, Kronecker, Christoffel, Frobenius, Lie, and more recently of Rosenow (Crelle, 108), Werner (1889), Killing (1890), and Scheffers (1891). The equivalence problem of non-quadratic forms has been studied by Christoffel. Schwarz (1872), Fuchs (1875–76), Klein (1877, 1884), Brioschi (1877), and Maschke (1887) have contributed to the theory of forms with linear transformations into themselves. Cayley (especially from 1870) and Sylvester (1877) have worked out the methods of denumeration by means of generating functions. Differential invariants have been studied by Sylvester, MacMahon, and Hammond. Starting from the differential invariant, which Cayley has termed the Schwarzian derivative, Sylvester (1885) has founded the theory of reciprocants, to which MacMahon, Hammond, Leudesdorf, Elliott, Forsyth, and Halphen have contributed. Canonical forms have been studied by Sylvester (1851), Cayley, and Hermite (to whom the term "canonical form" is due), and more recently by Rosanes (1873), Brill (1882), Gundelfinger (1883), and Hilbert (1886).

The Geometric Theory of Binary Forms may be traced to Poncelet and his followers. But the modern treatment has its origin in connection with the theory of elliptic modular functions, and dates from Dedekind's letter to Borchardt (Crelle, 1877). The names of Klein and Hurwitz are prominent in this connection. On the method of nets (réseaux), another geometric treatment of binary quadratic forms Gauss (1831), Dirichlet (1850), and Poincaré (1880) have written.

ART. 10. CALCULUS.

The Differential and Integral Calculus,* dating from Newton and Leibniz, was quite complete in its general range at

---

* Williamson, B., Infinitesimal Calculus, Encyclopædia Britannica, 9th edition; Cantor, M., Geschichte der Mathematik, Vol. III, pp. 150–316; Vivanti, G., Note sur l'histoire de l'infiniment petit, Bibliotheca Mathematica, 1894, p. 1; Mansion, P., Esquisse de l'histoire du calcul infinitésimal, Ghent, 1887. Le

the close of the eighteenth century. Aside from the study of first principles, to which Gauss, Cauchy. Jordan, Picard, Méray, and those whose names are mentioned in connection with the theory of functions, have contributed, there must be mentioned the development of symbolic methods, the theory of definite integrals, the calculus of variations, the theory of differential equations, and the numerous applications of the Newtonian calculus to physical problems. Among those who have prepared noteworthy general treatises are Cauchy (1821), Raabe (1839–47), Duhamel (1856), Sturm (1857–59), Bertrand (1864), Serret (1868), Jordan (2d ed., 1893), and Picard (1891–93). A recent contribution to analysis which promises to be valuable is Oltramare's Calcul de Généralization (1893).

Abel seems to have been the first to consider in a general way the question as to what differential expressions can be integrated in a finite form by the aid of ordinary functions, an investigation extended by Liouville. Cauchy early undertook the general theory of determining definite integrals, and the subject has been prominent during the century. Frullani's theorem (1821), Bierens de Haan's work on the theory (1862) and his elaborate tables (1867), Dirichlet's lectures (1858) embodied in Meyer's treatise (1871), and numerous memoirs of Legendre, Poisson, Plana, Raabe, Sohncke, Schlömilch, Elliott, Leudesdorf, and Kronecker are among the noteworthy contributions.

Eulerian Integrals were first studied by Euler and afterwards investigated by Legendre, by whom they were classed as Eulerian integrals of the first and second species, as follows:

$$\int_0^1 x^{n-1}(1-x)^{n-1}dx, \int_0^\infty e^{-x}x^{n-1}dx,$$ although these were not the exact forms of Euler's study. If n is integral, it follows that $\int_0^\infty e^{-x}x^{n-1}dx = n!$, but if n is fractional it is a transcendent function. To it Legendre assigned the symbol Γ, and it is

deux centième anniversaire de l'invention du calcul différentiel ; Mathesis, Vol. IV, p. 163.

now called the gamma function.   To the subject Dirichlet has contributed an important theorem (Liouville, 1839), which has been elaborated by Liouville, Catalan, Leslie Ellis, and others. On the evaluation of Γx and $\log \Gamma x$ Raabe (1843–44), Bauer (1859), and Gudermann (1845) have written.   Legendre's great table appeared in 1816.

Symbolic Methods may be traced back to Taylor, and the analogy between successive differentiation and ordinary exponentials had been observed by numerous writers before the nineteenth century.   Arbogast (1800) was the first, however, to separate the symbol of operation from that of quantity in a differential equation.   François (1812) and Servois (1814) seem to have been the first to give correct rules on the subject. Hargreave (1848) applied these methods in his memoir on differential equations, and Boole freely employed them.   Grassmann and Hankel made great use of the theory, the former in studying equations, the latter in his theory of complex numbers.

The Calculus of Variations * may be said to begin with a problem of Johann Bernoulli's (1696).   It immediately occupied the attention of Jakob Bernoulli and the Marquis de l'Hôpital, but Euler first elaborated the subject.   His contributions began in 1733, and his Elementa Calculi Variationum gave to the science its name.   Lagrange contributed extensively to the theory, and Legendre (1786) laid down a method, not entirely satisfactory, for the discrimination of maxima and minima.   To this discrimination Brunacci (1810), Gauss (1829), Poisson (1831), Ostrogradsky (1834), and Jacobi (1837) have been among the contributors.   An important general work is that of Sarrus (1842) which was condensed and improved by Cauchy (1844).   Other valuable treatises and memoirs have been written by Strauch (1849), Jellett (1850), Hesse (1857), Clebsch (1858), and Carll (1885), but perhaps the most

* Carll, L. B., Calculus of Variations, New York, 1885, Chap. V; Todhunter, I., History of the Progress of the Calculus of Variations, London, 1861 : Reiff, R., Die Anfänge der Variationsrechnung, Mathematisch-naturwissenschaftliche Mittheilungen, Tübingen, 1887, p. 90.

important work of the century is that of Weierstrass. His celebrated course on the theory is epoch-making, and it may be asserted that he was the first to place it on a firm and un-questionable foundation.

The Application of the Infinitesimal Calculus to problems in physics and astronomy was contemporary with the origin of the science. All through the eighteenth century these appli-cations were multiplied, until at its close Laplace and Lagrange had brought the whole range of the study of forces into the realm of analysis. To Lagrange (1773) we owe the introduc-tion of the theory of the potential* into dynamics, although the name "potential function" and the fundamental memoir of the subject are due to Green (1827, printed in 1828). The name "potential" is due to Gauss (1840), and the distinction between potential and potential function to Clausius. With its development are connected the names of Dirichlet, Rie-mann, Neumann, Heine, Kronecker, Lipschitz, Christoffel, Kirchhoff, Beltrami, and many of the leading physicists of the century.

It is impossible in this place to enter into the great variety of other applications of analysis to physical problems. Among them are the investigations of Euler on vibrating chords; Sophie Germain on elastic membranes; Poisson, Lamé, Saint-Venant, and Clebsch on the elasticity of three-dimensional bod-ies; Fourier on heat diffusion; Fresnel on light; Maxwell, Helm-holtz, and Hertz on electricity; Hansen, Hill, and Gyldén on astronomy; Maxwell on spherical harmonics; Lord Rayleigh on acoustics; and the contributions of Dirichlet, Weber, Kirchhoff, F. Neumann, Lord Kelvin, Clausius, Bjerknes, MacCullagh, and Fuhrmann to physics in general. The labors of Helm-holtz should be especially mentioned, since he contributed to the theories of dynamics, electricity, etc., and brought his great analytical powers to bear on the fundamental axioms of me-chanics as well as on those of pure mathematics.

* Bacharach, M., Abriss der Geschichte der Potentialtheorie, 1883. This contains an extensive bibliography.

ART. 11. DIFFERENTIAL EQUATIONS.

The Theory of Differential Equations [*] has been called by Lie [†] the most important of modern mathematics. The influence of geometry, physics, and astronomy, starting with Newton and Leibniz, and further manifested through the Bernoullis, Riccati, and Clairaut, but chiefly through d'Alembert and Euler, has been very marked, and especially on the theory of linear partial differential equations with constant coefficients. The first method of integrating linear ordinary differential equations with constant coefficients is due to Euler, who made the solution of his type, $\dfrac{d^n y}{dx^n} + A_1 \dfrac{d^{n-1} y}{dx^{n-1}} + \ldots + A_n y = 0$, depend on that of the algebraic equation of the nth degree, $F(z) = z^n + A_1 z^{n-1} + \ldots + A_n = 0$, in which z^k takes the place of $\dfrac{d^k y}{dx^k}$ $(k = 1, 2, \ldots n)$. This equation $F(z) = 0$, is the "characteristic" equation considered later by Monge and Cauchy.

The theory of linear partial differential equations may be said to begin with Lagrange (1779 to 1785). Monge (1809) treated ordinary and partial differential equations of the first and second order, uniting the theory to geometry, and introducing the notion of the "characteristic," the curve represented by $F(z) = 0$, which has recently been investigated by Darboux,

[*] Cantor, M., Geschichte der Mathematik, Vol. III, p. 429 ; Schlesinger, L., Handbuch der Theorie der linearen Differentialgleichungen, Vol. I, 1895, an excellent historical view ; review by Mathews in Nature, Vol. LII, p. 313; Lie, S., Zur allgemeinen Theorie der partiellen Differentialgleichungen, Berichte über die Verhandlungen der Gesellschaft der Wissenschaften zu Leipzig, 1895; Mansion, P., Theorie der partiellen Differentialgleichungen 1er Ordnung, German by Maser, Leipzig, 1892 excellent on history ; Craig, T., Some of the Developments in the Theory of Ordinary Differential Equations, 1878–1893, Bulletin New York Mathematical Society, Vol. II, p. 119; Goursat, E., Leçons sur l'intégration des équations aux dérivées partielles du premier ordre, Paris, 1891; Burkhardt, H., and Hefter, L., in Mathematical Papers of Chicago Congress, p. 13 and p. 96.

[†] "In der ganzen modernen Mathematik ist die Theorie der Differentialgleichungen die wichtigste Discipiin "

Levy, and Lie. Pfaff (1814, 1815) gave the first general method
of integrating partial differential equations of the first order, a
method of which Gauss (1815) at once recognized the value
and of which he gave an analysis. Soon after, Cauchy (1819)
gave a simpler method, attacking the subject from the analytical standpoint, but using the Monge characteristic. To him
is also due the theorem, corresponding to the fundamental
theorem of algebra, that every differential equation defines a
function expressible by means of a convergent series, a proposition more simply proved by Briot and Bouquet, and also by
Picard (1891). Jacobi (1827) also gave an analysis of Pfaff's
method, besides developing an original one (1836) which
Clebsch published (1862). Clebsch's own method appeared in
1866, and others are due to Boole (1859), Korkine (1869), and
A. Mayer (1872). Pfaff's problem has been a prominent subject of investigation, and with it are connected the names of
Natani (1859), Clebsch (1861, 1862), DuBois-Reymond (1869),
Cayley, Baltzer, Frobenius, Morera, Darboux, and Lie. The
next great improvement in the theory of partial differential
equations of the first order is due to Lie (1872), by whom the
whole subject has been placed on a rigid foundation. Since
about 1870, Darboux, Kovalevsky, Méray, Mansion, Graindorge, and Imschenetsky have been prominent in this line.
The theory of partial differential equations of the second and
higher orders, beginning with Laplace and Monge, was notably
advanced by Ampère (1840). Imschenetsky * has summarized
the contributions to 1873, but the theory remains in an
imperfect state.

The integration of partial differential equations with three
or more variables was the object of elaborate investigations by
Lagrange, and his name is still connected with certain subsidiary equations. To him and to Charpit, who did much to
develop the theory, is due one of the methods for integrating
the general equation with two variables, a method which now
bears Charpit's name.

* Grunert's Archiv für Mathematik, Vol. LIV.

The theory of singular solutions of ordinary and partial differential equations has been a subject of research from the time of Leibniz, but only since the middle of the present century has it received especial attention. A valuable but little-known work on the subject is that of Houtain (1854). Darboux (from 1873) has been a leader in the theory, and in the geometric interpretation of these solutions he has opened a field which has been worked by various writers, notably Casorati and Cayley. To the latter is due (1872) the theory of singular solutions of differential equations of the first order as at present accepted.

The primitive attempt in dealing with differential equations had in view a reduction to quadratures. As it had been the hope of eighteenth-century algebraists to find a method for solving the general equation of the *n*th degree, so it was the hope of analysts to find a general method for integrating any differential equation. Gauss (1799) showed, however, that the differential equation meets its limitations very soon unless complex numbers are introduced. Hence analysts began to study these equations as functions, thus opening a new and fertile field. Cauchy was the first to appreciate the importance of this view, and the modern theory may be said to begin with him. Thereafter the real question was to be, not whether a solution is possible by means of known functions or their integrals, but whether a given differential equation suffices for the definition of a function of the independent variable or variables, and if so, what are the characteristic properties of this function.

Within a half-century the theory of ordinary differential equations has come to be one of the most important branches of analysis, the theory of partial differential equations remaining as one still to be perfected. The difficulties of the general problem of integration are so manifest that all classes of investigators have confined themselves to the properties of the integrals in the neighborhood of certain given points. The new departure took its greatest inspiration from two memoirs by

Fuchs (Crelle, 1866, 1868), a work elaborated by Thomé and Frobenius. Collet has been a prominent contributor since 1869, although his method for integrating a non-linear system was communicated to Bertrand in 1868. Clebsch * (1873) attacked the theory along lines parallel to those followed in his theory of Abelian integrals. As the latter can be classified according to the properties of the fundamental curve which remains unchanged under a rational transformation, so Clebsch proposed to classify the transcendent functions defined by the differential equations according to the invariant properties of the corresponding surfaces $f = 0$ under rational one-to-one transformations.

Since 1870 Lie's † labors have put the entire theory of differential equations on a more satisfactory foundation. He has shown that the integration theories of the older mathematicians, which had been looked upon as isolated, can by the introduction of the concept of continuous groups of transformations be referred to a common source, and that ordinary differential equations which admit the same infinitesimal transformations present like difficulties of integration. He has also emphasized the subject of transformations of contact (Berührungstransformationen) which underlies so much of the recent theory. The modern school has also turned its attention to the theory of differential invariants, one of fundamental importance and one which Lie has made prominent. With this theory are associated the names of Cayley, Cockle, Sylvester, Forsyth, Laguerre, and Halphen. Recent writers have shown the same tendency noticeable in the work of Monge and Cauchy, the tendency to separate into two schools, the one inclining to use the geometric diagram, and represented by Schwarz, Klein, and Goursat, the other adhering to pure analysis, of which Weierstrass, Fuchs, and Frobenius are types. The work of Fuchs and the theory of elementary divisors has formed the basis of a late work by Sauvage (1895). Poincaré's

---

* Klein's Evanston Lectures, Lect. I.
† Klein's Evanston Lectures, Lect. II, III.

recent contributions are also very notable. His theory of
Fuchsian equations (also investigated by Klein) is connected
with the general theory. He has also brought the whole subject into close relations with the theory of functions. Appell
has recently contributed to the theory of linear differential
equations transformable into themselves by change of the function and the variable. Helge von Koch has written on infinite
determinants and linear differential equations. Picard has undertaken the generalization of the work of Fuchs and Poincaré
in the case of differential equations of the second order. Fabry
(1885) has generalized the normal integrals of Thomé, integrals
which Poincaré has called "intégrals anormales," and which
Picard has recently studied. Riquier has treated the question
of the existence of integrals in any differential system and
given a brief summary of the history to 1895.* The number of
contributors in recent times is very great, and includes, besides
those already mentioned, the names of Brioschi, Königsberger,
Peano, Graf, Hamburger, Graindorge, Schläfli, Glaisher, Lommel, Gilbert, Fabry, Craig, and Autonne.

ART. 12. INFINITE SERIES.

The Theory of Infinite Series† in its historical development has been divided by Reiff into three periods: (1) the
period of Newton and Leibniz, that of its introduction;
(2) that of Euler, the formal period; (3) the modern, that of
the scientific investigation of the validity of infinite series, a
period beginning with Gauss. This critical period begins with
the publication of Gauss's celebrated memoir on the series

$$1 + \frac{\alpha \cdot \beta}{1 \cdot \gamma} x + \frac{\alpha \cdot (\alpha + 1) \cdot \beta \cdot (\beta + 1)}{1 \cdot 2 \cdot \gamma \cdot (\gamma + 1)} x^2 + \dots, \text{ in } 1812. \text{ Euler}$$

---

* Riquier, C., Mémoire sur l'existence des intégrales dans un système différentiel quelconque, etc. Mémoires des Savants étrangers, Vol. XXXII, No. 3.

† Cantor, M., Geschichte der Mathematik, Vol. III, pp. 53. 71 : Reiff, R.,
Geschichte der unendlichen Reihen, Tübingen. 1889; Cajori, F., Bulletin
New York Mathematical Society, Vol. I, p. 184; History of Teaching of Mathe-
matics in United States, p. 361.

had already considered this series, but Gauss was the first to master it, and under the name "hypergeometric series" (due to Pfaff) it has since occupied the attention of Jacobi, Kummer, Schwarz, Cayley, Goursat, and numerous others. The particular series is not so important as is the standard of criticism which Gauss set up, embodying the simpler criteria of convergence and the questions of remainders and the range of convergence.

Gauss's contributions were not at once appreciated, and the next to call attention to the subject was Cauchy (1821), who may be considered the founder of the theory of convergence and divergence of series. He was one of the first to insist on strict tests of convergence; he showed that if two series are convergent their product is not necessarily so; and with him begins the discovery of effective criteria of convergence and divergence. It should be mentioned, however, that these terms had been introduced long before by Gregory (1668), that Euler and Gauss had given various criteria, and that Maclaurin had anticipated a few of Cauchy's discoveries. Cauchy advanced the theory of power series by his expansion of a complex function in such a form. · His test for convergence is still one of the most satisfactory when the integration involved is possible.

Abel was the next important contributor. In his memoir (1826) on the series $1 + \dfrac{m}{1}x + \dfrac{m(m-1)}{2!}x^2 + \dots$ he corrected certain of Cauchy's conclusions, and gave a completely scientific summation of the series for complex values of m and x. He was emphatic against the reckless use of series, and showed the necessity of considering the subject of continuity in questions of convergence.

Cauchy's methods led to special rather than general criteria, and the same may be said of Raabe (1832), who made the first elaborate investigation of the subject, of De Morgan (from 1842), whose logarithmic test DuBois-Reymond (1873) and Pringsheim (1889) have shown to fail within a certain region;

of Bertrand (1842), Bonnet (1843), Malmsten (1846, 1847, the latter without integration); Stokes (1847), Paucker (1852), Tchébichef (1852), and Arndt 1853). General criteria began with Kummer (1835), and have been studied by Eisenstein (1847), Weierstrass in his various contributions to the theory of functions, Dini (1867), DuBois-Reymond (1873), and many others. Pringsheim's (from 1889) memoirs present the most complete general theory.

The Theory of Uniform Convergence was treated by Cauchy (1821), his limitations being pointed out by Abel, but the first to attack it successfully were Stokes and Seidel (1847–48). Cauchy took up the problem again (1853), acknowledging Abel's criticism, and reaching the same conclusions which Stokes had already found. Thomé used the doctrine (1866), but there was great delay in recognizing the importance of distinguishing between uniform and non-uniform convergence, in spite of the demands of the theory of functions.

Semi-Convergent Series were studied by Poisson (1823), who also gave a general form for the remainder of the Maclaurin formula. The most important solution of the problem is due, however, to Jacobi (1834), who attacked the question of the remainder from a different standpoint and reached a different formula. This expression was also worked out, and another one given, by Malmsten (1847). Schlömilch (Zeitschrift, Vol. I, p. 192, 1856) also improved Jacobi's remainder, and showed the relation between the remainder and Bernoulli's function $F(x) = 1^n + 2^n + \ldots + (x-1)^n$. Genocchi (1852) has further contributed to the theory.

Among the early writers was Wronski, whose " loi suprême " (1815) was hardly recognized until Cayley (1873) brought it into prominence. Transon (1874), Ch. Lagrange (1884), Echols, and Dickstein* have published of late various memoirs on the subject.

Interpolation Formulas have been given by various writers

* Bibliotheca Mathematica, 1892–94; historical.

from Newton to the present time.   Lagrange's theorem is well
known, although Euler had already given an analogous form,
as are also Olivier's formula (1827), and those of Minding
(1830), Cauchy (1837), Jacobi (1845), Grunert (1850, 1853),
Christoffel (1858), and Mehler (1864).

Fourier's Series* were being investigated as the result of
physical considerations at the same time that Gauss, Abel,
and Cauchy were working out the theory of infinite series.
Series for the expansion of sines and cosines, of multiple arcs
in powers of the sine and cosine of the arc had been treated
by Jakob Bernoulli (1702) and his brother Johann (1701) and
still earlier by Viète.   Euler and Lagrange had simplified the
subject, as have, more recently, Poinsot, Schröter, Glaisher,
and Kummer.   Fourier (1807) set for himself a different problem, to expand a given function of x in terms of the sines or
cosines of multiples of x, a problem which he embodied in his
Théorie analytique de la Chaleur (1822).   Euler had already
given the formulas for determining the coefficients in the
series; and Lagrange had passed over them without recognizing their value, but Fourier was the first to assert and attempt to prove the general theorem.   Poisson (1820–23) also
attacked the problem from a different standpoint.   Fourier
did not, however, settle the question of convergence of his
series, a matter left for Cauchy (1826) to attempt and for
Dirichlet (1829) to handle in a thoroughly scientific manner.
Dirichlet's treatment (Crelle, 1829), while bringing the theory
of trigonometric series to a temporary conclusion, has been
the subject of criticism and improvement by Riemann (1854),
Heine, Lipschitz, Schläfli, and DuBois-Reymond.   Among
other prominent contributors to the theory of trigonometric
and Fourier series have been Dini, Hermite, Halphen, Krause,
Byerly and Appell.

* Historical Summary by Bôcher, Chap. IX of Byerly's Fourier's Series
and Spherical Harmonics, Boston, 1893; Sachse, A., Essai historique sur la
représentation d'une fonction.... par une serie trigonométrique. Bulletin
des Sciences mathematiques, Part I, 1880, pp. 43, 83.

ART. 13. THEORY OF FUNCTIONS.

The Theory of Functions * may be said to have its first development in Newton's works, although algebraists had already become familiar with irrational functions in considering cubic and quartic equations.   Newton seems first to have grasped the idea of such expressions in his consideration of symmetric functions of the roots of an equation.   The word was employed by Leibniz (1694), but in connection with the Cartesian geometry.   In its modern sense it seems to have been first used by Johann Bernoulli, who distinguished between algebraic and transcendent functions.   He also used (1718) the function symbol ϕ.   Clairaut (1734) used Πx, Φx, Δx, for various functions of x, a symbolism substantially followed by d'Alembert (1747) and Euler (1753).   Lagrange (1772, 1797, 1806) laid the foundations for the general theory, giving to the symbol a broader meaning, and to the symbols $f, \phi, F, \ldots,$ $f', \phi', F', \ldots$ their modern signification.   Gauss contributed to the theory, especially in his proofs of the fundamental theorem of algebra, and discussed and gave name to the theory of "conforme Abbildung," the "orthomorphosis" of Cayley.

Making Lagrange's work a point of departure, Cauchy so greatly developed the theory that he is justly considered one of its founders.   His memoirs extend over the period 1814–1851, and cover subjects like those of integrals with imaginary limits, infinite series and questions of convergence, the application of the infinitesimal calculus to the theory of complex

---

* Brill, A., and Noether, M., Die Entwickelung der Theorie der algebraischen Functionen in älterer und neuerer Zeit, Bericht erstattet der Deutschen Mathematiker-Vereinigung, Jahresbericht, Vol. II, pp. 107–566, Berlin, 1894; Königsberger, L., Zur Geschichte der Theorie der elliptischen Transcendenten in den Jahren 1826–29, Leipzig, 1879; Williamson, B., Infinitesimal Calculus, Encyclopædia Britannica; Schlesinger, L., Differentialgleichungen, Vol. I, 1895; Casorati, F., Teorica delle funzioni di variabili complesse, Vol. I, 1868: Klein's Evanston Lectures.   For bibliography and historical notes, see Harkness and Morley's Theory of Functions, 1893, and Forsyth's Theory of Functions, 1893: Eneström, G., Note historique sur les symboles. . . . Bibliotheca Mathematica, 1891, p. 89.

numbers, the investigation of the fundamental laws of mathematics, and numerous other lines which appear in the general theory of functions as considered to-day. Originally opposed to the movement started by Gauss, the free use of complex numbers, he finally became, like Abel, its advocate. To him is largely due the present orientation of mathematical research, making prominent the theory of functions, distinguishing between classes of functions, and placing the whole subject upon a rigid foundation. The historical development of the general theory now becomes so interwoven with that of special classes of functions, and notably the elliptic and Abelian, that economy of space requires their treatment together, and hence a digression at this point.

The Theory of Elliptic Functions* is usually referred for its origin to Landen's (1775) substitution of two elliptic arcs for a single hyperbolic arc. But Jakob Bernoulli (1691) had suggested the idea of comparing non-congruent arcs of the same curve, and Johann had followed up the investigation. Fagano (1716) had made similar studies, and both Maclaurin (1742) and d'Alembert (1746) had come upon the borderland of elliptic functions. Euler (from 1761) had summarized and extended the rudimentary theory, showing the necessity for a convenient notation for elliptic arcs, and prophesying (1766) that "such signs will afford a new sort of calculus of which I have here attempted the exposition of the first elements." Euler's investigations continued until about the time of his death (1783), and to him Legendre attributes the foundation of the theory. Euler was probably never aware of Landen's discovery.

It is to Legendre, however, that the theory of elliptic functions is largely due, and on it his fame to a considerable degree depends. His earlier treatment (1786) almost entirely substitutes a strict analytic for the geometric method. For forty years he had the theory in hand, his labor culminating in his

---

* Enneper, A., Elliptische Funktionen, Theorie und Geschichte, Halle, 1890; Königsberger, L., Zur Geschichte der Theorie der elliptischen Transcendenten in den Jahren 1826–29, Leipzig, 1879.

Traité des Fonctions elliptiques et des Intégrales Euleriennes (1825–28). A surprise now awaiting him is best told in his own words: "Hardly had my work seen the light—its name could scarcely have become known to scientific foreigners,—when I learned with equal surprise and satisfaction that two young mathematicians, MM. Jacobi of Königsberg and Abel of Christiania, had succeeded by their own studies in perfecting considerably the theory of elliptic functions in its highest parts." Abel began his contributions to the theory in 1825, and even then was in possession of his fundamental theorem which he communicated to the Paris Academy in 1826. This communication being so poorly transcribed was not published in full until 1841, although the theorem was sent to Crelle (1829) just before Abel's early death. Abel discovered the double periodicity of elliptic functions, and with him began the treatment of the elliptic integral as a function of the amplitude.

Jacobi, as also Legendre and Gauss, was especially cordial in praise of the delayed theorem of the youthful Abel. He calls it a "monumentum ære perennius," and his name "das Abel'sche Theorem" has since attached to it. The functions of multiple periodicity to which it refers have been called Abelian Functions. Abel's work was early proved and elucidated by Liouville and Hermite. Serret and Chasles in the Comptes Rendus, Weierstrass (1853), Clebsch and Gordan in their Theorie der Abel'sche Functionen (1866), and Briot and Bouquet in their two treatises have greatly elaborated the theory. Riemann's * (1857) celebrated memoir in Crelle presented the subject in such a novel form that his treatment was slow of acceptance. He based the theory of Abelian integrals and their inverse, the Abelian functions, on the idea of the surface now so well known by his name, and on the corresponding fundamental existence theorems. Clebsch, starting from

---

* Klein, Evanston Lectures, p. 3; Riemann and Modern Mathematics, translated by Ziwet, Bulletin of American Mathematical Society, Vol. I, p. 165; Burkhardt, H., Vortrag über Riemann, Göttingen, 1892.

an algebraic curve defined by its equation, made the subject
more accessible, and generalized the theory of Abelian integrals
to a theory of algebraic functions with several variables, thus
creating a branch which has been developed by Noether,
Picard, and Poincaré. The introduction of the theory of invariants and projective geometry into the domain of hyperelliptic and Abelian functions is an extension of Clebsch's
scheme. In this extension, as in the general theory of Abelian
functions, Klein has been a leader. With the development of
the theory of Abelian functions is connected a long list of
names, including those of Schottky, Humbert, C. Neumann,
Fricke, Königsberger, Prym, Schwarz, Painlevé, Hurwitz,
Brioschi, Borchardt, Cayley, Forsyth, and Rosenhain, besides
others already mentioned.

Returning to the theory of elliptic functions, Jacobi (1827)
began by adding greatly to Legendre's work. He created a
new notation and gave name to the "modular equations" of
which he made use. Among those who have written treatises
upon the elliptic-function theory are Briot and Bouquet,
Laurent, Halphen, Königsberger, Hermite, Durège, and Cayley.
The introduction of the subject into the Cambridge Tripos
(1873), and the fact that Cayley's only book was devoted to it,
have tended to popularize the theory in England.

The Theory of Theta Functions was the simultaneous and
independent creation of Jacobi and Abel (1828). Gauss's
notes show that he was aware of the properties of the theta
functions twenty years earlier, but he never published his investigations. Among the leading contributors to the theory
are Rosenhain (1846, published in 1851) and Göpel (1847), who
connected the double theta functions with the theory of Abelian
functions of two variables and established the theory of hyperelliptic functions in a manner corresponding to the Jacobian
theory of elliptic functions. Weierstrass has also developed
the theory of theta functions independently of the form of their
space boundaries, researches elaborated by Königsberger (1865)
to give the addition theorem. Riemann has completed the

investigation of the relation between the theory of the theta and the Abelian functions, and has raised theta functions to their present position by making them an essential part of his theory of Abelian integrals. H. J. S. Smith has included among his contributions to this subject the theory of omega functions. Among the recent contributors are Krazer and Prym (1892), and Wirtinger (1895).

Cayley was a prominent contributor to the theory of periodic functions. His memoir (1845) on doubly periodic functions extended Abel's investigations on doubly infinite products. Euler had given singly infinite products for sin x, cos x, and Abel had generalized these, obtaining for the elementary doubly periodic functions expressions for sn x, cn x, dn x. Starting from these expressions of Abel's Cayley laid a complete foundation for his theory of elliptic functions. Eisenstein (1847) followed, giving a discussion from the standpoint of pure analysis, of a general doubly infinite product, and his labors, as supplemented by Weierstrass, are classic.

The General Theory of Functions has received its present form largely from the works of Cauchy, Riemann, and Weierstrass. Endeavoring to subject all natural laws to interpretation by mathematical formulas, Riemann borrowed his methods from the theory of the potential, and found his inspiration in the contemplation of mathematics from the standpoint of the concrete. Weierstrass, on the other hand, proceeded from the purely analytic point of view. To Riemann* is due the idea of making certain partial differential equations, which express the fundamental properties of all functions, the foundation of a general analytical theory, and of seeking criteria for the determination of an analytic function by its discontinuities and boundary conditions. His theory has been elaborated by Klein (1882, and frequent memoirs) who has materially extended the theory of Riemann's surfaces. Clebsch, Lüroth, and later writers have based on this theory their researches on

---

* Klein, F., Riemann and Modern Mathematics, translated by Ziwet, Bulletin of American Mathematical Society, Vol. I, p. 165.

loops.   Riemann's speculations were not without weak points, and these have been fortified in connection with the theory of the potential by C. Neumann, and from the analytic standpoint by Schwarz.

In both the theory of general and of elliptic and other functions, Clebsch was prominent.   He introduced the systematic consideration of algebraic curves of deficiency 1, bringing to bear on the theory of elliptic functions the ideas of modern projective geometry.   This theory Klein has generalized in his Theorie der elliptischen Modulfunctionen, and has extended the method to the theory of hyperelliptic and Abelian functions.

Following Riemann came the equally fundamental and original and more rigorously worked out theory of Weierstrass. His early lectures on functions are justly considered a landmark in modern mathematical development.   In particular, his researches on Abelian transcendents are perhaps the most important since those of Abel and Jacobi.   His contributions to the theory of elliptic functions, including the introduction of the function $\mathfrak{p}(u)$, are also of great importance.   His contributions to the general function theory include much of the symbolism and nomenclature, and many theorems.   He first announced (1866) the existence of natural limits for analytic functions, a subject further investigated by Schwarz, Klein, and Fricke.   He developed the theory of functions of complex variables from its foundations, and his contributions to the theory of functions of real variables were no less marked.

Fuchs has been a prominent contributor, in particular (1872) on the general form of a function with essential singularities. On functions with an infinite number of essential singularities Mittag-Leffler (from 1882) has written and contributed a fundamental theorem.   On the classification of singularities of functions Guichard (1883) has summarized and extended the researches, and Mittag-Leffler and G. Cantor have contributed to the same result.   Laguerre (from 1882) was the first to discuss the "class" of transcendent functions, a subject to which Poincaré, Cesaro, Vivanti, and Hermite have also contributed. Automorphic functions, as named by Klein, have been investigated chiefly by Poincaré, who has established their general classification. The contributors to the theory include Schwarz, Fuchs, Cayley, Weber, Schlesinger, and Burnside.

The Theory of Elliptic Modular Functions, proceding from Eisenstein's memoir (1847) and the lectures of Weierstrass on elliptic functions, has of late assumed prominence through the influence of the Klein school. Schläfli (1870), and later Klein, Dyck, Gierster, and Hurwitz, have worked out the theory which Klein and Fricke have embodied in the recent Vorlesungen über die Theorie der elliptischen Modulfunctionen (1890–92). In this theory the memoirs of Dedekind (1877), Klein (1878), and Poincaré (from 1881) have been among the most prominent.

For the names of the leading contributors to the general and special theories, including among others Jordan, Hermite, Hölder, Picard, Biermann, Darboux, Pellet, Reichardt, Burkhardt, Krause, and Humbert, reference must be had to the Brill-Noether Bericht.

Of the various special algebraic functions space allows mention of but one class, that bearing Bessel's name. Bessel's functions [*] of the zero order are found in memoirs of Daniel Bernoulli (1732) and Euler (1764), and before the end of the eighteenth century all the Bessel functions of the first kind and integral order had been used. Their prominence as special functions is due, however, to Bessel (1816–17), who put them in their present form in 1824. Lagrange's series (1770), with Laplace's extension (1777), had been regarded as the best method of solving Kepler's problem (to express the variable quantities in undisturbed planetary motion in terms of the time or mean anomaly), and to improve this method Bessel's functions were first prominently used. Hankel (1869), Lommel (from 1868), F. Neumann, Heine, Graf (1893), Gray and

[*] Bôcher, M., A bit of mathematical history, Bulletin of New York Mathematical Society, Vol. II, p. 107.

Mathews (1895), and others have contributed to the theory. Lord Rayleigh (1878) has shown the relation between Bessel's and Laplace's functions, but they are nevertheless looked upon as a distinct system of transcendents. Tables of Bessel's functions were prepared by Bessel (1824), by Hansen (1843), and by Meissel (1888).

ART. 14. PROBABILITIES AND LEAST SQUARES.

The Theory of Probabilities and Errors* is, as applied to observations, largely a nineteenth-century development. The doctrine of probabilities dates, however, as far back as Fermat and Pascal (1654). Huygens (1657) gave the first scientific treatment of the subject, and Jakob Bernoulli's Ars Conjectandi (posthumous, 1713) and De Moivre's Doctrine of Chances (1718)†raised the subject to the plane of a branch of mathematics. The theory of errors may be traced back to Cotes's Opera Miscellanea (posthumous, 1722), but a memoir prepared by Simpson in 1755 (printed 1756) first applied the theory to the discussion of errors of observation. The reprint (1757) of this memoir lays down the axioms that positive and negative errors are equally probable, and that there are certain assignable limits within which all errors may be supposed to fall ; continuous errors are discussed and a probability curve is given. Laplace (1774) made the first attempt to deduce a rule for the combination of observations from the principles of the theory of probabilities. He represented the law of probability of errors by a curve $y = \phi(x)$, x being any error and y its probability, and laid down three properties of this curve : (1) It is symmetric as to the y-axis; (2) the x-axis is an asymptote, the probability of the error ∞ being 0; (3) the area enclosed is 1, it being certain that an error exists. He deduced a formula

---

* Merriman, M., Method of Least Squares, New York, 1884, p. 182 ; Transactions of Connecticut Academy, 1877, Vol. IV, p. 151, with complete bibliography; Todhunter, I., History of the Mathematical Theory of Probability, 1865; Cantor, M., Geschichte der Mathematik, Vol. III, p. 316.

† Eneström, G., Review of Cantor, Bibliotheca Mathematica, 1896, p. 20.

for the mean of three observations. He also gave (1781) a formula for the law of facility of error (a term due to Lagrange, 1774), but one which led to unmanageable equations. Daniel Bernoulli (1778) introduced the principle of the maximum product of the probabilities of a system of concurrent errors.

The Method of Least Squares is due to Legendre (1805), who introduced it in his Nouvelles méthodes pour la détermination des orbites des comètes. In ignorance of Legendre's contribution, an Irish-American writer, Adrain, editor of "The Analyst" (1808), first deduced the law of facility of error, $\phi(x) = ce^{-h^2x^2}$, c and h being constants depending on precision of observation. He gave two proofs, the second being essentially the same as Herschel's (1850). Gauss gave the first proof which seems to have been known in Europe (the third after Adrain's) in 1809. To him is due much of the honor of placing the subject before the mathematical world, both as to the theory and its applications.

Further proofs were given by Laplace (1810, 1812), Gauss (1823), Ivory (1825, 1826), Hagen (1837), Bessel (1838), Donkin (1844, 1856), and Crofton (1870). Other contributors have been Ellis (1844), De Morgan (1864), Glaisher (1872), and Schiaparelli (1875). Peter's (1856) formula for r, the probable error of a single observation, is well known.[*]

Among the contributors to the general theory of probabilities in the nineteenth century have been Laplace, Lacroix (1816), Littrow (1833), Quetelet (1853), Dedekind (1860), Helmert (1872), Laurent (1873), Liagre, Didion, and Pearson. De Morgan and Boole improved the theory, but added little that was fundamentally new. Czuber has done much both in his own contributions (1884, 1891), and in his translation (1879) of Meyer. On the geometric side the influence of Miller and The Educational Times has been marked, as also that of such contributors to this journal as Crofton, McColl, Wolstenholme, Watson, and Artemas Martin.

[*] Bulletin of New York Mathematical Society, Vol. II. p. 57.

ART. 15. ANALYTIC GEOMETRY.

The History of Geometry* may be roughly divided into the four periods: (1) The synthetic geometry of the Greeks, practically closing with Archimedes; (2) The birth of analytic geometry, in which the synthetic geometry of Guldin, Desargues, Kepler, and Roberval merged into the coordinate geometry of Descartes and Fermat; (3) 1650 to 1800, characterized by the application of the calculus to geometry, and including the names of Newton, Leibnitz, the Bernoullis, Clairaut, Maclaurin, Euler, and Lagrange, each an analyst rather than a geometer; (4) The nineteenth century, the renaissance of pure geometry, characterized by the descriptive geometry of Monge, the modern synthetic of Poncelet, Steiner, von Staudt, and Cremona, the modern analytic founded by Plücker, the non-Euclidean hypothesis of Lobachevsky and Bolyai, and the more elementary geometry of the triangle founded by Lemoine. It is quite impossible to draw the line between the analytic and the synthetic geometry of the nineteenth century, in their historical development, and Arts. 15 and 16 should be read together.

The Analytic Geometry which Descartes gave to the world in 1637 was confined to plane curves, and the various important properties common to all algebraic curves were soon discovered. To the theory Newton contributed three celebrated theorems on the Enumeratio linearum tertii ordinis † (1706), while others are due to Cotes (1722), Maclaurin, and Waring (1762, 1772,

* Loria, G., Il passato e il presente delle principali teorie geometriche. Memorie Accademia Torino, 1887; translated into German by F. Schütte under the title Die hauptsächlichsten Theorien der Geometrie in ihrer früheren und heutigen Entwickelung, Leipzig, 1888; Chasles. M., Aperçu historique sur l'origine et le développement des méthodes en Géométrie, 1889; Chasles, M., Rapport sur les Progrès de la Géométrie, Paris, 1870; Cayley, A., Curves, Encyclopædia Britannica; Klein, F., Evanston Lectures on Mathematics, New York, 1894; A. V. Braunmühl, Historische Studie über die organische Erzeugung ebener Curven, Dyck's Katalog mathematischer Modelle, 1892.

† Ball, W. W. R., On Newton's classification of cubic curves. Transactions of London Mathematical Society, 1891, p. 104.

etc.).　The scientific foundations of the theory of plane curves
may be ascribed, however, to Euler (1748) and Cramer (1750).
Euler distinguished between algebraic and transcendent curves,
and attempted a classification of the former.　Cramer is well
known for the " paradox " which bears his name, an obstacle
which Lamé (1818) finally removed from the theory.　To
Cramer is also due an attempt to put the theory of singularities of algebraic curves on a scientific foundation, although in
a modern geometric sense the theory was first treated by
Poncelet.

Meanwhile the study of surfaces was becoming prominent.
Descartes had suggested that his geometry could be extended
to three-dimensional space, Wren (1669) had discovered the
two systems of generating lines on the hyperboloid of one
sheet, and Parent (1700) had referred a surface to three coordinate planes.　The geometry of three dimensions began to
assume definite shape, however, in a memoir of Clairaut's (1731),
in which, at the age of sixteen, he solved with rare elegance
many of the problems relating to curves of double curvature.
Euler (1760) laid the foundations for the analytic theory of
curvature of surfaces, attempting the classification of those
of the second degree as the ancients had classified curves
of the second order.　Monge, Hachette, and other members of
that school entered into the study of surfaces with great zeal.
Monge introduced the notion of families of surfaces, and discovered the relation between the theory of surfaces and the
integration of partial differential equations, enabling each to be
advantageously viewed from the standpoint of the other.　The
theory of surfaces has attracted a long list of contributors in
the nineteenth century, including most of the geometers whose
names are mentioned in the present article.*

Möbius began his contributions to geometry in 1823, and
four years later published his Barycentrische Calcül.　In this
great work he introduced homogeneous coordinates with the

* For details see Loria, Il passato e il presente, etc.

attendant symmetry of geometric formulas, the scientific exposition of the principle of signs in geometry, and the establishment of the principle of geometric correspondence simple and multiple. He also (1852) summed up the classification of cubic curves, a service rendered by Zeuthen (1874) for quartics. To the period of Möbius also belong Bobillier (1827), who first used trilinear coordinates, and Bellavitis, whose contributions to analytic geometry were extensive. Gergonne's labors are mentioned in the next article.

Of all modern contributors to analytic geometry, Plücker stands foremost. In 1828 he published the first volume of his Analytisch-geometrische Entwickelungen, in which appeared the modern abridged notation, and which marks the beginning of a new era for analytic geometry. In the second volume (1831) he sets forth the present analytic form of the principle of duality. To him is due (1833) the general treatment of foci for curves of higher degree, and the complete classification of plane cubic curves (1835) which had been so frequently tried before him. He also gave (1839) an enumeration of plane curves of the fourth order, which Bragelogne and Euler had attempted. In 1842 he gave his celebrated "six equations" by which he showed that the characteristics of a curve (order, class, number of double points, number of cusps, number of double tangents, and number of inflections) are known when any three are given. To him is also due the first scientific dual definition of a curve, a system of tangential coordinates, and an investigation of the question of double tangents, a question further elaborated by Cayley (1847, 1858), Hesse (1847), Salmon (1858), and Dersch (1874). The theory of ruled surfaces, opened by Monge, was also extended by him. Possibly the greatest service rendered by Plücker was the introduction of the straight line as a space element, his first contribution (1865) being followed by his well-known treatise on the subject (1868–69). In this work he treats certain general properties of complexes, congruences, and ruled surfaces, as well as special properties of linear complexes and congruences, subjects also considered by Kummer and by Klein and others of the modern school. It is not a little due to Plücker that the concept of 4- and hence n-dimensional space, already suggested by Lagrange and Gauss, became the subject of later research. Riemann, Helmholtz, Lipschitz, Kronecker, Klein, Lie, Veronese, Cayley, d'Ovidio, and many others have elaborated the theory. The regular hypersolids in 4-dimensional space have been the subject of special study by Scheffler, Rudel, Hoppe, Schlegel, and Stringham.

Among Jacobi's contributions is the consideration (1836) of curves and groups of points resulting from the intersection of algebraic surfaces, a subject carried forward by Reye (1869). To Jacobi is also due the conformal representation of the ellipsoid on a plane, a treatment completed by Schering (1858). The number of examples of conformal representation of surfaces on planes or on spheres has been increased by Schwarz (1869) and Amstein (1872).

In 1844 Hesse, whose contributions to geometry in general are both numerous and valuable, gave the complete theory of inflections of a curve, and introduced the so-called Hessian curve as the first instance of a covariant of a ternary form. He also contributed to the theory of curves of the third order, and generalized the Pascal and Brianchon theorems on a spherical surface. Hesse's methods have recently been elaborated by Gundelfinger (1894).

Besides contributing extensively to synthetic geometry, Chasles added to the theory of curves of the third and fourth degrees. In the method of characteristics which he worked out may be found the first trace of the Abzählende Geometrie* which has been developed by Jonquières, Halphen (1875), and Schubert (1876, 1879), and to which Clebsch, Lindemann, and Hurwitz have also contributed. The general theory of correspondence starts with Geometry, and Chasles (1864) undertook

* Loria, G., Notizie storiche sulla Geometria numerativa. Bibliotheca Mathematica, 1888, pp. 39, 67; 1889, p. 23.

the first special researches on the correspondence of algebraic curves, limiting his investigations, however, to curves of deficiency zero. Cayley (1866) carried this theory to curves of higher deficiency, and Brill (from 1873) completed the theory.

Cayley's * influence on geometry was very great. He early carried on Plücker's consideration of singularities of a curve, and showed (1864, 1866) that every singularity may be considered as compounded of ordinary singularities so that the " six equations" apply to a curve with any singularities whatsoever. He thus opened a field for the later investigations of Noether, Zeuthen, Halphen, and H. J. S. Smith. Cayley's theorems on the intersection of curves (1843) and the determination of self-corresponding points for algebraic correspondences of a simple kind are fundamental in the present theory, subjects to which Bacharach, Brill, and Noether have also contributed extensively. Cayley added much to the theories of rational transformation and correspondence, showing the distinction between the theory of transformation of spaces and that of correspondence of loci. His investigations on the bitangents of plane curves, and in particular on the twenty-eight bitangents of a non-singular quartic, his developments of Plücker's conception of foci, his discussion of the osculating conics of curves and of the sextactic points on a plane curve, the geometric theory of the invariants and covariants of plane curves, are all noteworthy. He was the first to announce (1849) the twenty-seven lines which lie on a cubic surface, he extended Salmon's theory of reciprocal surfaces, and treated (1869) the classification of cubic surfaces, a subject already discussed by Schläfli. He also contributed to the theory of scrolls (skew-ruled surfaces), orthogonal systems of surfaces, the wave surface, etc., and was the first to reach (1845) any very general results in the theory of curves of double curvature, a theory in which the next great advance was made (1882) by Halphen and Noether. Among Cayley's other contributions to geometry is his theory of the Absolute, a figure in connection with which all metrical properties of a figure are considered.

* Biographical Notice in Cayley's Collected papers, Vol. VIII.

Clebsch * was also prominent in the study of curves and surfaces. He first applied the algebra of linear transformation to geometry. He emphasized the idea of deficiency (Geschlecht) of a curve, a notion which dates back to Abel, and applied the theory of elliptic and Abelian functions to geometry, using it for the study of curves. Clebsch (1872) investigated the shapes of surfaces of the third order. Following him, Klein attacked the problem of determining all possible forms of such surfaces, and established the fact that by the principle of continuity all forms of real surfaces of the third order can be derived from the particular surface having four real conical points. Zeuthen (1874) has discussed the various forms of plane curves of the fourth order, showing the relation between his results and those of Klein on cubic surfaces. Attempts have been made to extend the subject to curves of the nth order, but no general classification has been made. Quartic surfaces have been studied by Rohn (1887) but without a complete enumeration, and the same writer has contributed (1881) to the theory of Kummer surfaces.

Lie has adopted Plucker's generalized space element and extended the theory. His sphere geometry treats the subject from the higher standpoint of six homogeneous coordinates, as distinguished from the elementary sphere geometry with but five and characterized by the conformal group, a geometry studied by Darboux. Lie's theory of contact transformations, with its application to differential equations, his line and sphere complexes, and his work on minimum surfaces are all prominent.

Of great help in the study of curves and surfaces and of the theory of functions are the models prepared by Dyck, Brill, O. Henrici, Schwarz, Klein, Schönflies, Kummer, and others.†

The Theory of Minimum Surfaces has been developed along

---

* Klein, Evanston Lectures, Lect. I.

† Dyck, W., Katalog mathematischer und mathematisch-physikalischer Modelle, München, 1892 ; Deutsche Universitätsausstellung, Mathematical Papers of Chicago Congress, p. 49.

with the analytic geometry in general. Lagrange (1760–61)
gave the equation of the minimum surface through a given
contour, and Meusnier (1776, published in 1785) also studied
the question.   But from this time on for half a century little
that is noteworthy was done, save by Poisson (1813) as to certain imaginary surfaces.   Monge (1784) and Legendre (1787)
connected the study of surfaces with that of differential equations, but this did not immediately affect this question.   Scherk
(1835) added a number of important results, and first applied
the labors of Monge and Legendre to the theory.   Catalan
(1842), Björling (1844), and Dini (1865) have added to the
subject.   But the most prominent contributors have been
Bonnet, Schwarz, Darboux, and Weierstrass.   Bonnet (from
1853) has set forth a new system of formulas relative to the
general theory of surfaces, and completely solved the problem
of determining the minimum surface through any curve and
admitting in each point of this curve a given tangent plane.
Weierstrass (1866) has contributed several fundamental theorems, has shown how to find all of the real algebraic minimum
surfaces, and has shown the connection between the theory of
functions of an imaginaay variable and the theory of minimum
surfaces.

ART. 16.   MODERN GEOMETRY.

Descriptive,* Projective, and Modern Synthetic Geometry
are so interwoven in their historic development that it is even
more difficult to separate them from one another than from
the analytic geometry just mentioned.   Monge had been in
possession of his theory for over thirty years before the publication of his Géométrie Descriptive (1800), a delay due to the
jealous desire of the military authorities to keep the valuable
secret.   It is true that certain of its features can be traced
back to Desargues, Taylor, Lambert, and Frézier, but it was
Monge who worked it out in detail as a science, although

* Wiener, Chr., Lehrbuch der darstellenden Geometrie, Leipzig, 1884–87;
Geschichte der darstellenden Geometrie, 1884.

Lacroix (1795), inspired by Monge's lectures in the École Polytechnique, published the first work on the subject. After Monge's work appeared, Hachette (1812, 1818, 1821) added materially to its symmetry, subsequent French contributors being Leroy (1842), Olivier (from 1845), de la Gournerie (from 1860), Vallée, de Fourcy, Adhémar, and others. In Germany leading contributors have been Ziegler (1843), Anger (1858), and especially Fiedler (3d edn. 1883–88) and Wiener (1884–87). At this period Monge by no means confined himself to the descriptive geometry. So marked were his labors in the analytic geometry that he has been called the father of the modern theory. He also set forth the fundamental theorem of reciprocal polars, though not in modern language, gave some treatment of ruled surfaces, and extended the theory of polars to quadrics.*

Monge and his school concerned themselves especially with the relations of form, and particularly with those of surfaces and curves in a space of three dimensions. Inspired by the general activity of the period, but following rather the steps of Desargues and Pascal, Carnot treated chiefly the metrical relations of figures. In particular he investigated these relations as connected with the theory of transversals, a theory whose fundamental property of a four-rayed pencil goes back to Pappos, and which, though revived by Desargues, was set forth for the first time in its general form in Carnot's Géométrie de Position (1803), and supplemented in his Théorie des Transversales (1806). In these works he introduced negative magnitudes, the general quadrilateral and quadrangle, and numerous other generalizations of value to the elementary geometry of to-day. But although Carnot's work was important and many details are now commonplace, neither the name of the theory nor the method employed have endured. The present Geometry of Position (Geometrie der Lage) has little in common with Carnot's Géométrie de Position.

---

* On recent development of graphic methods and the influence of Monge upon this branch of mathematics, see Eddy, H. T., Modern Graphical Developments, Mathematical Papers of Chicago Congress (New York, 1896), p. 58.

Projective Geometry had its origin somewhat later than the period of Monge and Carnot. Newton had discovered that all curves of the third order can be derived by central projection from five fundamental types. But in spite of this fact the theory attracted so little attention for over a century that its origin is generally ascribed to Poncelet. A prisoner in the Russian campaign, confined at Saratoff on the Volga (1812–14), "privé," as he says, "de toute espèce de livres et de secours, surtout distrait par les malheurs de ma patrie et les miens propres," he still had the vigor of spirit and the leisure to conceive the great work which he published (1822) eight years later. In this work was first made prominent the power of central projection in demonstration and the power of the principle of continuity in research. His leading idea was the study of projective properties, and as a foundation principle he introduced the anharmonic ratio, a concept, however, which dates back to Pappos and which Desargues (1639) had also used. Möbius, following Poncelet, made much use of the anharmonic ratio in his Barycentrische Calcül (1827), but under the name "Doppelschnitt-Verhältniss" (ratio bisectionalis), a term now in common use under Steiner's abbreviated form "Doppelverhältniss." The name "anharmonic ratio" or "function" (rapport anharmonique, or fonction anharmonique) is due to Chasles, and "cross-ratio" was coined by Clifford. The anharmonic point and line properties of conics have been further elaborated by Brianchon, Chasles, Steiner, and von Staudt. To Poncelet is also due the theory of "figures homologiques," the perspective axis and perspective center (called by Chasles the axis and center of homology), an extension of Carnot's theory of transversals, and the "cordes idéales" of conics which Plücker applied to curves of all orders. He also discovered what Salmon has called "the circular points at infinity," thus completing and establishing the first great principle of modern geometry, the principle of continuity. Brianchon (1806), through his application of Desargues's theory of polars, completed the foundation which Monge had begun for Ponce-let's (1829) theory of reciprocal polars.

Among the most prominent geometers contemporary with Poncelet was Gergonne, who with more propriety might be ranked as an analytic geometer.   He first (1813) used the term " polar " in its modern geometric sense, although Servois (1811) had used the expression " pole."   He was also the first (1825–26) to grasp the idea that the parallelism which Maurolycus, Snell, and Viète had noticed is a fundamental principle.   This principle he stated and to it he gave the name which it now bears, the Principle of Duality, the most important, after that of continuity, in modern geometry.   This principle of geomet-ric reciprocation, the discovery of which was also claimed by Poncelet, has been greatly elaborated and has found its way into modern algebra and elementary geometry, and has recently been extended to mechanics by Genese.   Gergonne was the first to use the word "class" in describing a curve, explicitly defining class and degree (order) and showing the duality between the two.   He and Chasles were among the first to study scientifically surfaces of higher order.

Steiner (1832) gave the first complete discussion of the pro-jective relations between rows, pencils, etc., and laid the foun-dation for the subsequent development of pure geometry.   He practically closed the theory of conic sections, of the corre-sponding figures in three-dimensional space and of surfaces of the second order, and hence with him opens the period of special study of curves and surfaces of higher order.   His treat-ment of duality and his application of the theory of projective pencils to the generation of conics are masterpieces.   The theory of polars of a point in regard to a curve had been studied by Bobillier and by Grassmann, but Steiner (1848) showed that this theory can serve as the foundation for the study of plane curves independently of the use of coordinates, and introduced those noteworthy curves covariant to a given curve which now bear the names of himself, Hesse, and Cayley. This whole subject has been extended by Grassmann, Chasles,

Cremona, and Jonquières. Steiner was the first to make prominent (1832) an example of correspondence of a more complicated nature than that of Poncelet, Möbius, Magnus, and Chasles. His contributions, and those of Gudermann, to the geometry of the sphere were also noteworthy.

While Möbius, Plücker, and Steiner were at work in Germany, Chasles was closing the geometric era opened in France by Monge. His Aperçu Historique (1837) is a classic, and did for France what Salmon's works did for algebra and geometry in England, popularizing the researches of earlier writers and contributing both to the theory and the nomenclature of the subject. To him is due the name "homographic" and the complete exposition of the principle as applied to plane and solid figures, a subject which has received attention in England at the hands of Salmon, Townsend, and H. J. S. Smith.

Von Staudt began his labors after Plücker, Steiner, and Chasles had made their greatest contributions, but in spite of this seeming disadvantage he surpassed them all. Joining the Steiner school, as opposed to that of Plücker, he became the greatest exponent of pure synthetic geometry of modern times. He set forth (1847, 1856–60) a complete, pure geometric system in which metrical geometry finds no place. Projective properties foreign to measurements are established independently of number relations, number being drawn from geometry instead of conversely, and imaginary elements being systematically introduced from the geometric side. A projective geometry based on the group containing all the real projective and dualistic transformations, is developed, imaginary transformations being also introduced. Largely through his influence pure geometry again became a fruitful field. Since his time the distinction between the metrical and projective theories has been to a great extent obliterated,* the metrical properties

* Klein, F., Erlangen Programme of 1872, Haskell's translation, Bulletin of New York Mathematical Society, Vol. II, p. 215.

being considered as projective relations to a fundamental configuration, the circle at infinity common for all spheres.   Unfortunately von Staudt wrote in an unattractive style, and to
Reye is due much of the popularity which now attends the
subject.

Cremona began his publications in 1862.   His elementary
work on projective geometry (1875) in Leudesdorf's translation
is familiar to English readers.   His contributions to the theory
of geometric transformations are valuable, as also his works on
plane curves, surfaces, etc.

In England Mulcahy, but especially Townsend (1863), and
Hirst, a pupil of Steiner's, opened the subject of modern
geometry.   Clifford did much to make known the German
theories, besides himself contributing to the study of polars
and the general theory of curves.

ART. 17.   ELEMENTARY GEOMETRY.

Trigonometry and Elementary Geometry have also been
affected by the general mathematical spirit of the century.
In trigonometry the general substitution of ratios for lines in
the definitions of functions has simplified the treatment, and
certain formulas have been improved and others added.*
The convergence of trigonometric series, the introduction of
the Fourier series, and the free use of the imaginary have
already been mentioned.   The definition of the sine and cosine
by series, and the systematic development of the theory on
this basis, have been set forth by Cauchy (1821), Lobachevsky
(1833), and others.   The hyperbolic trigonometry,† already
founded by Mayer and Lambert, has been popularized and
further developed by Gudermann (1830), Hoüel, and Laisant
(1871), and projective formulas and generalized figures have

---

* Todhunter, I., History of certain formulas of spherical trigonometry,
Philosophical Magazine, 1873.

† Günther, S., Die Lehre von den gewöhnlichen und verallgemeinerten
Hyperbelfunktionen, Halle, 1881; Chrystal, G., Algebra, Vol. II, p. 288.

been introduced, notably by Gudermann, Möbius, Poncelet, and Steiner. Recently Study has investigated the formulas of spherical trigonometry from the standpoint of the modern theory of functions and theory of groups, and Macfarlane has generalized the fundamental theorem of trigonometry for three-dimensional space.

Elementary Geometry has been even more affected. Among the many contributions to the theory may be mentioned the following: That of Möbius on the opposite senses of lines, angles, surfaces, and solids; the principle of duality as given by Gergonne and Poncelet; the contributions of De Morgan to the logic of the subject; the theory of transversals as worked out by Monge, Brianchon, Servois, Carnot, Chasles, and others; the theory of the radical axis, a property discovered by the Arabs, but introduced as a definite concept by Gaultier (1813) and used by Steiner under the name of "line of equal power"; the researches of Gauss concerning inscriptible polygons, adding the 17- and 257-gon to the list below the 1000-gon; the theory of stellar polyhedra as worked out by Cauchy, Jacobi, Bertrand, Cayley, Möbius, Wiener, Hess, Hersel, and others, so that a whole series of bodies have been added to the four Kepler-Poinsot regular solids; and the researches of Muir on stellar polygons. These and many other improvements now find more or less place in the text-books of the day.

To these must be added the recent Geometry of the Triangle, now a prominent chapter in elementary mathematics. Crelle (1816) made some investigations in this line, Feuerbach (1822) soon after discovered the properties of the Nine-Point Circle, and Steiner also came across some of the properties of the triangle, but none of these followed up the investigation. Lemoine * (1873) was the first to take up the subject in a sys-

---

* Smith, D. E., Biography of Lemoine, American Mathematical Monthly, Vol. III, p. 29; Mackay, J. S., various articles on modern geometry in Proceedings Edinburgh Mathematical Society, various years; Vigarié, É., Géométrie du triangle. Articles in recent numbers of Journal de Mathématiques spéciales, Mathesis, and Proceedings of the Association française pour l'avancement des sciences.

tematic way, and he has contributed extensively to its development. His theory of "transformation continue" and his "géométrographie" should also be mentioned. Brocard's contributions to the geometry of the triangle began in 1877. Other prominent writers have been Tucker, Neuberg, Vigarié, Emmerich, M'Cay, Longchamps, and H. M. Taylor. The theory is also greatly indebted to Miller's work in The Educational Times, and to Hoffmann's Zeitschrift.

The study of linkages was opened by Peaucellier (1864), who gave the first theoretically exact method for drawing a straight line. Kempe and Sylvester have elaborated the subject.

In recent years the ancient problems of trisecting an angle, doubling the cube, and squaring the circle have all been settled by the proof of their insolubility through the use of compasses and straight edge.*

ART. 18. NON-EUCLIDEAN GEOMETRY.

The Non-Euclidean Geometry † is a natural result of the futile attempts which had been made from the time of Proklos to the opening of the nineteenth century to prove the fifth postulate (also called the twelfth axiom, and sometimes the

* Klein, F., Vorträge über ausgewählten Fragen; Rudio, F., Das Problem von der Quadratur des Zirkels. Naturforschende Gesellschaft Vierteljahrschrift, 1890; Archimedes, Huygens, Lambert, Legendre (Leipzig, 1892).

† Stäckel and Engel, Die Theorie der Parallellinien von Euklid bis auf Gauss, Leipzig, 1895; Halsted, G. B., various contributions: Bibliography of Hyperspace and Non-Euclidean Geometry, American Journal of Mathematics, Vols. I, II; The American Mathematical Monthly, Vol. I; translations of Lobachevsky's Geometry, Vasiliev's address on Lobachevsky, Saccheri's Geometry, Bolyai's work and his life; Non-Euclidean and Hyperspaces. Mathematical Papers of Chicago Congress, p. 92. Loria, G., Die hauptsächlichsten Theorien der Geometrie, p. 106; Karagiannides, A., Die Nichteuklidische Geometrie vom Alterthum bis zur Gegenwart, Berlin, 1893; McClintock, E., On the early history of Non-Euclidean Geometry, Bulletin of New York Mathematical Society, Vol. II, p. 144; Poincaré, Non-Euclidean Geom., Nature, 45:404; Articles on Parallels and Measurement in Encyclopædia Britannica, 9th edition; Vasiliev's address (German by Engel) also appears in the Abhandlungen zur Geschichte der Mathematik, 1895.

eleventh or thirteenth) of Euclid. The first scientific investigation of this part of the foundation of geometry was made by Saccheri (1733), a work which was not looked upon as a precursor of Lobachevsky, however, until Beltrami (1889) called attention to the fact. Lambert was the next to question the validity of Euclid's postulate, in his Theorie der Parallellinien (posthumous, 1786), the most important of many treatises on the subject between the publication of Saccheri's work and those of Lobachevsky and Bolyai. Legendre also worked in the field, but failed to bring himself to view the matter outside the Euclidean limitations.

During the closing years of the eighteenth century Kant's * doctrine of absolute space, and his assertion of the necessary postulates of geometry, were the object of much scrutiny and attack. At the same time Gauss was giving attention to the fifth postulate, though on the side of proving it. It was at one time surmised that Gauss was the real founder of the non-Euclidean geometry, his influence being exerted on Lobachevsky through his friend Bartels, and on Johann Bolyai through the father Wolfgang, who was a fellow student of Gauss's. But it is now certain that Gauss can lay no claim to priority of discovery, although the influence of himself and of Kant, in a general way, must have had its effect.

Bartels went to Kasan in 1807, and Lobachevsky was his pupil. The latter's lecture notes show that Bartels never mentioned the subject of the fifth postulate to him, so that his investigations, begun even before 1823, were made on his own motion and his results were wholly original. Early in 1826 he sent forth the principles of his famous doctrine of parallels, based on the assumption that through a given point more than one line can be drawn which shall never meet a given line coplanar with it. The theory was published in full in 1829-30, and he contributed to the subject, as well as to other branches of mathematics, until his death.

* Fink, E., Kant als Mathematiker, Leipzig, 1889.

Johann Bolyai received through his father, Wolfgang, some of the inspiration to original research which the latter had received from Gauss. When only twenty-one he discovered, at about the same time as Lobachevsky, the principles of non-Euclidean geometry, and refers to them in a letter of November, 1823. They were committed to writing in 1825 and published in 1832. Gauss asserts in his correspondence with Schumacher (1831–32) that he had brought out a theory along the same lines as Lobachevsky and Bolyai, but the publication of their works seems to have put an end to his investigations. Schweikart was also an independent discoverer of the non-Euclidean geometry, as his recently recovered letters show, but he never published anything on the subject, his work on the theory of parallels (1807), like that of his nephew Taurinus (1825), showing no trace of the Lobachevsky-Bolyai idea.

The hypothesis was slowly accepted by the mathematical world. Indeed it was about forty years after its publication that it began to attract any considerable attention. Hoüel (1866) and Flye St. Marie (1871) in France, Riemann (1868), Helmholtz (1868), Frischauf (1872), and Baltzer (1877) in Germany, Beltrami (1872) in Italy, de Tilly (1879) in Belgium, Clifford in England, and Halsted (1878) in America, have been among the most active in making the subject popular. Since 1880 the theory may be said to have become generally understood and accepted as legitimate.*

Of all these contributions the most noteworthy from the scientific standpoint is that of Riemann. In his Habilitationsschrift (1854) he applied the methods of analytic geometry to the theory, and suggested a surface of negative curvature, which Beltrami calls "pseudo-spherical," thus leaving Euclid's geometry on a surface of zero curvature midway between his own and Lobachevsky's. He thus set forth three kinds of

---

* For an excellent summary of the results of the hypothesis, see an article by McClintock, The Non-Euclidian Geometry, Bulletin of New York Mathematical Society, Vol. II, p. 1.

geometry, Bolyai having noted only two. These Klein (1871) has called the elliptic (Riemann's), parabolic (Euclid's), and hyperbolic (Lobachevsky's).

Starting from this broader point of view * there have contributed to the subject many of the leading mathematicians of the last quarter of a century, including, besides those already named, Cayley, Lie, Klein, Newcomb, Pasch, C. S. Peirce, Killing, Fiedler, Mansion, and McClintock. Cayley's contribution of his " metrical geometry " was not at once seen to be identical with that of Lobachevsky and Bolyai. It remained for Klein (1871) to show this, thus simplifying Cayley's treatment and adding one of the most important results of the entire theory. Cayley's metrical formulas are, when the Absolute is real, identical with those of the hyperbolic geometry ; when it is imaginary, with the elliptic ; the limiting case between the two gives the parabolic (Euclidean) geometry. The question raised by Cayley's memoir as to how far projective geometry can be defined in terms of space without the introduction of distance had already been discussed by von Staudt (1857) and has since been treated by Klein (1873) and by Lindemann (1876).

Art. 19.   Bibliography.

The following are a few of the general works on the history of mathematics in the nineteenth century, not already mentioned in the foot-notes.   For a complete bibliography of recent works the reader should consult the Jahrbuch über die Fortschritte der Mathematik, the Bibliotheca Mathematica, or the Revue Semestrielle, mentioned below.

Abhandlungen zur Geschichte der Mathematik (Leipzig).

Ball, W. W. R., A short account of the history of mathematics (London, 1893).

Ball, W. W. R., History of the study of mathematics at Cambridge (London, 1889).

Ball, W. W. R., Primer of the history of mathematics (London, 1895).

* Klein, Evanston Lectures, Lect. IX.

Bibliotheca Mathematica, G. Eneström, Stockholm. Quarterly. Should be consulted for bibliography of current articles and works on history of mathematics.

Bulletin des Sciences Mathématiqués (Paris, IIième Partie).

Cajori, F., History of Mathematics (New York, 1894).

Cayley, A., Inaugural address before the British Association, 1883. Nature, Vol. XXVIII, p. 491.

Dictionary of National Biography. London, not completed. Valuable on biographies of British mathematicians.

D'Ovidio, Enrico, Uno sguardo alle origini ed allo sviluppo della Matematica Pura (Torino, 1889).

Dupin, Ch., Coup d'œil sur quelques progrès des Sciences mathématiques, en France, 1830–35. Comptes Rendus, 1835.

Encyclopædia Britannica. Valuable biographical articles by Cayley, Chrystal, Clerke, and others.

Fink, K., Geschichte der Mathematik (Tübingen, 1890). Bibliography on p. 255.

Gerhardt, C. J., Geschichte der Mathematik in Deutschland (Munich, 1877).

Graf, J. H., Geschichte der Mathematik und der Naturwissenschaften in bernischen Landen (Bern, 1890). Also numerous biographical articles.

Günther, S., Vermischte Untersuchungen zur Geschichte der mathematischen Wissenschaften (Leipzig, 1876).

Günther, S., Ziele und Resultate der neueren mathematisch-historischen Forschung (Erlangen, 1876).

Hagen, J. G., Synopsis der höheren Mathematik. Two volumes (Berlin, 1891–93).

Hankel, H., Die Entwickelung der Mathematik in dem letzten Jahrhundert (Tübingen, 1884).

Hermite, Ch., Discours prononcé devant le président de la république le 5 août 1889 à l'inauguration de la nouvelle Sorbonne. Bulletin des Sciences mathématiques, 1890 : also Nature, Vol. XLI, p. 597. (History of nineteenth-century mathematics in France.)

Hoefer, F., Histoire des mathématiques (Paris, 1879).

Isely, L., Essai sur l'histoire des mathématiques dans la Suisse française (Neuchâtel, 1884).

Jahrbuch über die Fortschritte der Mathematik (Berlin, annually, 1868 to date).

Marie, M., Histoire des sciences mathématiques et physiques. Vols. X, XI, XII (Paris, 1887–88).

Matthiessen, L., Grundzüge der antiken und modernen Algebra der litteralen Gleichungen (Leipzig, 1878).

Newcomb, S., Modern mathematical thought.   Bulletin New York Mathematical Society, Vol. III, p. 95; Nature, Vol. XLIX, p. 325.

Poggendorff, J. C., Biographisch-literarisches Handwörterbuch zur Geschichte der exacten Wissenschaften.   Two volumes (Leipzig, 1863).

Quetelet, A., Sciences mathématiques et physiques chez les Belges au commencement du XIXᵉ siècle (Brussels, 1866).

Revue semestrielle des publications mathématiques rédigée sous les auspices de la Société mathématique d'Amsterdam. 1893 to date. (Current periodical literature.)

Roberts, R. A., Modern mathematics.   Proceedings of the Irish Academy, 1888.

Smith, H. J. S., On the present state and prospects of some branches of pure mathematics.   Proceedings of London Mathematical Society, 1876; Nature, Vol. XV, p. 79.

Sylvester, J. J., Address before the British Association.   Nature, Vol. I, pp. 237, 261.

Wolf, R., Handbuch der Mathematik.   Two volumes (Zurich, 1872).

Zeitschrift für Mathematik und Physik.   Historisch-literarische Abtheilung.   Leipzig.   The Abhandlungen zur Geschichte der Mathematik are supplements.

For a biographical table of mathematicians see Fink's Geschichte der Mathematik, p. 240.   For the names and positions of living mathematicians see the Jahrbuch der gelehrten Welt, published at Strassburg.

INDEX.